PRINCIPLES OF PHYSICS SERIES

ELECTRICITY AND MAGNETISM

by

FRANCIS WESTON SEARS

Department of Physics
Massachusetts Institute of Technology

ADDISON-WESLEY PUBLISHING COMPANY, Inc.
READING, MASSACHUSETTS, U.S.A.

Printed in the United States of America

Eleventh Printing—May 1958

PREFACE

This book is the second volume of a series of texts written for a two-year course in general physics. It is assumed that students using the book have completed a course in Calculus.

Except for brief mention of the electrostatic and electromagnetic systems, rationalized mks units are used throughout. The symbols and terminology, with a few exceptions, are those recommended by the Committee on Electric and Magnetic Units of the American Association of Physics Teachers in its report of June, 1938.

The author wishes to express his gratitude to Dr. Charles W. Sheppard, who wrote the sections on chemical emf's, and to Dr. Mark W. Zemansky for his assistance in preparing the manuscript for publication. Acknowledgement is also made to numerous contributors to the American Physics Teacher and the American Journal of Physics.

The entire staff of Addison-Wesley, and Mrs. Olga A. Crawford in particular, have been most cooperative in the task of seeing the series of texts through the press.

Cambridge, Mass. F.W.S.
March, 1946

ADDISON-WESLEY PHYSICS SERIES

Bonner and Phillips—PRINCIPLES OF PHYSICAL SCIENCE
Holton—INTRODUCTION TO CONCEPTS AND THEORIES IN PHYSICAL SCIENCE
Holton and Roller—FOUNDATIONS OF MODERN PHYSICAL SCIENCE
Knauss—DISCOVERING PHYSICS
Mitchell—FUNDAMENTALS OF ELECTRONICS
Sears and Zemansky—COLLEGE PHYSICS
Sears and Zemansky—UNIVERSITY PHYSICS

PRINCIPLES OF PHYSICS SERIES

Constant—THEORETICAL PHYSICS—MECHANICS
Constant—THEORETICAL PHYSICS—ELECTROMAGNETISM
Fowler—INTRODUCTION TO ELECTRIC THEORY
Randall—INTRODUCTION TO ACOUSTICS
Rossi—OPTICS
Sears—MECHANICS, HEAT, AND SOUND
Sears—ELECTRICITY AND MAGNETISM
Sears—OPTICS
Sears—MECHANICS, WAVE MOTION, AND HEAT
Sears—THERMODYNAMICS, THE KINETIC THEORY OF GASES, AND STATISTICAL MECHANICS
Symon—MECHANICS

ADDISON-WESLEY SERIES IN ADVANCED PHYSICS

Goldstein—CLASSICAL MECHANICS
Jauch and Rohrlich—THE THEORY OF PHOTONS AND ELECTRONS
Landau and Lifshitz—THE CLASSICAL THEORY OF FIELDS
Landau and Lifshitz—QUANTUM MECHANICS—NONRELATIVISTIC THEORY
Panofsky and Phillips—CLASSICAL ELECTRICITY AND MAGNETISM
Sachs—NUCLEAR THEORY

A-W SERIES IN NUCLEAR SCIENCE AND ENGINEERING

Goodman—INTRODUCTION TO PILE THEORY
(The Science and Engineering of Nuclear Power, I)
Goodman—APPLICATIONS OF NUCLEAR ENERGY
(The Science and Engineering of Nuclear Power, II)
Hughes—PILE NEUTRON RESEARCH
Kaplan—NUCLEAR PHYSICS

CONTENTS

CHAPTER 1

COULOMB'S LAW

1-1 The structure of the atom. The word atom is derived from the Greek *atomos*, meaning indivisible. It is scarcely necessary to point out that the term is inappropriate. All atoms are more or less complex arrangements of subatomic particles and there are many methods of splitting off some of these particles, either singly or in groups. The most spectacular example, of course, is the disruption of an atomic nucleus in the process of nuclear fission.

The subatomic particles, the building blocks out of which atoms are constructed, are of three different kinds: the electron, the proton, and the neutron. Other subatomic particles have been observed or postulated but their existence is transitory and they do not form a part of ordinary matter. The subatomic particles are arranged in the same general way in all atoms. The protons and neutrons always form a closely-packed group called the nucleus. The diameter of a nucleus, if we think of it as a sphere, is of the order of 10^{-12} cm. Outside the nucleus, but at relatively large distances from it, are the electrons. In the atomic model proposed by the Danish physicist Niels Bohr in 1913, the electrons were pictured as whirling about the nucleus in circular or elliptical orbits. We now believe this model is not entirely correct, but it is still useful in "visualizing" the structure of an atom. The diameters of the electronic orbits, which determine the size of the atom as a whole, are of the order of 2 or 3×10^{-8} cm, or about ten thousand times as great as the diameter of the nucleus. A Bohr atom is a solar system in miniature. The sun at the center corresponds to the nucleus, while the planets, at relatively large distances from the sun, correspond to the electrons.

The masses of a proton and a neutron are nearly equal, and the mass of each is about 1840 times as great as that of an electron. Practically all the mass of an atom is, therefore, concentrated in its nucleus. The masses of these subatomic particles can be computed as follows. We know that one gram-mole of monatomic hydrogen consists of 6.02×10^{23} particles (Avogadro's number) and that its mass is 1.008 gm. The mass of a single hydrogen atom is therefore, to three significant figures,

$$\frac{1.008}{6.02 \times 10^{23}} = 1.67 \times 10^{-24} \text{ gm} = 1.67 \times 10^{-27} \text{ kgm.}$$

The hydrogen atom (the question of isotopes will be discussed in the next paragraph) is the sole exception to the rule that all atoms are constructed of three kinds of subatomic particles. The nucleus of a hydrogen atom is a single proton, outside of which there is a single electron. Hence out of the total mass of the hydrogen atom, 1/1840th part is the mass of the electron and the remainder is the mass of a proton. Then to three significant figures,

$$\text{mass of electron} = \frac{1.67 \times 10^{-24}}{1840} = 9.11 \times 10^{-28}\ \text{gm} = 9.11 \times 10^{-31}\ \text{kgm},$$

$$\text{mass of proton} = 1.67 \times 10^{-24}\ \text{gm} = 1.67 \times 10^{-27}\ \text{kgm},$$

and since the masses of a proton and a neutron are nearly equal,

$$\text{mass of neutron} = 1.67 \times 10^{-24}\ \text{gm} = 1.67 \times 10^{-27}\ \text{kgm}.$$

Ordinary hydrogen is a mixture of three kinds of atoms called *isotopes*. In addition to the atom described in the preceding paragraph, which consisted of a single proton and a single electron, one finds atoms of "heavy hydrogen" or deuterium, and also atoms of tritium. The concentration of deuterium in ordinary hydrogen is only about one part in 5000, and that of tritium is even less. The nucleus of a deuterium atom (called a *deuteron*) consists of one proton and one neutron. The nucleus of a tritium atom contains one proton and two neutrons. In both dueterium and tritium there is one electron outside the nucleus

To distinguish between these forms of hydrogen, their chemical symbols are written as ${}_1H^1$, ${}_1D^2$ (a more consistent symbol would be ${}_1H^2$) and ${}_1H^3$. The number written below and to the left of the letter represents either the number of protons in the nucleus, or the number of electrons outside the nucleus, and it is called the *atomic number* of the atom. In every atom, provided it is in its normal or un-ionized state, the number of protons equals the number of electrons, a fact which is directly related to the electrical properties of the proton and the electron, as we shall see. The superscript, above and to the right of the letter, is the total number of particles in the nucleus, and is the nearest integer to the atomic weight. Evidently the mass of a deuterium atom is very nearly twice that of ${}_1H^1$, while the mass of a tritium atom is three times as great. The chemical properties of an atom depend chiefly on the number of extranuclear electrons and only to a small degree on the mass of the nucleus, so that chemically the isotopes of an element differ but little if at all.

The next element beyond hydrogen in the periodic table is helium. The most abundant isotope of helium is $_2He^4$. Its nucleus consists of two protons and two neutrons, and it has two extranuclear electrons. The nucleus of this isotope is called an alpha-particle, or α-particle. Helium also has an isotope, $_2He^3$, whose nucleus contains two protons and one neutron. Its concentration is about one part in 100,000. Lithium, the next element beyond helium, consists of about 92% of $_3Li^7$ and about 8% of $_3Li^6$. The arrangement of protons, neutrons, and electrons in these atoms is left for the reader to deduce.

As one proceeds through the periodic table the complexity of the atoms increases. The most complex atom of natural occurrence is uranium, a mixture of the three isotopes $_{92}U^{238}$, $_{92}U^{235}$, and $_{92}U^{234}$, in relative abundance 99.28%, 0.71%, and 0.006%. Even more complex atoms have been "manufactured" in the development of the atomic bomb. These are neptunium, $_{93}Np^{239}$, and plutonium, $_{94}Pu^{239}$. The plutonium nucleus consists of 94 protons and 145 neutrons, with 94 extranuclear electrons. This makes a grand total of 333 subatomic particles!

It is found that protons and electrons exert forces on one another, over and above the forces of gravitational attraction between them. These forces are "explained" by ascribing to protons and electrons a property known as *electricity* or *electric charge*, just as gravitational forces are "explained" by ascribing to matter the property of *gravitational mass*. There is this difference, however, that while in gravitation we find only forces of attraction, electrical forces of both attraction and repulsion exist. Protons exert forces of repulsion on other protons, electrons exert forces of repulsion on other electrons, while protons and electrons attract one another. There thus appear to be two kinds of electric charge, arbitrarily designated as positive (+) and negative (−). Protons have positive charge, electrons negative charge. The observed forces between protons and electrons then lead to the familiar statement that like charges repel one another, unlike charges attract one another.

All electrons have precisely the same negative charge, all protons have precisely the same positive charge. Furthermore, any system, such as a normal atom, containing equal numbers of protons and electrons, exhibits no net charge. This leads to the conclusion that the charges of a proton and an electron, while of opposite sign, are equal in magnitude.

No charges have ever been observed of smaller magnitude than those of a proton or electron. Hence these seem to be the ultimate, natural unit of electric charge.

In addition to the forces of attraction or repulsion between protons and electrons, which depend only on the separation of the particles, other forces are found to exist between them which depend on their relative motion. It is these forces which are responsible for *magnetic* phenomena. For many years, the apparent force of repulsion or attraction between a pair of bar magnets was explained on the theory that there existed magnetic entities similar to electric charges and called "magnetic poles." It is a familiar fact, however, that magnetic effects are also observed around a wire in which there is a current. But a current is simply a motion of electric charge, and it appears now that all magnetic effects come about as a result of the relative motion of electric charges. Hence magnetism and electricity are not two separate subjects, but are related phenomena arising from the properties of electric charges.

The preceding introductory statements summarize a small part of a tremendous body of experimental knowledge whose roots go back for hundreds of years, while its most recent aspects are still under investigation in many laboratories at the present day. Some of the experiments will be described in this book and a bibliography of reference books will be found at the end of Chap. 18.

1-2 Charging by contact. The atoms of ordinary matter, as stated in the preceding section, contain equal numbers of electrons and protons. Hence ordinary matter does not exhibit electrical effects and is said to be electrically *neutral* or *uncharged*. If by some means the balance between electrons and protons is upset, that is, if a body has an excess or deficiency of electrons, it is said to be *charged* and will be spoken of as a *charged body* or, for brevity, simply as a *charge*. There are many ways in which the normal balance between positive and negative charges can be altered. The oldest, historically, is the phenomenon of charging by friction, as it is usually called. Charging by contact is a better term, however, since close contact is all that is required. Vigorous rubbing of one surface against another simply serves to bring many points of the surfaces into good contact.

If a rod of hard rubber is rubbed with catskin and brought near a ball of pith suspended by a silk fiber, the pith ball is first attracted by the rubber, but after a moment of contact it is repelled. Two pith balls which have been similarly treated are found to repel one another. On the other hand, each ball is attracted by the catskin. The explanation of these effects is that when rubber and catskin are brought into contact there is a spontaneous transfer of electrons from the catskin to the rubber. The rubber therefore acquires an excess of electrons and becomes negatively charged, while the catskin, having lost some electrons, is positively charged.

The reason for the initial attraction of the *uncharged* pith ball by the rubber rod will be explained later (in Chap. 7). When the pith ball comes into contact with the negatively-charged rod some of the excess electrons on the rod transfer to the pith ball. The repulsion which is then observed between the rod and the pith ball, or between two charged pith balls, is evidence that bodies having like charges repel one another. The attraction between the pith balls and the catskin shows that bodies having unlike charges attract one another. All of the phenomena of *electrostatics* are simply manifestations of these forces between charged bodies or between the ultimate entities of charge, electrons and protons. The term "electrostatics" is somewhat of a misnomer, since the charges on which electrical forces act are often in motion.

Measurements show that the charge acquired by the rubber when it is brought into contact with the catskin is exactly equal and opposite to that acquired by the catskin. There is thus no *creation* of electric charge in the charging process, but simply a *transfer* of charge from one body to another. These experiments alone do not indicate which particles move. Other evidence shows it to be the electrons.

FIG. 1-1. The leaf electroscope.

A charged pith ball may be used as a test body to determine whether or not a second body is charged. A more sensitive test is afforded by the *leaf electroscope*, sketched in Fig. 1-1. Two strips of thin gold leaf or aluminum foil A are fastened at the end of a metal rod B which passes through a support C of rubber, amber or sulfur. The surrounding case D is provided with windows through which the leaves can be observed and which serve to protect them from air currents. When the knob of the electroscope is touched by a charged body, the leaves acquire charges of the same sign and repel one another, their divergence being a measure of the quantity of charge they have received.

An instrument that is better adapted to quantitative measurements is the *string electrometer*, illustrated in principle in Fig. 1-2. A fine fibre A (the "string") is stretched under a small tension between two metal plates B and C, which are, respectively, positively and negatively charged. If a positive charge, for example, is now given to the fiber it is repelled by the positive charges on B and attracted by the negative charges on C. As a result, it bows slightly toward the right and the displacement of its center, which can be observed with a measuring microscope, is a measure of the charge it has acquired.

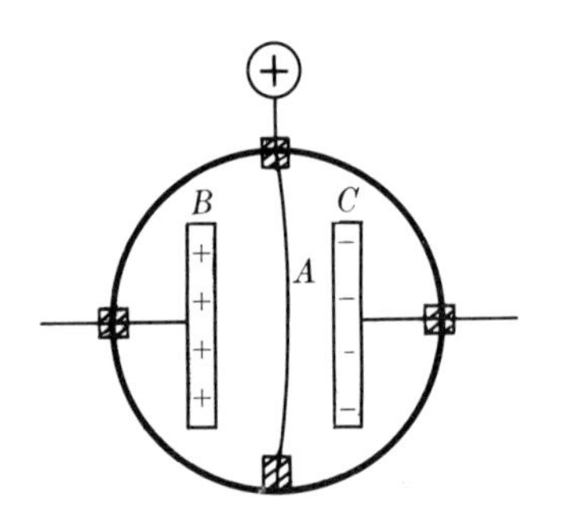

FIG. 1-2. The string electrometer.

If one terminal of a "B" battery of a few hundred volts potential difference is connected to the knob of an electroscope and the other terminal to the electroscope case, the leaves will diverge just as if they had been charged from a body electrified by contact. There is no difference between "kinds" of charge given the leaves in the two processes and, in general, there is no distinction between "static electricity" and "current electricity." The term "current" refers to a flow of charge, while "electrostatics" is concerned for the most part with interactions between charges at rest. The charges themselves in either case are those of electrons or protons.

1-3 Conductors and insulators. Let one end of a copper wire be attached to the knob of an electroscope, its other end being supported by a glass rod as in Fig. 1-3. If a charged rubber rod is touched to the far end of the wire, the electroscope leaves will immediately diverge. There has, therefore, been a transfer of charge along or through the wire, and the wire is called a *conductor*. If the experiment is repeated, using a silk thread or rubber band in place of the wire, no such deflection of the electroscope occurs and the thread or rubber is called an *insulator* or *dielectric*. The motion of charge through a material substance will be studied in more detail in Chap. 4, but for our present purposes it is

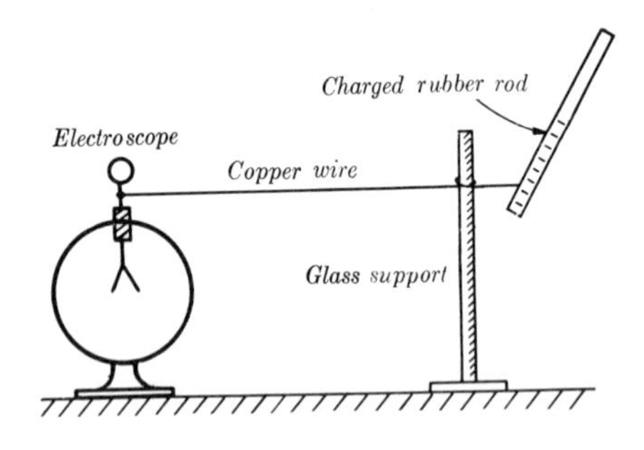

FIG. 1-3. Copper is a conductor of electricity.

sufficient to state that most substances fall into one or the other of the two classes above. Conductors permit the passage of charge through them while insulators do not.

Metals in general are good conductors, while nonmetals are insulators. The positive valency of metals, and the fact that they form positive ions in solution, indicate that the atoms of a metal will part readily with one or more of their outer electrons. Within a metallic conductor, such as a copper wire, a few outer electrons become detached from each atom and can move freely throughout the metal. These are called *free* electrons. The positive nuclei and the remainder of the electrons remain fixed in position. Within an insulator, on the other hand, there are none (or at any rate very few) of these free electrons.

The phenomenon of charging by contact is not limited to rubber and catskin, or indeed to insulators in general. Any two dissimilar substances exhibit the effect to a greater or less extent, but evidently a conductor must be supported on an insulating handle or the charges developed on it will at once leak away.

1-4 Quantity of charge. Coulomb's law. A charged body is one having an excess number of electrons or protons, and the magnitude of the body's net charge might be described by a statement of this number. In practice, however, the charge of a body is expressed in terms of a unit much larger than the charge of an individual electron or proton. We shall use the letter q or Q to represent the excess quantity of $+$ or $-$ charge on a body, postponing for a moment the definition of the unit charge.

The first quantitative investigation of the law of force between charged bodies was carried out by Charles Augustin Coulomb (1736–1806) in 1784–1785, utilizing for the measurement of forces a torsion balance of the type employed 13 years later by Cavendish in measuring gravitational forces. Within the limits of accuracy of his measurements, Coulomb showed that the force of attraction or repulsion between two charged bodies followed an inverse square law,

$$F \propto \frac{1}{r^2}.$$

The concept of quantity of charge was not clearly appreciated in Coulomb's time, and no unit of charge or method of measuring it had been devised. Later work has shown that at a given separation, the force between two charged bodies is proportional to the product of their individual charges, q and q'. That is,

$$F \propto \frac{qq'}{r^2}. \tag{1-1}$$

This proportion may be converted to an equation by multiplying by a constant k.

$$F = k\frac{qq'}{r^2}. \tag{1-2}$$

The magnitude of the proportionality constant k depends on the units in which force, charge, and distance are expressed. Either of the relations (1-1) or (1-2) is called *Coulomb's law*, which may be stated:

The force of attraction or repulsion exerted on one charged body by another is proportional to the product of their charges and inversely proportional to the square of their separation.

Subject to certain generalizations which will be described later on, Coulomb's law in the form given above is restricted to point charges, that is, charges or charged bodies whose dimensions are small compared with their separation. If there is matter in the space between the charges, for example if the charged bodies are in air, the force between them is altered because charges are "induced" in the molecules of the intervening medium. This effect will also be described later on. As a practical matter, the law can be used as it stands for point charges in air, since even at atmospheric pressure the effect of the air is to alter the force from its value in vacuum by only about one part in two thousand.

The same law of force holds whatever may be the sign of the charges q and q'. If the charges are of like sign the force is a repulsion, if the charges are of opposite sign the force is an attraction. Forces of the same magnitude, but in opposite directions, are exerted on each of the charges.

1-5 Verification of Coulomb's law. Rutherford's nuclear atom. The nearest thing to a truly geometrical point charge, in nature, is the positively charged nucleus of an atom, and we have direct experimental evidence that the force of repulsion between a pair of atomic nuclei obeys Coulomb's law. The experiments were performed not for the primary purpose of verifying this law, but to learn, if possible, how the positive and negative electricities are arranged within an atom, and it was these very experiments that first showed the atom to consist of a small, massive nucleus, surrounded by electrons. The experiments were carried out in 1910–11 by Sir Ernest Rutherford and two of his students, Hans Geiger and Ernest Marsden, at Cambridge, England.

The electron had been "discovered" in 1897 by Sir J. J. Thomson, and by 1910 its mass and charge were quite accurately known. It had also been well established that, with the sole exception of hydrogen, all atoms contained more than one electron. Thomson had proposed an atomic model consisting of a relatively large sphere of positive charge (about 2 or 3×10^{-8} cm in diameter) within which were embedded, like plums in a pudding, the electrons.

What Rutherford and his co-workers did was to project other particles at the atoms under investigation, and from observations of the way in which the projected particles were deflected or *scattered*, draw conclusions about the distribution of charge within the atoms at which the particles had been projected. Fortunately, a supply of projectiles was available. In certain radioactive disintegrations, there are emitted alpha-particles (or α-particles). These particles, which we now know are the

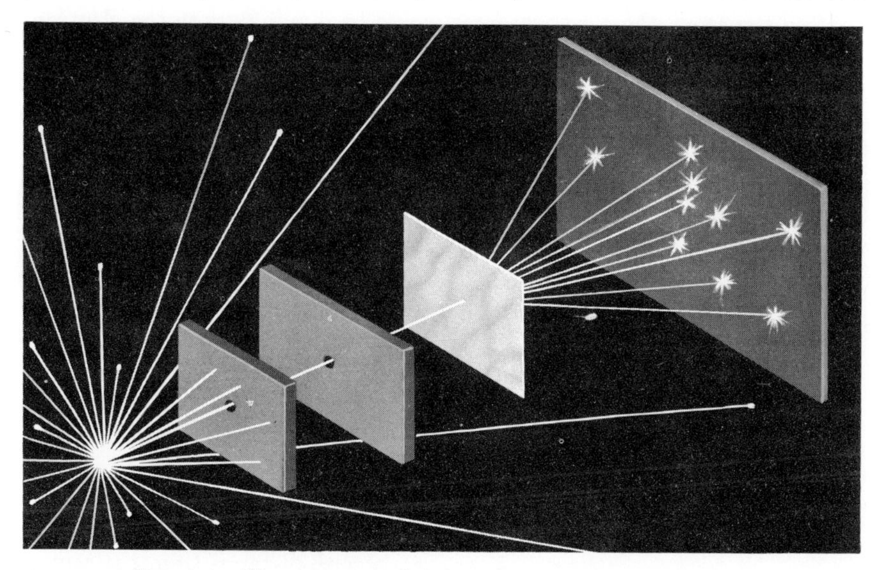

FIG. 1-4. The scattering of alpha particles by a thin metal foil.

nuclei of helium atoms, are ejected with velocities of the order of 10^9 cm/sec. The normal helium atom has two electrons. Its nucleus, the α-particle, therefore has a double positive charge and consists of two protons and two neutrons. Because of the tremendous speeds of α-particles, they are capable of traveling several centimeters in air, or a few tenths or hundredths of a millimeter through solids, before they are brought to rest by collisions.

The experimental setup is shown schematically in Fig. 1-4. A radioactive source at the left emits α-particles. Thick lead screens stop all particles except those in a narrow beam defined by small holes. The beam then passes through a thin metal foil (gold, silver, and copper were used) and strikes a plate coated with zinc sulphide. A momentary flash or scintillation can be observed on the screen whenever it is struck by an α-particle, and the number of particles that have been deflected through any angle from their original direction can therefore be determined.

According to the Thomson model, the atoms of a solid are packed together like marbles in a box. The experimental fact that an α-particle can pass right through a sheet of metal foil forces one to conclude, if this model is correct, that the α-particle is capable of actually penetrating the spheres of positive charge. Granted that this is possible, one can compute the deflection it would undergo. The Thomson atom is electrically neutral, so outside the atom no force would be exerted on the α-particle. Within the atom, the electrical force would be due in part to the electrons and in part to the sphere of positive charge. However, the mass of an α-particle is about 7400 times that of an electron, and from momentum considerations it follows that the α-particle can suffer only a negligible scattering as a consequence of forces between it and the much less massive electrons. It is only interactions with the positive charge, which makes up most of the atomic mass, that can deviate the α-particle.

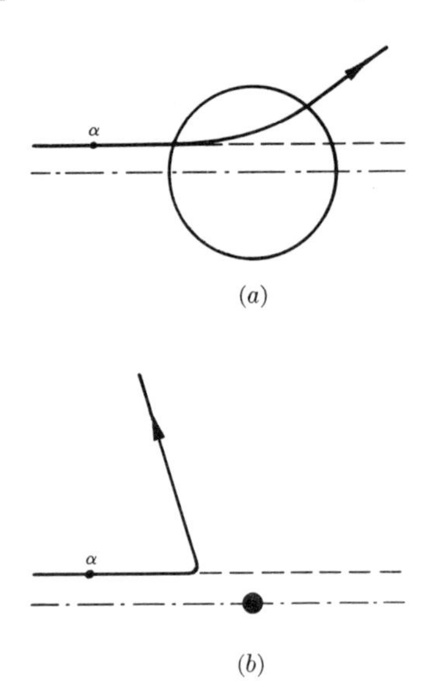

FIG. 1-5. (a) Alpha particle scattered through a small angle by the Thomson atom. (b) Alpha particle scattered through a large angle by the Rutherford nuclear atom.

The electrical force on an α-particle within a sphere of positive charge is like the gravitational force on a mass point within a sphere, except that gravitational forces are attractive, while the force between two positive charges is a repulsion. It will be recalled that the gravitational intensity within a sphere is zero at the center and increases linearly with distance from the center, because that part of the mass of the sphere lying *outside* any radius exerts no force at interior points. The α-particle is therefore *repelled* from the center of the sphere with a force proportional to its distance from the center (see problem 13 at end of Chap. 2), and its trajectory can be computed for any initial direction of approach such as that in Fig. 1-5(a.) On the basis of such calculations,

Rutherford predicted the number of α-particles that should be scattered at any angle with their original direction.

The experimental results did not agree with the calculations based on the Thomson atom. In particular, many more particles were scattered through large angles than were predicted. To account for the observed large-angle scattering, Rutherford concluded that the positive charge, instead of being spread through a sphere of atomic dimensions (2 or 3×10^{-8} cm) was concentrated in a much smaller volume whose dimensions were only about one ten-thousandth as great or about 10^{-12} cm, and which he called the *nucleus*. When an α-particle approaches a charge of these dimensions, the entire charge exerts a repelling effect on it down to extremely small separations, with the consequence that much larger deviations can be produced. Fig. 1-5(b) shows the trajectory of an α-particle deflected by a Rutherford nuclear atom, for the same original path as that in part (a) of the figure.

Rutherford again computed the expected number of particles scattered through any angle, assuming an inverse square law of force between the α-particle (itself a nucleus also) and a nucleus whose dimensions were less than 10^{-12} cm. Within the limits of experimental accuracy, the computed and observed results were in complete agreement. We may therefore say that these experiments verify the Coulomb law for "point" charges, and at the same time measure the size of the nucleus.

Every serious student of Physics should read Rutherford's account of these researches in the *Philosophical Magazine*, Vol. 21, p. 669 (1911).

1-6 Systems of units. The reader will recall that many of the equations of mechanics can be written more simply if the units of *force*, *mass*, and *acceleration* are chosen of such magnitude that the proportionality constant k in Newton's second law ($F = kma$) is equal to unity. The proportionality constant can then be dropped from many equations in which it would otherwise appear. A similar simplification in some of the equations of electricity and magnetism can be attained if the same procedure is adopted with regard to Coulomb's law, that is, if the units of force, distance, and *quantity of charge* are so chosen that the proportionality constant in Coulomb's law is equal to unity.

While some of the common systems of electrical units are set up in this way, there are others in which the units of force, distance, and charge are defined independently of Coulomb's law, with the consequence that the proportionality constant is not equal to unity and must be carried along in all equations which derive from this law.

At the present time, the formulation of the equations of electricity and magnetism is in a stage of transition, with no general agreement on symbols or terminology. One may choose a half-dozen books or articles at random, and find a different system of notation in each. All have something in their favor, and no matter what system we were to adopt in this book the reader would be certain to encounter other systems in his more advanced study and in his reference to material already published.

For many years, the equations of electrostatics have been written using units of the *electrostatic* system, while equations relating to magnetic phenomena used the *electromagnetic* system. Units of the same quantity are not of the same magnitude in the two systems. For example, the electromagnetic unit of charge is 3×10^{10} times as great as the electrostatic unit. Furthermore, the units of the *practical* system, such as the familiar volt, ampere, and ohm, are different from those in both of the above systems. Hence it has been customary to carry along three sets of units and the necessary factors for converting from one to the other.

The mechanical units in both the electrostatic and the electromagnetic systems are those of the cgs system, i.e., the gram, centimeter, dyne, etc. It was pointed out comparatively recently, however, that, with a few minor changes, the units of the familiar practical system would fit nicely into a single system adapted to both electricity and magnetism if, instead of the cgs units, one adopted the mks mechanical units, i.e., the kilogram, meter, newton, etc. There is little doubt that this mks system, in one form or another, will eventually receive world-wide adoption and for that reason it is the one we shall use in this book. When, if ever, general agreement is attained on the precise form the equations should take, it is safe to say they will differ from ours only in the symbols and names used for various quantities, and perhaps in the location of the factor 4π.

Although we shall refer to it only occasionally, some familiarity with the electrostatic system is necessary because of the large number of books, articles, and reference tables already in existence in which this system is used. In the electrostatic system, forces are expressed in dynes, distances in centimeters, and the unit of charge is chosen of such magnitude that the proportionality constant in Coulomb's law is equal to unity. This law, with $k = 1$, then *defines* the unit charge in this system. This unit is called *one statcoulomb*, the prefix "stat" being applied in general to quantities in the electrostatic system. *One statcoulomb* may accordingly be defined as *that charge which repels an equal charge of the same sign with a force of one dyne when the charges are separated by one centimeter*, the charges being carried by bodies whose dimensions are small compared with one centimeter. Coulomb's law takes the simple form

$$F = \frac{qq'}{r^2}. \tag{1-3}$$

In the mks system, forces are expressed in newtons and distances in meters. Quantity of charge is defined not by Coulomb's law, but in terms of the unit of current, the ampere. (See Sec. 11-4.) The mks unit of charge is called *one coulomb* and is defined as the quantity of charge which in one second crosses a section of a conductor in which there is a constant current of one ampere. The coulomb is very nearly 3×10^9 times as large as the statcoulomb. (The precise value is

$$1 \text{ coulomb} = 2.99592 \times 10^9 \text{ statcoulombs.)}$$

The coulomb is a relatively large quantity of charge and for many purposes the *microcoulomb* ($= 10^{-6}$ coulomb) is a unit of more convenient size.

The "natural" unit of electric charge is the charge carried by an electron or proton. The most precise measurements which have been made up to the present time find this charge to be

$$\begin{aligned} e &= 1.6029 \times 10^{-19} \text{ coulombs} \\ &= 4.8021 \times 10^{-10} \text{ statcoulombs.} \end{aligned}$$

One statcoulomb therefore represents the aggregate charge carried by 2.08×10^9 electrons, or about two billion electrons, and one coulomb the charge of about 6×10^{18} electrons. By way of comparison, the total population of the earth is estimated to be about 2×10^9 persons, while on the other hand, a cube of copper one centimeter on a side contains about 8×10^{22} free electrons.

Since the units of force, charge, and distance, in the mks system, are defined independently of Coulomb's law, the numerical value of the proportionality constant k in this system must be found by experiment. In principle, the experiment would consist of measuring the force, in newtons, between two point charges whose charges had been measured in coulombs, at a measured separation expressed in meters. Eq. (1-2) could then be solved for k. In practice, the magnitude of k is determined indirectly. The best value to date is, numerically,

$$k = 8.98776 \times 10^9.$$

For most purposes, the approximation $k = 9 \times 10^9$ is sufficiently accurate.

In any system of units, the units of k are those of force $\times$ distance^2 $\div$ charge^2. Hence in the electrostatic system

$$k = 1 \frac{\text{dyne-cm}^2}{\text{statcoulomb}^2},$$

and in the mks system,

$$k = 9 \times 10^9 \frac{\text{newton-m}^2}{\text{coulomb}^2}.$$

For the purpose of avoiding the appearance of the factor 4π in other equations derived from Coulomb's law, and which are used more frequently than is Coulomb's law itself, we now define a new constant ϵ_0 by the relation

$$\epsilon_0 = \frac{1}{4\pi k},$$

or

$$k = \frac{1}{4\pi\epsilon_0},$$

and write Coulomb's law as

$$\boxed{F = \frac{1}{4\pi\epsilon_0}\frac{qq'}{r^2}.} \tag{1-4}$$

From the numerical value of k given above it follows that in the mks system

$$\boxed{\epsilon_0 = \frac{1}{4\pi k} = \frac{1}{4\pi \times 9 \times 10^9} = 8.85 \times 10^{-12} \frac{\text{coulomb}^2}{\text{newton-meter}^2}}$$

while the factor $1/4\pi\epsilon_0$ is

$$\boxed{\frac{1}{4\pi\epsilon_0} = k = 9 \times 10^9 \frac{\text{newton-meter}^2}{\text{coulomb}^2}}$$

When the factor 4π is written out explicitly in the definition of ϵ_0, as we have done here, one obtains the so-called *rationalized mks system.* Many writers choose to define ϵ_0 by the equation

$$\epsilon_0 = \frac{1}{k},$$

instead of

$$\epsilon_0 = \frac{1}{4\pi k}.$$

Coulomb's law then becomes

$$F = \frac{1}{\epsilon_0} \frac{qq'}{r^2}$$

and the numerical value of ϵ_0 defined in this way is

$$\epsilon_0 = \frac{1}{k} = \frac{1}{9 \times 10^9} = 1.11 \times 10^{-10}.$$

This method of defining ϵ_0 leads to the so-called *nonrationalized mks system*. One must be very careful when reading other texts or articles to ascertain whether the author is using the rationalized or unrationalized system. We shall use only the former in this book.

Example. Compute the electrostatic force of repulsion between two α-particles at a separation of 10^{-11} cm, and compare with the force of gravitational attraction between them.

Each α-particle has a charge of $+2e$, or $2 \times 1.60 \times 10^{-19} = 3.20 \times 10^{-19}$ coul. The force of repulsion at a separation of 10^{-11} cm or 10^{-13} m is

$$\begin{aligned} F &= \frac{1}{4\pi\epsilon_0} \frac{qq'}{r^2} \\ &= 9 \times 10^9 \frac{(3.20 \times 10^{-19})^2}{(10^{-13})^2} \\ &= 9.18 \times 10^{-2} \text{ newtons,} \end{aligned}$$

and since 1 newton = 10^5 dynes,

$$F = 9{,}180 \text{ dynes.}$$

This is a sizable force, equal to the weight of nearly 10 grams!

The mass of an α-particle (2 protons + 2 neutrons) is

$$4 \times 1.67 \times 10^{-24} = 6.68 \times 10^{-24} \text{ gm} = 6.68 \times 10^{-27} \text{ kgm.}$$

The gravitational constant, γ, is

$$\gamma = 6.67 \times 10^{-11} \frac{\text{newton-m}^2}{\text{kgm}^2}.$$

The force of gravitational attraction is

$$\begin{aligned} F &= \gamma \frac{mm'}{r^2} \\ &= 6.67 \times 10^{-11} \frac{(6.68 \times 10^{-27})^2}{(10^{-13})^2} \\ &= 2.97 \times 10^{-37} \text{ newtons.} \end{aligned}$$

The gravitational force is evidently negligible in comparison with the electrostatic force.

Problems—Chapter 1

(1) (a) A point charge of $+6$ statcoulombs is 5 cm from a point charge of -8 statcoulombs. What force, in dynes, is exerted on each charge by the other? (b) A point charge of $+10$ statcoulombs is 4 cm from the positive charge in part (a) and 3 cm from the negative charge. What is the resultant force exerted on it?

(2) Two very small spheres, each weighing 3 dynes, are attached to silk fibres 5 cm long and hung from a common point. When the spheres are given equal quantities of negative charge, each supporting fibre makes an angle of 30° with the vertical. Find the magnitude of the charges.

(3) It will be shown in the next chapter that when a conducting sphere 1 cm in radius, in air, is given a charge of more than about 3×10^{-8} coulombs, the air around the sphere becomes conducting and the charge "leaks" away. Use the figure above as the maximum charge a sphere of this radius can retain, in air, and compute the maximum force between two such spheres at a center-to-center distance of 10 cm.

(4) One gram of monatomic hydrogen contains 6.02×10^{23} atoms, each atom consisting of a positively-charged nucleus and a single outer electron. If all the electrons in one gram of hydrogen could be concentrated at the north pole of the earth and all the nuclei concentrated at the south pole, what would be the force of attraction between them, in tons? The polar radius of the earth is 6357 km.

(5) Rutherford's experiments showed that Coulomb's law held for the force of repulsion of two nuclei, down to a separation of about 10^{-12} cm. The gold nucleus consists of 118 neutrons and 79 protons, and the helium nucleus (an α-particle) consists of two neutrons and two protons. (a) What is the force of repulsion, in newtons, between a gold nucleus and an α-particle at a separation of 10^{-12} cm? (b) Find the acceleration of an α-particle when acted on by this force.

(6) In the Bohr model of the hydrogen atom, an electron revolves in a circular orbit around a nucleus consisting of a single proton. If the radius of the orbit is 5.28×10^{-9} cm, find the number of revolutions of the electron per second. The force of electrostatic attraction between proton and electron provides the centripetal force.

CHAPTER 2

THE ELECTRIC FIELD

2-1 The electric field. Fig. 2-1(a) represents two positively charged bodies, A and B, between which there is an electrical force of repulsion, F. Like the force of gravitational attraction, this force is of the action-at-a-distance type, making itself felt without the presence of any material connection between A and B. No one knows "why" this is possible—it is an experimental fact that charged bodies behave in this way. It is useful, however, to think of each of the charged bodies as modifying the state of affairs in the space around it, so that this state is different in some way from whatever it may be when the charged bodies are not present. Thus, let body B be removed. Point P, Fig. 2-1(b), is the point of space at which B was formerly located. The charged body A is said to produce or to set up an *electric field* at the point P, and if the charged body B is now placed at P, one considers that a force is exerted on B *by the field*, rather than by body A directly. Since a force would be experienced by body B at all points of space around body A, the whole of space around A is an electric field.

One can equally well consider that body B sets up a field, and that a force is exerted on body A by the field of B.

The experimental test for the existence of an electric field at any point is simply to place a charged body, which will be called a *test charge*, at

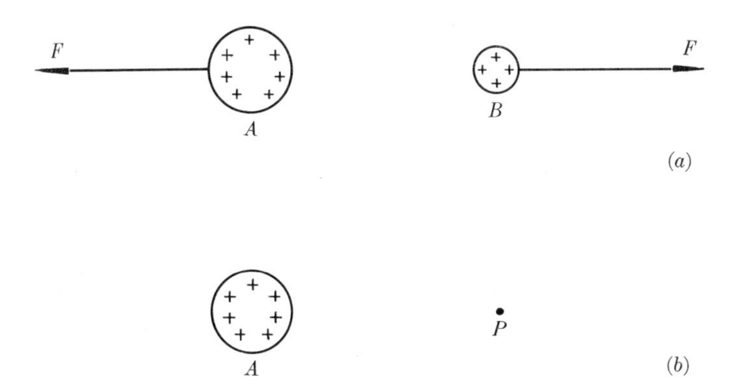

FIG. 2-1. The space around a charged body is an electric field.

the point. If a force (of electrical origin) is exerted on the test charge then an electric field exists at the point.

An electric field is said to exist at a point if a force of electrical origin is exerted on a charged body placed at the point.

Whether or not the force is of electrical origin can be determined by comparing the forces on the test body when it is charged and when it is uncharged. Any force that may be observed when the test body is charged, and which did not exist when it was uncharged, is a force of electrical origin. Thus an uncharged pith ball experiences a force because of its weight, even if no other charged bodies are near. This force is not due to electrical causes, since the ball weighs no more when charged. However, if a second charged body is placed near the pith ball, the force on the pith ball is different in its charged and uncharged states. The change in the force indicates the existence of an electric field, due to the presence of the second charged body.

Since force is a vector quantity, the electric field is a vector quantity also, having both magnitude and direction. The *magnitude* of the field at any point, represented by E, is defined as *the quotient obtained when the force F on a test charge placed at the point, is divided by the quantity of charge q on the test charge.*

$$E = \frac{F}{q} \tag{2-1}$$

In other words, the magnitude of an electric field is the *force per unit charge.*

The *direction* of an electric field at any point is *the direction of the force on a positive test charge placed at the point.* The force on a negative charge, such as an electron, is therefore opposite to the direction of the field.

We shall refer to the electric field vector at a point as the *electric intensity* at the point. Several other terms are commonly used, for example, electric field strength, electric field intensity, or merely field intensity. The words "intensity" or "strength" are really unnecessary and it would be sufficient to speak of the vector merely as the *electric field.*

In the mks system, where forces are expressed in newtons and charges in coulombs, the unit of electric intensity is *one newton per coulomb.* Other units in which electric intensity may be expressed will be defined later.

Eq. (2-1) may be written

$$F = qE. \tag{2-2}$$

That is, the force exerted on a charge q at a point where the electric intensity is E equals the product of the electric intensity and the charge.

The significance of many of the equations of electricity and magnetism is much more evident if they are written in such a form as to bring out the vector or scalar nature of the terms involved. Although we shall develop all of our equations without using vector notation, some of the more important relations will be given in vector form. For example, since electric intensity and force are vectors, while charge q is a scalar, Eq. (2-2) is written in vector notation

$$\vec{F} = q\vec{E}. \tag{2-3}$$

This shows at once that the vectors $\vec{F}$ and $\vec{E}$ are in the same direction if q is positive, and in opposite directions if q is negative.

Examples. We shall show in the next chapter that when the terminals of a 100-volt battery are connected to two parallel plates 1 cm apart, the electric intensity in the space between the plates is 10^4 newtons/coulomb. Suppose we have a field of this intensity whose direction is vertically upward.

(1) Compute the force on an electron in this field and compare with the weight of the electron.

Electronic charge $e = 1.60 \times 10^{-19}$ coul.
Electronic mass $m = 9.1 \times 10^{-28}$ gm $= 9.1 \times 10^{-31}$ kgm.

$$F_{\text{electrical}} = eE = 1.60 \times 10^{-19} \times 10^4 = 1.60 \times 10^{-15} \text{ newtons.}$$

$$F_{\text{gravitational}} = mg = 9.1 \times 10^{-31} \times 9.8 = 8.9 \times 10^{-30} \text{ newtons.}$$

The ratio of the electrical to the gravitational force is therefore

$$\frac{1.6 \times 10^{-15}}{8.9 \times 10^{-30}} = 1.8 \times 10^{14}!$$

It will be seen that the gravitational force is negligible.

(2) If released from rest, what velocity will the electron acquire while traveling 1 cm? What will then be its kinetic energy? How long a time is required?

The force is constant so the electron moves with a constant acceleration of

$$a = \frac{F}{m} = \frac{eE}{m} = \frac{1.6 \times 10^{-19} \times 10^4}{9.1 \times 10^{-31}} = 1.8 \times 10^{15} \text{ m/sec.}^2$$

Its velocity after traveling 1 cm or 10^{-2} m is

$$v = \sqrt{2ax} = \sqrt{2 \times 1.8 \times 10^{15} \times 10^{-2}} = 6.0 \times 10^6 \text{ m/sec.}$$

Its kinetic energy is

$$\tfrac{1}{2}mv^2 = \tfrac{1}{2} \times 9.1 \times 10^{-31} \times (6.0 \times 10^6)^2 = 16 \times 10^{-18} \text{ joules.}$$

The time is

$$t = \frac{v}{a} = \frac{6.0 \times 10^6}{1.8 \times 10^{15}} = 3.3 \times 10^{-9} \text{ sec.}$$

(3) If the electron is projected into the field with a horizontal velocity, find the equation of its trajectory. (Fig. 2-2.)

The direction of the field is upward in Fig. 2-2, so the force on the electron is downward. The initial velocity is along the positive X-axis. The X-acceleration is zero, the Y-acceleration is $-\frac{eE}{m}\cdot$ Hence after a time t,

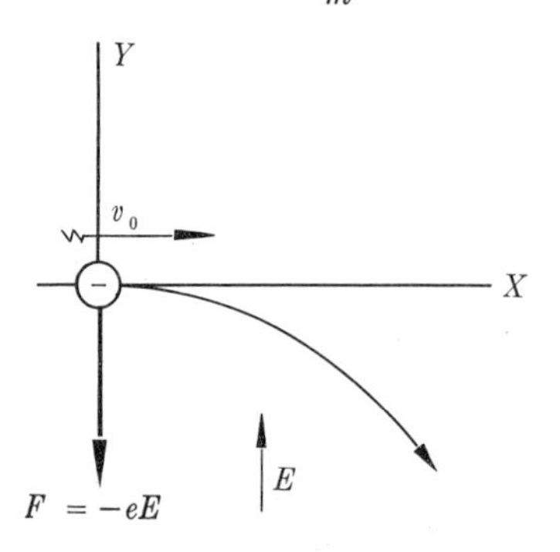

FIG. 2-2. Trajectory of an electron in an electric field.

$$x = v_0 t, \quad y = -\frac{1}{2}\frac{eE}{m}t^2.$$

Elimination of t gives

$$y = -\frac{eE}{2mv_0^{\,2}}x^2,$$

which is the equation of a parabola. The motion is the same as that of a body projected horizontally in the earth's gravitational field. The deflection of electrons by an electric field is used to control the direction of an electron stream in many electronic devices such as the cathode-ray oscillograph.

2-2 Calculation of electric intensity. The preceding section describes an experimental method of measuring the electric intensity at a point. The method consists of placing a test charge at the point, measuring the force on it, and taking the ratio of the force to the charge. The electric intensity at a point may also be computed from Coulomb's law if the magnitudes and positions of all charges contributing to the field are known. Thus to find the electric intensity at a point of space P, distant r from a point charge q, imagine a test charge q' to be placed at P. The force on the test charge, by Coulomb's law, is

$$F = \frac{1}{4\pi\epsilon_0}\frac{qq'}{r^2},$$

and hence the electric intensity at P is

$$E = \frac{F}{q'} = \frac{1}{4\pi\epsilon_0}\frac{q}{r^2}.$$

The direction of the field is away from the charge q if the latter is positive, toward q if it is negative.

If a number of point charges q_1, q_2, etc., are at distances, r_1, r_2, etc., from a given point P, as in Fig. 2-3, each exerts a force on a test charge q' placed at the point and the resultant force on the test charge is the vector sum of these forces.

$$F = \frac{1}{4\pi\epsilon_0}\frac{q_1 q'}{r_1^2} + \frac{1}{4\pi\epsilon_0}\frac{q_2 q'}{r_2^2} + \cdots \text{ (vector sum)}$$

$$= \frac{1}{4\pi\epsilon_0} q' \left(\frac{q_1}{r_1^2} + \frac{q_2}{r_2^2} + \cdots\right) = \frac{1}{4\pi\epsilon_0} q' \, \Sigma \frac{q}{r^2} \text{ (vector sum)}$$

The electric intensity at the point,

$$E = \frac{F}{q'},$$

is therefore

$$\boxed{E = \frac{1}{4\pi\epsilon_0} \Sigma \frac{q}{r^2}} \quad \text{(vector sum).} \tag{2-4}$$

If the charges setting up the field are not point charges but are distributed over bodies of finite size, the charges may be subdivided into infinitesimal elements dq and the obvious generalization of Eq. (2-4) is

$$\boxed{E = \frac{1}{4\pi\epsilon_0} \int \frac{dq}{r^2}} \quad \text{(vector sum)} \tag{2-5}$$

Either Eq. (2-1) or Eq. (2-5) may be considered the definition of the electric intensity at a point. The former defines the intensity at a point of space in terms of experimental measurements that might be made at that point; it does not require any knowledge of the magnitude or position of the charges setting up the field. The latter enables one to *compute* the intensity at a point without actually going to the point to make measurements, provided the distribution of charges setting up the field is known.

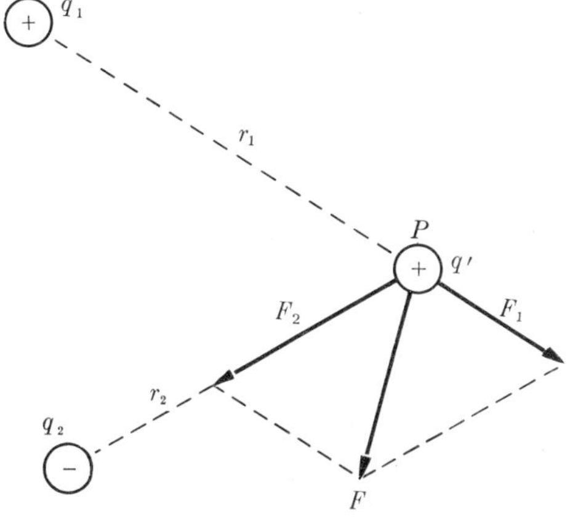

FIG. 2-3. The resultant force F on a test charge is the vector sum of the individual forces F_1 and F_2.

Notice carefully that the charge

q in Eq. (2-1) refers to the *test* charge that was placed at point P to determine the magnitude and direction of the field, while in Eq. (2-4) or (2-5), q refers to *the charges that set up the field* and which are, of course, located at points other than P.

Eq. (2-5) is written as follows in vector notation. Let $\vec{r}$ be the vector from any element of charge dq to the point at which we wish to compute the field. Then

$$\vec{E} = \frac{1}{4\pi\epsilon_0}\int\frac{dq\,\vec{r}}{r^3}. \tag{2-6}$$

The absolute magnitude of $\vec{r}/r^3$ is $1/r^2$, so the numerical value of $dq\,\vec{r}/r^3$ is equal to dq/r^2.

It will be recognized that Coulomb's law for the force between point charges is of precisely the same form as Newton's law for the gravitational force between point masses,

$$F = \gamma\frac{mm'}{r^2}.$$

Also, the definitions of the intensity of an electric field are the same as the corresponding definitions of the intensity of a gravitational field

$$G = \frac{F}{m'}, \quad \text{or} \quad G = \gamma\int\frac{dm}{r^2}.$$

It would appear from the form of these equations that the gravitational and electrical problems were identical in principle. There are, however, important differences between them that render the concept of field intensity, in the electrical case, of deeper significance than in the gravitational case, and make its definition as the ratio of force to charge preferable to its definition by Eq. (2-5). The difference arises from the fact that while gravitational effects appear to propagate through space with an infinite velocity, electrical (or, more correctly, electromagnetic) effects do not.

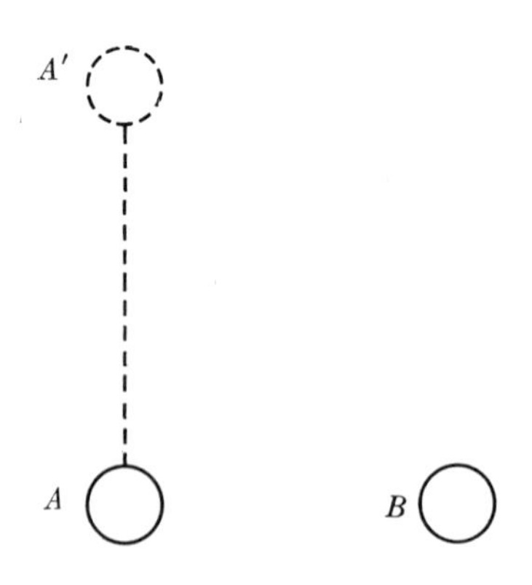

FIG. 2-4. Interaction of body B and a moving body A.

Let A and B in Fig. 2-4 represent two masses which occupy at some instant the positions shown by the solid circles. The gravitational force exerted on B by A is directed toward A. Now let A move along the dotted line toward A'. The instant A is displaced from its original position the direction of the force on B changes, and in such a way that this force is directed at every instant toward the position occupied by A at that instant.

Now let A and B represent two electric charges. If A and B are at rest, the force exerted on B by A is along the line joining them and is given by Coulomb's law. If A starts to move toward A', the force on B *does not alter its direction instantaneously*, but a certain time elapses before the changes in A's field have propagated to the point occupied by B. The consequence is that if charge A is in motion, the force exerted by it on charge B does not depend on where charge A is *now*, but on where it was *a moment ago*. The electric intensity at B, at any instant, can be defined as the ratio of the force on B at that instant to B's charge, but it can not be computed from Coulomb's law [or Eq. (2-5)] giving to r a value equal to the instantaneous distance between A and B. Hence the definition of electric intensity as force per unit charge,

$$E = \frac{F}{q},$$

may be considered more fundamental than its definition in terms of Coulomb's law,

$$E = \frac{1}{4\pi\epsilon_0}\int\frac{dq}{r^2}.$$

Still another reason for preferring the first definition becomes evident when the phenomena of *induced electromotive force* and *electromagnetic radiation* are studied. We shall see that it is quite possible to have a force exerted on a charge even in the absence of any other charged bodies, provided only that there is present a magnetic field which varies with time.

The velocity with which electromagnetic effects propagate in free space is the same as the velocity of light, or 3×10^8 m/sec. Hence for slowly moving charges the time lag in the propagation of electric effects can be ignored and either definition of E can be used. For further discussion of this question a more advanced text should be consulted.

One more point should be mentioned in connection with the definitions of electric intensity. Suppose we wish to know the electric intensity at a point of empty space. The field, in purely electrostatic phenomena at any rate, is due to the presence of charged bodies in the vicinity of the point. But if one places an actual test charge q at the point, the forces exerted by the test charge on the original charges will, in general, cause them to move from their original positions.

For example, let body A in Fig. 2-5(a) be a spherical conductor with positive charge uniformly distributed over its surface. We wish to know the electric intensity at point P, a point in empty space. If we bring a positive point test charge q to point P, the repulsion between this charge and the charges on the sphere will bring about the non-uniform distribution of charge on the sphere indicated in Fig. 2-5(b). While the ratio of the observed force on the test charge to the magnitude of this charge gives (correctly!) the electric intensity at P *while the test charge is at this point*, the field is obviously *not* the same as it was *before* the test charge was brought up to the point.

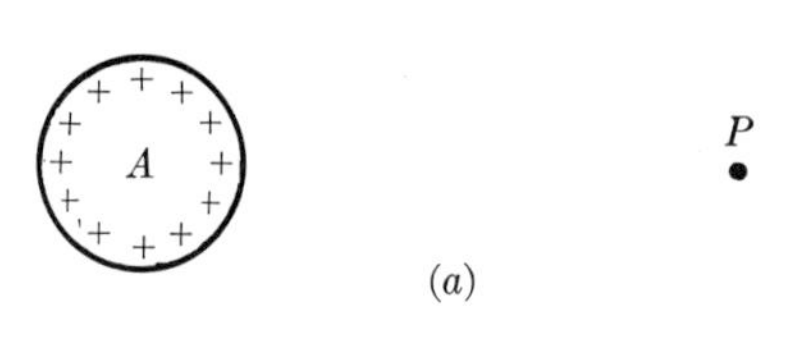

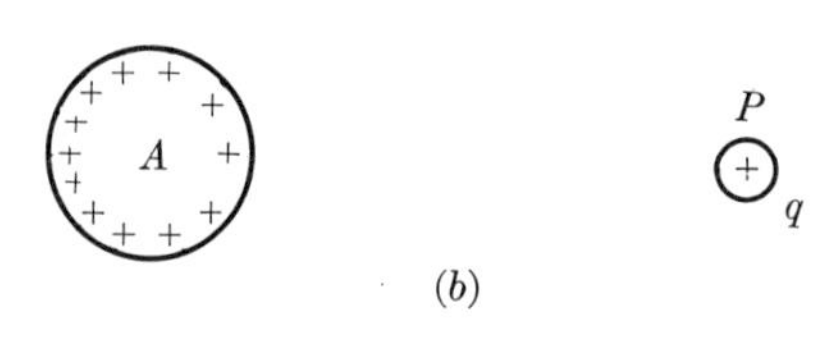

FIG. 2-5. The introduction of a test charge into an electric field may alter the field.

Hence, if we wish to define the electric intensity at a point of *empty space* by the equation

$$E = \frac{F}{q},$$

we must either postulate (a) that the charges setting up the field are assumed not to be displaced by the presence of the test charge, or (b) that an *infinitesimal* test charge shall be used. In the latter case the test charge produces only a negligible alteration in the original charge distribution.

On the other hand, if the electric intensity at a point is defined by the equation

$$E = \frac{1}{4\pi\epsilon_0}\int\frac{dq}{r^2},$$

the difficulty above does not occur, since no use is made of a test charge.

Examples. (1) Compute the electric intensity at the following distances from a positive point charge of magnitude 10^{-9} coulombs: 1 mm, 1 cm, 10 cm, 100 cm. If the field at any point is represented by a vector on a scale of 1 cm = 10,000 newtons/coulomb, find the length of the vector in each instance.

The electric intensity at a distance r from a single point charge is

$$E = \frac{1}{4\pi\epsilon_0}\frac{q}{r^2} = 9 \times 10^9 \frac{q}{r^2}.$$

When $r = 1$ mm $= 10^{-3}$ m,

$$E = 9 \times 10^9 \times \frac{10^{-9}}{(10^{-3})^2} = 9 \times 10^6 \text{ newtons/coulomb.}$$

When $r = 1$ cm, $E = 9 \times 10^4$ newtons/coulomb.
When $r = 10$ cm, $E = 900$ newtons/coulomb.
When $r = 100$ cm, $E = 9$ newtons/coulomb.

The corresponding lengths of the field vectors are: 9 m, 9 cm, 0.9 mm, 0.009 mm.

A few of the vectors representing the electric intensity at points around a positive point charge are indicated in Fig. 2-6. The figure shows only the field vectors at points in the plane of the diagram. In reality, of course, the field is three-dimensional. Since the field has a value at *all* points of space around the charge it would be necessary to completely fill this space with vectors to show the field in its entirety.

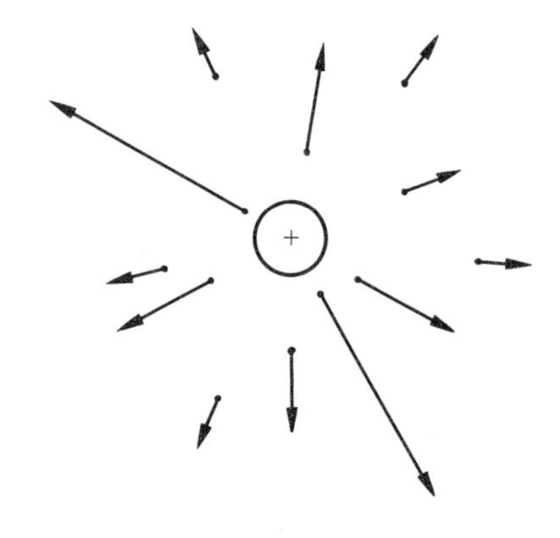

FIG. 2-6. Electric intensity at points in the field of a point charge.

(2) Point charges q_1 and q_2 of $+12 \times 10^{-9}$ and -12×10^{-9} coulombs are placed 10 cm apart as in Fig. 2-7. Compute the electric intensities due to these charges at points a, b, and c.

We must evaluate the vector sum $\frac{1}{4\pi\epsilon_0} \Sigma \frac{q}{r^2}$ at each point.

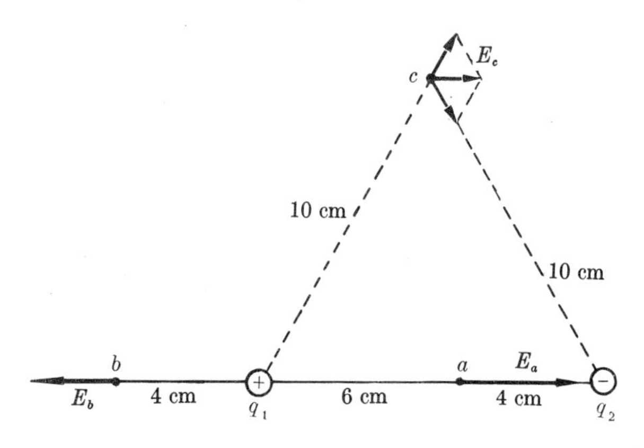

FIG. 2-7. Electric intensity at three points, a, b, c, in the field due to charges q_1 and q_2.

At point a, the vector due to the positive charge is directed toward the right and its magnitude is

$$9 \times 10^9 \times \frac{12 \times 10^{-9}}{(.06)^2} = 3.00 \times 10^4 \text{ newtons/coulomb.}$$

The vector due to the negative charge is also directed toward the right and its magnitude is

$$9 \times 10^9 \times \frac{12 \times 10^{-9}}{(.04)^2} = 6.75 \times 10^4 \text{ newtons/coulomb.}$$

Hence at point a,

$$E_a = (3.00 + 6.75) \times 10^4 = 9.75 \times 10^4 \text{ newtons/coulomb, directed toward the right.}$$

At point b, the vector due to the positive charge is directed toward the left and its magnitude is

$$9 \times 10^9 \times \frac{12 \times 10^{-9}}{(.04)^2} = 6.75 \times 10^4 \text{ newtons/coulomb.}$$

The vector due to the negative charge is directed toward the right and its magnitude is

$$9 \times 10^9 \times \frac{12 \times 10^{-9}}{(.14)^2} = 0.55 \times 10^4 \text{ newtons/coulomb.}$$

Hence at point b

$$E_b = (6.75 - 0.55) \times 10^4 = 6.20 \times 10^4 \text{ newtons/coulomb, directed toward the left.}$$

At point c, the magnitude of each vector is

$$9 \times 10^9 \times \frac{12 \times 10^{-9}}{(.10)^2} = 1.08 \times 10^4 \text{ newtons/coulomb.}$$

The directions of these vectors are shown in the figure and their resultant E is easily seen to be

$$E_c = 1.08 \times 10^4 \text{ newtons/coulomb, directed toward the right.}$$

2-3 Field of a dipole. The field set up by two charges of equal magnitude but of opposite sign is of sufficient importance to warrant further discussion. The pair of equal and opposite charges is called a *dipole*. Fig. 2-8 shows a dipole consisting of the point charges $+q$ and $-q$, separated by a distance l.

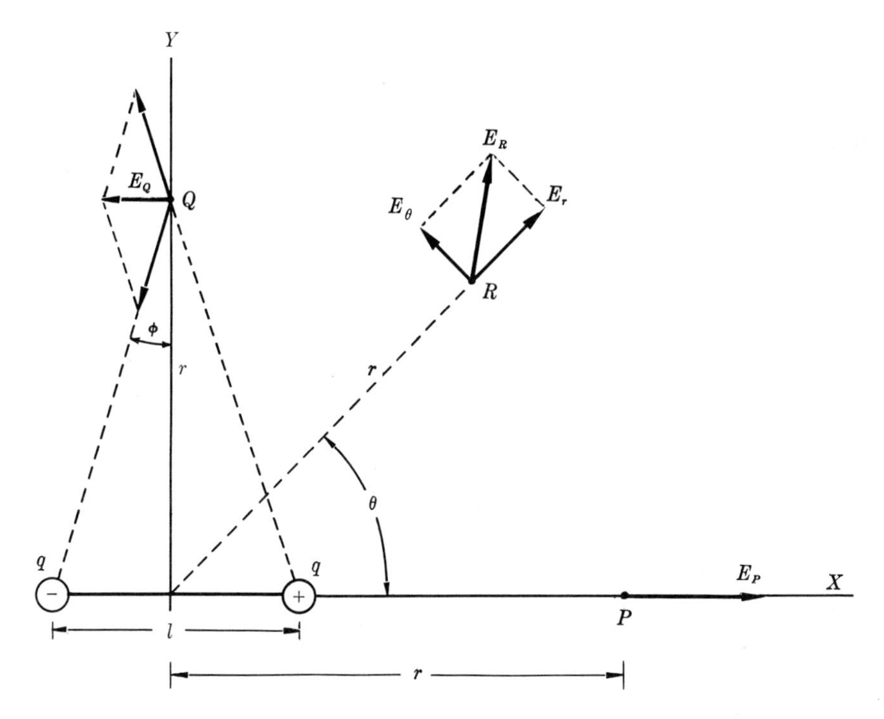

FIG. 2-8. Electric intensity at points in the field of a dipole.

Consider first the field at a point P on the prolonged axis of the dipole, at a distance r from the midpoint of the dipole. From Eq. (2-4) we find

$$E_P = \frac{1}{4\pi\epsilon_0}\left(\frac{q}{\left(r - \frac{l}{2}\right)^2} - \frac{q}{\left(r + \frac{l}{2}\right)^2}\right)$$

$$= \frac{1}{4\pi\epsilon_0} q \frac{(r + l/2)^2 - (r - l/2)^2}{\left(r^2 - \frac{l^2}{4}\right)^2}$$

$$= \frac{1}{4\pi\epsilon_0} q \frac{2rl}{(r^2 - l^2/4)^2}.$$

If the distance between the charges, l, is small in comparison with r, the term $l^2/4$ in the denominator may be neglected in comparison with r^2 and we get

$$E_P = \frac{1}{4\pi\epsilon_0}\frac{2ql}{r^3}.$$

We see that the electric intensity is proportional to the product ql. This product is called the *electric moment* of the dipole. If we abbreviate ql by p, then

$$E_P = \frac{1}{4\pi\epsilon_0}\frac{2p}{r^3}. \tag{2-7}$$

Consider next the point Q, on the perpendicular bisector of the axis of the dipole. The magnitude of the field set up at Q by each charge is

$$E = \frac{1}{4\pi\epsilon_0}\frac{q}{r^2 + \frac{l^2}{4}}.$$

The components of these fields perpendicular to the axis of the dipole evidently cancel one another. The component of each parallel to the axis is $E \sin \phi$, and the resultant electric intensity E_Q is

$$\begin{aligned} E_Q &= 2E \sin \phi \\ &= \frac{1}{4\pi\epsilon_0}\frac{2q}{r^2 + \frac{l^2}{4}}\frac{l/2}{\left(r^2 + \frac{l^2}{4}\right)^{\frac{1}{2}}}. \end{aligned}$$

If r is sufficiently large in comparison with l we may neglect the terms $l^2/4$ in the denominator, and the expression reduces to

$$E_Q = \frac{1}{4\pi\epsilon_0}\frac{ql}{r^3},$$

or, in terms of the electric moment p,

$$E_Q = \frac{1}{4\pi\epsilon_0}\frac{p}{r^3}. \tag{2-8}$$

Hence we see that at both points P and Q the intensity is proportional to the electric moment and inversely proportional to the *cube* of the distance from the center of the dipole (at distances sufficiently large for the approximations to be justified).

The electric intensity at any point such as R may be evaluated by the method described above, but it is much simpler to use another method that will be explained in Sec. 3-5. It turns out to be more convenient to use polar coordinates, r and θ, and to evaluate the intensity in terms of its rectangular components E_r, in the direction of increasing r, and E_θ, in the direction of increasing θ. The final expressions are

$$E_r = \frac{1}{4\pi\epsilon_0}\frac{2p\cos\theta}{r^3}, \tag{2-9}$$

$$E_\theta = \frac{1}{4\pi\epsilon_0}\frac{p\sin\theta}{r^3}, \tag{2-10}$$

where $r \gg l$.

It is left as a problem to show that these equations reduce to Eq. (2-7) or Eq. (2-8) when the point lies on the X- or Y-axis.

2-4 Field due to continuous distribution of charge. Thus far we have computed only the fields set up by one or more point charges. In practice, one is nearly always concerned with fields due to charges on bodies of finite size. The calculation of such fields is extremely complicated, if not impossible, except for bodies of simple geometrical shape such as wires, planes, cylinders or spheres. Furthermore, if the charged bodies are immersed in a liquid or gas, or if other uncharged bodies are nearby, so-called "induced" charges are developed within or at the surfaces of these bodies which further complicates the problem. Hence we shall for the present consider only charged bodies of simple geometrical shape in otherwise empty space.

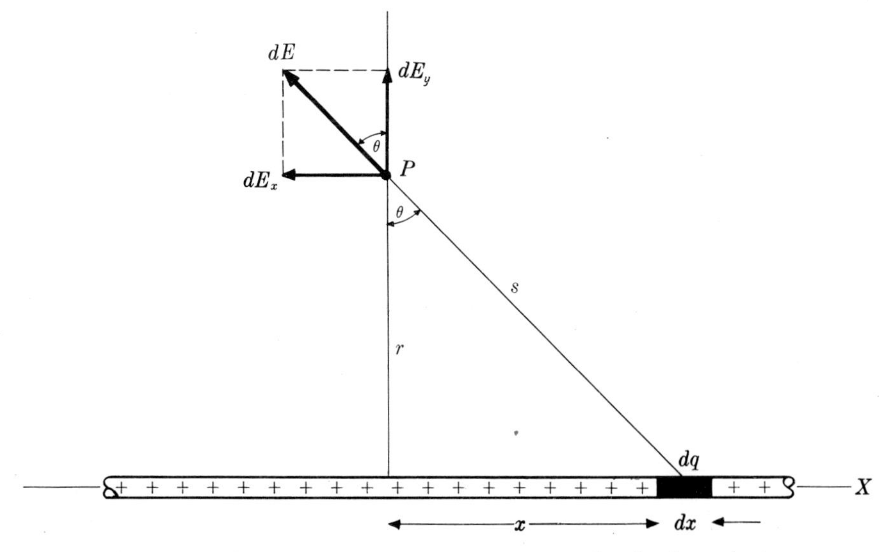

FIG. 2-9. Electric intensity at a point near a uniformly charged wire.

Let us begin with the example of a long fine wire, having a positive charge uniformly distributed along its length. We wish to find the magnitude and direction of the electric intensity at a point P, Fig. 2-9, at a perpendicular distance r from the wire. The radius of the wire itself is small enough so that it may be considered a geometrical line.

Let the wire be divided into short segments of length dx, each of which may be treated as a point charge dq. The resultant intensity at any point is then to be found by summing (vectorially) the fields set up by all of these point charges. At point P, the element dq sets up a field of magnitude

$$dE = \frac{1}{4\pi\epsilon_0}\frac{dq}{s^2}.$$

Since the fields of other elements are not in the same direction as dE, the resultant intensity cannot be found by integration of this equation. (Integration is a process of *algebraic*, not *vector*, summation.) The X- and Y-components of dE can, however, be integrated separately. That is,

$$E_x = \int dE_x = \int dE \sin\theta$$

$$E_y = \int dE_y = \int dE \cos\theta.$$

Let λ represent the charge per unit length on the wire. The charge on an element of length dx is

$$dq = \lambda\, dx,$$

and

$$dE = \frac{1}{4\pi\epsilon_0}\lambda\frac{dx}{s^2}.$$

The integration may be simplified if θ replaces x as the independent variable. From Fig. 2-9,

$$x = r\tan\theta, \quad dx = r\sec^2\theta d\theta, \; s = r\sec\theta.$$

Making these substitutions, we obtain

$$E_x = \frac{1}{4\pi\epsilon_0}\frac{\lambda}{r}\int \sin\theta d\theta,$$

$$E_y = \frac{1}{4\pi\epsilon_0}\frac{\lambda}{r}\int \cos\theta d\theta.$$

If the wire is infinitely long, the limits of integration are from $\theta = -\pi/2$ to $\theta = +\pi/2$, and we obtain finally

$$E_x = 0, \quad E_y = \frac{1}{2\pi\epsilon_0}\frac{\lambda}{r}.$$

The resultant intensity is therefore at right angles to the wire, as might have been expected by symmetry. Notice that the intensity is inversely proportional to the *first* power of the distance from the wire.

As a second example, let Fig. 2-10 represent a circular ring lying in the X-Z plane, with its center at the origin. The radius of the ring is a and it carries a positive charge q. We wish to evaluate the electric intensity at a point P on the axis of the ring, at a distance b from its center.

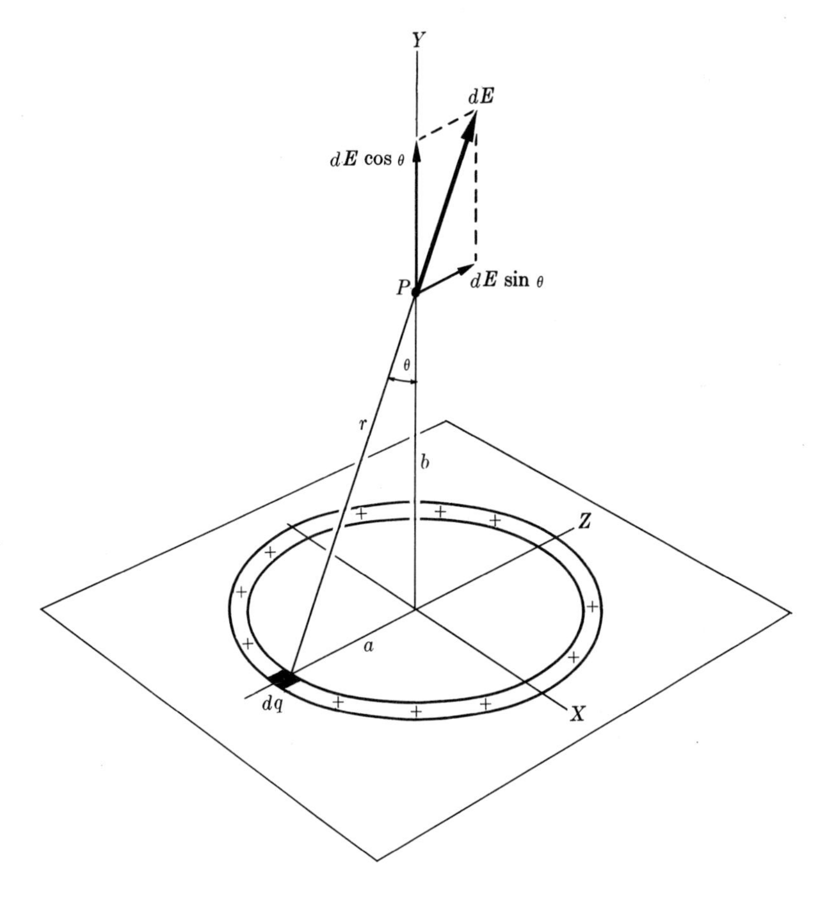

FIG. 2-10. Electric intensity at a point on the axis of a uniformly charged ring.

A small element of the ring whose charge is dq sets up at P a field dE, of magnitude

$$dE = \frac{1}{4\pi\epsilon_0}\frac{dq}{r^2}.$$

It will be seen that when all of the elements of the ring are considered, the components $dE \sin\theta$, perpendicular to the axis of the ring, will cancel and the resultant field at P is the sum of the components $dE \cos\theta$. Hence

$$E = \int dE \cos\theta = \frac{1}{4\pi\epsilon_0}\frac{\cos\theta}{r^2}\int dq,$$

or,

$$E = \frac{1}{4\pi\epsilon_0}\frac{q\cos\theta}{r^2} = \frac{1}{4\pi\epsilon_0}\, q\,\frac{b}{(a^2 + b^2)^{\frac{3}{2}}}.$$

At the center of the ring, where $b = 0$, $E = 0$, as is obvious by symmetry. At distances such that $b \gg a$, a^2 may be neglected in comparison with b^2 and

$$E = \frac{1}{4\pi\epsilon_0}\frac{q}{b^2}.$$

In other words, at sufficiently great distances the field is the same as that of a point charge q.

2-5 Lines of force. The concept of lines of force was introduced by Michael Faraday (1791–1867) as an aid in visualizing electric (and magnetic) fields. A *line of force* (in an electric field) is *an imaginary line drawn in such a way that its direction at any point* (i.e., the direction of its tangent) *is the same as the direction of the field at that point.* (See Fig. 2-11.) Since, in general, the direction of a field varies from point to point, lines of force are usually curves.

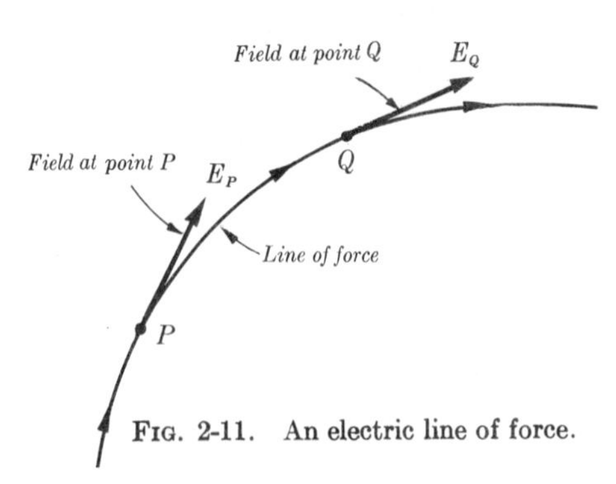

FIG. 2-11. An electric line of force.

Fig. 2-12 shows some of the lines of force around a single positive charge; around two equal charges, one positive and one negative; and around two equal positive charges. Computation such as that in Secs. 2-3 and 2-4 will show that the direction of the resultant intensity at every point in each diagram

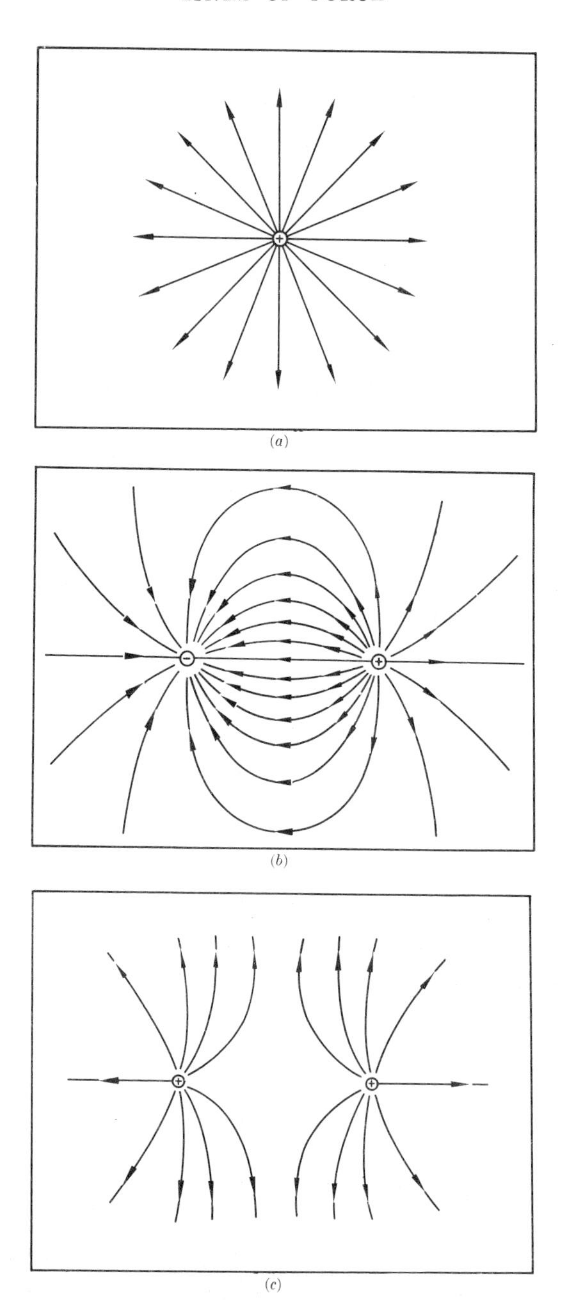

FIG. 2-12. The mapping of an electric field with the aid of lines of force.

is along the tangent to the line of force passing through the point. Arrowheads on the lines indicate the direction in which the tangent is to be drawn.

It is, of course, possible to draw a line of force through every point of an electric field, but if this were done, the whole of space and the entire surface of a diagram would be filled with lines, and no individual line could be distinguished. By suitably limiting the number of force lines which one draws to represent a field, the lines can be used to indicate the magnitude of a field as well as its direction. This is accomplished by spacing the lines in such a way that the number per unit area crossing a surface at right angles to the direction of the field, is at every point numerically equal to the electric intensity. That is, if at some point in an electric field the intensity is 10^4 newtons/coulomb parallel to the X-axis toward the right, one constructs at that point 10^4 lines per square meter crossing a plane perpendicular to the X-axis. In a region where the intensity is large, the lines of force will be closely spaced, and, conversely, in a region where the intensity is small the lines will be widely separated.

If the electric intensity, and hence the number of lines per unit area, is the same at all points of a surface of finite extent perpendicular to the field, the total number of lines N threading through the surface is

$$N = EA,$$

where E is the uniform electric intensity and A is the surface area. The condition above is fulfilled at all points of a spherical surface of any radius concentric with an isolated point charge, since the intensity at a distance r from a point charge q is

$$E = \frac{1}{4\pi\epsilon_0}\frac{q}{r^2},$$

and the direction of the field is everywhere radially outward from the charge or perpendicular to the spherical surface. The total number of lines threading through the sphere is therefore

$$N = EA = \frac{1}{4\pi\epsilon_0}\frac{q}{r^2} \times 4\pi r^2 = \frac{1}{\epsilon_0}q. \tag{2-11}$$

This number is independent of the radius of the sphere, from which it follows that the same number of lines cross *every* sphere concentric with the charge. In other words, no lines of force originate or terminate in the space surrounding the sphere. This conclusion is found to be true in all instances. Every line of force in an electrostatic field is a continuous line terminated by a positive charge at one end and a negative charge at the other. While sometimes for convenience we speak of an "isolated" charge, and draw its field as in Fig. 2-12(a), this simply means that the charges on which the lines terminate are at large distances from the charge under consideration. For example, if the charged body in Fig. 2-12(a) is a small sphere suspended by a thread from the laboratory ceiling, the negative charges on which its force lines terminate would be found on the walls, floor, or ceiling, or on other objects in the laboratory.

At any one point in an electric field the field can have but one direction. Hence only one line of force can pass through each point of the field. In other words, lines of force never intersect.

If the electric intensity varies from point to point on a surface, and if the surface is not everywhere at right angles to the field, the number of lines crossing the surface can be expressed as follows. In Fig. 2-13, dA is an infinitesimal element of area whose normal makes an angle ϕ with the direction of an electric field. The projected area of dA perpendicular to the field is $dA \cos \phi$, and the number of lines passing through dA is $E\,dA \cos \phi$, where E is the intensity at the point where dA is located. The total number of lines through a finite area is then

$$N = \int E\,dA \cos \phi, \quad (2\text{-}12)$$

where the limits of integration must be chosen so as to include the entire surface.

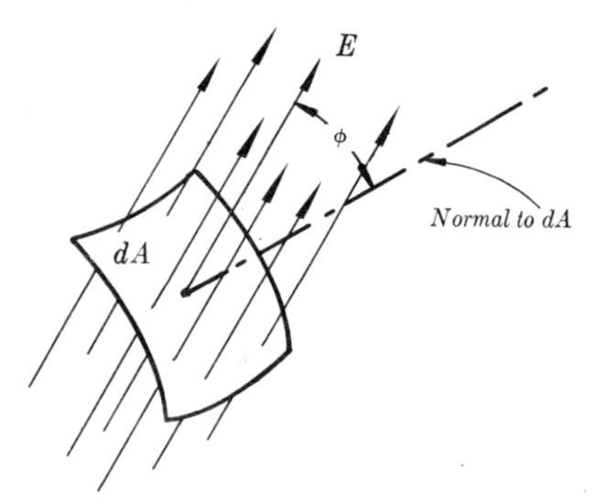

Fig. 2-13. Electric field at an angle to an element of a surface.

The term $\cos \phi$ in Eq. (2-12) can equally well be associated with E, that is, $E \cos \phi$ is the component of E normal to the surface.

Since $1/\epsilon_0 = 36\pi \times 10^9$, it follows from Eq. (2-11) that $36\pi \times 10^9$ lines of force emanate from a positive charge of one coulomb.

2-6 Gauss's law. We showed in the preceding section that a total of $(1/\epsilon_0)q$ lines of force cross every spherical surface concentric with a point charge q. Since the lines are continuous, it is evident that the same number will cross a closed surface of any shape enclosing the same charge. *Gauss's law* is a generalization of this result, and it states that *if a closed surface of any shape is constructed in an electric field, the total number of lines of force crossing it in an outward direction is equal to* $1/\epsilon_0$ *times the net positive charge within the surface,*[1] regardless of how the charge may

[1] But see Sec. 7-1 for a generalization if the surface passes through matter.

be distributed. Making use of Eq. (2-12) as the general expression for the total number of lines crossing a surface in terms of the field intensity at the surface, Gauss's law is

$$\boxed{\int E \cos \phi \, dA = \frac{1}{\epsilon_0} q,} \tag{2-13}$$

where it is understood that the integral extends over a closed surface, and where q is the *net* positive charge within the surface.

As a corollary, if there is no net charge within a closed surface, the net number of lines crossing the surface is zero.

An element of area, dA, can be represented vectorially by a vector of length proportional to dA, and at right angles to dA. Then Gauss's law, in vector notation, is

$$\int \vec{E} \cdot d\vec{A} = \frac{1}{\epsilon_0} q.$$

Example. Fig. 2-14 is a schematic diagram of a number of charged bodies, showing a few of the force lines of their field, and the traces with the plane of the diagram of four closed surfaces, A, B, C, and D. The actual field, and the closed surfaces, are, of course, three dimensional. Each + and − sign represents one coulomb, but only one of the $36\pi \times 10^9$ lines which leave or terminate on each unit charge is shown.

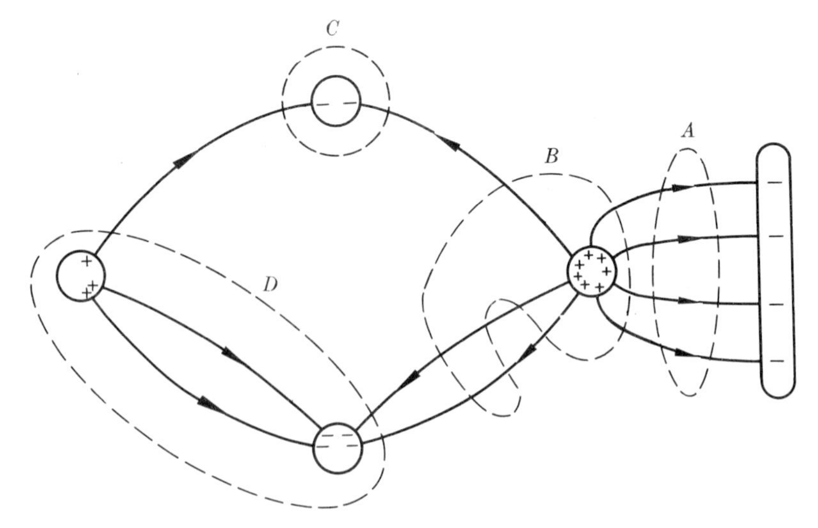

FIG. 2-14. Gauss's law.

The number of lines crossing each surface can be found by inspection without resorting to integration. Consider first surface A. Four lines leave it and four lines enter it. No line can terminate within the surface, since there is no charge within it. The net number of lines crossing the surface in an outward direction (zero) equals $36\pi \times 10^9$ times the net charge within the surface (also zero).

Consider next surface B, which encloses 7 units of positive charge. The surface is intersected at 11 points, at 9 of which the lines cross it in an outward direction and at 2 of which the crossing is inward. The net number of lines crossing in an outward direction is 7, and since $1/36\pi \times 10^9$th of the lines are shown, the total number crossing the surface is $252\pi \times 10^9$, or $36\pi \times 10^9$ times the positive charge within the surface.

Similar analyses of surfaces C and D are left as an exercise.

2-7 Field and charge within a conductor. A conductor is a material within which there are free charges, that is, charges which are free to move provided a force is exerted on them by an electric field. If by any means an electric field is maintained within a conductor, there will be a continuous motion of its free charges. (This motion is called a current.) On the other hand, if there is no field within a conductor there will be no motion of its free charges.[1] We may now reverse the reasoning and conclude that if the charges in a conductor are known to be at rest, the field within the conductor must be zero. With the help of this fact and Gauss's law we may now show that when a conductor is charged, the *excess* charge is confined entirely to the surface of the conductor.

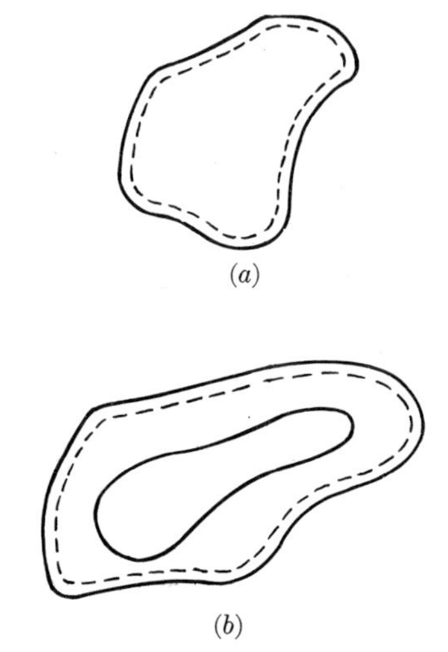

FIG. 2-15. Application of Gauss's law to a conductor.

Fig. 2-15(a) represents a solid conductor of irregular shape, to which has been given a charge q. Construct a surface (shown dotted in the figure) lying in the material of the conductor, just inside its surface. According to Gauss's law, the

[1] The free electrons within a metal share in the thermal energy of the atoms, but their thermal motion is a random one like that of the atoms in a gas and does not result in a net transfer of charge across any surface. For our present purposes this random electronic motion can be ignored.

total number of lines of force crossing this surface equals $1/\epsilon_0$ times the net charge enclosed by the surface. But if the charges are at rest, the electric intensity at all points within the conductor is zero. The total number of lines crossing the dotted surface is therefore zero and, from Gauss's law, the net charge within this surface is zero also. All of the charge on the conductor must therefore lie *outside* the dotted surface. But this surface is only an infinitesimal distance inside the actual surface of the conductor. Hence all of the excess charge on the conductor lies on its surface.

The result derived above is in agreement with what one would expect from the forces of repulsion between like charges. Suppose that in some way an excess positive or negative charge had been distributed throughout the interior of the conductor. The mutual forces of repulsion would cause the charges to separate as far as possible, and the greatest separation would be obtained with all of the charge on the outside surface.

If the conductor is not solid but is hollow as in Fig. 2-15(b) the same result holds true. Gauss's law, applied to the dotted surface in Fig. 2-15(b), shows that there can be no charge within this surface, and hence no charge and no field within the cavity. The fact that the field within a closed conductor is zero is the basis of what is known as electrical or electrostatic *shielding*. Any device such as a radio tube or an electroscope can be isolated from the influence of other charges by surrounding it by a conductor. This purpose is served in the modern metal-walled vacuum tube by the walls of the tube itself. An interesting experiment is to surround an electroscope by a shield of wire mesh. The shield may be connected directly to one terminal of a static machine, and the electroscope will show practically no deflection when the machine is in operation. The wire mesh is not a continuous closed surface, but its shielding is nearly as effective as if it were.

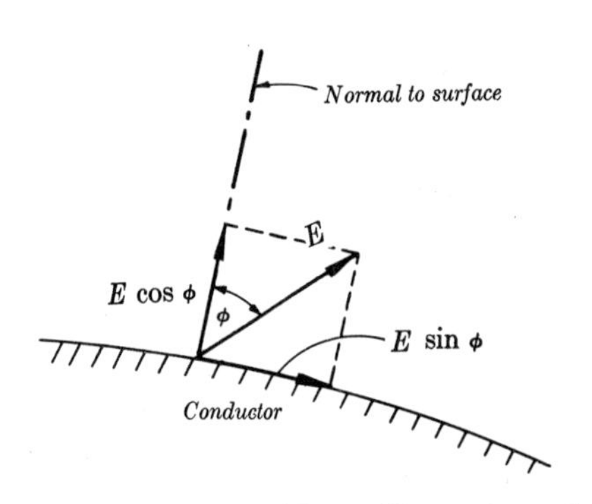

FIG. 2-16. The field outside a charged conductor is perpendicular to the surface.

The direction of the field just outside the surface of a charged conductor is perpendicular to the surface, if the charges on the conductor are at rest. To show that this is true, assume, as in Fig. 2-16, that the field at the surface of a conductor makes an angle ϕ with the normal to the surface. The field can then be resolved into components perpendicular and parallel to the surface. Under the influence of the parallel component, $E \sin \phi$, charges would move along the surface of the conductor, which is contrary to the hypothesis that the charges are at rest. Hence $E \sin \phi = 0$, and the field is normal to the surface.

2-8 Application of Gauss's law. As an illustration of the utility of Gauss's law, we shall compute with its aid the electric intensity due to a few commonly encountered distributions of charge.

Field of a long charged wire. This problem has already been worked out in Sec. 2-4, where E was computed from the defining equation

$$E = \frac{1}{4\pi\epsilon_0} \int \frac{dq}{r^2} \text{ (vector sum).}$$

The same problem will now be solved, making use of Gauss's law. Construct a closed "gaussian" surface as in Fig. 2-17, in the form of a right cylinder of radius r and length b, coaxial with the wire. At points not too near its ends, the lines of force around the wire are, by symmetry, perpendicular to the wire and to the curved surfaces of the cylinder. Also, by symmetry, the field has the same intensity at all points of the curved surface. The total number of lines of force crossing the cylinder in an outward direction is therefore the product of the electric intensity E at its curved surface and the area of this surface. (No lines cross the ends of the cylinder.) The quantity of charge within the cylinder is λb, where λ is the charge per unit length. Therefore

$$E \times 2\pi r b = \frac{1}{\epsilon_0} \lambda b,$$

$$E = \frac{1}{2\pi\epsilon_0} \frac{\lambda}{r} \cdot$$

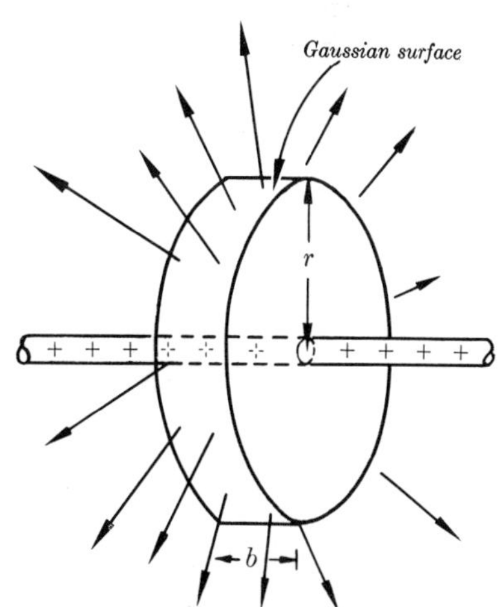

FIG. 2-17. Gaussian surface for a long charged wire.

This example illustrates the simplification made possible by the use of Gauss's law in a problem where symmetry considerations show the field to be constant over a surface of simple geometrical shape.

Field around a cylinder. Fig. 2-18 shows a long cylinder of radius a, having a uniform surface density of charge σ. (The surface density of charge is the charge per unit area, and is expressed in coulombs/meter^2.) Construct a gaussian surface as in Fig. 2-18. By the same reasoning as in the previous example,

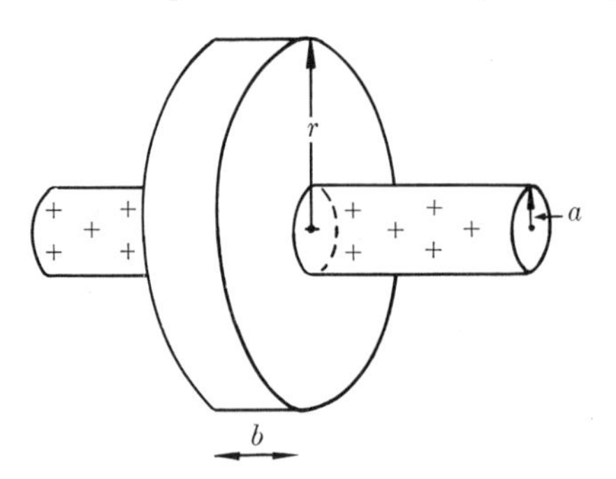

FIG. 2-18. Gaussian surface for a long charged cylinder.

$$E \times 2\pi rb = \frac{1}{\epsilon_0} \times \sigma \times 2\pi ab,$$

$$E = \frac{1}{\epsilon_0}\frac{\sigma a}{r}. \tag{2-14}$$

The charge per unit length of the cylinder is $2\pi a\sigma$. Calling this λ as before, we have

$$\lambda = 2\pi a\sigma, \quad \sigma a = \lambda/2\pi,$$

and Eq. (2-14) becomes

$$E = \frac{1}{2\pi\epsilon_0}\frac{\lambda}{r}. \tag{2-15}$$

This is the same as the expression for the field around a "line" of charge. Hence the field *outside* a charged cylinder is the same as though the charge on the cylinder were concentrated along its axis. Of course the field inside the cylinder is zero, and Eqs (2-14) and (2-15) are true only for values of r equal to or greater than a.

The field in the space between two oppositely charged coaxial cylinders, having equal charges per unit length, is also given by Eq. (2-15). The quantity λ represents the charge per unit length on either cylinder.

Field around a charged sphere. It will be left as an exercise to show by Gauss's law that the field at a distance r from the center of a sphere of radius a, on which the surface density of charge is uniform and given by σ, is

$$E = \frac{1}{\epsilon_0}\frac{\sigma a^2}{r^2}, \tag{2-16}$$

where r is equal to or greater than a.

Since the total charge q on the sphere is

$$q = 4\pi a^2\sigma,$$

Eq. (2-16) can be written

$$E = \frac{1}{4\pi\epsilon_0}\frac{q}{r^2},$$

and the field outside a uniformly charged conducting sphere is the same as though all of the charge were concentrated at the center of the sphere. The field inside the sphere is zero.

Eq. (2-16) also gives the electric intensity at points in the space between two concentric spherical conductors.

Field between parallel plates. When two plane parallel conducting plates, having the size and spacing shown in Fig. 2-19, are given equal and opposite charges, the field between and around them is approximately as shown in Fig. 2-19(a). While most of the charge accumulates at the opposing faces of the plates, and the field is essentially uniform in the space between them, there is a small quantity of charge on the outer surfaces of the plates and a certain spreading or "fringing" of the field at the edges of the plates.

As the plates are made larger and the distance between them diminished, the fringing becomes relatively less. Such an arrangement of two oppositely charged plates separated by a distance small compared with their linear dimensions is encountered in many pieces of electrical equipment, notably in capacitors (see Sec. 8-3). In many instances the fringing is entirely negligible, and even if it is not, it is usually neglected for simplicity in computation. We shall therefore assume that the field between two

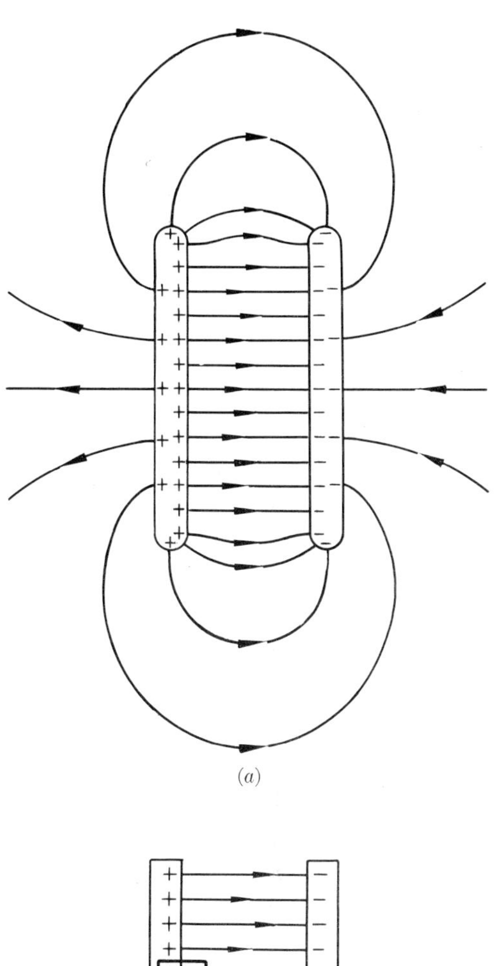
(a)

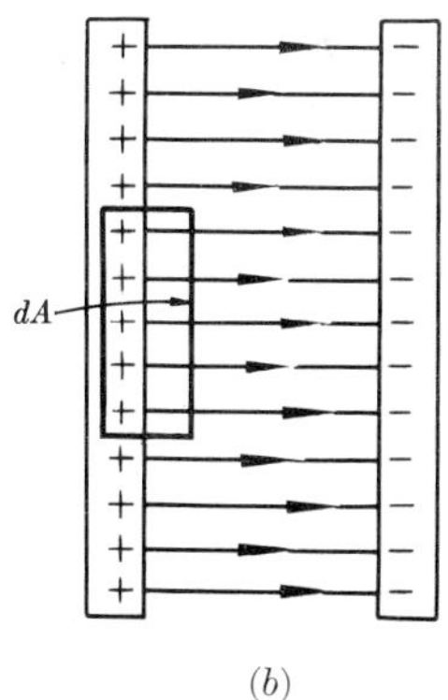

(b)

FIG. 2-19. Electric field between oppositely-charged parallel plates.

oppositely charged plates is uniform as in Fig. 2-19(b), and that the charges are distributed uniformly over the opposing surfaces.

The electric intensity at a point in the space between the plates can be computed from the definition

$$E = \frac{1}{4\pi\epsilon_0}\int\frac{dq}{r^2}$$

by performing a double integration over both plates, but it is much simpler to compute it from Gauss's law. The small rectangle in Fig. 2-19(b) is a side view of a closed surface shaped like a pillbox. Its ends, of area dA, are perpendicular to the plane of the figure. One of them lies within the left conductor, the other in the field. Let E represent the (uniform) electric intensity between the plates, and σ the surface density of charge (charge per unit area) on either plate. Lines of force cross the surface of the pillbox only over the end in the space between the plates, since the field within the conductor is zero. The number of lines crossing this end is $E\,dA$. The charge within the pillbox is $\sigma\,dA$. Then from Gauss's law

$$E\,dA = \frac{1}{\epsilon_0}\sigma\,dA,$$

and

$$\boxed{E = \frac{1}{\epsilon_0}\sigma} \tag{2-17}$$

In practice, electric fields are much more commonly set up by charges distributed over parallel plates, than by some arrangement of point charges. If we had *not* introduced the factor 4π in our original formulation of Coulomb's law, we would have found it appearing in Eq. (2-17). The *rationalized* system therefore relegates the factor 4π to equations which, while fundamental, are less frequently encountered.

We have shown in Sec. 2-7 that the field *just outside* the surface of any charged conductor is at right angles to the surface. If a small pillbox is constructed as in Fig. 2-19(b), enclosing a small portion of any charged surface, it follows that the electric intensity just outside the surface is also given by Eq. (2-17). In general, of course, the field varies as one proceeds away from the surface. In the special case of two plane parallel plates, the field is the same at all points between the plates.

Note that Eqs. (2-14) and (2-16) reduce to

$$E = \frac{1}{\epsilon_0}\sigma$$

at the surfaces of the cylinder and the sphere, i.e., when $r = a$.

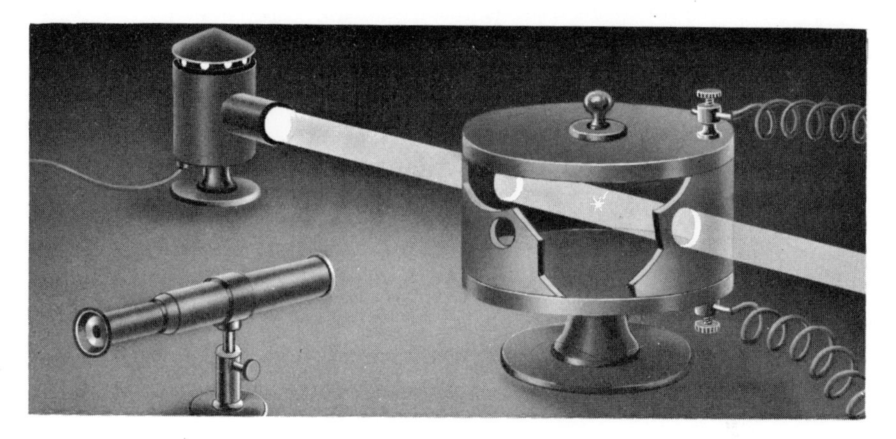

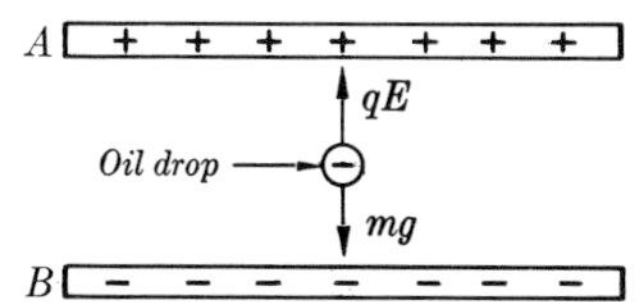

FIG. 2-20. Millikan's oil drop experiment.

2-9 The Millikan oil drop experiment. We have now developed the theory of electrostatics to a point where one of the classical experiments of all time can be described—the measurement of the charge of an individual electron by Robert Andrews Millikan.

Millikan's apparatus is shown in Fig. 2-20. A and B are two accurately parallel horizontal metal plates. Oil is sprayed in fine droplets from an atomizer above the upper plate and a few of the droplets allowed to fall through a small hole in this plate. A beam of light is directed between the plates and a telescope set up with its axis transverse to the light beam. The oil drops, illuminated by the light beam, appear like tiny bright stars when viewed through the telescope, falling slowly under the combined influence of their weight, the buoyant force of the air, and the viscous force opposing their motion.

It is found that the oil droplets in the spray from an atomizer are electrically charged, presumably because of frictional effects. This charge is usually negative, meaning that the drops have acquired one or more excess electrons. If now the upper plate is positively charged and the lower plate negatively charged, the region between the plates becomes a uniform electric field. By adjusting the electric intensity, the force exerted on a negatively charged drop by the field may be made just equal to the weight of the drop less the buoyant force of the air, so that the drop can be held stationary between the plates. When the field has been adjusted to this value we have

$$qE = \text{weight} - \text{buoyant force}$$
$$= \tfrac{4}{3}\pi a^3 g(\rho - \rho'), \tag{2-18}$$

where q is the charge of the drop, E the electric intensity, g the acceleration of gravity, a the radius of the drop, ρ its density, and ρ' the density of air. Hence, if E, g, a, ρ, and ρ' are known, the charge may be found.

The radii of the drops are about 10^{-5} cm, much too small to be measured directly. Millikan devised an extremely ingenious method for determining these radii. When the field is removed, the drop falls slowly and its rate of fall may be measured by timing it as it passes reference cross hairs in the telescope. We have seen that a body falling in a viscous medium accelerates until a terminal velocity is reached such that the net downward force (weight minus buoyant force) equals the viscous force as given by Stokes' law.

$$\tfrac{4}{3}\pi a^3 g(\rho - \rho') = 6\pi\eta a v, \tag{2-19}$$

where η is the viscosity of the medium and v the terminal velocity. Since g, ρ, ρ' and η are known and v may be measured, a can be found. Hence q can be computed from Eq. (2-18).

During the course of experiments carried on over the period 1909–1913, Millikan and his co-workers measured the charges of some thousands of drops, with the following remarkable result. Every drop observed was found to have a charge equal to some small integral multiple of a basic quantity of (negative) charge e. That is, drops were observed having charges of e, or exactly $2e$ or exactly $3e$, and so forth, but never such values as $0.76e$ or $2.49e$. The evidence is conclusive that electric charge is not something which can be divided indefinitely, but that it exists in nature only in units of magnitude e. When a drop is observed with charge e we conclude it has acquired one extra electron, if its charge is $2e$, it has two extra electrons, and so on.

Measurements of the electronic charge by Millikan's method are still being carried on in a number of laboratories. The best results to date give[1]

$$e = 1.6065 \times 10^{-19} \text{ coulomb}$$
$$= 4.8130 \times 10^{-10} \text{ statcoulomb}$$

[1] This figure differs slightly from that quoted in Sec. 1-16, which is based on the results of *all* measurements of e, including those made by other methods.

The brief description above cannot do justice to Millikan's work. For a more complete description of his experiments see Millikan's book, "*Electrons + and −.*"

2-10 Dielectric strength. An insulator, or dielectric, is a substance within which there are no (or at least relatively few) charged particles free to move continuously under the influence of an electric field. For every dielectric there exists, however, a certain limiting electric intensity above which the substance loses its insulating properties and becomes a conductor. The maximum electric intensity which a dielectric can withstand without breakdown is called its *dielectric strength.* For example, the dielectric strength of air at atmospheric pressure is about 100 dynes per statcoulomb or 3×10^6 newtons per coulomb, while that of glass is two or three times as great. (See Sec. 3-5 for other units in which dielectric strengths are more commonly expressed.)

Since the maximum electric intensity that can exist in air at atmospheric pressure without breakdown is about 3×10^6 newtons/coulomb, the maximum surface density of charge on a conductor in air can be computed from Eq. (2-17), setting $E = 3 \times 10^6$. One finds

$$\begin{aligned}\sigma = \epsilon_0 E = \frac{3 \times 10^6}{36\pi \times 10^9} &= 27 \times 10^{-6}\ \text{coulombs/m}^2\\ &= 27\ \text{microcoulombs/m}^2.\end{aligned}$$

Hence the maximum *total* charge that can be retained by a metal sphere 1 cm in radius in air is

$$\begin{aligned}q = 4\pi a^2 \sigma &= 3.3 \times 10^{-8}\ \text{coulombs}\\ &= .033\ \text{microcoulombs}.\end{aligned}$$

It is of interest to compare this charge with the total charge of all the free electrons in the sphere. Assume it to be of copper, of density 8.8 gm/cm^3. Its volume is $4\pi a^3/3$ or 4.2 cm^3, its mass is 37 gm, and it therefore contains $37/63 = 0.59$ mole of copper, or 3.5×10^{23} atoms. If each atom contributes one free electron, of charge 1.6×10^{-19} coulombs, the total free electronic charge is 5.6×10^4 coulombs. Hence an excess charge of 3.3×10^{-8} coulombs means an excess (or deficiency) of only about one free electron in 10^{12}. The sphere may be "highly charged," but the proportional change in its normal electron population is negligible.

Actually, the conditions under which a spark will jump from one conductor to another in air are much more complex than is implied by the simple statement that the electric intensity cannot exceed 3×10^6 newtons/coulomb. The size and shape of the conductors, their separation, and many other factors must be considered. A text on the conduction of electricity in gases should be consulted for further details.

Problems—Chapter 2

(1) The electric intensity in the region between the deflecting plates of a certain cathode ray oscilloscope is 30,000 newtons/coulomb. (a) What is the force on an electron in this region? (b) What is the acceleration of an electron when acted on by this force?

(2) A uniform electric field exists in the region between two oppositely-charged plane parallel plates. An electron is released from rest at the surface of the negatively-charged plate and strikes the surface of the opposite plate, 2 cm distant from the first, in a time interval of 1.5×10^{-8} sec. (a) Find the electric intensity. (b) Find the velocity of the electron when it strikes the second plate.

(3) An electron is projected with an initial velocity $v_0 = 10^7$ m/sec into the uniform field between the parallel plates in Fig. 2-21. The direction of the field is vertically downward, and the field is zero except in the space between the plates. The electron enters the field at a point midway between the plates. (a) If the electron just misses the upper plate as it emerges from the field, find the magnitude of the electric intensity. (b) Find the direction of the velocity of the electron as it emerges from the field.

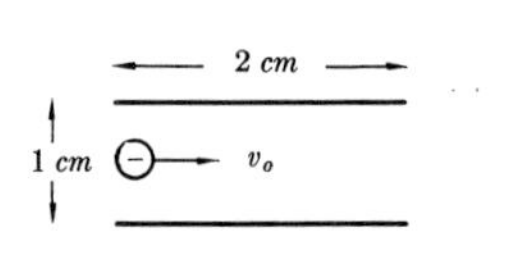

FIG. 2-21.

(4) An electron is projected into a uniform electric field of intensity 5000 newtons/coulomb. The direction of the field is vertically upward. The initial velocity of the electron is 10^7 m/sec, at an angle of 30° above the horizontal. (a) Find the maximum distance the electron rises vertically above its initial elevation. (b) After what horizontal distance does the electron return to its original elevation? (c) Sketch the trajectory of the electron.

(5) In a rectangular coordinate system, two positive point charges of 10^{-8} coulomb each are fixed at the points $x = +0.1$ m, $y = 0$, and $x = -0.1$ m, $y = 0$. Find the magnitude and direction of the electric intensity at the following points: (a) the origin; (b) $x = 0.2$ m, $y = 0$; (c) $x = 0.1$ m, $y = 0.15$ m; (d) $x = 0$, $y = 0.1$ m.

(6) Same as problem 5, except that one of the point charges is positive and the other negative.

(7) The maximum charge that can be retained by one of the spherical terminals of a large Van de Graaff generator is about 10^{-3} coulomb. Assume a positive charge of this magnitude, distributed uniformly over the surface of a sphere in otherwise empty space. (a) Compute the magnitude of the electric intensity at a point 5 m from the center of the sphere. (b) If an electron were released at this point, what would be the magnitude and direction of its initial acceleration?

(8) (a) What is the electric intensity in the field of a gold nucleus, at a distance of 10^{-12} cm from the nucleus? (See problem 5 in Chap. 1.) (b) What is the electric intensity in the field of a proton, at a distance of 5.28×10^{-9} cm from the proton? (See problem 6 in Chap. 1.)

(9) A dipole consisting of an electron and a proton, 4×10^{-8} cm apart, is located with its midpoint at the origin, the dipole axis along the X-axis, and the electron at the left of the origin. Compute and show in a diagram the components of electric intensity of the dipole field, E_r and E_θ, at the following points: (a) $r = 10^{-6}$ cm, $\theta = 0$; (b) $r = 10^{-6}$ cm, $\theta = \pi/2$; (c) $r = 10^{-6}$ cm, $\theta = \pi/6$.

(10) A circular ring of fine wire carries a uniformly distributed positive charge q. (a) Find the magnitude and direction of the electric intensity at the center of the ring, due to the charge on a portion of the ring subtending an angle θ at the center.

(11) A circular ring 5 cm in radius has a positive charge of 100 statcoulombs. (a) Compute the magnitude and direction of the electric intensity at points on the axis of the ring, at the following distances from the center of the ring: 0, 2 cm, 3 cm, 4 cm, 6 cm, 10 cm. Use the electrostatic system of units in which $1/4\pi\epsilon_0 = 1$, and E is expressed in dynes/statcoulomb. Show the results in a graph. (b) From the general expression for the electric intensity at any point on the axis of a ring of charge, find by differentiation the expression for the distance from the center of a ring of radius a at which the intensity is a maximum. Does the answer agree with the graph plotted in part (a)?

(12) Two circular metal plates, parallel to one another and separated by a distance small in comparison with their radii, have equal and opposite surface densities of charge σ. Consider each plate to be subdivided into a number of concentric zones. (a) Find the expression for the electric intensity, at a point in the space between the plates on a line joining their centers, due to the charge on a single zone of radius r and width dr. (b) Integrate this expression to find the resultant electric intensity.

(13) Prove with the aid of Gauss's law that the electric intensity at points outside a sphere having a uniform surface density of charge is the same as if all the charge on the sphere were concentrated at its center.

(14) Suppose that positive charge is distributed uniformly throughout a spherical volume of radius R, the charge per unit volume being ρ. Prove with the help of Gauss's law that a positive point charge q, at a distance r from the center of the sphere, is repelled from the center with a force

$$F = \frac{\rho q r}{3\epsilon_0},$$

where $r \leqq R$.

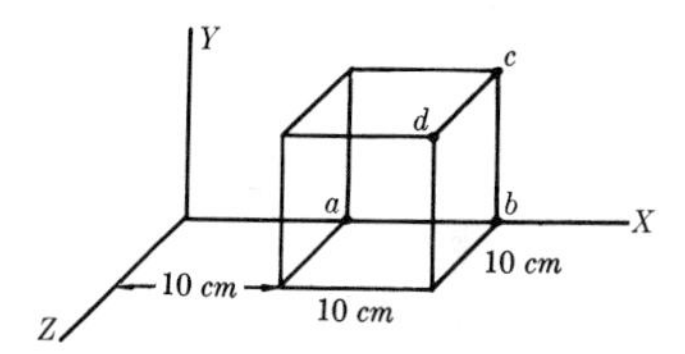

FIG. 2-22.

(15) The electric intensity in the region between a pair of oppositely-charged plane parallel plates, each 100 cm² in area, is 10 newtons/coulomb. What is the charge on each plate? Neglect edge effects.

(16) The three rectangular components of electric intensity throughout a certain region are: $E_x = 10^3\, x^{1/3}$, $E_y = 0$, $E_z = 0$. (E is in newtons/coulomb, x is in meters.) Find (a) the number of lines of force threading through the surface of the cubical volume element in Fig. 2-22, in an outward direction; (b) the net positive charge within the element.

(17) A negative point charge $-q$ is located at the point $x = -a$, $y = z = 0$, in a rectangular coordinate system. An equal positive charge is located at the point $x = a$, $y = z = 0$. How many lines of force pass through a circular area lying in the Y-Z plane with its center at the origin? A line drawn from either charge to the circumference of the circle makes an angle β with the X-axis. Discuss the special case when $\beta = \pi/2$.

(18) Evaluate $\int E \cos \phi \, dA$ over the surface of a sphere concentric with the origin, where E is the field set up by a dipole at the origin. The radius of the sphere is

large compared with the length of the dipole. The expression for an element of area on the surface of a sphere, in spherical coordinates, is

$$dA = r^2 \sin \theta \, d\theta \, d\phi.$$

(See Phillips Analytical Geometry and Calculus, p. 449.) How many lines of force thread through the surface of the sphere in an outward direction? What is the net charge within the sphere?

(19) A small sphere whose mass is 0.1 gm carries a charge of 3×10^{-9} coulombs and is attached to one end of a silk fibre 5 cm long. The other end of the fibre is attached to a large vertical conducting plate which has a surface density of charge of 25×10^{-7} coulombs/m^2. Find the angle which the fibre makes with the vertical.

(20) Fig. 2–23 is an end view of two long coaxial cylinders of radii a and b. The inner cylinder has a negative charge per unit length of λ, the outer cylinder an equal positive charge per unit length. A positively-charged particle of charge q and mass m revolves in a circular path of radius r in the evacuated space between the cylinders. Find the expression for the speed of the particle in terms of the given quantities.

(21) A charged oil drop, in a Millikan oil drop apparatus, is observed to fall through a distance of 1 mm in a time of 27.4 sec, in the absence of any external field. The same drop can be held stationary in a field of 2.37×10^4 newtons/coulomb. How many excess electrons has the drop acquired? The viscosity of air is 180×10^{-7} newton-sec/m^2. The density of the oil is 824 kgm/m^3, and the density of air is 1.29 kgm/m^3.

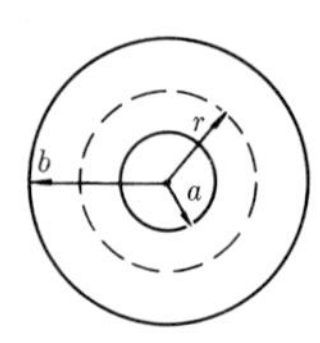

FIG. 2-23.

CHAPTER 3

POTENTIAL

3-1 Electrostatic potential energy. Many problems in mechanics can be greatly simplified with the help of energy considerations. For example, the velocity of a body sliding down a frictionless surface of any arbitrary shape can be found by setting the increase in the kinetic energy of the body equal to the decrease in its potential energy. The principle of conservation of energy is, of course, equally applicable to the motion of an electron, an ion, or any charged body in an electric field. In practically all cases of interest the gravitational force on a charge is so small in comparison with the electrical force that gravitational potential energy is negligible. However, every charged particle in an electric field has electrical potential energy which arises from the work done in moving it against electrical forces, just as gravitational potential energy arises from the work done in lifting a body against gravitational forces.

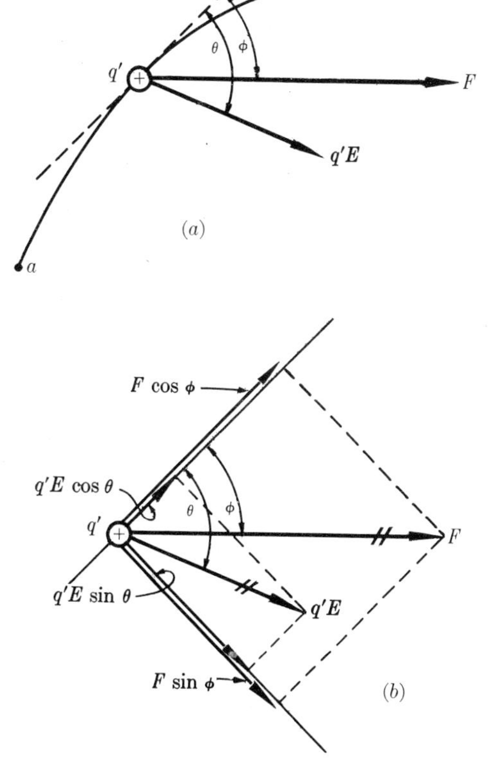

FIG. 3-1. Forces acting on a charge q' in an electrostatic field.

The region in Fig. 3-1(a) is an electric field. The curved line passing through points a and b represents a path of arbitrary shape between these points. A positive test charge q' (for example, a small charged metal sphere on an insulating rod) is to be moved along this path from a to b. The magnitude and direction of the field may vary from point to point of the path. The charges setting up the field are not shown. Gravitational and friction forces are neglected.

The vector $q'E$, in Fig. 3-1(a), is the force exerted on the charge by the field at some arbitrary point along the line. The vector F is the external, nonelectrical force exerted on the charge. That is, it is the force exerted by one's hand if one is holding the insulating rod and moving the charge along the line. Let θ be the angle between the tangent to the path and the force $q'E$, and ϕ the angle between the tangent and the external force F. In Fig. 3-1(b) the forces $q'E$ and F are resolved into components parallel and normal to the path. The resultant normal force, ΣF_n, is

$$\Sigma F_n = F \sin\phi + q'E \sin\theta.$$

The resultant tangential force, ΣF_t, is

$$\Sigma F_t = F \cos\phi + q'E \cos\theta.$$

The normal force, at right angles to the path, is a centripetal force. It alters the direction but not the magnitude of the velocity of the charge. The tangential force results in an acceleration of the charge along its path. The magnitude of the acceleration is given by Newton's second law,

$$F \cos\phi + q'E \cos\theta = ma,$$

where m is the mass of the charge.

Now

$$a = \frac{dv}{dt} = \frac{dv}{dt}\frac{ds}{ds} = \frac{ds}{dt}\frac{dv}{ds} = v\frac{dv}{ds},$$

where ds is an element of length along the path. Then

$$F \cos\phi + q'E \cos\theta = mv\frac{dv}{ds}$$

and

$$F \cos\phi \, ds + q'E \cos\theta \, ds = mv \, dv,$$

or,

$$F \cos\phi \, ds = mv \, dv - q'E \cos\theta \, ds. \tag{3-1}$$

The three terms in this equation have the following significance. The first term, $F \cos\phi \, ds$, is evidently the work done on the charge by the external force F in the displacement ds. Let us call this work dW.

$$dW = F \cos\phi \, ds. \tag{3-2}$$

The next term, $mv\,dv$, can be written as $d(\frac{1}{2}mv^2)$ and is therefore the increase in kinetic energy of the charge, $d(KE)$.

$$d(KE) = mv\,dv = d(\tfrac{1}{2}mv^2). \tag{3-3}$$

The last term, $-q'E\cos\theta\,ds$, is the work done against the electrical force $q'E$ exerted on the charge by the field. (The negative sign means that work is done *against* the electrical force; the work done *by* the force $q'E$ would be $+q'E\cos\theta\,ds$.) Hence the last term represents the increase in potential energy of the charge, $d(PE)$.

$$d(PE) = -q'E\cos\theta\,ds. \tag{3-4}$$

Eq. (3-1) is therefore the form taken by the work-energy equation when a charged body is moved in an electric field, and may be written

$$dW = d(KE) + d(PE).$$

When Eq. (3-1) is integrated along the path from a to b we get

$$\int_a^b F\cos\phi\,ds = \int_{v_a}^{v_b} mv\,dv - \int_a^b q'E\cos\theta\,ds. \tag{3-5}$$

The first integral is evidently the total work W done on the charge by the external force F. It is called a *line integral* (See Phillips' Analytical Geometry and Calculus, p. 427) and it means that at every element of the path, of length ds, the component of force $F\cos\phi$ is to be multiplied by ds and these products summed over all elements of the path between a and b. The limits a and b are not limits in the usual sense but merely serve to indicate the end points of the path. Of course, in order to evaluate the integral, the path must be specified and we must know how the force varies in magnitude and direction at all points of the path.

$$\int_a^b F\cos\phi\,ds = W. \tag{3-6}$$

The first integral on the right side of Eq. (3-5) can be evaluated at once, regardless of how the force varies. The limits v_a and v_b are the velocities of the charge at points a and b.

$$\int_{v_a}^{v_b} mv\,dv = \tfrac{1}{2}mv_b^2 - \tfrac{1}{2}mv_a^2 = KE_b - KE_a. \tag{3-7}$$

That is, this integral is the total increase in kinetic energy of the charge.

The last integral is also a line integral. It evidently represents the total work done against the force exerted by the field, or the total increase in potential energy, $PE_b - PE_a$.

$$-\int_a^b q'E \cos\theta\, ds = PE_b - PE_a. \tag{3-8}$$

In the special case in which the magnitude of the velocity is the same at points a and b, the increase in kinetic energy is zero and all the external work is accounted for by an increase in potential energy. That is, in this special case,

$$\int_a^b F \cos\phi\, ds = -\int_a^b q'E \cos\theta\, ds = PE_b - PE_a. \tag{3-9}$$

In the special case (but a very common one) in which the external force is zero and the charge moves solely under the influence of the force exerted on it by the field, the first integral in Eq. (3-5) is zero and

$$0 = KE_b - KE_a + PE_b - PE_a,$$

or

$$KE_b + PE_b = KE_a + PE_a = \text{constant}. \tag{3-10}$$

That is, when the only force is that exerted by the field, the sum of the kinetic and potential energies is the same at all points of the path.

Eq. (3-8) is the general expression for the difference between the potential energies of the test charge q' at points a and b in an electrostatic field. Before we can speak of *the* potential energy of the charge at any one point it is necessary to agree on some reference point at which the potential energy is arbitrarily considered zero. Mathematical simplicity often results if this point is chosen at infinity. That is, the potential energy of the test charge is considered zero when it is far removed from the other charges setting up the field. Then if the test charge is brought from infinity to any point in the field, the work done against the force exerted on it by the field equals its potential energy at the point. If we let point a be the point at infinity and set $PE_a = 0$, Eq. (3-8) becomes

$$PE_b = -\int_\infty^b q'E \cos\theta\, ds. \tag{3-11}$$

Point b can, of course, be any point in the field. It is therefore customary to drop the subscript and the limits in the integral, and write merely

$$PE = -\int q'E \cos\theta\, ds, \tag{3-12}$$

where it is understood that the integral is a line integral along a line from infinity to the point in question. *The potential energy of a test charge at a point in an electric field* may therefore be defined as *the work done against the force exerted on it by the field, when the charge is brought from infinity to the point.* We shall show later that the work done is independent of the path followed; if this were not so, the potential energy would have no unique value and the concept of potential energy would be meaningless. It should also be pointed out that the potential energy is a property of the *system* of charges (i.e., the test charge and the charges setting up the field) and, strictly speaking, should not be assigned to the test charge alone. Nevertheless, for convenience, one usually speaks of the potential energy of the test charge.

3-2 Potential. The electric intensity at a point in an electric field is the ratio of the force on a test charge at the point to the magnitude of the charge, or it is the force per unit charge. Similarly, the *potential* at a point in an electric field is defined as *the ratio of the potential energy of a test charge to the magnitude of the charge,* or as *the potential energy per unit charge.* The potential at a point is considered to have a value even though there may be no charge at the point. One places (in imagination) a test charge at the point, computes its potential energy, and takes the ratio of potential energy to charge. As in Sec. 2-2, it must be assumed that the original charge distribution is not altered by the introduction of the test charge. This will be true if the test charge is sufficiently small.

Potential will be represented by the letter V, or by V_a or V_b, for example, if we wish to call attention to the fact that it refers to a specific point a or b.

$$\text{Potential at point } a = V_a = \frac{PE \text{ of charge } q' \text{ at point } a}{q'}. \tag{3-13}$$

It follows from this definition that the potential energy of a charge q' at a point a in an electric field where the potential is V_a, is the product of the potential at the point and the charge q'.

$$PE \text{ of charge } q' \text{ at point } a = q'V_a. \tag{3-14}$$

Since energy is a scalar quantity, potential is a scalar also. It has magnitude but not direction, and in this respect differs from electric intensity, which is a vector.

From Eqs. (3-12) and (3-13), the potential at a point is

$$V = \frac{PE}{q'} = \frac{-\int q'E \cos\theta \, ds}{q'},$$

or

$$\boxed{V = -\int E \cos\theta \, ds} \tag{3-15}$$

where it is understood that the line integral is taken along a line from infinity to the point in question. In mathematical language, *the potential at a point equals the negative of the line integral of the electric intensity from infinity to the point.* Physically, the potential at a point may be defined as *the work done per unit charge against the force exerted by the field, when a charge is brought from infinity to the point.*

From its definition as potential energy per unit charge, potential is expressed in the mks system in *joules per coulomb.* (The corresponding unit in the electrostatic system is one erg per statcoulomb.) A potential of one joule per coulomb is called *one volt.* The volt is named in honor of Alessandro Volta (1745–1827), an Italian scientist. He was the inventor of the "voltaic pile," the first electrolytic cell. The volt may be defined as follows:

The potential at a point in an electrostatic field is one volt if one joule of work per coulomb is done against electrical forces when a charge is brought from infinity to the point. An alternate definition is: *The potential at a point in an electrostatic field is one volt if the ratio of the potential energy of a charge at the point, to the magnitude of the charge, is one joule per coulomb.*

One one-thousandth of a volt is called one *millivolt* (mv), and one one-millionth of a volt is called a *microvolt* (μv). One thousand volts is a *kilovolt* (kv) and one million volts is a *megavolt* (Mv).

In the electrostatic system, a potential of one erg per statcoulomb is called a *statvolt.* One statvolt = 300 volts (very nearly).

3-3 Potential difference. The difference between the potential at two points in an electrostatic field is called the *potential difference* between the points. When both sides of Eq. (3-8) are divided by q' we get

$$\frac{PE_b}{q'} - \frac{PE_a}{q'} = -\int_a^b E \cos\theta \, ds. \tag{3-16}$$

But

$$PE_b/q' = V_b \quad \text{and} \quad PE_a/q' = V_a,$$

where V_b and V_a are the potentials at points b and a. Hence the potential difference between points b and a is

$$V_b - V_a = -\int_a^b E \cos\theta \, ds. \tag{3-17}$$

The potential difference between points b and a equals the negative of the line integral of the electric intensity from point a to point b, or, it equals *the work done per unit charge against electrical forces when a charge is transported from a to b.*

Since potentials are expressed in volts, potential differences are in volts also. *The potential difference between points b and a is one volt if one joule of work per coulomb is done against electrical forces when a charge is moved from point a to point b.*

Point b is said to be at a higher potential than point a if work is done against electrical forces when a positive charge is moved from a to b. That is, b is at a higher potential than a if the potential energy of a *positive* charge is greater at b than it is at a.

The potential *at a point* can be considered as the potential difference between the point and a point at infinity where the potential is arbitrarily assumed zero.

The concept of potential difference is an extremely important one, both in electrostatics and in electric circuits. Electrical engineers commonly refer to it as "voltage."

For example, the potential difference between the terminals of a common automobile storage battery is about 6 volts, the terminal at the higher potential being designated by a + sign, that at the lower potential by a − sign. The + terminal is positively charged, the − terminal negatively charged. There exists, therefore, an electric field in the space around and between the battery terminals, and the statement that the potential difference between the terminals is 6 volts simply means that if one were to move a positively-charged body from the negative to the positive terminal (say by transporting a small positively-charged metal sphere attached to an insulating handle from one point to the other), the work done against the electrical forces of the field between the terminals would be 6 joules per coulomb of charge transported.

The potential difference between two points, say a and b, or p and q, will be abbreviated from $V_a - V_b$ or $V_p - V_q$ to V_{ab} or V_{pq}.

$$V_{ab} \equiv V_a - V_b, \quad V_{pq} \equiv V_p - V_q, \text{ etc.}$$

If this difference is a positive quantity, the first point (a or p) is at a *higher* potential than the second (b or q), and external work must be done to move positive charge *from* b, *to* a, or *from* q, to p. Conversely, if the difference is negative, the second point is at a higher potential than the first. Also, since $V_{ab} \equiv V_a - V_b$, and $V_{ba} \equiv V_b - V_a$, it follows that

$$V_{ab} \equiv -V_{ba}.$$

This simply means that if work must be done against the field to move positive charge from b to a, work can be obtained from the field when positive charge is allowed to move from a to b.

Differences of potential may be measured by electroscopes, electrometers, and voltmeters. If the knob of a leaf electroscope is connected to a body at one potential and the case of the instrument to a body at a different potential, the divergence of the electroscope leaves is a rough measure of the potential difference between the bodies. Since there is no simple relation between the divergence of its leaves and the potential difference, the leaf electroscope is rarely used in quantitative work, some form of electrometer being preferred. The familiar pivoted coil voltmeter also measures the potential difference between its terminals, but being a current-operated instrument, it is not suitable for purely electrostatic measurements.

One more general relation should be derived at this point. We have shown that the negative of the line integral of the electric intensity along any path in an electrostatic field equals the potential difference between the end points of the path. If the path closes on itself, so that the end points coincide, the potential difference between them is obviously zero and the line integral is therefore zero also. In mathematical language,

$$\oint E \cos \theta \, ds = 0, \tag{3-18}$$

where the symbol $\oint$ means that the line integral extends over a closed path.

The scalar product (or "dot product") of two vectors is defined as a scalar quantity equal to the product of the magnitudes of the two vectors multiplied by the cosine of the angle between them. Hence in vector notation, Eq. (3–15) is written

$$V = -\int \vec{E} \cdot ds.$$

Eq. (3-17) becomes

$$V_b - V_a = -\int_a^b \vec{E} \cdot d\vec{s},$$

and Eq. (3-18) is

$$\oint \vec{E} \cdot d\vec{s} = 0$$

Examples. (1) If the electric intensity E is constant in magnitude and direction within a certain region, as, for example, between a pair of oppositely-charged plane parallel plates, the right side of Eq. (3-17) may readily be integrated. Let the X-axis be taken in the direction of the field, and let $ds = dx$. (Fig. 3-2.) Then $\cos\theta = 1$, and

$$V_b - V_a = -\int_{x_1}^{x_2} E\,dx = -E(x_2 - x_1) = -Ed.$$

Hence

$$V_a = V_b + Ed,$$

and

$$E = \frac{V_a - V_b}{d} = \frac{V_{ab}}{d}. \qquad (3\text{-}19)$$

That is, the electric intensity equals the potential difference between the plates divided by their separation. Note that $V_a > V_b$ (the potential of the positive plate is greater than that of the negative plate) so V_{ab} is positive and E is positive, i.e., the field is directed from left to right.

Eq. (3-19) is a more useful expression for the electric intensity between parallel plates than Eq. (2-17), $E = \sigma/\epsilon_0$, since the potential difference V_{ab} can be determined experimentally more readily than the surface density of charge.

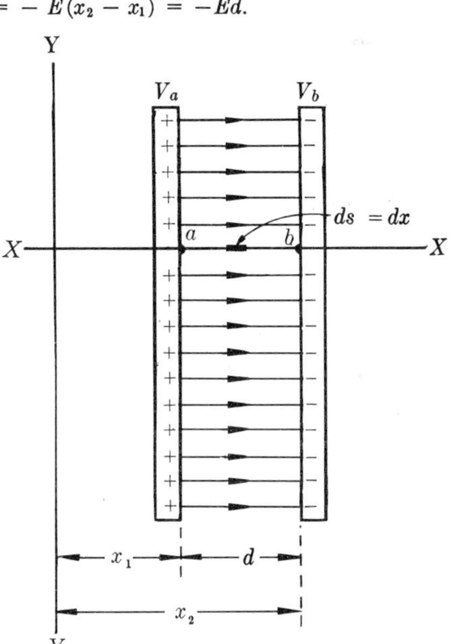

FIG. 3-2. The electric intensity between two parallel plates equals the ratio V_{ab}/d.

(2) A simple type of vacuum tube known as a *diode* consists essentially of two electrodes within a highly evacuated enclosure. One electrode, the *cathode*, is maintained at a high temperature and emits electrons from its surface (see Sec. 18-1). A potential difference of a few hundred volts is maintained between the cathode and the other electrode, known as the *anode* or plate, with the anode at the higher potential. Suppose that in a certain diode the plate potential is 250 volts above that of the cathode, and an electron is emitted from the cathode with no initial velocity. What is its velocity when it reaches the anode?

Let V_k and V_p represent the cathode and anode potentials. Since the electron has a negative charge of magnitude e, its potential energy at the cathode is, from Eq. (3-14),

$$PE_k = -eV_k,$$

and at the anode its potential energy is

$$PE_p = -eV_p.$$

Let v_p be the velocity of the electron at the anode. Its kinetic energy at the anode is then

$$KE_p = \tfrac{1}{2}mv_p^2.$$

Its kinetic energy at the cathode is zero (given).

The only force on the electron is that exerted by the field, so from Eq. (3-10),

$$\begin{aligned} KE_p + PE_p &= KE_k + PE_k \\ \tfrac{1}{2}mv_p^2 - eV_p &= -eV_k \\ \tfrac{1}{2}mv_p^2 &= e(V_p - V_k) = eV_{pk}\,. \end{aligned}$$

Hence

$$\begin{aligned} v_p &= \sqrt{\frac{2eV_{pk}}{m}} \\ &= \sqrt{\frac{2 \times 1.6 \times 10^{-19} \times 250}{9.1 \times 10^{-31}}} \\ &= 9.4 \times 10^6 \text{ m/sec.} \end{aligned}$$

Notice that the shape or separation of the electrodes need not be known. The final velocity depends only on the potential difference through which the electron has "fallen" (or perhaps "risen" is better, since the electron, having a negative charge, moves from points of lower to points of higher potential). Of course the *time* of transit from cathode to anode depends on the geometry of the tube.

3-4 Potential and charge distribution. The general considerations of the preceding sections show that the potential difference between points b and a in an electrostatic field equals the negative of the line integral of the electric intensity along a line from a to b.

$$V_b - V_a = -\int_a^b E \cos\theta \, ds. \tag{3-17}$$

We now evaluate this integral for the special case in which the field is set up by a single point charge q (Fig. 3-3). Let r represent the distance from the charge q to any point of an arbitrary path between a and b. The electric intensity E at the point is

$$E = \frac{1}{4\pi\epsilon_0}\frac{q}{r^2}.$$

In a short distance ds along the path the distance r increases by dr. It can be seen from the diagram that

$$dr = ds \cos\theta.$$

Hence Eq. (3-17) becomes

$$V_b - V_a = -\frac{1}{4\pi\epsilon_0} q \int_{r_a}^{r_b} \frac{dr}{r^2}$$

$$= -\frac{1}{4\pi\epsilon_0} q \left[-\frac{1}{r}\right]_{r_a}^{r_b},$$

or,

$$V_b - V_a = \frac{1}{4\pi\epsilon_0}\frac{q}{r_b} - \frac{1}{4\pi\epsilon_0}\frac{q}{r_a}, \quad (3\text{-}20)$$

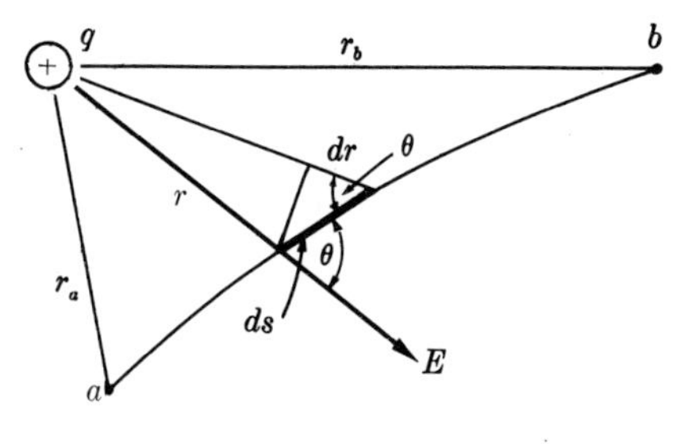

FIG. 3-3. Evaluating the potential difference between points a and b in the field of a single point charge q.

where r_b and r_a are the distances from the charge q to the points b and a respectively.

To find the potential *at* a point, relative to a point at infinity, we may either return to Eq. (3-15) and evaluate the negative of the line integral of E from infinity to any arbitrary point, or make use of Eq. (3-20), setting $r_a = \infty$ and $V_a = 0$. Either method leads to the same result, namely,

$$V_b = \frac{1}{4\pi\epsilon_0}\frac{q}{r_b}.$$

Since b can be any point in the field, the subscripts are usually dropped and we have as the expression for the potential at a point at a distance r from a point charge q, relative to a point at infinity,

$$\boxed{V = \frac{1}{4\pi\epsilon_0}\frac{q}{r}.} \quad (3\text{-}21)$$

The potential is in volts if q is in coulombs and r in meters.

The distances r_a, r_b, and r, in Eqs. (3-20) and (3-21), are measured outward from the charge q and are considered positive always. The charge q may, however, be positive or negative. The potential therefore has the same algebraic sign as the charge q.

In general, the field at a point is set up by a number of charges, rather than by a single point charge. A consideration of the simple case of the field of two point charges will indicate how Eq. (3-21) is to be generalized.

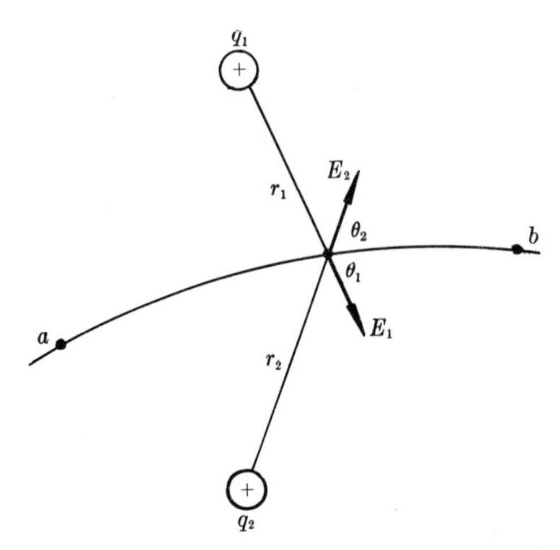

FIG. 3-4. The potential at a point in an electric field is the algebraic sum of the potentials due to the charges setting up the field.

In Fig. 3-4, the positive point charges q_1 and q_2 set up a field. The vectors E_1 and E_2 are the electric intensities due to the respective charges at any point of an arbitrary path between a and b. If an external force F at an angle ϕ with the path (not shown in Fig. 3-4) acts on a test charge q' at the point, the resultant accelerating force is

$$\Sigma F_t = F \cos \phi + q'E_1 \cos \theta_1 + q'E_2 \cos \theta_2.$$

By the same reasoning as in Secs. 3-1 and 3-2 we obtain

$$\int_a^b F \cos \phi \, ds = \int_{v_a}^{v_b} mv \, dv - \int_a^b q'E_1 \cos \theta_1 \, ds - \int_a^b q'E_2 \cos \theta_2 \, ds.$$

The increase in potential energy of the test charge is

$$PE_b - PE_a = -\int_a^b q'E_1 \cos \theta_1 \, ds - \int_a^b q'E_2 \cos \theta_2 \, ds.$$

The potential difference $V_b - V_a$ is

$$V_b - V_a = -\int_a^b E_1 \cos \theta_1 \, ds - \int_a^b E_2 \cos \theta_2 \, ds, \tag{3-22}$$

and the potential at a point is

$$V = -\int E_1 \cos \theta_1 \, ds - \int E_2 \cos \theta_2 \, ds, \tag{3-23}$$

where the integrals are line integrals from infinity to the point in question. But from Fig. 3-4,

$$E_1 = \frac{1}{4\pi\epsilon_0}\frac{q_1}{r_1^2}, \quad E_2 = \frac{1}{4\pi\epsilon_0}\frac{q_2}{r_2^2},$$

$$\cos\theta_1\, ds = dr_1, \quad \cos\theta_2\, ds = dr_2.$$

Hence from Eq. (3-23),

$$V = \frac{1}{4\pi\epsilon_0}\frac{q_1}{r_1} + \frac{1}{4\pi\epsilon_0}\frac{q_2}{r_2}, \tag{3-24}$$

where V is the potential at any point and r_1 and r_2 are the distances from q_1 and q_2 respectively to this point. We see that the potential at a point is the sum of the potentials due to the charges q_1 and q_2 respectively. The sum is *algebraic, not vectorial,* since potential is a scalar quantity.

If any number of point charges contribute to the field, the obvious generalization of Eq. (3-24) is

$$\boxed{V = \frac{1}{4\pi\epsilon_0}\Sigma\frac{q}{r}.} \tag{3-25}$$

Eq. (3-25) applies equally well to the potential at a point due to uniformly charged spheres, since the field at points outside a sphere having a uniform surface density of charge is the same as though all of the charge were concentrated at the center of the sphere.

If the charges are distributed over surfaces or throughout volumes, Eq. (3-25) becomes

$$V = \frac{1}{4\pi\epsilon_0}\int\frac{dq}{r}. \tag{3-26}$$

It will be seen from Eqs. (3-24), (3-25), and (3-26) that the potential at a point in an electrostatic field depends only on the magnitudes of the charges setting up the field and their distances from the chosen point. The potential is therefore independent of the particular path from infinity that is used in evaluating the line integral of the electric intensity.

Examples. (1) Compute the potentials at the following distances from a positive point charge of 10^{-9} coulombs: 1 mm, 1 cm, 10 cm, 100 cm. (Compare Example (1) in Sec. 2-2.)

$$V = \frac{1}{4\pi\epsilon_0}\frac{q}{r} = 9 \times 10^9 \frac{q}{r}.$$

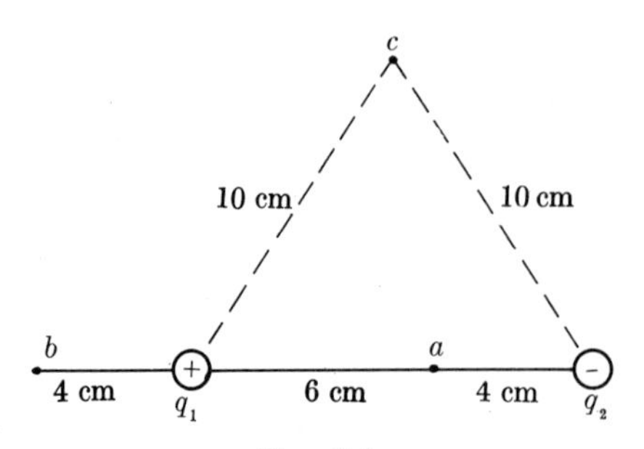

FIG. 3-5.

When $r = 1$ mm $= 10^{-3}$ m,

$$V = 9 \times 10^9 \times \frac{10^{-9}}{10^{-3}}$$

$$= 9000 \text{ volts.}$$

Similarly,

at 1 cm, $V = 900$ volts,
at 10 cm, $V = 90$ volts,
at 100 cm, $V = 9$ volts.

(2) Point charges of $+12 \times 10^{-9}$ and -12×10^{-9} coulombs are placed 10 cm apart as in Fig. 3-5. Compute the potentials at points, a, b, and c. (Compare Example (2) in Sec. 2-2.)

We must evaluate the *algebraic* sum $\frac{1}{4\pi\epsilon_0} \Sigma \frac{q}{r}$ at each point. At point a, the potential due to the positive charge is

$$9 \times 10^9 \times \frac{12 \times 10^{-9}}{.06} = +1800 \text{ volts,}$$

and the potential due to the negative charge is

$$9 \times 10^9 \times \frac{(-12 \times 10^{-9})}{.04} = -2700 \text{ volts.}$$

Hence

$$V_a = +1800 - 2700 = -900 \text{ volts.}$$

At point b, the potential due to the positive charge is $+2700$ volts and that due to the negative charge is -770 volts. Hence

$$V_b = +2700 - 770 = +1930 \text{ volts.}$$

At point c, the potential is

$$V_c = +1080 - 1080 = 0.$$

(3) Compute the potential energy of a point charge of $+4 \times 10^{-9}$ coulombs if placed at points a, b, and c in Fig. 3-5.

$$PE = qV.$$

Hence at point a,

$$PE = qV_a = 4 \times 10^{-9} \times (-900) = -36 \times 10^{-7} \text{ joules} = -36 \text{ ergs.}$$

At point b,

$$PE = qV_b = 4 \times 10^{-9} \times 1930 = 77 \times 10^{-7} \text{ joules} = 77 \text{ ergs.}$$

At point c,

$$PE = qV_c = 0.$$

(All relative to a point at infinity.)

(4) What is the potential difference or voltage between points a and b? b and a? b and c? How much work would be required to take a point charge of $+4 \times 10^{-9}$ coulombs from a to b, with no increase in its kinetic energy? from c to a?

The potential difference between points a and b is $V_a - V_b$ or V_{ab}.

$$V_{ab} = V_a - V_b = -900 - 1930 = -2830 \text{ volts},$$
$$V_{ba} = V_b - V_a = 1930 - (-900) = +2830 \text{ volts},$$
$$V_{bc} = V_b - V_c = 1930 - 0 = +1930 \text{ volts}.$$

The work required to take a charge q from a to b is qV_{ba}.

$$W_{a \to b} = q \times V_{ba} = 4 \times 10^{-9} \times 2830 = 113 \times 10^{-7} \text{ joules}$$
$$W_{c \to a} = q \times V_{ac} = 4 \times 10^{-9} \times (-900) = -36 \times 10^{-7} \text{ joules}$$

3-5 Potential gradient. The general expression for the potential difference between two points is

$$V_b - V_a = -\int_a^b E \cos\theta \, ds.$$

When the points are only an infinitesimal distance apart, the potential difference between them becomes dV and from the preceding equation,

$$dV = -E \cos\theta \, ds. \tag{3-27}$$

The same result follows from Eq (3-4) when both sides are divided by q'.

Eq. (3-27) can be written

$$\boxed{E \cos\theta = -\frac{dV}{ds}.} \tag{3-28}$$

The ratio dV/ds, or the rate of change of potential with distance in the direction of ds, is called the *potential gradient.* Since θ is the angle between the electric intensity E and the element of distance ds, the product $E \cos\theta$ is the component of the electric intensity in the direction of ds. Hence we have the important relation: *The component of the electric intensity in any direction is equal to the negative of the potential gradient in that direction.*

In particular, if the direction of ds is the same as that of the electric intensity, the angle $\theta = 0$, $\cos\theta = 1$, and *the electric intensity is equal to the negative of the potential gradient in the direction of the field.*

Potential gradients, in the mks system, are expressed in *volts per meter.*

Since the component of electric intensity in any direction is numerically equal to the potential gradient in that direction (disregarding the negative sign) it follows that the units of electric intensity and potential gradient are equivalent. That is 1 volt/m = 1 newton/coulomb, 100 volts/m = 100 newtons/coulomb, etc. In fact, it is accepted practice to use the volt per meter rather than the newton per coulomb as the common unit in which to express electric intensities.

Other units sometimes used are the volt/cm and the volt/mil. (1 mil = 0.001 inch.)

The field around any assemblage of charges is of course three-dimensional, and, in general, the potential V at any point is some function of the coordinates of the point, say x, y, and z. We may then take the direction of ds in Eq. (3-28), first parallel to the X-axis, then parallel to the Y-axis, and finally parallel to the Z-axis. The three potential gradients then give the three rectangular components of E, say E_x, E_y, and E_z. That is,

$$E_x = -\frac{dV}{dx}, \quad E_y = -\frac{dV}{dy}, \quad E_z = -\frac{dV}{dz}. \tag{3-29}$$

(Strictly speaking, these should all be written as partial derivatives.)

It is easily seen that throughout a region in which the potential has the same value at all points, the three derivatives in Eq. (3-29) are all zero. Hence the three components of the electric intensity are zero and the region is one of zero field. Conversely, if the electric intensity throughout a region is zero, the potential is the same at all points. We have shown that the electric intensity within a conductor is zero when the charges on the conductor are at rest. Hence the potential has the same value at all points within such a conductor.

One of the useful properties of the concept of potential gradient is that it is often much simpler to compute the electric intensity at a point by first finding an expression for the potential at the point and then using Eq. (3-29), than it is to compute the intensity directly from Eq. (2-5). The simplification arises from the fact that potential, a scalar, involves only algebraic summation or integration, rather than vector summation. See Examples 2 and 3 at the end of this section.

The *gradient* of a scalar function of x, y, and z, such as the potential V, is defined as a vector with components equal to the partial derivatives of the function with respect to x, y, and z. (See Phillips' Analytical Geometry and Calculus, p. 424.)

$$\text{grad } V = \vec{i}\,\frac{\partial V}{\partial x} + \vec{j}\,\frac{\partial V}{\partial y} + \vec{k}\,\frac{\partial V}{\partial z},$$

where $\vec{i}$, $\vec{j}$, and $\vec{k}$ are unit vectors along the X-, Y-, and Z-axes respectively. It follows that

$$\vec{E} = -\text{grad } V,$$

which is equivalent to the three equations (3-29).

Examples. (1) We have shown that the potential due to a single point charge q is

$$V = \frac{1}{4\pi\epsilon_0}\frac{q}{r}.$$

By symmetry, the field of the charge is radial. Hence if we take ds, in Eq. (3-28), in the radial direction, the component of E obtained will be the electric intensity itself. We accordingly let ds equal dr, and obtain

$$E = -\frac{dV}{ds} = -\frac{dV}{dr} = -\frac{d}{dr}\left(\frac{1}{4\pi\epsilon_0}\frac{q}{r}\right) = \frac{1}{4\pi\epsilon_0}\frac{q}{r^2},$$

which agrees with the expression previously derived for the electric intensity due to a point charge.

The analytical method above is represented graphically as follows. In Fig. 3-6(a), the potential of a positive point charge q is plotted as a function of r. The slope of the curve, at any point, is the potential gradient in a radial direction, dV/dr. The potential gradient can be found graphically by constructing a tangent to the curve at any point and measuring its slope. The electric intensity at the point then equals the negative of this slope. Evidently, from Fig. 3-6(a), the slope of the graph of V vs r is negative, so the electric intensity is positive, i.e., radially outward.

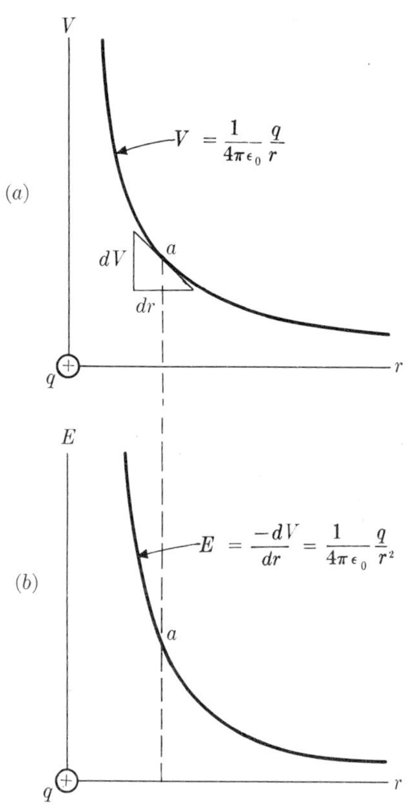

FIG. 3-6. Potential V and intensity E as a function of the distance r from a point charge q.

Fig. 3-6(b) is the graph of E as a function of r. The *ordinate* of this graph, at every point, equals the negative of the *slope* of the upper graph at the same point.

(2) To illustrate the simplification made possible by the use of potential gradients, let us compute the electric intensity on the axis of a ring of charge, a problem that has been worked out already in Sec. 2-4. Although the total charge q on the ring is dis

tributed over its length, every portion of the charge is at the same distance $\sqrt{x^2 + a^2}$ from an axial point P (Fig. 3-7(a)). Hence the potential at P is

$$V = \frac{1}{4\pi\epsilon_0} \frac{q}{\sqrt{x^2 + a^2}}.$$

(The general expression for the potential at points not on the axis is, of course, more complex.)

By symmetry, the field at an axial point is in the direction of the axis, so that the negative of the potential gradient dV/dx will give the field at such a point.

$$E = -\frac{dV}{dx} = -\frac{d}{dx}\left(\frac{1}{4\pi\epsilon_0} \frac{q}{\sqrt{x^2 + a^2}}\right)$$
$$= \frac{1}{4\pi\epsilon_0} \frac{qx}{(x^2 + a^2)^{\frac{3}{2}}}.$$

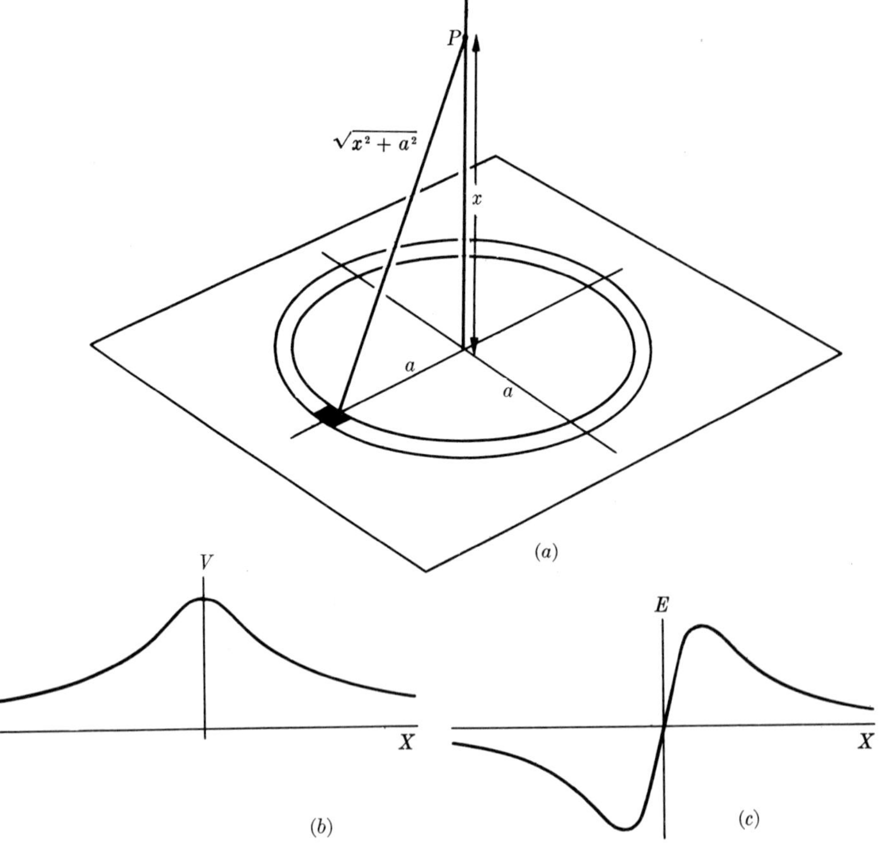

FIG. 3-7. (a) Point P on the axis of a charged ring. (b) Potential and intensity on the axis of a charged ring as a function of the distance from the center.

Except for a difference in notation, this is the same as the result obtained in Sec. 2-4, and the computations are evidently much simpler. Fig. 3-7(b) is a graph of V vs x, and Fig. 3-7(c) a graph of E vs x.

(3) Consider next the potential and field of a dipole (Fig. 3-8). Let us use polar coordinates, r, θ, with origin at the center of the dipole, to specify the position of a point P. With P as center, strike the arcs indicated by the dotted lines. If P is sufficiently far from the dipole the arcs can be considered straight lines, at right angles both to r and the lines joining P and the two charges. The distance from the charge $+q$ to P is very nearly $r - \frac{l}{2}\cos\theta$, and from the charge $-q$ to P it is $r + \frac{l}{2}\cos\theta$. The potential at P is

$$V = \frac{1}{4\pi\epsilon_0}\left(\frac{q}{r - \frac{l}{2}\cos\theta} + \frac{-q}{r + \frac{l}{2}\cos\theta}\right)$$

$$= \frac{1}{4\pi\epsilon_0} q \left(\frac{l\cos\theta}{r^2 - \frac{l^2}{4}\cos^2\theta}\right)$$

$$= \frac{1}{4\pi\epsilon_0}\frac{ql\cos\theta}{r^2},$$

if we neglect $\frac{l^2}{4}\cos^2\theta$ in comparison with r^2.

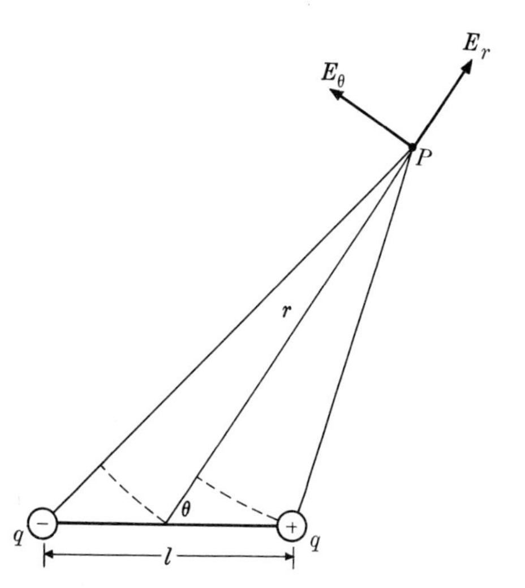

FIG. 3-8. Components of electric intensity in the field of a dipole.

In polar coordinates, a distance ds in a radial direction is simply dr, while a distance ds at right angles to r is $r\,d\theta$. Hence

$$\left(\frac{dV}{ds}\right)_{\text{radially}} = \frac{dV}{dr}, \quad \left(\frac{dV}{ds}\right)_{\text{tangentially}} = \frac{dV}{r d\theta} = \frac{1}{r}\frac{dV}{d\theta},$$

and

$$E_r = -\frac{dV}{dr}, \quad E_\theta = -\frac{1}{r}\frac{dV}{d\theta}.$$

Thus

$$E_r = -\frac{d}{dr}\left(\frac{1}{4\pi\epsilon_0}\frac{ql\cos\theta}{r^2}\right),$$

$$= \frac{1}{4\pi\epsilon_0}\frac{2ql\cos\theta}{r^3}.$$

$$E_\theta = -\frac{1}{r}\frac{d}{d\theta}\left(\frac{1}{4\pi\epsilon_0}\frac{ql\cos\theta}{r^2}\right),$$

$$= \frac{1}{4\pi\epsilon_0}\frac{ql\sin\theta}{r^3}.$$

These expressions for E_r and E_θ were given without proof in Sec. 2-3.

(4) A long conducting cylinder of radius a is surrounded by a coaxial hollow cylinder of inner radius b. The cylinders are given equal and opposite charges. Find the expression for the electric intensity at a point of space between the cylinders, in terms of the potential difference between them.

We have shown with the help of Gauss's law (see Sec. 2-8) that in terms of the charge per unit length, λ, the electric intensity between the cylinders is

$$E = \frac{1}{2\pi\epsilon_0}\frac{\lambda}{r},$$

and since

$$E = -dV/dr,$$

$$-\frac{dV}{dr} = \frac{1}{2\pi\epsilon_0}\frac{\lambda}{r}$$

$$-\int_{V_a}^{V_b} dV = \frac{1}{2\pi\epsilon_0}\lambda\int_a^b \frac{dr}{r}, \quad V_a - V_b = \frac{1}{2\pi\epsilon_0}\lambda \ln\frac{b}{a},$$

$$\lambda = 2\pi\epsilon_0\frac{V_{ab}}{\ln\frac{b}{a}}.$$

When this expression for λ is inserted in the first equation we get

$$E = \frac{1}{r}\frac{V_{ab}}{\ln\frac{b}{a}}. \tag{3-30}$$

This is a more useful expression for the intensity, since the potential difference between the cylinders can be measured more readily than their charge per unit length. End effects have been neglected.

3-6 Potential of a charged spherical conductor. The electric intensity *outside* a charged spherical conductor is the same as though all of the charge on the conductor were concentrated at its center. It follows that the potential at points outside the conductor is given by the same expression as that for a point charge, namely,

$$V = \frac{1}{4\pi\epsilon_0}\frac{q}{r}, \tag{3-31}$$

where r is equal to or greater than the radius of the sphere. At points *inside* the sphere the electric intensity is zero. The potential is therefore the same at all internal points and is equal to the potential at the surface, namely,

$$V = \frac{1}{4\pi\epsilon_0}\frac{q}{a}, \tag{3-32}$$

where a is the radius of the sphere.

Hence Eq. (3-31) gives the potential at points outside and Eq. (3-32) the potential at points inside the sphere.

The potential and electric intensity due to a positively-charged sphere are shown graphically in Fig. 3-9. At internal points, where V is constant, the slope of the graph of V *vs* r is zero and E is zero. Just outside the surface of the sphere the slope of the graph of V *vs* r is a maximum and is negative. Hence E is a maximum, and is positive, just outside the surface. As the distance from the center increases, V decreases as $1/r$ and E as $1/r^2$.

It has been shown previously that the maximum charge that can be retained by a conductor in air is limited by the fact that the air itself becomes conducting at an electric intensity of about 3×10^6 volts/meter. In general, if E_m represents the upper limit of electric intensity, the maximum charge that can be retained by a spherical conductor in air is

$$q_m = 4\pi\epsilon_0 a^2 E_m.$$

Hence the maximum *potential* to which a spherical conductor in air can be raised is, from Eq. (3-32),

$$V_m = aE_m.$$

For a sphere one centimeter in radius,

$$V_m = .01 \times 3 \times 10^6 = 30{,}000 \text{ volts},$$

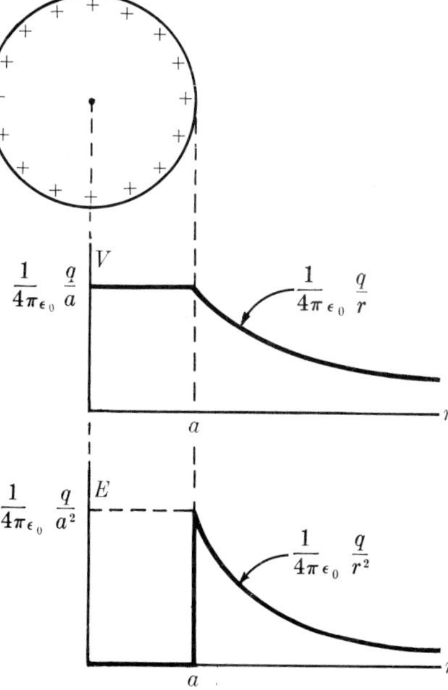

FIG. 3-9. Potential and intensity inside and outside of a charged sphere.

and no amount of "charging" could raise the potential of a sphere of this size, in air, higher than about 30,000 volts.

It is this fact which necessitates the use of such large spherical terminals on high voltage machines of the Van de Graaff type. If we make $a = 2$ meters, then

$$V_m = 2 \times 3 \times 10^6 = 6 \text{ million volts}.$$

At the other extreme is the effect produced by sharp points, a point being a portion of a surface of very small radius of curvature. Since the maximum potential is directly proportional to the radius, even relatively small potentials applied to sharp points in air will produce sufficiently high fields just outside the point to result in ionization of the surrounding air.

Example. Two conducting spheres A and B, each of 10 cm radius, are placed with their centers 1 meter apart (Fig. 3-10). If sphere A is given a charge of $+30 \times 10^{-9}$ coulombs and sphere B a charge of -60×10^{-9} coulombs, compute the potential of each.

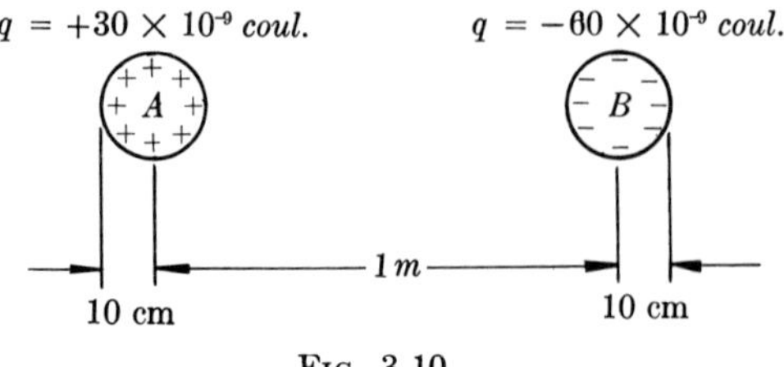

FIG. 3-10.

Since the potential is the same at all points within each sphere and equal to that at the surface of the sphere, let us compute the potential at the center of each. At the center of sphere A, the potential due to A's own charge is

$$V = \frac{1}{4\pi\epsilon_0}\frac{q}{a} = 9 \times 10^9 \times \frac{30 \times 10^{-9}}{.10} = 2700 \text{ volts.}$$

At points outside of sphere B, all of the charge on B can be considered as concentrated at its center, or at a distance of 1 meter from the center of A. Hence the potential at the center of A due to B's charge is

$$V = 9 \times 10^9 \times \frac{-60 \times 10^{-9}}{1} = -540 \text{ volts.}$$

The potential at the center of A, which is the same as the potential at its surface, is therefore

$$V_A = 2700 - 540 = +2160 \text{ volts.}$$

Similarly, at the center of sphere B the potential is

$$V_B = -5400 + 270 = -5130 \text{ volts.}$$

The potential difference V_{AB} between the spheres is

$$V_{AB} = V_A - V_B = 2160 - (-5130) = +7290 \text{ volts,}$$

with A at the higher potential.

The preceding calculation is not strictly correct, since the forces of attraction between the charges on A and B will result in an accumulation of charge on the surfaces of the spheres which face one another. The resulting distribution is therefore not spherically symmetrical and at outside points the field and potential are not the same as if all of the charge were concentrated at the centers of the spheres. The potential at the center of each sphere due to the sphere's own charge is, however, equal to the charge divided by the radius whether the surface density is uniform or not, since each element of charge is at a distance of one radius from the center.

The exact treatment of the problem is beyond the scope of this book, but if the separation of the spheres is large compared with their radii, the asymmetry is not great and the method above gives substantially correct results.

3-7 Poisson's and Laplace's equations. Poisson's and Laplace's equations are useful mathematical relations in the solution of many problems in electrostatics. We shall derive them for the one-dimensional case and then generalize for three dimensions. Let A and B in Fig. 3-11 denote two large plane parallel conducting plates between which there is an electric field directed from A toward B. Suppose that there is electric charge in the space between the plates. Such a condition would exist between the cathode and anode of a vacuum tube, where the space would be occupied by electrons streaming from the cathode to the anode. The distributed charge is called a *space charge*. Let ρ represent the space charge density, or the charge per unit volume. We shall assume that the charge density may vary from one point of space to another, but that it is uniform throughout any thin layer parallel to the plates A and B. In other words, if we take the X-axis perpendicular to the plates, the charge density is a function of x only. The electric intensity may also vary from point to point, but it also is assumed to depend only on x. The electric intensity is indicated in Fig. 3-11(a) by constructing one line of force from each positive to each negative charge. (A positive space charge has been assumed.) It will be seen that the number of lines per unit area, and hence the electric intensity, increases with increasing x because of the space charge.

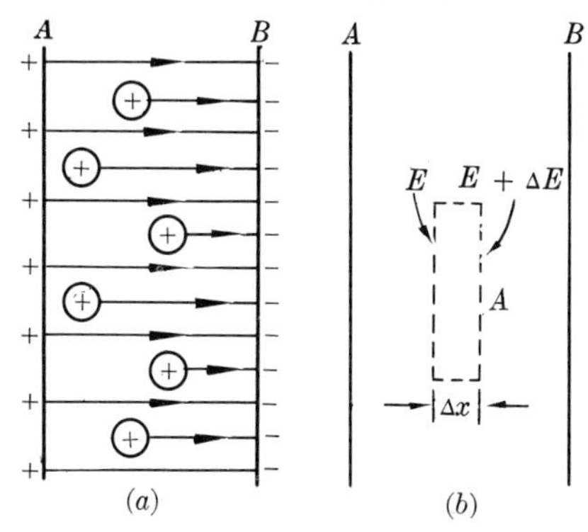

FIG. 3-11. Positive space charge between two plates.

We now apply Gauss's law to a thin volume element of cross sectional area A and thickness Δx, shown in Fig. 3-11(b). Let E represent the electric intensity at the left face of the element. By hypothesis, E has the same value at all points of this face. The electric intensity at the right face of the slab we shall call $E + \Delta E$. If dE/dx is the rate at which the electric intensity increases with increasing x, the increase in intensity ΔE, in the distance Δx, is

$$\Delta E = \frac{dE}{dx}\,\Delta x.$$

Hence

$$E + \Delta E = E + \frac{dE}{dx}\,\Delta x.$$

The surface integral of E over the right face of the element is

$$\left(E + \frac{dE}{dx}\,\Delta x\right) \times A.$$

The surface integral of E over the left face is

$$-E \times A.$$

This integral is negative, since E is directed inward over this face. The surface integral of E over the entire surface of the element is

$$\left(E + \frac{dE}{dx}\,\Delta x\right) \times A - E \times A = A\Delta x\,\frac{dE}{dx}\cdot$$

Gauss's law states that the surface integral of the electric intensity over a closed surface equals $1/\epsilon_0$ times the charge enclosed within the surface. The volume of the element is $A\Delta x$ and the net charge enclosed within it is $\rho A\Delta x$. Hence

$$A\Delta x\,\frac{dE}{dx} = \frac{1}{\epsilon_0}\,\rho A\Delta x.$$

After cancelling $A\Delta x$ from both sides, this reduces to

$$\frac{dE}{dx} = \frac{1}{\epsilon_0}\,\rho. \tag{3-33}$$

But the electric intensity equals the negative of the potential gradient, or

$$E = -\frac{dV}{dx}\cdot$$

When both sides are differentiated with respect to x we obtain

$$\frac{dE}{dx} = -\frac{d^2V}{dx^2}\cdot \tag{3-34}$$

Eq. (3-34) may now be combined with Eq.(3-33) to give

$$\boxed{\frac{d^2V}{dx^2} = -\frac{1}{\epsilon_0}\,\rho.} \tag{3-35}$$

This is *Poisson's equation* in one dimension, where the potential varies only with x. Its general form, where V may depend on x, y, and z, is found by the same method as that above to be

$$\frac{d^2V}{dx^2} + \frac{d^2V}{dy^2} + \frac{d^2V}{dz^2} = -\frac{1}{\epsilon_0}\rho. \tag{3-36}$$

Strictly speaking, the three derivatives should be written as partial derivatives.

In the special case in which the charge density is zero, Eq. (3-35) becomes

$$\boxed{\frac{d^2V}{dx^2} = 0,} \tag{3-37}$$

of which the general form is

$$\frac{d^2V}{dx^2} + \frac{d^2V}{dy^2} + \frac{d^2V}{dz^2} = 0, \tag{3-38}$$

Eq. (3-38) is *Laplace's equation.*

3-8 Electric intensity, potential, and charge distribution. An electric field may be completely described by giving either the charge distribution, or the electric intensity at all points, or the potential at all points. If any one of these is known, the others may be computed (except for possible mathematical difficulties).

If the charge distribution is given, the electric intensity and the potential at any point can be found from

$$E = \frac{1}{4\pi\epsilon_0}\int\frac{dq}{r^2}, \quad V = \frac{1}{4\pi\epsilon_0}\int\frac{dq}{r}.$$

If the electric intensity is given, the charge within any volume (and hence at any point) can be found from Gauss's law. The potential can then be computed from the known charge distribution and the equation

$$V = \frac{1}{4\pi\epsilon_0}\int\frac{dq}{r},$$

or it can be found directly from

$$V = -\int E\cos\theta\, ds.$$

If the potential is given at all points, the electric intensity can be computed from

$$E_x = -\frac{dV}{dx}, \quad E_y = -\frac{dV}{dy}, \quad E_z = -\frac{dV}{dz}$$

and since the electric intensity is now known, the charge distribution can be found by Gauss's law.

3-9 Equipotential surfaces. The potential distribution in an electric field may be represented graphically by *equipotential surfaces.* An equipotential surface is one at all points of which the potential has the same value. While an equipotential surface may be constructed through every point of an electric field, it is customary to show only a few of the equipotentials in a diagram.

Since the potential energy of a charged body is the same at all points of a given equipotential surface, it follows that no (electrical) work is needed to move a charged body over such a surface. Hence the equipotential surface through any point must be at right angles to the direction of the field at that point. If this were not so, the field would have a component lying in the surface and work would have to be done against electrical forces to move a charge in the direction of this component. The lines of force and the equipotential surfaces thus form a mutually perpendicular network. In general, the lines of force of a field are curves and the equipotentials are curved surfaces. For the special case of a uniform field, where the lines of force are straight and parallel, the equipotentials are parallel planes perpendicular to the lines of force.

It has been shown (see Sec. 2-7) that the lines of force at the surface of a charged conductor are at right angles to the conductor if the charges on it are at rest. Hence the surface of such a conductor is an equipotential. Furthermore, since the field within a charged conductor is zero, the entire interior of the conductor is an equipotential *volume*, at the same potential as the surface of the conductor.

Fig. 3-12 shows the same arrangements of charges as in Fig. 2-12 in Sec. 2-5. The lines of force have been drawn dotted and the intersections of the equipotential surfaces with the plane of the diagram are shown by the full lines. The actual field is, of course, three-dimensional.

Suppose that a given electric field has been mapped by its network of lines of force and equipotential surfaces, with the (electrical) spacing between the equipotentials equal to some constant difference ΔV such

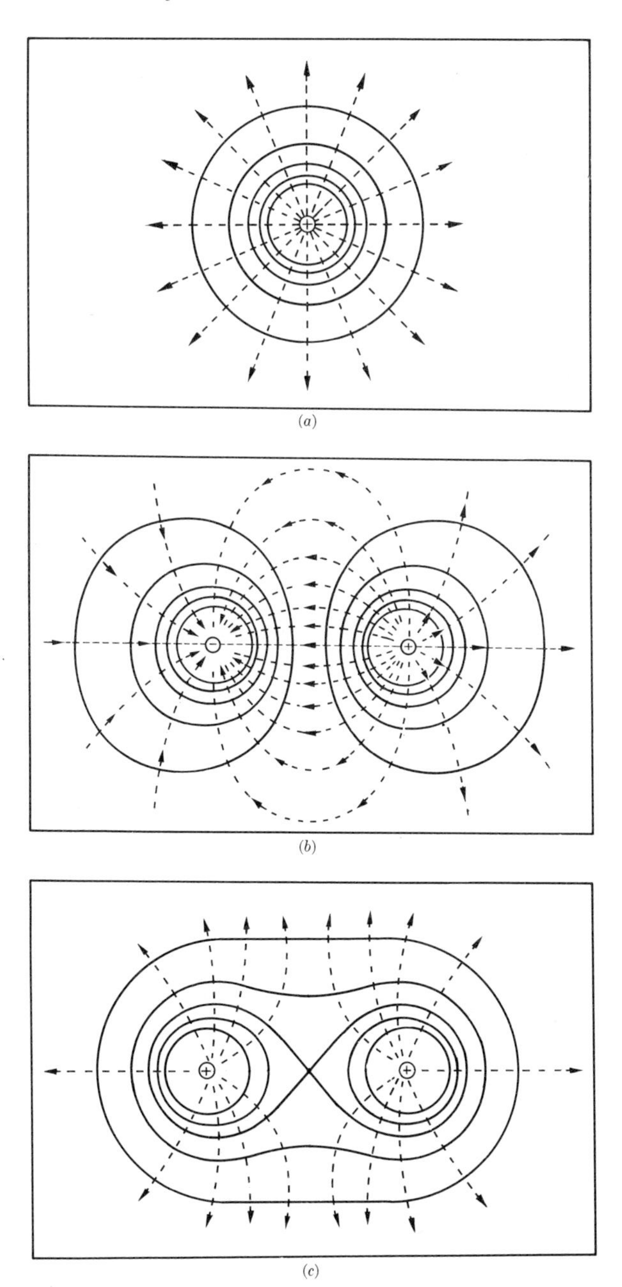

Fig. 3-12. Equipotential surfaces (solid lines) and lines of force (dotted lines) in the neighborhood of point charges.

as 1 volt or 100 volts. Let Δs represent the perpendicular distance between two equipotentials. Then Δs is in the direction of the field and it follows that

$$E = -\frac{\Delta V}{\Delta s} \text{ (approximately)}$$

or

$$\Delta s = -\frac{\Delta V}{E}.$$

That is. the greater the electric intensity E, the smaller the perpendicular distance Δs between the equipotentials. The equipotentials are therefore crowded close together in a strong field and are more widely separated in a weak field.

The geometrical relations between lines of force and equipotential surfaces frequently enable one to obtain a qualitative idea of the charge distribution and electric intensity in an arrangement of conductors where it would be difficult if not impossible to obtain an exact analytic expression for the desired result. For example, suppose that the conductor in Fig. 3-13 is given a positive charge, and we wish to know how the charge will be distributed over its surface. We know that the surface of the conductor is an equipotential, and also that at distances large compared with the dimensions of the body the equipotentials are spherical surfaces,

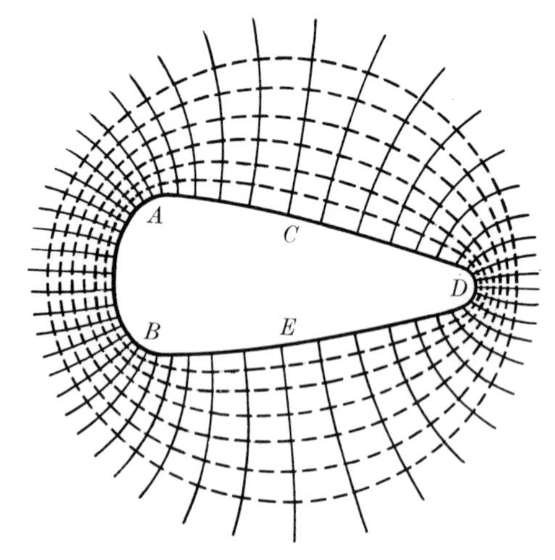

FIG. 3-13. Equipotential surfaces (dotted lines) and lines of force (solid lines) about a conductor of irregular shape.

since at a sufficiently great distance the body may be considered a point charge. Traces of the equipotentials may then be sketched, starting with one just outside the body and gradually altering the shape so that the trace approaches a circle. It will be found that the smaller the radius of curvature of the surface, the more closely must the equipotentials be spaced. The closest spacing occurs near the point D, with a somewhat larger spacing at A and B and the greatest spacing outside the surfaces C and E. The potential gradient, or the electric intensity, is therefore greatest near the sharply curved point D, and least near C and E.

The lines of force may next be sketched in, making them everywhere perpendicular to the equipotentials, of which the surface of the conductor is one, and spacing them more closely together the smaller the separation of the equipotentials.

Finally, since the electric intensity just outside a surface is proportional to the surface density of charge, it follows that the surface density is greatest at regions of small radius of curvature, least where the radius of curvature is large.

Notice that although the surface density of charge is far from uniform over the surface of a body of irregular shape, the *potential* is necessarily constant over the entire surface.

3-10 Sharing of charge by conductors. When a charged conductor is brought into electrical contact with one which is uncharged, the original charge is shared between the two. That this should happen is evident from the mutual forces of repulsion between the component parts of the original charge. The question as to precisely how much charge will be transferred has not yet been answered, but we can now see that it must be such as to bring all points of both conductors to the same potential. We shall discuss two examples only, (1) two spheres at a separation large compared with their radii, contact being made by a wire, and (2) a charged conductor making internal contact with a second hollow conductor. The principles will be illustrated by numerical examples.

Examples. (1) Suppose that charge is to be transferred from a conducting sphere A 1 cm in radius, supported on an insulated rod, to a 10 cm sphere B similarly supported, connection being made by a fine wire (Fig. 3-14). The centers of the spheres are 50 cm apart and the wire is fine enough so that the charge on it can be neglected. A charge of 10×10^{-9} coulomb is given initially to the smaller sphere.

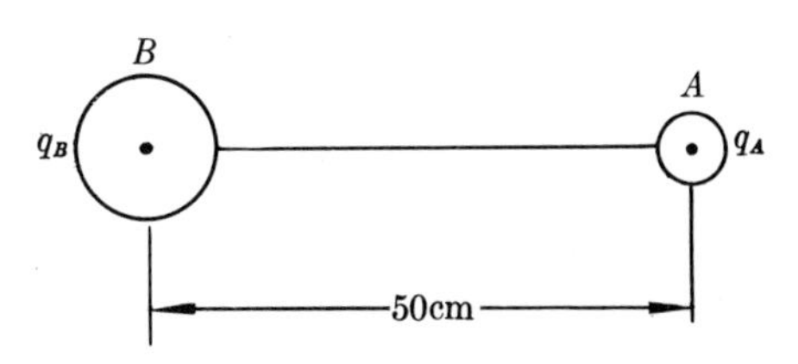

FIG. 3-14. Transfer of a charge from one sphere to another through a fine wire.

When the spheres are connected together the charge of 10×10^{-9} coulomb will distribute itself in such a way that both spheres and the connecting wire come to the same potential. Let q_A and q_B represent the charges on the spheres, and let us first write the expression for the potential at the center of sphere B. This is

$$V_B = 9 \times 10^9 \left(\frac{q_B}{.10} + \frac{q_A}{.50}\right).$$

Similarly, at the center of the other sphere,

$$V_A = 9 \times 10^9 \left(\frac{q_A}{.01} + \frac{q_B}{.50}\right).$$

But since both spheres are at the same potential, $V_A = V_B$, or

$$\frac{q_B}{.10} + \frac{q_A}{.50} = \frac{q_A}{.01} + \frac{q_B}{.50},$$

and since $q_A + q_B = 10 \times 10^{-9}$ coulomb (given), we find

$$\begin{aligned} q_B &= 9.25 \times 10^{-9} \text{ coulomb} \\ q_A &= 0.75 \times 10^{-9} \text{ coulomb} \\ V &= 845 \text{ volts.} \end{aligned}$$

The surface density of charge on the larger sphere is

$$\sigma_B = \frac{9.25 \times 10^{-9}}{4\pi \times (.10)^2} = 7.35 \times 10^{-8} \text{ coul/m}^2,$$

while on the smaller sphere it is

$$\sigma_A = \frac{.75 \times 10^{-9}}{4\pi \times (.01)^2} = 59.7 \times 10^{-8} \text{ coul/m}^2,$$

another illustration of the accumulation of charge at those portions of a conductor of small radius of curvature.

(2) Let an opening be made in the 10 cm sphere large enough to admit the 1 cm sphere. Assume for simplicity that the smaller sphere (A) is at the center of the larger sphere (B). (Fig. 3-15.) Let sphere B have a charge q_B to begin with, and let q_A be the charge on sphere A. The potential of B, neglecting any effect of the small opening, is

$$V_B = \frac{1}{4\pi\epsilon_0}\left(\frac{q_B}{.10} + \frac{q_A}{.10}\right),$$

and the potential of A is

$$V_A = \frac{1}{4\pi\epsilon_0}\left(\frac{q_B}{.10} + \frac{q_A}{.01}\right).$$

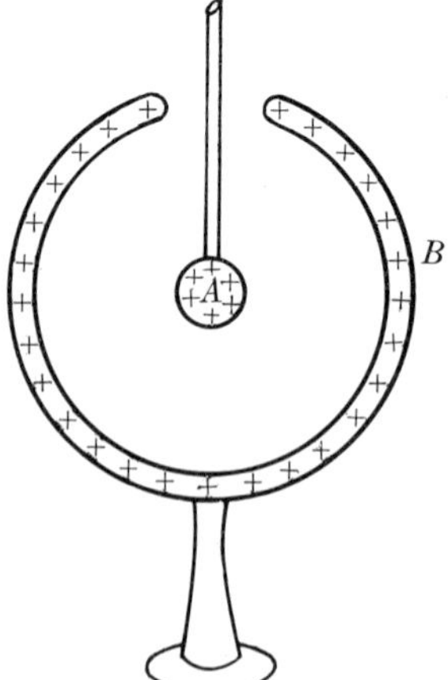

FIG. 3-15. Small charged sphere completely enclosed by a larger charged sphere.

The potential difference between A and B is therefore

$$V_{AB} = \frac{1}{4\pi\epsilon_0}\left(\frac{q_B}{.10} + \frac{q_A}{.01} - \frac{q_B}{.10} - \frac{q_A}{.10}\right)$$

$$= \frac{1}{4\pi\epsilon_0} q_A \left(\frac{1}{.01} - \frac{1}{.10}\right)$$

$$= \frac{1}{4\pi\epsilon_0} 90\, q_A. \tag{3-39}$$

Hence V_{AB} is positive and A is at a higher potential than B. If contact is now made between the spheres, charge will flow from A to B until $V_{AB} = 0$, or, from Eq. (3-39), until $q_A = 0$, which leads to the unexpected conclusion that *all* of the charge on sphere A transfers to sphere B, *regardless of the initial values of B's charge and potential.* It is this fact which is made use of in the Van de Graaff generator, described in the next section.

In the preceding analysis the small sphere was assumed to be at the center of the larger one for simplicity in computing the potentials. But since the charge on the larger sphere produces no field at internal points, no work is required to move the smaller sphere from one point to another within the larger sphere. The potential of the smaller sphere is therefore the same at all internal points, and regardless of its position its potential will be higher than that of the surrounding conductor as long as it has any charge at all. Furthermore, it is not necessary that the bodies be spheres. A conductor of any shape, inserted through a small opening in a hollow conductor of any shape, will give up all of its charge to the other conductor when internal contact is made between the two.

3-11 The Van de Graaff generator. The Van de Graaff generator makes use of the principle described in Sec. 3-10, namely, that if a charged conductor is brought into *internal* contact with a second hollow conductor, *all* of its charge transfers to the hollow conductor no matter how high the potential of the latter may be. Thus, were it not for insulation difficulties, the charge and hence the potential of a hollow conductor could be raised to any desired value by successively adding charges to it by internal contact. Actually, since the conductor must be supported in some way, its maximum potential will be limited to that at which the rate of leakage of charge from it through its supports or through the surrounding air equals the rate at which charge is delivered to it.

The Van de Graaff generator is shown schematically in Fig. 3-16 and a photograph of the installation at M.I.T. in Fig. 3-17. Referring to Fig. 3-16, A is a hollow spherical conductor supported on an insulating hollow column B. A belt C passes over pulleys D, the lower pulley being driven by a motor, while the upper is an idler. Terminal E, which consists of a number of sharp points projecting from a horizontal rod, is maintained at a negative potential of some tens of thousands of volts relative to ground by the auxiliary *DC* source *F*. Terminal *G*, similarly constructed, is maintained at a positive potential with respect to the spherical terminal *A*. The air around the points becomes ionized and a "spray" of ions is repelled from each set of points, negative ions from *E* and positive ions from *G*. Some of these attach themselves to the surface of the moving belt, so that the upward moving half becomes negatively and the other half positively charged. The ion current at terminals *E* and *G* is adjusted so that first, the charge carried toward either terminal is neutralized and second an equal charge of opposite sign is deposited on the belt.

The installation shown in Fig. 3-16 is enclosed in the left column of Fig. 3-17. The right column in Fig. 3-17 encloses a giant x-ray tube some 20 ft long. The cathode of the tube is inside the sphere at the top of this column. Electrons are driven down the tube by the potential difference between spheres and ground (about 2.5 million volts), and strike a metal anode mounted in a chamber below ground level. The anode then becomes a source of x-rays of short wave length and great penetrating power.

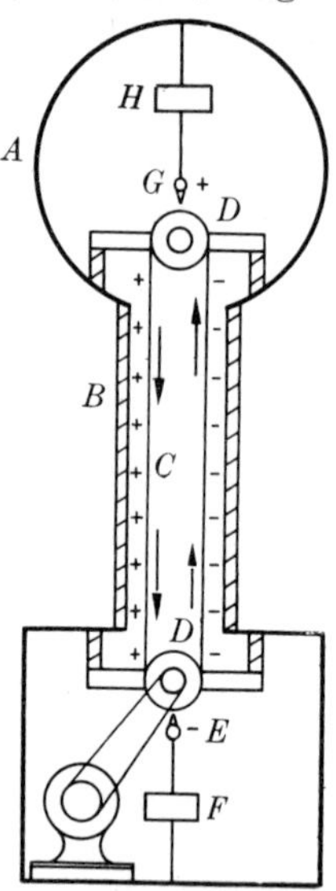

FIG. 3-16. Schematic diagram of Van de Graaff generator.

FIG. 3-17. Photograph of Van de Graaff generator at the Massachusetts Institute of Technology. (*Courtesy M.I.T. News Service*)

Problems—Chapter 3

(1) A dipole of small dimensions, and of dipole moment 10^{-10} coulomb-meters, is located at the origin with its axis along the X-axis. (a) Evaluate the line integral of the electric intensity along a line at 45° with the X-axis in the first quadrant, from infinity to a point 2 cm from the origin, to find the potential at this point. Use the expressions for the components of electric intensity in the field of a dipole given in Sec. 2-3. (b) How much work would be required to bring a positive point charge of 10^{-9} coulombs from infinity to this point? (c) What is the potential energy of a charge of 10^{-9} coulombs, at this point?

(2) Refer to problem 16 and Fig. 2-22 in Chap. 2. (a) Evaluate the line integral of the electric intensity from point b to point a, along the X-axis, and compute the potential difference V_{ab}. (b) Find the potential differences V_{bc} and V_{cd}. (c) Which faces of the cube are equipotential surfaces?

(3) Suppose a uniform electric field exists in the room in which you are working, with the lines of force horizontal and at right angles to one wall. As you walk toward the wall from which the lines of force emerge into the room, are you walking toward points of higher and higher, or lower and lower potential?

(4) In an apparatus for measuring the electronic charge e by Millikan's method, an electric intensity of 6.34×10^4 volts/m is required to just support a charged oil drop. If the plates are 1.5 cm apart, what potential difference between them is required?

(5) (a) Prove that when a particle of constant mass and charge is accelerated from rest in an electric field, its final velocity is proportional to the square root of the potential difference through which it is accelerated. (b) Find the magnitude of the proportionality constant if the particle is an electron, the velocity is in m/sec, and the potential difference is in volts. (c) At velocities much greater than $1/10$ of the velocity of light the mass of a body becomes appreciably larger than its "rest mass" and cannot be considered constant. (The velocity of light is 3×10^8 m/sec.) Through what voltage must an electron be accelerated, assuming no change in its mass, to acquire a velocity $1/10$th that of light? (d) What would be the velocity of a deuteron, accelerated through the same voltage? Express the velocity as a fraction of the velocity of light.

(6) An *electron-volt* is a unit of energy, equal to the kinetic energy of an electron that has been accelerated from rest through a potential difference of 1 volt. (a) Express this energy in joules and in ergs. (b) What is the velocity of an electron whose kinetic energy is 1 electron-volt? 100 electron-volts? (c) What is the velocity of a deuteron whose kinetic energy is 100 electron-volts?

(7) A vacuum diode consists of a cylindrical cathode 0.05 cm in radius, mounted coaxially within a cylindrical anode 0.45 cm in radius. The potential of the anode is 300 volts above that of the cathode. An electron leaves the surface of the cathode with zero initial velocity. Find its velocity (a) when it has travelled 2 mm toward the anode, (b) when it strikes the anode.

(8) Two point charges, $q_1 = +40 \times 10^{-9}$ coulombs and $q_2 = -30 \times 10^{-9}$ coulombs, are 10 cm apart. Point A is midway between them, point B is 8 cm from q_1 and 6 cm from q_2. Find (a) the potential at point A; (b) the potential at point B; (c) the work required to carry a charge of 25×10^{-9} coulombs from point B to point A.

(9) Refer to problem 5 in Chap. 2. Find the potential at the points a, b, c, and d.

(10) Refer to problem 6 in Chap. 2. Find the potential at the points a, b, c, and d.

(11) How much work would be required to bring a positive point charge of 10^{-9} coulombs from infinity to each of the points a, b, and c, in Fig. 3-5? What is the significance of the negative sign?

(12) In the Bohr model of the hydrogen atom a single electron revolves around a single proton in a circle of radius 5.28×10^{-9} cm. (a) What is the electrostatic potential energy of the atom? (b) Through what potential difference would an electron have to be accelerated to gain this amount of energy?

(13) The potentials at points a, b, c, d, and e, in Fig. 3-18, on the circumference of a circle 1 cm in radius, are: $V_a = 100^{\mathrm{v}}$, $V_b = 90^{\mathrm{v}}$, $V_c = 80^{\mathrm{v}}$, $V_d = 70^{\mathrm{v}}$, $V_e = 60^{\mathrm{v}}$. The potential at the center of the circle is 100^{v}. Approximate dV/ds by $\Delta V/\Delta s$ and compute the components of electric intensity in the directions of the radii Oa, Ob, Oc, Od, and Oe.

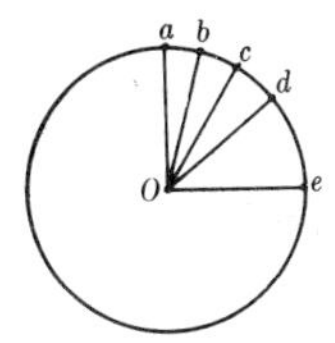

FIG. 3-18.

(14) (a) What is the potential gradient, in volts/m, at a distance of 10^{-12} cm from the nucleus of a gold atom? (See problem 5 in Chap. 1.) (b) What is the potential gradient at a distance of 5.28×10^{-9} cm from a proton? (See problem 6 in Chap. 1.)

(15) The potential at points in a plane is given by

$$V = \frac{a \cos \theta}{r^2} + \frac{b}{r},$$

where r and θ are the polar coordinates of a point in the plane and a and b are constants Find the components E_r and E_θ of the electric intensity at any point.

(16) The potential at points in a plane is given by

$$V = \frac{ax}{(x^2 + y^2)^{3/2}} + \frac{b}{(x^2 + y^2)^{1/2}},$$

where x and y are the rectangular coordinates of a point and a and b are constants. Find the components E_x and E_y of the electric intensity at any point.

(17) Refer to Problem 20 in Chap. 2 and Fig. 2-23. If the revolving particle is a deuteron traveling with a speed of 10^6 m/sec and the radii of the cylinders are 5 cm and 6 cm respectively, find the required potential difference between the cylinders.

(18) Two plane parallel plates are separated by 2 cm. The potential of one plate is +500 volts and that of the other is −500 volts. Show in a diagram the equipotential surfaces between the plates at potentials of +250 volts, 0, and −250 volts.

(19) The potential of a positively-charged conducting sphere of radius 5 cm is 5000 volts. No other charged bodies are nearby. (a) Show in a diagram the equipotential surfaces at potentials of 4000 volts, 3000 volts, and 2000 volts. (b) Construct a graph of potential as a function of radial distance from the center of the sphere, from $r = 0$ to $r = 15$ cm.

(20) Fig. 3–19 is an end view of two long coaxial hollow cylinders. (a) Prove that the potential V_r at the point P is given by

$$V_r = V_a - V_{ab}\frac{\log \frac{r}{a}}{\log \frac{b}{a}},$$

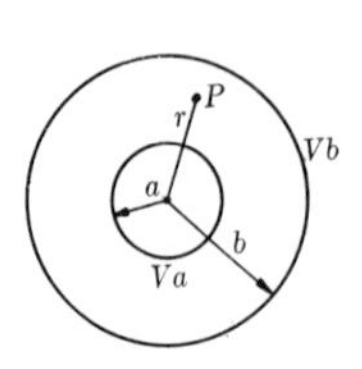

FIG. 3-19.

where V_a and V_b are the potentials of the inner and outer cylinders and a and b are their radii. (b) Let $V_a = +400$ volts, $V_b = -400$ volts, $a = 1$ cm, $b = 10$ cm. Find the radii of the equipotential surfaces at potentials of +200 volts, 0, and −200 volts. Draw a diagram. (c) Compute the electric intensities at each of these surfaces, and also at the surfaces of the cylinders. (d) What would be the electric intensity in the region between two large parallel plates 9 cm apart, the potentials of the plates being +400 volts and −400 volts?

(21) Two thin-walled conducting tubes of radii 1 cm and 10 cm respectively are mounted coaxially and insulated from one another. The outer tube is at ground potential. (a) What is the maximum potential to which the inner tube may be raised if the electric intensity in the region between the tubes is not to exceed 5×10^2 volts/m at any point? (b) When the inner tube is at this potential, find the potential and the electric intensity at a point 5 cm from the axis of the system. (c) Find the charge per unit length on each cylinder.

(22) The electrode structure of a certain vacuum tube consists of two coaxial cylinders of radii 1 mm and 3 mm respectively. The inner electrode is at ground potential and the outer electrode at a potential of +150 volts. Electrons are emitted from the surface of the inner electrode, or cathode, and are accelerated toward the outer electrode, the plate or anode. Assume the electrons to leave the cathode with zero initial velocity and neglect any forces exerted on one electron by others in the space between the electrodes. (a) What is the force, in dynes, on an electron just as it leaves the cathode? (b) Halfway between cathode and plate? (c) Just as it reaches the plate? (d) What is the acceleration of the electron at each of these points? (e) What is the speed of the electron when it strikes the plate? (f) What is its energy, in electron-volts?

(23) A conducting sphere 2 cm in radius has a charge of +100 statcoulombs. Compute the potential and the electric intensity, at points along a radial line, at the following distances from the center of the sphere: 0, 1 cm, 2 cm, 3 cm, 5 cm, 10 cm. Use the electrostatic system in which $1/4\pi\epsilon_0 = 1$. Construct graphs of V and E. What is the potential of the sphere, in volts?

(24) Two conducting spheres, each 2 cm in radius, are each given a charge of +60 statcoulombs and placed with their centers 12 cm apart. Assume the surface density of charge on each to be uniform. Compute the potential and electric intensity at points spaced 2 cm apart along a line joining the centers of the spheres. Make a graph of the results. Use the electrostatic system of units in which $1/4\pi\epsilon_0 = 1$.

(25) Refer to problem 16 in Chap. 2. (a) Find the space charge density as a function of x. (b) How many electrons per cubic millimeter would be required to give a space charge density equal to that at the plane perpendicular to the X-axis, passing through point a?

(26) A vacuum triode may be idealized as follows. A plane surface (the cathode) emits electrons with negligible initial velocities. Parallel to the cathode and 3 mm away from it is an open grid of fine wire at a potential of 18 volts above the cathode. A second plane surface (the anode) is 12 mm beyond the grid and is at a potential of 15 volts above the cathode. Assume that the plane of the grid is an equipotential surface, and that the potential gradients between cathode and grid, and between grid and anode, are uniform. Assume also that the structure of the grid is sufficiently open for electrons to pass through it freely. (a) Draw a diagram of potential vs distance, along a line from cathode to anode. (b) With what velocity will electrons strike the anode? (c) If the anode is made 18 volts negative with respect to the cathode, how far will electrons travel beyond the grid?

(27) Two conducting spheres of radii 1 cm and 2 cm respectively are each given a charge of 10^{-8} coulomb. If the distance between the centers of the spheres is 1 meter, what will be the final charge on each, and the potential of each, when they are connected by a fine wire?

(28) Suppose the potential difference between the spherical terminal of a Van de Graaff generator and the point at which charges are sprayed onto the upward moving belt, is 2 million volts. If the belt delivers negative charge to the sphere at the rate of 2×10^{-3} coulombs/sec and removes positive charge at the same rate, what horsepower must be expended to drive the belt against electrical forces?

CHAPTER 4

CURRENT, RESISTANCE, RESISTIVITY

4-1 Current. In purely electrostatic problems such as those considered in the preceding chapters, one is concerned chiefly with the forces between charges, the final, steady-state distribution of charge brought about by these forces, and the motion of charged particles in empty space. We are next to discuss the motion of charge in a conductor when an electric field is maintained within the conductor. This motion constitutes a *current.*

A conductor, it will be recalled, is a material within which there are "free" charges which will move when a force is exerted on them by an electric field. The free charges in a *metallic* conductor are negative electrons. The free charges in an *electrolyte* are ions, both positive and negative. A *gas* under the proper conditions, as in a neon sign or a fluorescent lamp, is also a conductor and its free charges are positive and negative ions and negative electrons.

We have seen that when an isolated conductor is placed in an electric field the charges within the conductor rearrange themselves so as to make the interior of the conductor a field-free region throughout which the potential is constant. The motion of the charges in the rearranging process constitutes a current, but it is of short duration only and is called a *transient* current. If we wish to maintain a continuous current in a conductor, we must continuously maintain a field or a potential gradient within it. If the field is always in the same direction, even though it may fluctuate in magnitude, the current is called *direct.* If the field reverses direction periodically, the flow of charge reverses also and the current is *alternating.* Direct and alternating currents are abbreviated to DC and AC respectively.

There are a number of electrical devices that will be discussed in more detail later on, which have the property of maintaining their terminals continuously at different potentials. The most familiar of these are the dry cell, the storage battery, and the generator. If the ends of a metal wire are connected to the terminals of any one of these devices, a potential gradient or an electric field will be maintained within the wire and there will be a continuous motion of charge through it. To be specific, if the ends of a copper wire one meter long are connected to the terminals of a 6-volt storage battery, a potential gradient or electric field of intensity 6 volts/m or 6 newtons/coulomb is established and maintained along the wire.

Under the influence of this field the free electrons in the wire experience a force in the opposite direction to the field and accelerate in the direction of this force. (The other electrons, as well as the positive nuclei, are also acted on by the field but are prevented from accelerating by the binding forces which hold these electrons to the nuclei and hold the nuclei together to form a solid body.) Collisions with the stationary particles in the metal very soon slow down the free electrons or stop them, after which they again accelerate, and so on. Their motion is thus a succession of accelerations and decelerations, but they will acquire a certain average velocity in the opposite direction to the field and we may assume that they all move steadily with this average velocity. The free electrons also share in the thermal energy of the conductor, but their thermal motion is a random one and for our present purposes may be ignored.

Fig. 4-1 illustrates a portion of a wire in which there is a field toward the left and consequently a motion of free electrons toward the right. Each electron is assumed to move with the same constant velocity v,

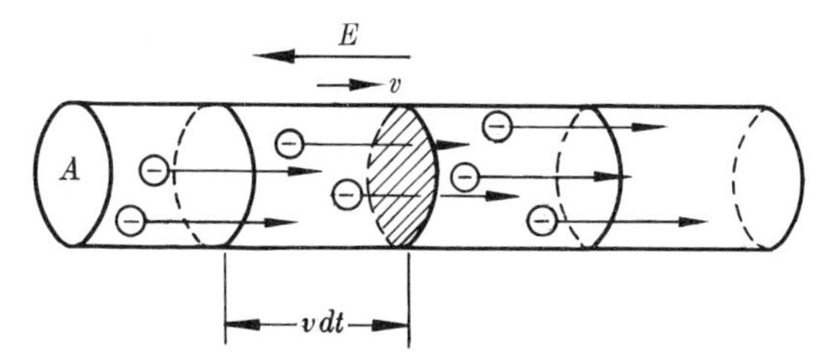

FIG. 4-1. Motion of free electrons in a wire.

and in time dt each advances a distance $v\,dt$. In this time, the number of electrons crossing any plane such as the one shown shaded, is the number contained in a section of the wire of length $v\,dt$ or volume $Av\,dt$, where A is the cross section of the wire. If there are n free electrons per unit volume, the number crossing the plane in time dt is $nAv\,dt$, and if e represents the charge of each, the total charge crossing the area in time dt is

$$dq = nevA\,dt. \tag{4-1}$$

The rate at which charge is transported across a section of the wire, or dq/dt, is called the current in the wire. Current is represented by the letter i (or I).

$$\boxed{i = \frac{dq}{dt}} \tag{4-2}$$

From Eqs. 4-1 and 4-2,

$$i = nevA. \tag{4-3}$$

The mks unit of current is *one coulomb per second* and is called one *ampere*, in honor of the French scientist André Marie Ampere (1775–1836) who developed many of the concepts of electricity and magnetism. Small currents are more conveniently expressed in milliamperes (ma) (1 milliampere $= 10^{-3}$ amp.) or in microamperes (μa) (1 microampere $= 10^{-6}$ amp.).

Since a current is a flow of charge, the common expression "flow of current" should be avoided, since literally it means "flow of flow of charge."

When there is a current in a conductor in which free charges of both signs are present, as in an electrolyte, negative charge crosses a surface in one direction and positive charge in the other. In general, if any number of different kinds of charged particles are present, in different concentrations and moving with different velocities, the net charge crossing a surface in time dt is

$$dq = Adt\,[n_1q_1v_1 + n_2q_2v_2 + \cdots]$$

and the current is

$$i = A\,\Sigma nqv$$

All of the nqv products will have the same sign, since charges of opposite sign move in opposite directions.

The distribution of charge in a wire carrying a current should not be confused with the steady-state charge distribution on an isolated conductor having an *excess* charge which, as we have seen, is confined to the surface of the conductor. There is no *excess* charge on a wire carrying a current, the positive and negative charge per unit volume being equal. (The positive charges are not shown in Fig. 4-1.) The free electrons in a wire carrying a current are distributed uniformly throughout the wire and the current in a wire of constant cross section is distributed uniformly across any section. (Except that if the current is alternating, there is a tendency for it to concentrate at the surface.)

The *current density* in the wire, represented by J, is defined as the ratio of the current to the cross-sectional area.

$$J = \frac{i}{A} = nev \tag{4-4}$$

Strictly speaking, Eq. (4-4) defines the average current density over the area A. If the current is not uniformly distributed, one considers an infinitesimal area dA across which the current is di, and defines the current density as

$$J = \frac{di}{dA} \tag{4-5}$$

It is of interest to estimate the average velocity of the free electrons in a conductor in which there is a current. Consider a copper conductor 1 cm in diameter in which the current is 200 amp. The current density is

$$J = \frac{i}{A} = \frac{200}{\frac{1}{4}\pi(.01)^2} = 2.54 \times 10^6 \text{ amp/m}^2.$$

Previous calculations (see Sec. 2-10) have shown that there are in copper about 8.5×10^{22} free electrons/cm^3 or 8.5×10^{28} free electrons/m^3. Then since $J = nev$,

$$v = \frac{J}{ne} = \frac{2.54 \times 10^6}{8.5 \times 10^{28} \times 1.6 \times 10^{-19}} = 1.9 \times 10^{-4} \text{ m/sec}$$

or about .02 cm/sec. The velocity is therefore quite small.

The average velocity of the free electrons in a conductor should not be confused with the velocity of propagation of electromagnetic waves in free space, which is 3×10^8 m/sec or 186,000 mi/sec.

4-2 The direction of a current. It would seem logical to define the direction of a current as the direction of motion of the free charges. We immediately come up against the difficulty, however, that in an electrolytic or gaseous conductor free charges of both signs are in motion in opposite directions. Whichever direction was assigned to the current, we would find charges moving in the opposite direction. Since some convention must be adopted, it has been agreed to speak of the direction of any current as if the carriers were all positive charges. Hence *in a metallic conductor the electrons move in the opposite direction to the conventional current.* In Fig. 4-1, for example, the direction of the current is said to be from right to left, although the electrons move from left to right.

From now on we shall adopt this convention, and indeed shall often speak of the current in a metal as though it consisted of the motion of positive charge.

4-3 The complete circuit. When the ends of a wire are connected to two points maintained at fixed, but different, potentials, such as the terminals of a cell or generator, there will be a current in the wire but the potential of each point of the wire remains constant in time. Consider any short element of the wire bounded by two transverse planes and shown shaded in Fig. 4-2. If the electrons in the wire are moving from left to right, negative charge is entering the element at its left face and leaving it at its right face. The quantity of charge entering the element in any time interval must be exactly equal to that leaving for, if it were not, the quantity of charge in the element, and hence the potential of the element, would change. We conclude that the current must be the same at both faces of the element, and therefore must be the same at *all* sections of the wire. In this respect a flow of electrons is like the flow of an incompressible fluid.

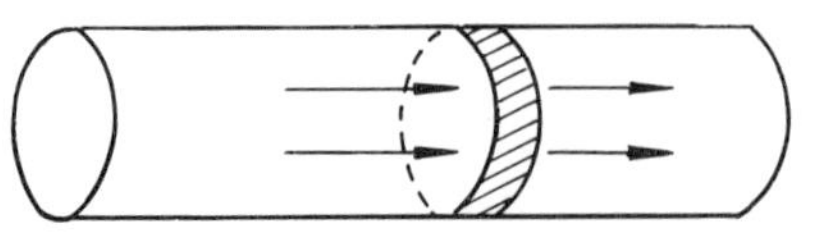

FIG. 4-2. Electric charge entering and leaving a small element of a wire.

The conducting wire and the cell or generator to which its ends are connected are said to form a complete circuit or a *closed circuit.* An electrolytic cell is represented by the symbol $\dashv\vdash^{+}$ and a generator by $-\!\!\bigcirc\!\!-^{+}$, the + sign denoting the terminal which is normally at the higher potential. Fig. 4-3 is a diagram of such a closed circuit. The light line marked with arrowheads indicates the direction of the conventional current. The electrons in the wire circulate in the opposite direction. The positive ions within the cell move in the same direction as the conventional current, the negative ions in the opposite direction.

A number of cross sections of the circuit are indicated by dotted lines. The current is the same at all of these sections, including the one through the cell. Current is not something which squirts out of the positive terminal of a cell and gets all used up by the time it reaches the negative terminal.

A steady current can exist only in a closed circuit equivalent to that of Fig. 4-3. If the wire is disconnected from the cell at either end, or if a break is made at any other

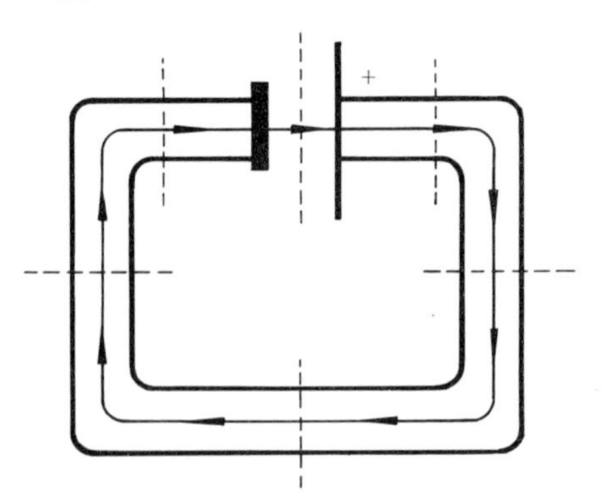

FIG. 4-3. A closed circuit consisting of an electrolytic cell and a wire.

point of the circuit, the current immediately ceases. The circuit is then said to be *open*.

Notice carefully that the direction of the conventional current is "from plus to minus" *in the external circuit only*. Within the cell, the direction is from minus to plus.

4-4 Electrical conductivity. In order to maintain a current in a conductor there must be an electric field or a potential gradient within it. Conducting materials differ from one another, however, in the magnitude of the current density established by a given electric field. Fig. 4-4 represents a portion of a conductor of cross section A within which there is a (conventional) current i and a current density $J = \frac{i}{A}$. The electric intensity at the shaded section is E. The ratio of the current density to the electric intensity, or the current density per unit electric intensity, is called the *electrical conductivity* of the material and is represented by σ.

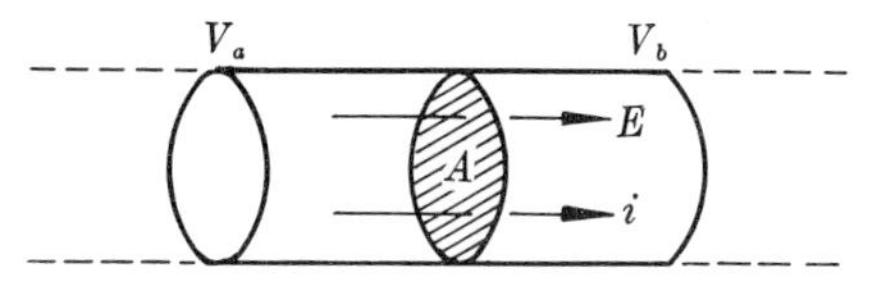

FIG. 4-4. Current within a portion of a conductor.

$$\sigma = \frac{J}{E}, \quad \text{or} \quad J = \sigma E \tag{4-6}$$

The greater the conductivity, the greater the current density for a given electric intensity. The conductivity of a given material varies with temperature, and to a small degree with other physical conditions which will be mentioned later. For many materials, notably the metals, the conductivity is independent of the current density. On the other hand, there are substances for which the conductivity varies markedly with current density.

Current density is a vector, in the same direction as the field. The vector form of Eq. (4-6) is therefore

$$\vec{J} = \sigma\vec{E}$$

Eq. (4-6) may be put in the following form. If we replace E by $-\frac{dV}{dx}$, and J by $\frac{i}{A}$, we obtain

$$\frac{i}{A} = \sigma\left(-\frac{dV}{dx}\right)$$

$$i = -\sigma A \frac{dV}{dx}. \tag{4-7}$$

It is interesting to note that this equation is exactly analogous to the equation of steady-state heat conduction,

$$H = -KA\frac{dt}{dx}.$$

The electrical current i corresponds to the heat current H, the electrical conductivity σ to the thermal conductivity K, and the potential gradient $\frac{dV}{dx}$ to the temperature gradient $\frac{dt}{dx}$.

The relation between the conduction of electricity and the conduction of heat is more than a mere mathematical analogy. The free electrons which are the carriers of charge in electrical conduction also play an important role in the conduction of heat and hence a correlation can be expected between electrical and thermal conductivity. It is a familiar fact that good electrical conductors, such as the metals, are good conductors of heat, while poor electrical conductors are also poor conductors of heat.

Theoretical reasoning which we cannot go into here predicts the following relation, known as the *Wiedemann-Franz law*, between the thermal and electrical conductivities of metals:

$$\frac{K}{\sigma} = 3\left(\frac{k}{e}\right)^2 T$$

where k is the Boltzmann constant, e the electronic charge, and T the Kelvin temperature.

4-5 Resistance and resistivity. Ohm's law. While the relation $J = \sigma E$ is the fundamental equation of electrical conduction, it is more convenient in practice to work with currents and potential differences than with current densities and electric intensities. For this purpose we start with the equivalent form

$$i = -\sigma A\frac{dV}{dx}, \tag{4-7}$$

and consider a conductor of length L and constant cross section A (Fig. 4-5) in which there is a current i. Let V_a and V_b be the potentials at its terminals. We shall restrict the discussion to materials for which σ is independent of J and shall assume the temperature (and any other factors affecting σ) to be the same at all points of the conductor. Under these

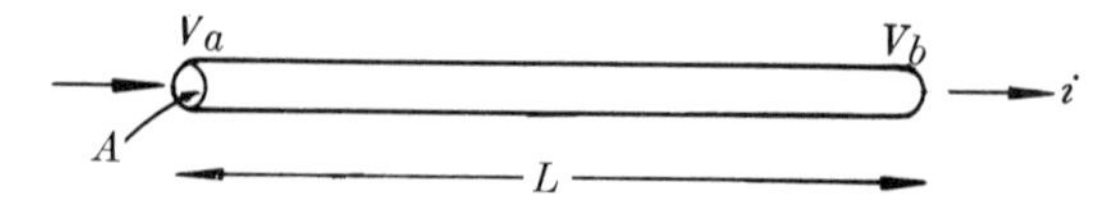

FIG. 4-5. Conductor of length L and constant cross section.

conditions i, σ, and A are all constants and Eq. (4-7) can readily be integrated. Take the X-axis along the wire with origin at end a. Then

$$i\,dx = -\sigma A\,dV$$

$$i\int_0^L dx = -\sigma A \int_{V_a}^{V_b} dV$$

$$iL = \sigma A(V_a - V_b)$$

$$i = \frac{\sigma A}{L}(V_a - V_b). \tag{4-8}$$

This is the desired relation between the current in the conductor and the potential difference between its terminals. The factor $\dfrac{\sigma A}{L}$ is called the *conductance* of the wire. The greater the conductance, the greater the current for a given potential difference.

In practice, the reciprocal of the conductance, called the *resistance*, is almost universally used. Resistance is represented by the letter R (or r). Evidently, for a homogeneous conductor of constant cross section and for which σ is constant,

$$R = \frac{L}{\sigma A}. \tag{4-9}$$

The conducting properties of materials are commonly tabulated in terms of the reciprocal of the conductivity, called the *resistivity* and denoted by ρ.

$$\rho = \frac{1}{\sigma}.$$

In terms of resistivity Eq. (4-9) becomes

$$\boxed{R = \frac{\rho L}{A},} \tag{4-10}$$

and in terms of resistance Eq. (4-8) becomes

$$i = \frac{V_a - V_b}{R} = \frac{V_{ab}}{R}$$

or

$$\boxed{V_{ab} = Ri.} \tag{4-11}$$

Note carefully that this relation between V_{ab}, R, and i, applies only when the path between a and b is a so-called "pure resistance," that is, it does not contain any batteries, motors, generators, etc. The general expression for the potential difference between any two points in a DC circuit will be derived in the next chapter.

It must also be remembered that in deriving Eq. (4-11) from the basic equation $J = \sigma E$ we assumed the constancy of the conductivity σ. If (and only if) σ is independent of J, the resistance R is a constant independent of i, and the potential difference V_{ab} between the terminals of a conductor is a *linear* function of the current in the conductor. This direct proportionality between the current in a metallic conductor and the potential difference between its terminals was first discovered experimentally by the German scientist, Georg Simon Ohm (1789–1854), and it is known as *Ohm's law.*

Materials for which V_{ab} is a linear function of i (or for which R is a constant) are called *linear conductors;* others are called *nonlinear.* Fortunately, for simplicity of computation, the metals, which are the most widely used conductors, are linear conductors and their resistance is independent of the current. But since there are many materials which do not obey Ohm's law, this law describes a special property of certain materials and not a general property of all matter.

The mks unit of resistance, from Eq. (4-11), is the *volt per ampere.* A resistance of one volt per ampere is called one *ohm.* That is, *the resistance of a conductor is one ohm if the potential difference between the terminals of the conductor is one volt when the current in the conductor is one ampere.*

The Greek letter ω or Ω is used to designate a resistance in ohms. Thus a resistance of 4.5 volts/amp or 4.5 ohms is written

$$4.5^{\omega} \quad \text{or} \quad 4.5\,\Omega.$$

Large resistances are more conveniently expressed in megohms (1 megohm = 10^6 ohms) and small resistances in microhms (1 microhm = 10^{-6} ohms).

Resistance units constructed to introduce into a circuit lumped resistances large compared with those of leads and contacts are called *resistors.* A resistor is represented by the symbol

Portions of a circuit of negligible resistance are shown by straight lines.

An adjustable resistor is called a *rheostat.* A common type is constructed by winding resistance wire on a porcelain or enamel tube, and making connections to one end and to a sliding contact. A rheostat is represented by

or simply

4-6 Standard resistors. The ohm as defined in the preceding section—i.e., the resistance of a conductor in which the current is one ampere when the potential difference between its terminals is one volt—is known as the *absolute ohm.* For practical convenience, the ohm has been defined by international agreement in terms of a physical standard. At the time of its adoption, in 1908, this *international ohm* was believed to be identical with the absolute ohm. Later measurements have shown the two to differ slightly, the best value at present being

1 int. ohm = 1.00048 abs. ohm.

The international ohm is defined as the resistance offered to an unvarying electric current by a column of mercury at the temperature of melting ice, 14.4521 gm in mass, of a constant cross-sectional area and of a length of 106.300 cm.

The international standard consists of a mercury column in a glass tube and is obviously not suited for a working standard in the laboratory. So-called "standard resistors" are, however, available commercially with resistances ranging from 0.001 ohm to 10,000 ohms, accurate to within a few hundredths of one per cent. They are constructed of wire, usually manganin because of its low temperature coefficient of resistivity (see Sec. 4-7), and are mounted in a metal case through which oil can be circulated to maintain constant temperature (see Fig. 4-6). Heavy "current terminals' are provided for leading the current into and out of the resistor, together with a pair of "potential terminals" between which the resistance has its specified value and which are used for measuring the potential difference across the resistor.

FIG. 4-6. Standard resistor. (*Courtsey of Leeds and Northrup*)

4-7 Calculation of resistance. The resistance of a homogeneous conductor of constant cross section is given by Eq. (4-10),

$$R = \frac{\rho L}{A}.$$

The resistance is proportional to the length of the conductor and inversely proportional to its cross section. If the conductor is of unit length and unit cross section, the ratio $\frac{L}{A}$ is unity and the resistance R and the resistivity ρ are numerically equal. Hence the resistivity of a material is numerically equal to the resistance of a specimen of the material of unit length and unit cross-sectional area.

In the mks system, where the unit of length is the meter and the unit of area the square meter, the unit of resistivity is one *ohm-meter*. The resistivity of a material in ohm-meters is numerically equal to the resistance, in ohms, between opposite faces of a cube of the material one meter on a side.

If the centimeter is used as a unit of length and the square centimeter as a unit of area, resistivities are expressed in *ohm-centimeters*.

In engineering work it is customary to express the length of a conductor in feet and its cross-sectional area in circular mils.[1] The resistivity of a material in terms of these units is expressed in ohm-circular mils per foot, usually abbreviated to the "mil-foot" resistivity. The mil-foot resistivity is numerically equal to the resistance in ohms of a wire of circular cross section, one foot long and one one-thousandth of an inch in diameter.

It is evident from the significance of resistivity that materials having large resistivities are poor conductors or good insulators. Conversely, substances of small resistivity are good conductors. No perfect insulator ($\rho = \infty$) exists, nor does a perfect conductor[2] ($\rho = 0$). There is, however, a wide difference in the resistivities of different materials so that in general they can be grouped into two classes, conductors and insulators. See Table 4-1.

[1] A circular mil (abbreviated CM) is a unit of *area*, equal to the area of a circle whose diameter is one one-thousandth of an inch or one mil. Since the areas of two circles are proportional to the squares of their diameters, the area of any circle in circular mils is equal to the square of its diameter in mils.

$$\text{Area (in CM)} = [D \text{ (in mils)}]^2$$

Since most wires are circular in cross section, the use of a unit of area which is itself circular eliminates the factor π in calculations of area.

It follows at once from its definition that

$$1 \text{ CM} = \frac{\pi}{4{,}000{,}000} \textit{ square } \text{inches.}$$

[2] At extremely low temperatures, near absolute zero, some conductors lose the last vestige of resistance, a phenomenon known as *superconductivity*. A current once started in such a conductor will continue of itself for hours or even days.

TABLE 4-1

RESISTIVITIES AND TEMPERATURE COEFFICIENTS

(Approximate values near room temperature)

(From Handbook of Chemistry and Physics, 26th Ed.)

Material	ρ (ohm-meters)	α (°C^{-1})
Aluminum	2.63×10^{-8}	.0039
Brass	6-8	.0020
Carbon	3500	$-.0005$
Constantan (Cu 60, Ni 40)	49	$+.000002$
Copper (Commercial annealed)	1.72	.00393
Iron	10	.0050
Lead	22	.0043
Manganin (Cu 84, Mn 12, Ni 4)	44	.000000
Mercury	94	.00088
Nichrome	100	.0004
Silver	1.47	.0038
Tungsten	5.51	.0045
Amber	5×10^{14}	
Bakelite	$2 \times 10^{5} - 2 \times 10^{14}$	
Glass	$10^{10} - 10^{14}$	
Hard rubber	$10^{13} - 10^{16}$	
Mica	$10^{11} - 10^{15}$	
Quartz (fused)	75×10^{16}	
Sulfur	10^{15}	
Wood	$10^{8} - 10^{11}$	

Multiply ρ in ohm-meters by 10^2 to get ρ in ohm-cm.
Multiply ρ in ohm-meters by 6.02×10^8 to get ρ in the mil-foot system.

The resistivity of a pure metal in a definite crystalline state and at a definite temperature is a quantity characteristic of the metal. Changes in the crystalline structure due to heat treatment or mechanical strain, or impurities alloyed with the metal even in minute quantities, may have a pronounced effect on its resistivity. For example, the resistivity of commercial annealed copper at 20° C. is 1.72×10^{-8} ohm-meter, that of hard drawn copper is 1.77×10^{-8} ohm-meter.

Hydrostatic pressure applied to a metal modifies its resistivity very slightly. The resistivity of bismuth is increased if the bismuth is in a magnetic field, and this change in resistivity affords a convenient method of measuring such fields. The resistivity of the metal selenium is decreased when the metal is illuminated. The light liberates photoelectrons within the metal, rendering it a better conductor.

The resistivity of all conducting materials is affected by their temperature. A plot of resistivity vs temperature, for a metallic conductor, is given in Fig. 4-7. The curve may be satisfactorily represented by an equation of the form

$$\rho = \rho_0 + at + bt^2 + \cdots$$

where ρ_0 is the resistivity at 0° C; a, b, etc., are constants characteristic of a particular material, and t is the Centigrade temperature. For temperatures which are not too great, the terms in t^2 and higher powers may be neglected and one may write

$$\rho = \rho_0 + at. \tag{4-12}$$

It is convenient to put Eq. (4-12) in the form

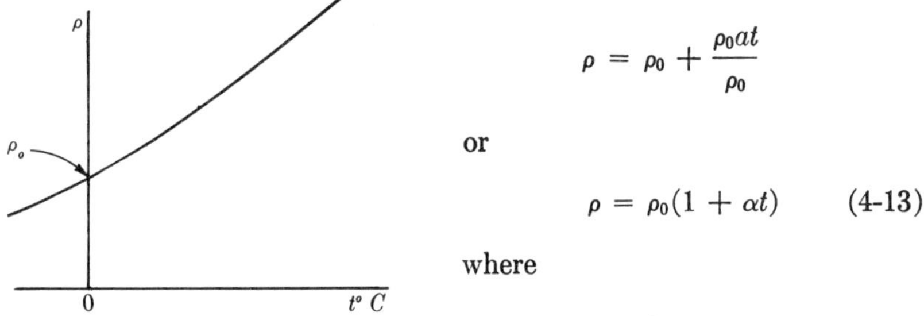

$$\rho = \rho_0 + \frac{\rho_0 at}{\rho_0}$$

or

$$\rho = \rho_0(1 + \alpha t) \tag{4-13}$$

where

$$\alpha = \frac{a}{\rho_0} = \frac{\rho - \rho_0}{\rho_0 t}.$$

FIG. 4-7. Variation of resistivity of a metal with temperature.

The quantity α is called the *temperature coefficient of resistivity* and is the fractional increase in resistivity per degree increase in temperature. Its units are deg^{-1} or "reciprocal degrees."

Since the resistance of a given conductor is proportional to its resistivity, Eq. (4-13) may also be written

$$R = R_0(1 + \alpha t). \tag{4-14}$$

R_0 is the resistance at 0° C and R is the resistance at t° C.

The resistivities of nonmetals decrease with increasing temperature, and their temperature coefficients of resistivity are negative. Electrolytes also have negative temperature coefficients of resistivity. In other words, they conduct more readily at high temperatures, which is probably because their decreased viscosity affords less opposition to the transport of ions through them. The temperature coefficients of resistivity of some common materials are listed in Table 4-1.

Examples. (1) What is the electric intensity in the copper conductor referred to in the example in Sec. 4-1?

From Table 4-1 the resistivity of copper is 1.72×10^{-8} ohm-meters and hence its conductivity is $\sigma = \dfrac{1}{1.72 \times 10^{-8}} = 5.83 \times 10^{7}$ reciprocal ohm-meters. The current density, from the example on page 89 is 2.54×10^{6} amp m². Hence

$$E = \frac{J}{\sigma} = \frac{2.54 \times 10^{6}}{5.83 \times 10^{7}} = 4.37 \times 10^{-2} \text{ volts/m.}$$

(2) What is the potential difference between two points on the wire 100 m apart?

(a) Since the electric intensity or potential gradient in the wire is 4.37×10^{-2} volts/m, the potential difference between two points 100 m apart is

$$V_{ab} = 4.37 \times 10^{-2} \times 100 = 4.37 \text{ volts.}$$

(b) The more common method of solution would be to compute first the resistance of the wire from $R = \rho \dfrac{l}{A}$ and then find the potential difference from $V_{ab} = Ri$. We have

$$R = 1.72 \times 10^{-8} \times \frac{100}{\frac{1}{4}\pi(.01)^2} = .0218 \text{ ohms.}$$

$$i = 200 \text{ amp. (given)}$$

$$V_{ab} = Ri = .0218 \times 200 = 4.37 \text{ volts.}$$

which is the same as the previous answer.

(3) Fig. 4-8 represents an armored submarine cable consisting of a central metallic conductor of radius a surrounded by an insulator of outer radius b and encased in a protecting metal sheath. Under operating conditions, the central conductor and the sheath are at different potentials. Hence there will be a potential gradient between conductor and sheath and a radial leakage current through the "insulator" in addition to the current along the cable. What is the resistance of the insulator to this radial current, in a length L of cable?

Let the potentials of the inner conductor and the sheath be V_a and V_b respectively.

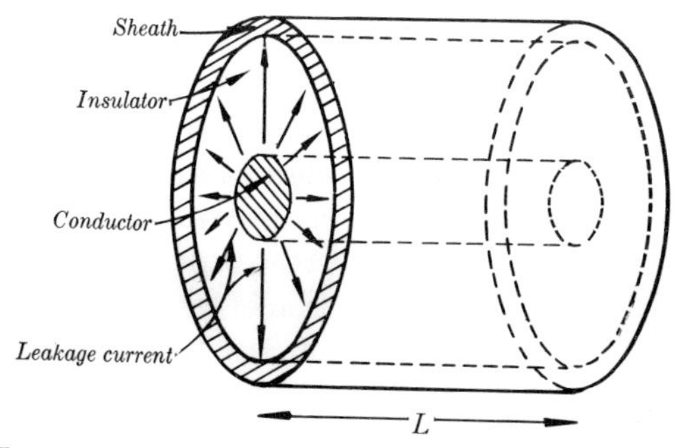

FIG. 4-8. Radial current in an armored submarine cable.

We shall assume them constant along the length L. It has been shown (see Sec. 3-5) that the electric intensity at a point between two concentric cylinders between which the potential difference is V_{ab} is

$$E = \frac{1}{r}\frac{V_{ab}}{\ln b/a}.$$

Hence

$$J = \sigma E = \frac{\sigma}{r}\frac{V_{ab}}{\ln b/a}$$

$$i = AJ = 2\pi r L \frac{\sigma}{r}\frac{V_{ab}}{\ln b/a}$$

and

$$R = \frac{V_{ab}}{i} = \frac{\ln b/a}{2\pi L\sigma} = \rho\frac{\ln b/a}{2\pi L}$$

The similarity between this problem and that of computing the radial flow of heat in a cylindrical "heat insulator" should be noted.

4-8 Measurement of current, potential difference, and resistance. Currents are measured by instruments called *galvanometers* or *ammeters.* The most common type makes use of the interaction between a current-carrying conductor and a magnetic field and is described in Chap. 10. For our present purposes it is sufficient to know that such instruments exist.

To measure the current at a point such as a, b, or c in Fig. 4-9(a), the circuit must be opened and the ammeter inserted at that point so that the current to be measured passes *through* the ammeter as in Fig. 4-9(b). An ammeter is a low resistance instrument, representative values being a few hundredths or thousandths of an ohm.

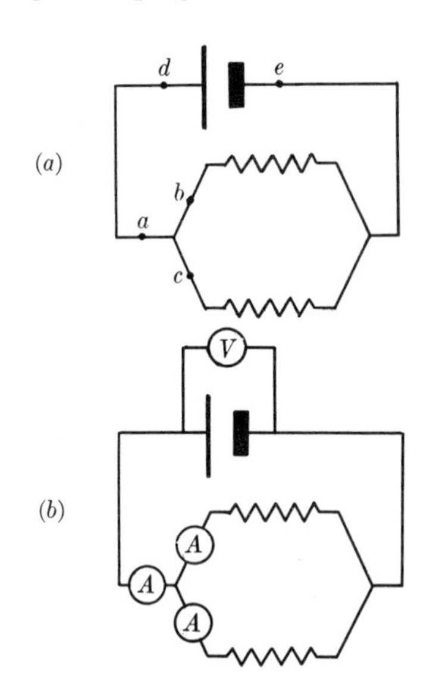

FIG. 4-9. Ammeter and voltmeter connections.

The potential difference between two points of a circuit might be measured with an electroscope or electrometer (see Sec. 1-2). It is more convenient, however, to use some type of *voltmeter*, the construction of which is described more fully in Sec. 10-3. Most voltmeters, unlike electroscopes and electrometers, are current operated. The voltmeter terminals are connected to the points between which the potential difference is to be measured. Fig. 4-9(b) shows a voltmeter V connected so as to measure the potential difference between the terminals of the cell. If the details of its construction are disregarded, a voltmeter may be treated as a resistor which automatically indicates the potential difference between its terminals. Typical resistances, for a 100-volt instrument, are from 10,000 to 100,000 ohms.

The resistance of a conductor is the ratio of the potential difference between its terminals to the current in it. The most straightforward method of measuring resistance is therefore to measure these two quantities and divide one by the other. The ammeter-voltmeter method for so doing is illustrated in Fig. 4-10. In circuit (a) the ammeter measures the current i_R in the resistor, but the voltmeter reads V_{ac} and not the potential difference V_{ab} between the terminals of the resistor. In circuit (b) the voltmeter reads V_{ab} but the ammeter reads the *sum* of the currents in the resistor and in the voltmeter. Hence, whichever circuit is used, corrections must be made to the reading of one meter or the other, unless these corrections can be shown to be negligible. See Sec. 5-5.

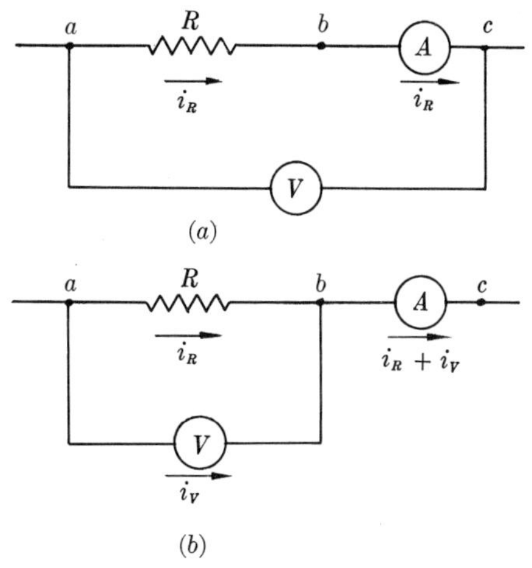

FIG. 4-10. Ammeter-voltmeter methods of measuring resistance.

4-9 Joule's law. While we have for simplicity been speaking of the current in a conductor as if all of the free electrons moved with the same constant velocity v, the electronic motion may more correctly be described as a series of accelerations, each of which is terminated by a collision with one of the fixed particles of the conductor. The electrons gain kinetic energy in the free paths between collisions, and give up to the fixed particles, in each collision, the same amount of energy they have gained. The energy acquired by the fixed particles (which are "fixed" only in the sense that their *mean* position does not change) increases their amplitude of vibration. In other words, it is converted to thermal energy or heat.

To derive the expression for the rate of development of heat in a conductor, we first work out the general expression for the power input to any portion of an electric circuit. The rectangle in Fig. 4-11 represents a portion of a circuit in which there is a (conventional) current i from left to right. V_a and V_b are the potentials at terminals a and b. The nature of the circuit between a and b is immaterial—it may be a conductor, a motor, generator, battery, or any combination of these. The power input, as we shall show, depends only on the magnitudes and relative directions of the current and the terminal potential difference.

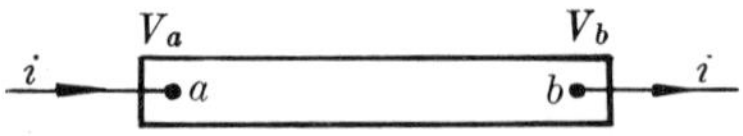

FIG. 4-11. Transfer of charge in a portion of a circuit.

In a time interval dt, a quantity of charge $dq = i\,dt$ enters the portion of the circuit under consideration at terminal a, and in the same time an equal quantity of charge leaves at terminal b. There has thus been a transfer of charge dq from a potential V_a to a potential V_b. The energy dW given up by the charge is

$$dW = dq(V_a - V_b) = i\,dtV_{ab},$$

and the *rate* at which energy is given up, or the power input P, is

$$\boxed{P = \frac{dW}{dt} = iV_{ab}.} \tag{4-15}$$

That is, the power is equal to the product of the current and the potential difference. If the current is in amperes, or coulombs/sec, and the potential difference in volts, or joules/coulomb, the power is in joules/sec or watts, since

$$\frac{\text{coulombs}}{\text{sec}} \times \frac{\text{joules}}{\text{coulomb}} = \frac{\text{joules}}{\text{sec}} = \text{watts}$$

Eq. (4-15) is a perfectly general relation that holds whatever the nature of the circuit elements between a and b. We shall return to it later on.

In the special case in which the circuit between a and b is a pure resistance R, all of the energy supplied is converted to heat, and in this special case the potential difference V_{ab} is given by

$$V_{ab} = iR.$$

Hence

$$P = iV_{ab} = i \times iR,$$

or

$$P = i^2R. \tag{4-16}$$

To bring out more explicitly that in this special case the energy appears as heat, we may set $P = dH/dt$, where dH is the heat developed in time dt. Eq. (4-16) then becomes

$$\boxed{\frac{dH}{dt} = i^2R} \tag{4-17}$$

If the conductor is linear, i.e., if R is a constant independent of i, Eq. (4-17) states that *the rate of development of heat is directly proportional to the square of the current.* This fact was discovered experimentally by Joule in the course of his measurements of the mechanical equivalent of heat and it is known as *Joule's law.* Of course, it is a law only in the same sense as is Ohm's law, that is, it expresses a special property of certain materials rather than a general property of all matter. A material which obeys Ohm's law necessarily obeys Joule's law also, and the two are not independent relations.

If the resistance, in Eq. (4-17), is expressed in ohms and the current in amperes, the rate of development of heat is in joules/sec or watts. This can readily be converted to cal/sec from the relation 1 cal = 4.186 joules.

Notice that the rate of development of heat in a conductor is not the same thing as the rate of increase of temperature of the conductor. The latter depends on the heat capacity of the conductor and the rate at which heat can escape from the conductor by conduction, convection, and radiation. The rate of loss of heat increases as the temperature of the conductor increases, and the temperature of a current-carrying conductor will rise until the rate of loss of heat equals the rate of development of heat, after which the temperature remains constant. Thus when the circuit is closed through an incandescent lamp, the temperature of the filament rises rapidly until the rate of heat loss (chiefly by radiation) equals the rate of development of heat, Ri^2. On the other hand, a fuse is constructed so that when the current in it exceeds a certain predetermined value, the fuse melts before its final equilibrium temperature can be attained.

4-10 Average and effective values of a current. The current at a section of a circuit is defined by the equation $i = dq/dt$. It follows at once that

$$dq = i\,dt,$$

and hence the total charge q crossing a section in a finite time t is

$$q = \int_0^t i\,dt. \tag{4-18}$$

If the current is constant,

$$q = it, \tag{4-19}$$

and the quantity of charge crossing *any* section of a circuit in which there is a constant current equals the product of the current and the time interval. This equation has already been used, by implication, in the definition of the coulomb (in Sec. 1-6) as the quantity of charge which in one second crosses a section of a conductor in which there is a constant current of one ampere.

It will be seen that the term *ampere-second* is equivalent to the coulomb, and that Eq. (4-19) may be used to define other units of charge. One *ampere-hour*, for example, is the charge which in one hour crosses a section of a circuit in which there is a constant current of one ampere. Evidently, 1 amp-hr = 3600 amp-sec = 3600 coulombs.

The average value (i_{av} or $\bar{\imath}$) of a varying current over a time interval is defined as the constant current which will transport the same charge in the same time as does the varying current. The charge q transported by the constant current i_{av} is

$$q = i_{av}t,$$

and since this must equal the charge transported by the varying current,

$$i_{av}t = \int_0^t i\,dt, \quad \text{or} \quad i_{av} = \frac{1}{t}\int_0^t i\,dt, \tag{4-20}$$

which is the familiar definition of the average value of a varying quantity. (See Phillips' Analytical Geometry and Calculus, p. 450.)

The average value of a varying current is a measure of the electrolytic effect it can produce, since the number of ions that can be neutralized at an electrode and hence the mass of material deposited, are directly proportional to the quantity of charge delivered to the electrode.

It is found experimentally that 96,514 coulombs are required to liberate one gram-equivalent weight of material at an electrode of an electrolytic cell. This quantity of charge is called *one faraday.*

$$1 \text{ faraday} = 96{,}514 \text{ coulombs.}$$

(A gram-equivalent weight is the atomic or molecular weight of an ion divided by its valency. That is, the gram-equivalent weight of H^+ is one gm, of Cu^{++} it is 63.57/2 = 31.79 gm, of SO_4^{--} it is 96.06/2 = 48.03 gm, etc.).

The faraday can be determined quite apart from any theory regarding the "atomicity" of electricity. However, if the reasonable assumption is made that all ions of the same kind carry the same charge, the quantity of charge of any monovalent ion can be found by dividing the faraday by Avogadro's number. Hence the charge on an H^+ ion is

$$\frac{96{,}514}{6.0228 \times 10^{23}} = 1.60248 \times 10^{-19} \text{ coulombs.}$$

This is in excellent agreement with the Millikan value of the electronic charge e. (See Sec. 2-9.)

The average current in a circuit can be determined experimentally by measuring the charge passing any point in the circuit during a measured time interval. One experimental method is to pass the current through an electrolytic cell and compute the quantity of charge from the valency and mass of material liberated at an electrode. In fact, the *international* ampere is defined in this way, namely, as *that unvarying current which will deposit 0.001118 grams of silver per second* from a solution of silver salt of specified nature in a cell of certain specifications. The International Electrotechnical Commission has recommended, however, that this standard be abandoned in favor of one based on the magnetic forces between current-carrying conductors. See Sec. 11-4.

We consider next another kind of average of a varying current, defined in terms of the heating effect of the current. From Eq. (4-17), the heat dH developed in a conductor in time dt is

$$dH = Ri^2\, dt,$$

and in a finite time interval from 0 to t the heat is

$$H = \int_0^t Ri^2\, dt.$$

If the current and resistance are constant,

$$H = Ri^2 t.$$

The *effective value* of a varying current over a specified time interval is defined as the constant current which will develop the same heat in any resistance in the same specified time. If we represent the constant current by i_{eff}, the heat developed by it in a resistance R in time t is

$$H = i_{\text{eff}}^2 Rt.$$

Since this equals the heat developed by the varying current,

$$i_{\text{eff}}^2 Rt = \int_0^t Ri^2\,dt,$$

$$i_{\text{eff}} = \sqrt{\frac{1}{t}\int_0^t i^2\,dt}. \tag{4-21}$$

But $\dfrac{1}{t}\displaystyle\int_0^t i^2\,dt$ is the average or mean value of the *square* of the current, and the effective value equals the square-root-of-the-mean-value-of-the-square-of-the-current. This is usually abbreviated to the root-mean-square or the rms current, i_{rms}. Note that the rms value of a varying current is not the same as the average value, which is defined as

$$i_{\text{av}} = \frac{1}{t}\int_0^t i\,dt.$$

As an example, consider the most common type of varying current, an alternating current which varies sinusoidally with time according to the relation

$$i = I_m \sin 2\pi ft$$

where I_m is the maximum current and f is the frequency. (See Fig. 4-12.)

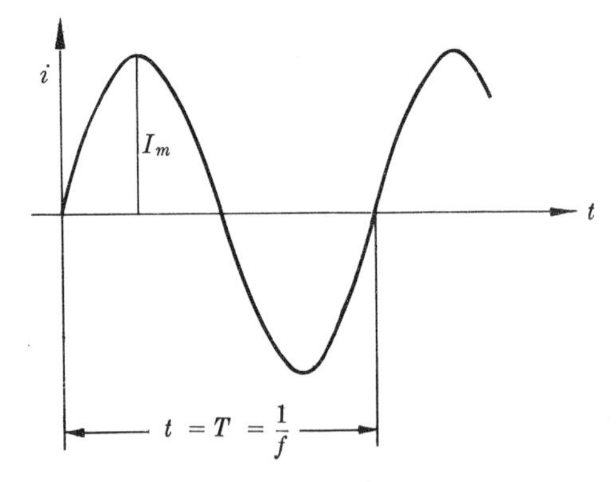

FIG. 4-12. Sinusoidal alternating current.

The average current during one cycle is

$$i_{\text{av}} = \frac{1}{T}\int_0^T i\,dt = \frac{1}{T}\int_0^T I_m \sin 2\pi ft\,dt.$$

Let us represent $2\pi ft$ by θ, and integrate with respect to θ rather than t. In the integrand, we replace $2\pi ft$ by θ and dt by $\dfrac{1}{2\pi f}\,d\theta$ or $\dfrac{T}{2\pi}\,d\theta$. The limits of integration, since θ is now the variable, are 0 and 2π. Then

$$\begin{aligned} i_{\text{av}} &= \frac{1}{2\pi}\int_0^{2\pi} I_m \sin\theta\,d\theta \\ &= \frac{I_m}{2\pi}\Big[-\cos\theta\Big]_0^{2\pi} \\ &= 0 \end{aligned}$$

and the average current is zero as is evident from the diagram, since the area above the axis equals that below it.

The root-mean-square current is

$$
\begin{aligned}
i_{\text{rms}} = i_{\text{eff}} &= \sqrt{\frac{1}{T}\int_0^T i^2\,dt} \\
&= \sqrt{\frac{1}{T}\int_0^T (I_m \sin 2\pi f t)^2\,dt} \\
&= \sqrt{\frac{I_m^2}{2\pi}\int_0^{2\pi} \sin^2\theta\,d\theta} \\
&= \sqrt{\frac{I_m^2}{2\pi}\left[\frac{\theta}{2} - \frac{\sin 2\theta}{4}\right]_0^{2\pi}} \\
&= \sqrt{\frac{I_m^2}{2}} \\
&= \frac{I_m}{\sqrt{2}} = 0.707 I_m.
\end{aligned}
$$

The heating effect of a sinusoidal alternating current is therefore the same as that of a steady current .707 times as great as the maximum value of the alternating current. In practice, alternating currents are usually described in terms of their effective or rms values.

Problems—Chapter 4

(1) A wire carries a constant current of 10 amperes. How many coulombs pass a cross section of the wire in 20 seconds? How many electrons?

(2) A sensitive galvanometer will measure a current as small as 10^{-10} ampere, or one ten-thousandth of a microampere. (a) How many million electrons cross a section of a conductor, per second, within which there is a current of this magnitude? (b) What is the average drift velocity of the free electrons in the conductor, if its cross section is 1 mm²? Assume there are 8.5×10^{28} free electrons per cubic meter in the conductor. (c) How long a time, on the average, is required for an electron to advance a distance of 1 cm along the conductor?

(3) A vacuum diode can be approximated by a plane cathode and a plane anode, parallel to one another and 5 mm apart. The area of both cathode and anode is 2 cm². In the region between cathode and anode the current is carried solely by electrons. If the electron current is 50 ma, and the electrons strike the anode surface with a velocity of 1.2×10^7 m/sec, find the number of electrons per cubic millimeter in the space just outside the surface of the anode.

(4) The belt of a Van de Graaff generator is one meter wide and travels with a speed of 25 m/sec. (a) Neglecting leakage, at what rate in coulombs/sec must charge be sprayed on one face of the belt to correspond to a current of 10^{-4} ampere into the collecting sphere? (b) Compute the surface density of charge on the belt, assuming it to be uniform.

(5) Charge is removed from a sphere through a wire. The charge on the sphere at any time is given by

$$q = 10^{-3}\epsilon^{-2t},$$

where q is in coulombs and t is in seconds. Find the current in the wire at times $t = 0$ and $t = 5$ sec.

(6) (a) Show that the Wiedemann-Franz law predicts that the ratio of thermal to electrical conductivity should be the same for all metals at the same temperature. (b) Look up the thermal and electrical conductivities of at least four metals and test the validity of the prediction above.

(7) What electric intensity is necessary to set up the same current density in an aluminum conductor as in the copper conductor in example (1) at the end of Sec. 4–7?

(8) What is the current density in No. 10 rubber-covered copper wire when carrying the maximum current allowable by the National Board of Fire Underwriters? See the table on page 109. Express your answer in amperes/circular mil, amperes/square inch, and amperes/square meter. (b) What is the electric intensity or potential gradient in the wire, in volts/meter? (c) What is the potential difference between two points on the wire, 1 km apart?

The National Board of Fire Underwriters sets the following allowable limits to the "carrying capacities" of copper wire.

Size, B & S Gauge,	Diameter, mils	Maximum current (amp)	
		Rubber Insulation	Other Insulation
0000	460.0	225	325
0	324.9	125	200
10	101.9	25	30
18	40.3	3	5

(9) It is found that when the potential difference between the terminals of a resistor is 10 volts, the current in the resistor is 2 amperes. (a) What will be the current if the potential difference is 20 volts? 100 volts. (b) What will be the potential difference if the current is 10 amperes? 0.10 amperes? (c) What is the resistance of the resistor?

(10) (a) The following measurements of current and potential difference were made on a resistor constructed of Nichrome wire:

i (amp)	V_{ab} (volts)
0.5	2.18
1.0	4.36
2.0	8.72
4.0	17.44

Make a graph of V_{ab} as a function of i. Does the nichrome obey Ohm's law? What is the resistance of the resistor, in ohms?

(b) The following measurements were made on a "thyrite" resistor.

i (amp)	V_{ab} (volts)
0.5	4.76
1.0	5.81
2.0	7.05
4.0	8.56

Make a graph of V_{ab} as a function of i. Does "thyrite" obey Ohm's law? What is the resistance of the resistor?

(11) From the definition of the international ohm, compute the resistivity of mercury at 0° C. The specific gravity of mercury is 13.60.

(12) (a) Compute the resistance of 1 mile of No. 10 copper conductor. See the table above. (b) Find the required diameter, in mils, of an aluminum conductor of circular cross section and having the same resistance per mile.

(13) (a) Find the resistance of 1 km of No. 0 copper wire. See the table above. (b) Find the potential difference between two points 1 km apart on a No. 0 copper wire, when the current in the wire is 25 amp.

(14) The conductors in the transmission line from Boulder Dam are copper tubes, inside diameter 1.203 inches, outside diameter 1.400 inches. Find the resistance of a one-mile length of conductor.

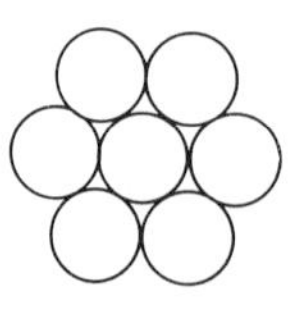

FIG. 4-13.

(15) Each conductor in a seven-strand copper cable (see Fig. 4-13) is 0.10 inch in diameter. Find the resistance of 1000 ft of the cable.

(16) The cross section of a copper wire is a square, one-tenth inch on a side. What is the cross section in circular mils?

(17) A solid copper conductor 0.5 cm in radius is surrounded by a concentric insulating sheath 1 cm in outside radius. The insulated conductor is protected by an outer lead sheath. Compute: (a) The resistance of the insulator to radial flow, in a length of 1 km, of 2 km; (b) the leakage current in a length of 1 km, if the potential difference between the central conductor and the sheath is 100 volts; (c) the current density at points of a cylindrical surface of radius 0.8 cm; (d) the current across this surface in a length of 1 km. How should the answers to (b) and (d) compare? The resistivity of the insulator is 10^{10} ohm-meters.

(18) A spherical silver electrode 8 cm in radius is concentric with a spherical metal shell of inner radius 10 cm. The interspace contains an insulating material of resistivity 10^6 ohm-meters. Compute the resistance of the insulator.

(19) (a) Show that if R_0, R, R_1 and R_2 are the resistances of a conductor at the Centigrade temperatures zero, t, t_1 and t_2, that

$$R_2 = R_1 \frac{1 + \alpha t_2}{1 + \alpha t_1},$$

where

$$\alpha = \frac{R - R_0}{R_0 t}.$$

(b) The resistance of the field windings of a motor was measured at 20° C and found to be 54 ohms. The resistance was measured again after the motor had been running and was found to be 60 ohms. To what temperature have the windings risen? The windings are of copper wire.

(20) The rate of development of heat in a resistor is 20 watts when the current in it is 4 amp. (a) What will be the rate of development of heat when the current is 8 amp? 2 amp? (b) What is the resistance of the resistor?

(21) A "660-watt" electric heater is designed to operate from 120 volt lines. (a) What is its resistance? (b) What current does it draw? (c) What is the rate of development of heat, in calories/sec? (d) If the line voltage drops to 110 volts, what power does the heater take, in watts? (Assume the resistance constant. Actually it will change somewhat because of the change in temperature.)

(22) Express the rate of development of heat in a resistor in terms of (a) potential difference and current; (b) resistance and current; (c) potential difference and resistance.

(23) The current in a wire varies with time according to the relation

$$i = 4 + 2t^2,$$

where i is in amperes and t in seconds. (a) How many coulombs pass a cross section of the wire in the time interval between $t = 5$ sec and $t = 10$ sec? (b) What is the average current in this time interval? (c) What is the rms current?

(24) The current in a wire varies according to the relation

$$i = 20 \sin 377t,$$

where i is in amperes, t in seconds, and ($377t$) in radians. (a) How many coulombs pass a cross section of the wire in the time interval between $t = 0$ and $t = 1/120$ sec? (b) In the interval between $t = 0$ and $t = 1/60$ sec? (c) What is the average current during each of the intervals above? (d) What is the rms current?

(25) Fig. 4-14 is a graph of a half-wave rectified alternating current. During the intervals from $t = 0$ to $t = T/2$, T to $3T/2$, etc., the current is given by

$$i = I_m \sin \frac{2\pi t}{T}.$$

During the intervals from $t = T/2$ to $t = T$, etc., the current is zero. (a) Find the average current during the interval from 0 to T. (b) Find the rms current during the same interval.

(26) The "wave-form" of a cyclic current is shown in Fig. 4-15. Find the average and rms values of this current during an interval of 0.05 sec.

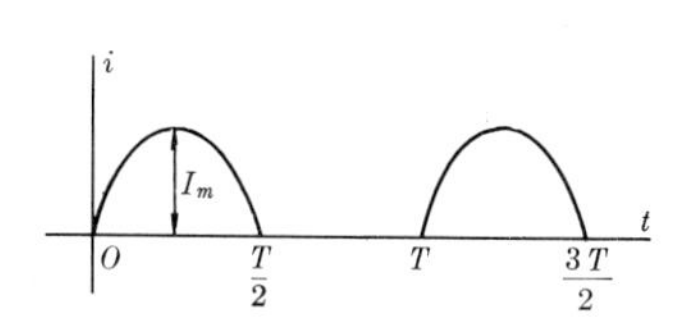

Fig. 4-14.

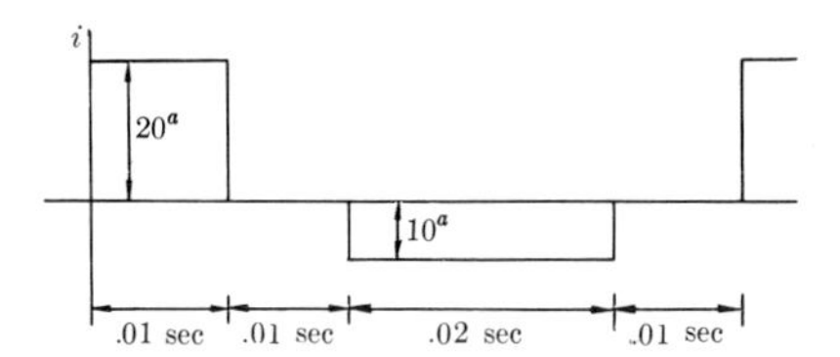

Fig. 4-15.

CHAPTER 5

D. C. CIRCUITS

5-1 Electromotive force. We have shown in the preceding chapter that in order to have a current in a conductor an electric field, or a potential gradient, must be maintained in the conductor. We have also shown that a continuous energy input is necessary to maintain the current, the energy being converted to heat in the conductor. The development of heat in a conductor is an "irreversible" process in the thermodynamic sense. That is, while energy is conserved in the process, the heat developed can not be reconverted to electrical energy except under the restrictions imposed by the second law of thermodynamics.

We now wish to consider another type of energy transformation, one which is reversible in the thermodynamic sense. Devices in which such reversible transformations occur are the storage battery, the generator, and the motor. (There are other devices also, but they need not concern us at present.) Consider a storage battery which is operating the starting motor of an automobile. The chemical energy of the materials in the battery is decreasing, and is converted to mechanical form available at the shaft of the starter motor. After the automobile engine starts, the starting motor is disconnected and the generator sends a reversed, charging current through the storage battery. The chemical reactions in the battery then proceed in the reversed direction and chemical energy is developed at the expense of mechanical work that must be done by the automobile engine to drive the generator. Hence, depending on the direction of the current through it, we may have in the storage battery either of these transformations:

Chemical energy ⟶ electrical energy
Electrical energy ⟶ chemical energy

As contrasted with the *irreversible* conversion of electrical energy into heat via i^2R heating in a resistor, the conversion of electrical energy into chemical energy in a storage battery is *reversible* in the sense that the chemical energy can be completely recovered and converted to electrical form, with no outstanding changes in other bodies.[1]

[1] The fact that the recovery is not 100% in an actual storage battery under the conditions in which it is used does not vitiate the possibility of complete recovery under the proper conditions.

The mechanism of energy transformations in a motor or generator will be discussed in more detail in succeeding chapters. It will suffice for the present to state that in these devices we have conversion from electrical to mechanical (rather than chemical) energy or vice versa, and that the conversion may be 100% in either direction.

Any device in which a reversible transformation between electrical energy and some other form of energy can take place, is called a *seat of electromotive force.*

The magnitude of the electromotive force of a seat may be defined quantitatively as the energy converted from electrical to nonelectrical form or vice versa (exclusive of energy converted irreversibly to heat) per unit of charge passing across a section through the seat. More briefly, electromotive force may be defined as the *work per unit charge.* Electromotive force will be represented by the symbol $\mathcal{E}$, and is often abbreviated by emf.[1] Hence if dq is the charge crossing a section through the seat in time dt, and dW is the energy transformed in this time, the emf is

$$\boxed{\mathcal{E} = \frac{dW}{dq}} \tag{5-1}$$

Since electromotive force is work per unit charge, the unit of emf in the mks system is *one joule per coulomb.* This is the same as the unit of potential and has been abbreviated to one volt. Hence emf's can be expressed in volts. It should be noted, however, that although emf and potential are expressible in the same unit, they relate to different concepts. The distinction will be made clearer as we proceed.

The work done by the seat in time dt is

$$dW = \mathcal{E}\, dq, \tag{5-2}$$

and the rate of doing work, or the power, is

$$P = \frac{dW}{dt} = \mathcal{E}\frac{dq}{dt} = \mathcal{E}i. \tag{5-3}$$

For example, the emf of the common automobile storage battery is about 6 volts or 6 joules/coulomb. This means that for every coulomb passing across a section through the battery (or across any section of a circuit to which the battery is connected) 6 joules of chemical energy are converted to electrical form if the battery is discharging, or 6 joules of

[1] The use of the word "force" in connection with this concept is unfortunate, since an emf is not a force but work per unit charge. It has been suggested that "electromotance" would be a better term.

chemical energy are developed at the expense of electrical energy if the battery is being charged. If the current in the battery is 10 amperes, or 10 coulombs/sec, the rate of energy conversion is

$$P = \mathcal{E}i = 6\,\frac{\text{joules}}{\text{coulomb}} \times 10\,\frac{\text{coulombs}}{\text{sec}} = 60\,\frac{\text{joules}}{\text{sec}} = 60 \text{ watts.}$$

5-2 The circuit equation. Let us consider now a simple circuit (Fig. 5-1) in which a resistor of resistance R is connected by leads of negligible resistance to the terminals of a seat of emf such as a dry cell or storage battery. Let i represent the current in the circuit. When a number of circuit elements are connected as in Fig. 5-1 so as to provide only a single conducting path they are said to be *in series*. Other types of connection will be considered later.

One minor point should be clarified at the start. Points p and a, in Fig. 5-1(a), are connected by a resistanceless conductor. The potential difference between two points of a conductor equals the product of the current and the resistance between the points. But since the resistance between p and a is zero, the potential difference between them is also zero, or in other words, points p and a are at the same potential. Hence p and a, and, in fact, all points of the conductor connecting them, are *electrically* equivalent and separate letters need not be assigned to them. The same is evidently true of points b and q. The circuit is therefore relettered in Fig. 5-1(b).

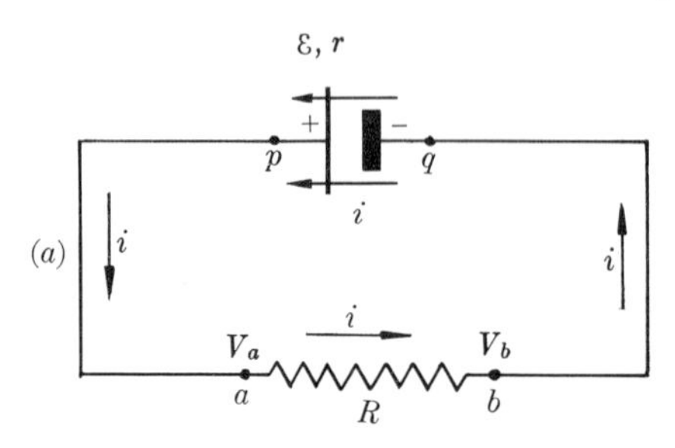

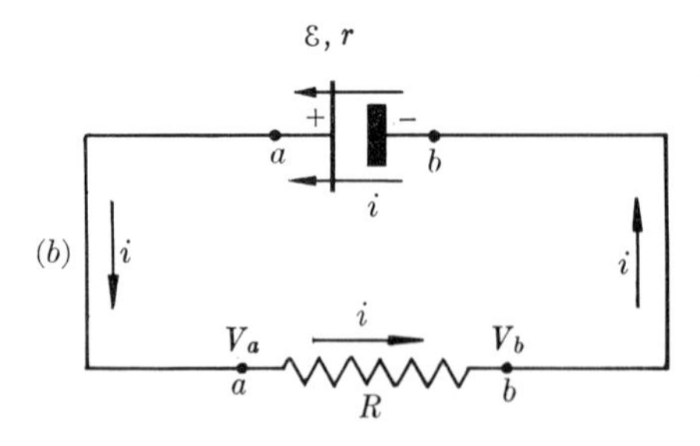

FIG. 5-1. A seat of emf connected in series with a resistor by means of resistanceless connecting wires.

If the direction of the current in the resistor is from a toward b, the potential V_a is higher than the potential V_b. The + sign at the terminal a of the cell indicates this fact.

Although strictly speaking emf is not a vector quantity, it is useful to assign to it a direction (or better, a sense). We shall arbitrarily consider the direction or sense of an emf to be from the − toward the + terminal of the seat, within the seat. This directed emf is indicated by an arrow in Fig. 5-1.

We now set down the expressions for the rates of energy transformation in the various parts of the circuit. Heat is developed in the external resistor at a rate $\frac{dH_R}{dt} = Ri^2$. It is also found, as might be expected, that every seat of emf has resistance. This is called the *internal resistance* of the seat and we shall represent it by r. (The symbols B and R_B are also commonly used for internal resistance.) Hence the rate of development of heat in the cell is $\frac{dH_r}{dt} = ri^2$. The sum of Ri^2 and ri^2 is the rate at which energy is given up by the circulating charge in the form of heat, and the seat of emf must supply energy at an equal rate. Then from Eq. (5-3),

$$\mathcal{E}i = Ri^2 + ri^2. \tag{5-4}$$

After cancelling and rearranging terms we get

$$\boxed{i = \frac{\mathcal{E}}{R + r}} \tag{5-5}$$

The principle of conservation of energy therefore leads to this exceedingly useful relation between the current, the emf, and the resistance in a simple series circuit. It may be called the "circuit equation."

Consider next a circuit that contains a seat of emf within which work is done *by* the circulating charge. Examples are a motor, where the work done by the charge appears as mechanical energy, or a storage battery being "charged" where the work appears as chemical energy. Fig. 5-2 represents a circuit in which a motor, represented by the symbol —⊏◯⊐— is being run by a battery. A resistance R is included for generality.

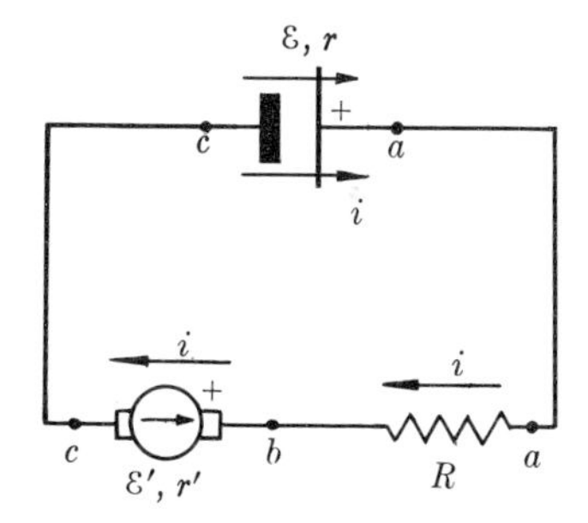

FIG. 5-2. A series circuit with two seats of emf.

The current in the motor is directed from right to left, or from b to c. Since the circulating charge gives up energy in passing through the motor, the potential V_b must be higher than the potential V_c. Hence a + sign is attached to terminal b of the motor and the sense of its emf is from c toward b.

Let us represent the emf of the motor by $\mathcal{E}'$ and its internal resistance by r'. From the definition of electromotive force, the rate of conversion of energy to mechanical form by the motor is $\mathcal{E}'i$ and the rate of conversion to heat in the motor is $r'i^2$. The rates of development of heat in the external resistor and in the battery are Ri^2 and ri^2. The rate at which work is done on the circulating charge by the battery is $\mathcal{E}i$. Therefore

$$\mathcal{E}i = \mathcal{E}'i + r'i^2 + Ri^2 + ri^2$$

and

$$i = \frac{\mathcal{E} - \mathcal{E}'}{R + r + r'}, \tag{5-6}$$

or

$$i = \frac{\Sigma\mathcal{E}}{\Sigma R}. \tag{5-7}$$

This equation is evidently a generalization of Eq. (5-5) and may be stated: the current (in a series circuit) equals the algebraic sum of the emf's in the circuit divided by the sum of the resistances in the circuit.

It is necessary that a convention of signs be adopted when using Eq. (5-7). The simplest rule is to consider the direction of the current the positive direction in the circuit. Then emfs are positive if they are in the same direction as the current, negative if in the opposite direction. Resistances are considered positive always. It will be seen that the signs in Eq. (5-6) are consistent with this convention.

The *relative* directions of current and emf in the two seats in Fig. 5-2 lead to the following generalizations. When the current in a seat of emf is in the *same* direction as the emf, as it is in the battery, work is done *on* the circulating charge at the expense of energy of some other form. When the current in a seat of emf is *opposite* in direction to the emf, as in the motor, work is done *by* the circulating charge and energy of some other form appears. The emf in the latter case is often called a *back electromotive force.*

5-3 Alternate definition of electromotive force. A seat of emf is a portion of a circuit in which nonelectrical energy is converted reversibly to electrical energy, or vice versa. In purely mechanical problems, whenever work is done on a body it is always possible to attribute this work to a force which acts on the body while it moves a certain distance. In some seats of emf, notably those of a chemical nature, the mechanism by which the charges acquire energy at the expense of chemical energy is certainly not as simple as would be implied by the statement that "chemical forces" do work on the charges in moving them from one electrode to the other. However, since a definite quantity of work is, in fact, done on each unit charge which passes through a cell, one could speak of an "equivalent" force of such magnitude that the work done by it per unit of charge transferred equals the emf. Since a force on a charge implies an electric field, one can equally well speak of an "equivalent" electric field within the cell.

In other seats of emf, notably induced emf's (see Sec. 12-2) the work done on the circulating charge can definitely be ascribed to an electric field, which, however, is set up not by positive and negative charges but by a changing magnetic field.

Fig. 5-3 illustrates in a schematic way a seat of emf. The *electrostatic* field set up by the charges on the positive and negative terminals is shown by light lines. The heavy lines represent the "equivalent" field, that is, the field which does work on the charge passing through the seat. Note that its sense is opposite to that of the electrostatic field within the seat, and is the same as the sense of the emf, i.e., from the negative toward the positive terminal. The dotted line is an arbitrary closed path passing through the seat.

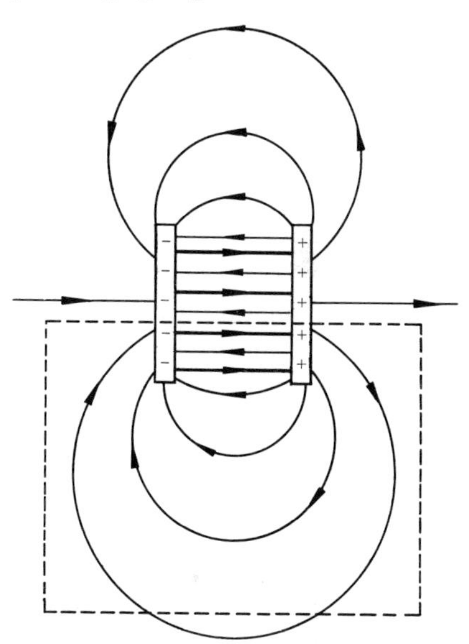

FIG. 5-3. Schematic diagram of a seat of emf.

Let us now evaluate the line integral of the electric intensity around this closed path. The resultant intensity E at any point is the (vector) sum of the electrostatic intensity E_0 and the "equivalent" intensity E'. The line integral of the resultant intensity is then the sum of two terms,

$$\oint E \cos\theta \, ds = \oint E_0 \cos\theta \, ds + \oint E' \cos\theta \, ds.$$

But we have shown in Sec. 3-3 that in a purely electrostatic field the line integral of the electric intensity around a closed path is zero, so the first term on the right drops out. This appears reasonable, since if the dotted path is traversed in a clockwise direction we are moving *with* the electrostatic field outside the seat but *against* it within the seat, and the longer path outside the seat is compensated by the weaker field in this region. The "equivalent" field, however, exists only within the seat and is zero at external points. The second term on the right therefore reduces to the line integral of E' from one terminal of the seat to the other. But E' is the "equivalent" force per unit charge, so that $\oint E' \cos\theta\, ds$ is the work done per unit charge by the forces of the "equivalent" field. This work, however, is just what we have called the electromotive force of the seat. Hence one may say, that *the emf of a circuit equals the line integral of the electric intensity around the circuit,* where the intensity includes *all* forces acting on a charge at any point of the circuit.

$$\mathcal{E} = \oint E \cos\theta\, ds \tag{5-8}$$

This equation is sometimes made the definition of the emf of a circuit. We shall need it later in deriving the expression for the velocity of electromagnetic waves.

5-4 Potential difference between points in a circuit. We next deduce the general expression for the difference in potential between any two points in a series circuit. Fig. 5-4 shows a portion of such a circuit in which the direction of the current i is from a toward b. The rate at which the circulating charge gives up energy to the portion of the circuit between a and b is iV_{ab}. In other words, this is the power input to this portion of the circuit, supplied by the seat or seats of emf in the remainder of the circuit (not shown). Power is also supplied *by* the first seat of emf (since i and $\mathcal{E}$ are in the same direction) at a rate $\mathcal{E}i$. Power is supplied *to* the second seat of emf (i and $\mathcal{E}'$ are in opposite directions) at a rate $\mathcal{E}'i$. Heat is developed in the resistor and in the seats of emf at a total rate $(R + r + r')i^2$. Then if we equate power input to power output,

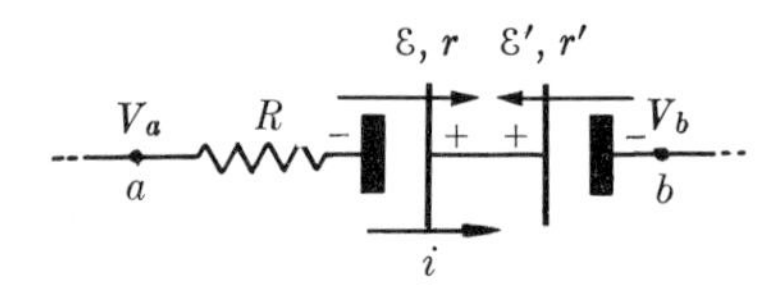

FIG. 5-4. Portion of a circuit with two seats of emf.

$$iV_{ab} + \mathcal{E}i = \mathcal{E}'i + (R + r + r')i^2,$$

$$V_{ab} = (R + r + r')i - (\mathcal{E} - \mathcal{E}'), \tag{5-9}$$

$$V_{ab} = \Sigma Ri - \Sigma\mathcal{E} \tag{5-10}$$

This is the desired general expression for the potential difference $V_{ab} \equiv V_a - V_b$ between any two points a and b of a series circuit.

Careful attention must be paid to algebraic signs. The simplest rule is to consider the direction from a toward b as positive. Then currents and emf's are positive if their direction is from a toward b, negative if their direction is from b toward a. Resistances are always positive. The potential difference V_{ab} is positive if a is at a higher potential than b, negative if b is at a higher potential than a. It will be seen that the signs in Eq. (5-9) are consistent with these conventions.

Eq. (5-10) reduces, of course, to the familiar relation $V_{ab} = Ri$ for the special case where a and b are the terminals of a pure resistance. It also contains implicitly the general circuit equation, Eq. (5-7). That is, if points a and b coincide,

$$V_{ab} = 0 = \Sigma Ri - \Sigma \mathcal{E}, \quad \text{and} \quad i = \frac{\Sigma \mathcal{E}}{\Sigma R}.$$

Examples. (1) Given that in Fig. 5-5, $\mathcal{E}_1 = 12^{\text{v}}$, $r_1 = 0.2^{\omega}$; $\mathcal{E}_2 = 6^{\text{v}}$, $r_2 = 0.1^{\omega}$; $R_3 = 1.4^{\omega}$; $R_4 = 2.3^{\omega}$; compute (a) the current in the circuit, in magnitude and direction, and (b) the potential difference V_{ac}.

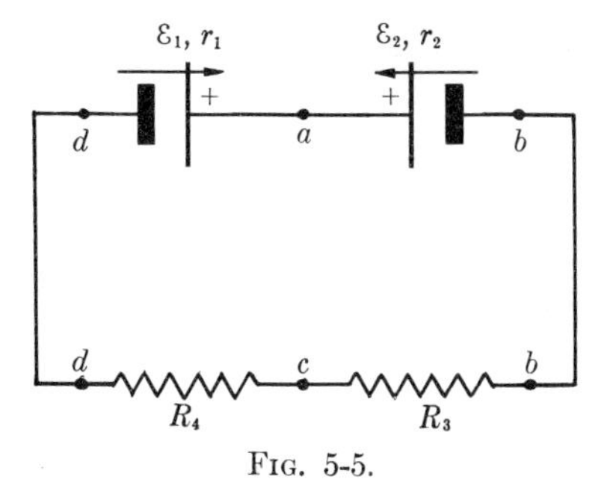

FIG. 5-5.

(a) It seems obvious that the current will be clockwise, since $\mathcal{E}_1 > \mathcal{E}_2$. However, to illustrate that one need not rely on intuition to find its direction, let us assume it to be counterclockwise. Then with the help of the sign conventions on page 116,

$$\Sigma \mathcal{E} = -12 + 6 = -6 \text{ volts}, \quad \Sigma R = 4 \text{ ohms}$$

$$i = \frac{\Sigma \mathcal{E}}{\Sigma R} = \frac{-6}{4} = -1.5 \text{ amp.}$$

The significance of the negative sign is merely that the incorrect assumption was made regarding the direction of the current. Had we assumed the opposite direction we would have obtained

$$\Sigma \mathcal{E} = 12 - 6 = +6 \text{ volts}, \quad \Sigma R = 4 \text{ ohms},$$

$$i = \frac{6}{4} = +1.5 \text{ amp.}$$

So the same numerical value is obtained in either case, and the direction of the current need not be known in advance.

(b) The potential difference between a and c is $V_{ac} = \Sigma Ri - \Sigma \mathcal{E}$. One can proceed from a to c by two paths and either may be used to compute V_{ac}. Let us first use path abc. We have already shown that the current is clockwise. Use the sign conventions above. The direction of the current in the path abc is from a toward c, hence i is positive and

$$\Sigma Ri = (.1 + 1.4)(+1.5) = +2.25^{\text{v}}.$$

The direction of $\mathcal{E}_2$ is from b toward a, so

$$\Sigma\mathcal{E} = -6^{\text{V}}.$$

Hence

$$V_{ac} = +2.25 - (-6) = +8.25^{\text{V}},$$

and since

$$V_{ac} \equiv V_a - V_c,$$

$$V_a - V_c = 8.25^{\text{V}}, \quad V_a = V_c + 8.25^{\text{V}}.$$

That is, the potential at point a is 8.25 volts above that at point c.

If we proceed from a to c along path adc, the current term has a negative sign, since its direction is from c toward a. Hence

$$\Sigma Ri = (.2 + 2.3)(-1.5) = -3.75^{\text{V}}.$$

The direction of $\mathcal{E}_1$ is from c toward a, so

$$\Sigma\mathcal{E} = -12^{\text{V}}.$$

Hence

$$V_{ac} = -3.75 - (-12) = +8.25^{\text{V}}$$

and the same answer is obtained whichever path is used.

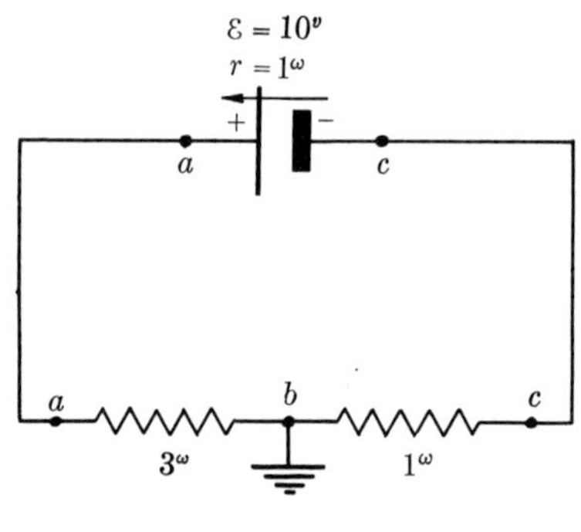

FIG. 5-6.

(2) In practically all circuits, whether power distribution systems or small units such as radio amplifiers, one or more points of the circuit are connected to the earth or to "ground." In contrast to the reference level of potential used in general field theory (that of a point at infinity), when dealing with circuits the potential of the grounded points is assumed to be zero and the potential of any other point of the circuit is expressed relative to this reference level. Consider the simple circuit in Fig. 5-6, grounded at point b. Compute the potentials of points a and c.

The current in the circuit is

$$i = \frac{10}{5} = 2^{\text{a}}$$

and its direction is counterclockwise. The potential differences V_{ab} and V_{bc} are

$$V_{ab} = Ri = 3 \times 2 = 6^{\text{V}} = V_a - V_b,$$
$$V_{bc} = Ri = 1 \times 2 = 2^{\text{V}} = V_b - V_c.$$

Since $V_b = 0$, it follows that

$$V_a = +6^{\text{V}}, \quad V_c = -2^{\text{V}}.$$

That is, point a is 6^{V} *above* ground and point c is 2^{V} *below* ground. The potential difference V_{ac} can now be found by subtraction.

$$V_{ac} \equiv V_a - V_c = 6 - (-2) = +8^{\text{V}}.$$

As a check, we may proceed from a to c through the cell,

$$V_{ac} = \Sigma Ri - \Sigma\mathcal{E} = -1 \times 2 - (-10) = +8^{\text{V}}.$$

5-5 Terminal voltage of a seat of emf. It is convenient to develop the special form of Eq. (5-10) when points a and b are the terminals of a seat of electromotive force. There can be but two possibilities, shown in Fig. 5-7(a) and (b), namely, when the current and emf are in the same or opposite directions. Let us agree that point a shall always refer to the + terminal and point b to the − terminal of the seat. Then in Fig. 5-7(a), according to the sign convention in Sec. 5-2, both $\mathcal{E}$ and i are negative and

$$V_{ab} = V_{+-} = \Sigma Ri - \Sigma\mathcal{E} = -ri - (-\mathcal{E}) = \mathcal{E} - ri.$$

In Fig. 5-7(b), $\mathcal{E}$ is negative but i is positive. Hence

$$V_{ab} = V_{+-} = ri - (-\mathcal{E}) = \mathcal{E} + ri.$$

The potential difference *between the + and − terminals* of a seat of emf is therefore equal to the emf of the seat, diminished by the product of the internal resistance and the current when emf and current are in the same direction, and increased by this product when the two are in opposite directions. (In the latter case we have a seat of back emf.) When the current in a seat of emf is zero, the terminal voltage equals the electromotive force. Therefore the emf of a seat may be found experimentally by measuring its terminal voltage on open circuit.

If the terminal voltage is measured with the common type of voltmeter, care must be taken that the current drawn by the voltmeter does

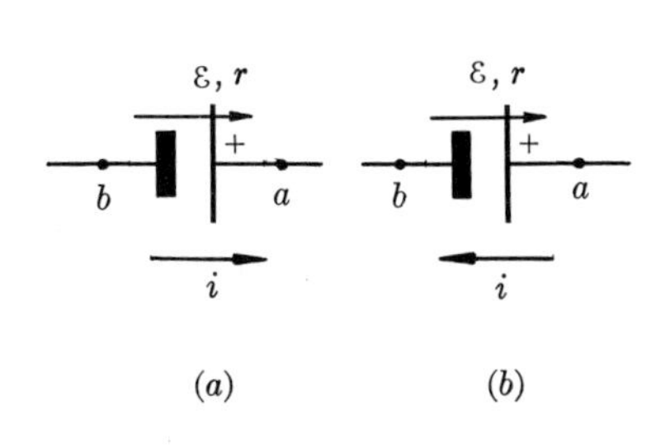

FIG. 5-7. (a) Current and emf in the same direction. (b) Current and emf in opposite directions.

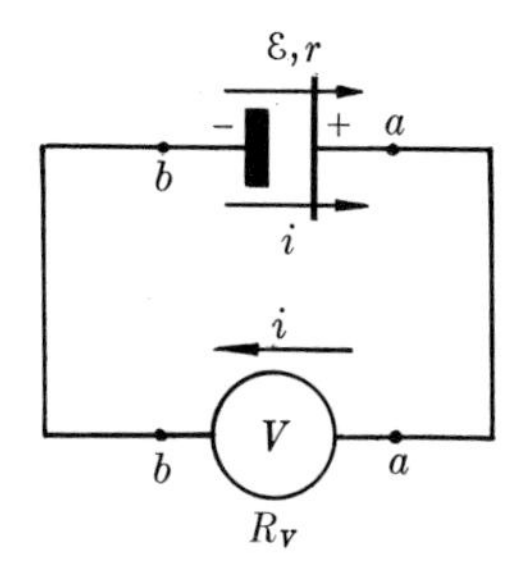

FIG. 5-8. Measurement of the terminal voltage of a seat of emf by means of a voltmeter.

not result in any appreciable lowering of the terminal voltage below the emf. Fig. 5-8 shows a cell of emf $\mathcal{E}$ and internal resistance r with a voltmeter of resistance R_V connected to its terminals. Since the voltmeter provides a conducting path, we have a closed circuit as soon as it is connected to the cell. The current in the circuit is

$$i = \frac{\mathcal{E}}{R_V + r}.$$

The terminal voltage is

$$V_{ab} = \mathcal{E} - ri = \mathcal{E} - r \cdot \frac{\mathcal{E}}{R_V + r} = \mathcal{E}\left(\frac{1}{1 + \frac{r}{R_V}}\right).$$

If $R_V \gg r$, the term in parentheses is nearly unity and the voltmeter reading (V_{ab}) is nearly equal to the emf. It is evident, therefore, that a voltmeter should be a high resistance instrument.

5-6 The potentiometer. The potentiometer is an instrument which can be used to measure the emf of a seat without drawing any current from the seat. It also has a number of other useful applications. Essentially, it balances an unknown potential difference against an adjustable, measurable potential difference.

The principle of the potentiometer is shown schematically in Fig. 5-9. A resistance wire ab is permanently connected to the terminals of a seat of emf $\mathcal{E}_1$. A sliding contact c is connected through the galvanometer G to a second cell whose emf $\mathcal{E}_2$ is to be measured. Contact c is moved along the wire until a position is found at which the galvanometer shows no deflection. (This necessitates that $V_{ab} \geq \mathcal{E}_2$.) If we then write the expressions for V_{cb} for both paths between these points, and remember that there is no current in the lower path,

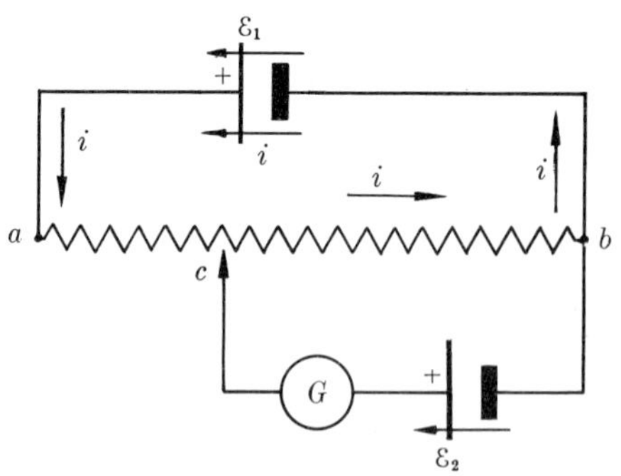

FIG. 5-9. Principle of the potentiometer.

$$\text{(Upper path)} \quad V_{cb} = iR_{cb},$$
$$\text{(Lower path)} \quad V_{cb} = \mathcal{E}_2.$$

Hence iR_{cb} is *exactly* equal to the emf $\mathcal{E}_2$, and $\mathcal{E}_2$ can be computed if i and R_{cb} are known. No correction need be made for the "ir" term since the current in $\mathcal{E}_2$ is zero.

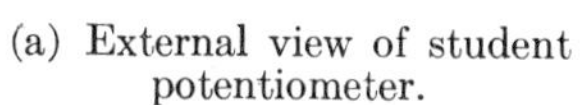

(a) External view of student potentiometer.

(b) Inside view of student potentiometer.

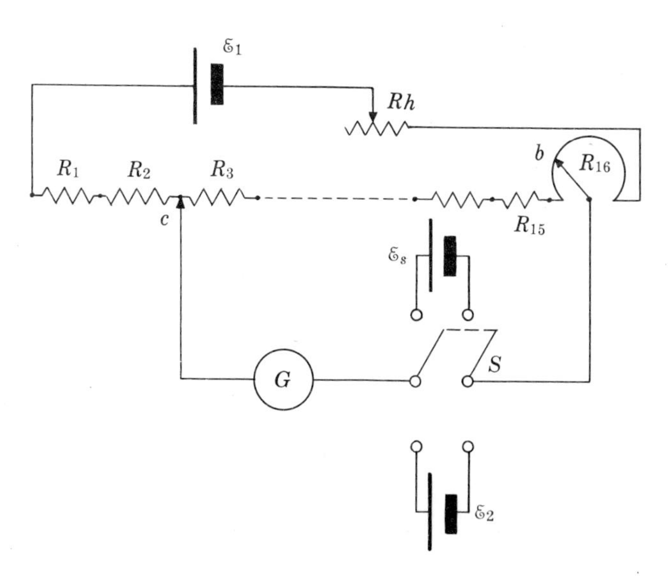

(c) Circuit diagram of student potentiometer.

Fig. 5 10.

(*Courtesy of Leeds & Northrup*)

In practice, the simple potentiometer circuit is modified as shown in Fig. 5-10. The following description relates specifically to the Leeds and Northrup student type potentiometer. Still other refinements are introduced in high precision instruments. Figs. 5-10(a) and (b) are photographs of the instrument, Fig. 5-10(c) is a circuit diagram.

The single slide wire of Fig. 5-9 is replaced by sixteen resistors R_1, R_2, R_3, $\cdots$ R_{15}, R_{16} in series, each of resistance 10 ohms. R_{16} alone is a slide wire. The movable contacts c and b may include any whole number of resistors plus any fraction of the slide wire. The potentiometer is always to be used with a current of 0.010 amp. in the upper circuit. The drop across each 10-ohm resistor is therefore 0.100 volt, and the instrument may be calibrated to read volts directly.

To insure that the current in the upper circuit has its proper value, switch S is first thrown in the "up" position, connecting the standard cell of known emf $\mathcal{E}_s$ in the lower circuit. Contacts c and b are set at the points where the circuit *should* be in balance when the current in the upper circuit has its correct value, and this circuit is then adjusted by the rheostat Rh until the galvanometer G shows no deflection. Then, without changing the setting of Rh, switch S is thrown to the "down" position and the galvanometer current again brought to zero by adjusting c and b. The unknown emf $\mathcal{E}_2$ may then be read directly from the positions of c and b.

During preliminary adjustments a protective resistor is inserted *in series* with the galvanometer to prevent damage both to it and to the standard cell.

The potentiometer described above can be read to within 10^{-4} volt. High precision instruments read to 10^{-6} volts.

5-7 Series and parallel connection of resistors. Most electrical circuits consist not merely of a single seat of emf and a single external resistor, but comprise a number of emf's, resistors, or other elements such as capacitors, motors, etc., interconnected in a more or less complicated manner. The general term applied to such a circuit is a *network*. We shall next consider a few of the simpler types of network.

Fig. 5-11 illustrates four different ways in which three resistors having resistances R_1, R_2, and R_3 might be connected. In Fig. 5-11(a), the resistors provide only a single path between the end-points a and b, and are said to be connected in *series* between these points. Any number of circuit elements such as resistors, cells, motors, etc., are similarly said to be in series with one another between two points if connected as in (a) so as to provide only a single path between the points.

The resistors in Fig. 5-11(b) are said to be in *parallel* between points a and b. Each resistor provides an alternate path between the end-points, and any number of circuit elements similarly connected are in parallel with one another.

In Fig. 5-11(c), resistors R_2 and R_3 are in parallel with one another, and this combination is in series with the resistor R_1. In Fig. 5-11(d), R_2 and R_3 are in series, and this combination is in parallel with R_1.

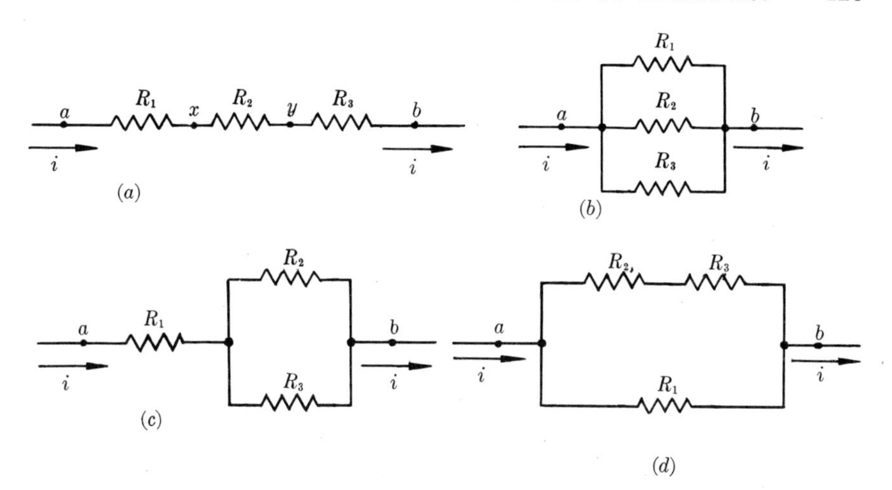

FIG. 5-11. Four different ways of connecting three resistors.

It is always possible to find a single resistor which could replace a combination of resistors in any given circuit and leave unaltered the potential difference between the terminals of the combination and the current in the rest of the circuit. The resistance of this single resistor is called the *equivalent* resistance of the combination. If any one of the networks in Fig. 5-11 were replaced by its equivalent resistance R, one could write

$$V_{ab} = Ri, \quad \text{or} \quad R = V_{ab}/i,$$

where V_{ab} is the potential difference between the terminals of the network and i is the current at the point a or b. Hence the method of computing an equivalent resistance is to assume a potential difference V_{ab} across the actual network, compute the corresponding current i (or vice versa) and take the ratio of one to the other. The simple series and parallel connections of resistors are sufficiently common so that it is worth while to develop formulas for these two special cases.

If the resistors are in series as in Fig. 5-11(a), the current in each must be the same and equal to the line current i. Hence

$$V_{ax} = iR_1, \quad V_{xy} = iR_2, \quad V_{yb} = iR_3$$

$$V_{ax} + V_{xy} + V_{yb} = i(R_1 + R_2 + R_3)$$

But $$V_{ax} = V_a - V_x, V_{xy} = V_x - V_y, V_{yb} = V_y - V_b.$$

Hence
$$V_{ax} + V_{xy} + V_{yb} = V_a - V_b = V_{ab}.$$
$$V_{ab} = i(R_1 + R_2 + R_3), \text{ and}$$
$$\frac{V_{ab}}{i} = R_1 + R_2 + R_3.$$

But V_{ab}/i is by definition the equivalent resistance R. Therefore

$$R = R_1 + R_2 + R_3. \tag{5-11}$$

Evidently the equivalent resistance of any number of resistors in series equals the sum of their individual resistances.

If the resistors are in parallel as in Fig. 5-11(b), the potential difference between the terminals of each must be the same and equal to V_{ab}. If the currents in each are denoted by i_1, i_2, and i_3 respectively,

$$i_1 = \frac{V_{ab}}{R_1}, \quad i_2 = \frac{V_{ab}}{R_2}, \quad i_3 = \frac{V_{ab}}{R_3}.$$

Now charge is delivered to point a by the line current i, and removed from a by the currents i_1, i_2, and i_3. Since charge is not accumulating at a, it follows that

$$i = i_1 + i_2 + i_3, \text{ or}$$
$$i = \frac{V_{ab}}{R_1} + \frac{V_{ab}}{R_2} + \frac{V_{ab}}{R_3}, \text{ or}$$
$$\frac{i}{V_{ab}} = \frac{1}{R_1} + \frac{1}{R_2} + \frac{1}{R_3}.$$

Since
$$\frac{i}{V_{ab}} = \frac{1}{R},$$
$$\frac{1}{R} = \frac{1}{R_1} + \frac{1}{R_2} + \frac{1}{R_3} \tag{5-12}$$

Evidently, for any number of resistors in parallel, the *reciprocal* of the equivalent resistance equals the *sum of the reciprocals* of their individual resistances.

For the special case of two resistors in parallel,

$$\frac{1}{R} = \frac{1}{R_1} + \frac{1}{R_2} = \frac{R_2 + R_1}{R_1 R_2}, \quad \text{and}$$
$$R = \frac{R_1 R_2}{R_1 + R_2}$$

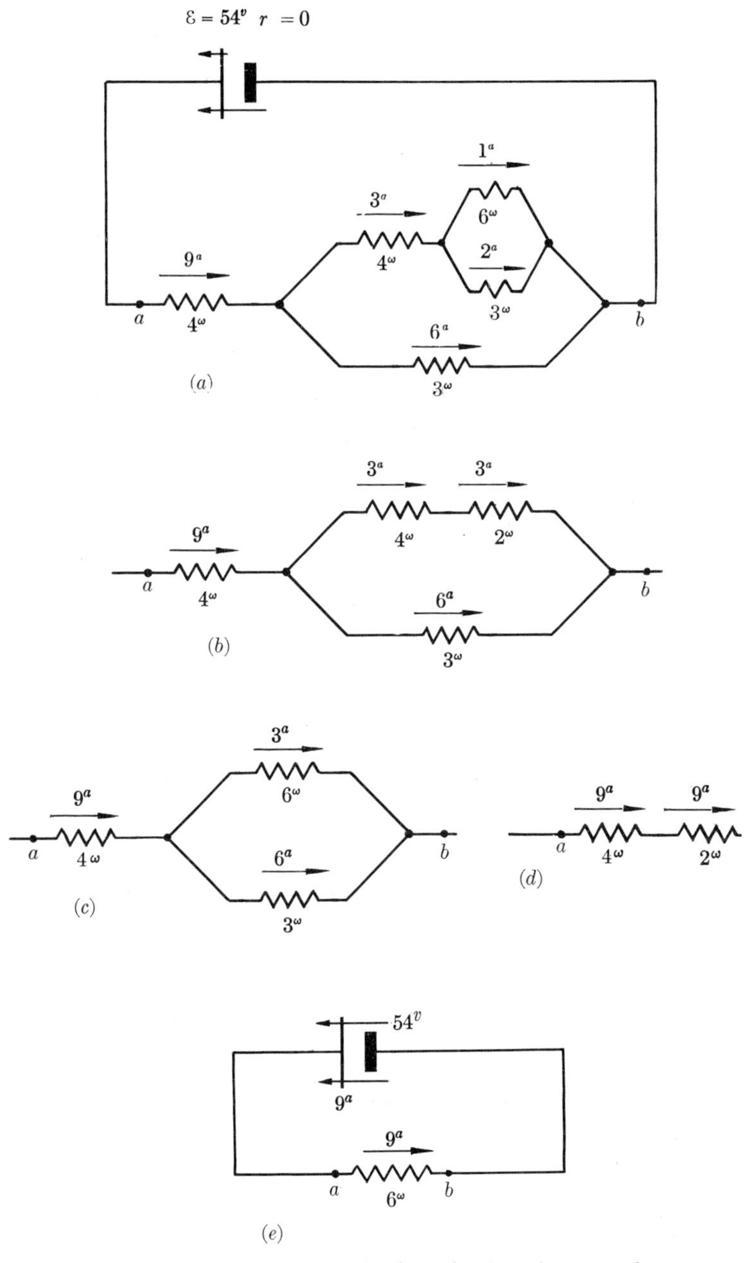

Fig. 5-12. Successive stages in the reduction of a network to a single equivalent resistance

Also, since $V_{ab} = i_1R_1 = i_2R_2$,

$$\frac{i_1}{i_2} = \frac{R_2}{R_1},$$

and the currents carried by two resistors in parallel are inversely proportional to their resistances.

The equivalent resistances of the networks in Fig. 5-11(c) and (d) could be found by the same general method, but it is simpler to consider them as combinations of series and parallel arrangements. Thus in (c) the combination of R_2 and R_3 in parallel is first replaced by its equivalent resistance, which then forms a simple series combination with R_1. In (d), the combination of R_2 and R_3 in series forms a simple parallel combination with R_1. Not all networks, however, can be reduced to simple series-parallel combinations and special methods for handling such networks will be given later.

Example. Compute the equivalent resistance of the network in Fig. 5-12, and find the current in each resistor.

Successive stages in the reduction to a single equivalent resistance are shown in parts (b) to (e) of the figure. For simplicity, the seat of emf has been omitted from all except the first and last diagrams. The 6^ω and the 3^ω resistor in parallel in (a) are equivalent to the single 2^ω resistor which replaces them in (b). In (c) the 4^ω and the 2^ω resistor in series are replaced by their equivalent 6^ω resistor. Repetition of this process leads to the single equivalent 6^ω resistor in (e).

We now work back up the series of diagrams and compute the currents. In the simple series circuit of (e) the current is $i = \dfrac{\Sigma\mathcal{E}}{\Sigma R} = \dfrac{54}{6} = 9^a$. Since the 6^ω resistor in (e) replaces the 4^ω and the 2^ω resistors in series, the current in each of these in (d) is also 9^a. In (c), the current in the 4^ω resistor remains 9^a but the currents in the 6^ω and 3^ω resistors divide as shown. Repetition of the process gives the current distribution in (a).

5-8 Networks containing seats of emf. The arrangement of two or more seats of emf is not uniquely described by the statement that the seats are in series or parallel. Two cells in series, for example, might be connected either as in Fig. 5-13(a) or (b). The connection in (a) is denoted as "series aiding"; that in (b) as "series opposing." A combination such as that in Fig. 5-13(c) is described as "in parallel, like poles together," and that in (d) as "in parallel, unlike poles together."

Note that although the cells in Fig. 5-13(c) and (d) are in parallel as far as the rest of the circuit is concerned, each of these arrangements in itself forms a closed circuit with the cells in series opposing in the first case and in series aiding in the second. Hence the terms "series" and "parallel" are not mutually exclusive, nor do they by themselves completely describe a network. Very frequently a more explicit statement is required.

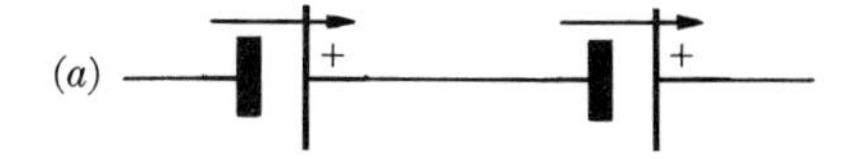

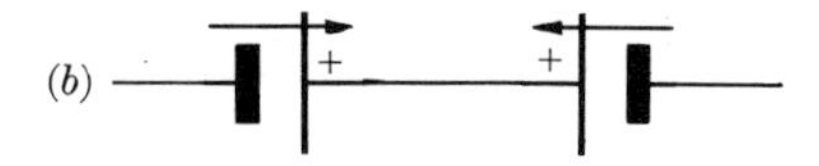

The expression for the current in a *series* circuit containing a number of seats of emf either aiding or opposing can be written

$$i = \frac{\Sigma \mathcal{E}}{R + \Sigma r}.$$

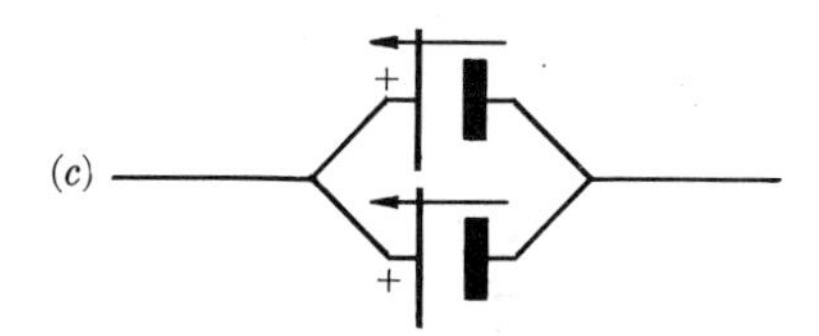

It is obvious that the equivalent emf is the algebraic sum of the individual emf's and the equivalent internal resistance is the arithmetic sum of the internal resistances.

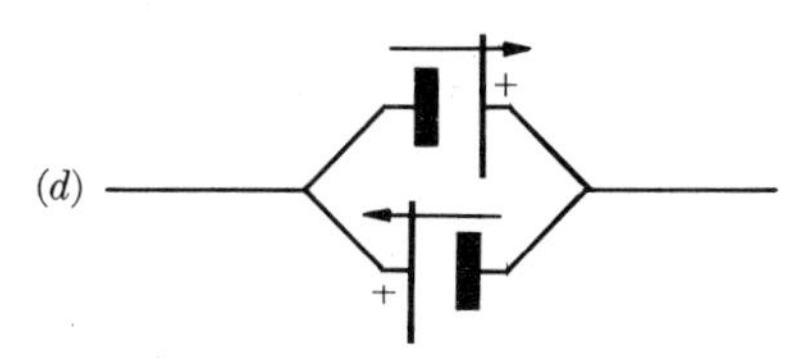

Fig. 5-13. Four ways of connecting two seats of emf.

When any number of seats whose emf's are equal are connected in parallel with like poles together as in Fig. 5-13(c), the equivalent emf is equal to that of a single seat and the equivalent internal resistance is computed by the usual method for resistances in parallel. When the seats are of unequal emf, or are connected as in Fig. 5-13(d), the problem becomes more complicated and methods for handling it will be described in the next section.

5-9 Kirchhoff's rules. Networks in which the resistors are not in simple series or parallel groupings, or in which there are seats of emf in parallel paths, can not in general be solved by the method of equivalent resistances. Two rules first formally stated by Gustav Robert Kirchhoff (1824–1887) enable such problems to be handled systematically.

We first define two terms. A *branch point* is a point of a network at which three (or more) conductors are joined. A *loop* is any closed conducting path. For example, points a, b, c, and d, in Fig. 5-14, are branch points, and paths $abca$, $abdca$, $abcd\mathcal{E}a$, etc., are loops.

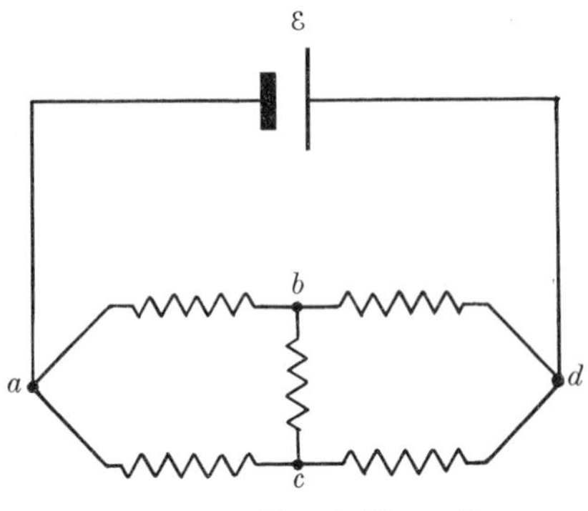

FIG. 5-14. Circuit illustrating branch points and loops.

Kirchhoff's rules may be stated as follows:

Point rule: The algebraic sum of the currents toward any branch point of a network is zero.

$$\Sigma i = 0.$$

Loop rule: The algebraic sum of the emf's in any loop of a network equals the algebraic sum of the Ri products in the same loop.

$$\Sigma \mathcal{E} = \Sigma Ri.$$

The first rule merely states formally that no charge accumulates at a branch point of the network. The second rule follows from the generalized expression for the potential difference between two points of a circuit, $V_{ab} = \Sigma Ri - \Sigma \mathcal{E}$. This equation was developed for a simple series circuit in which the current was the same at all points. The currents in various parts of a loop will, in general, differ from one another, and the term ΣRi must be understood to mean the sum of the products of each resistance and the current *in that particular resistance.* Then, if one continues completely around a loop so that the second point coincides with the first, the potential difference is zero and $\Sigma \mathcal{E} = \Sigma Ri$.

The first step in applying Kirchhoff's rules is to assign a magnitude and direction to all unknown currents and emf's, and a magnitude to all unknown resistances. These, as well as the known quantities, are represented in a diagram with directions carefully shown. The solution is then carried through on the basis of the assumed directions. If in the numerical solution of the equations a negative value is found for a current or an emf, its correct direction is opposite to that assumed. The correct *numerical* value is obtained in any case. Hence the rules provide a method for ascertaining the *directions* as well as the magnitudes of currents and emf's and it is not necessary that these directions be known in advance.

The expressions Σi, ΣRi, and $\Sigma \mathcal{E}$ are algebraic sums. When applying the point rule, a current is considered positive if its direction is toward a branch point, negative if away from the point. (Of course the opposite convention may also be used.) When applying the loop rule, some direction around the loop (i.e., clockwise or counterclockwise) must be chosen as the positive direction. All currents and emf's in this direction are positive, those in the opposite direction are negative. Note that a current which has a positive sign in the point rule may have a negative sign in the term in which it appears in the loop rule. Note also that the direction around the loop which is considered positive is immaterial, the result of choosing the opposite direction being merely to obtain the same equation with signs reversed. There is a tendency to assume that the "correct" direction to consider positive is that of the current in the loop, but in general such a choice is not possible, since the currents in some elements of a loop

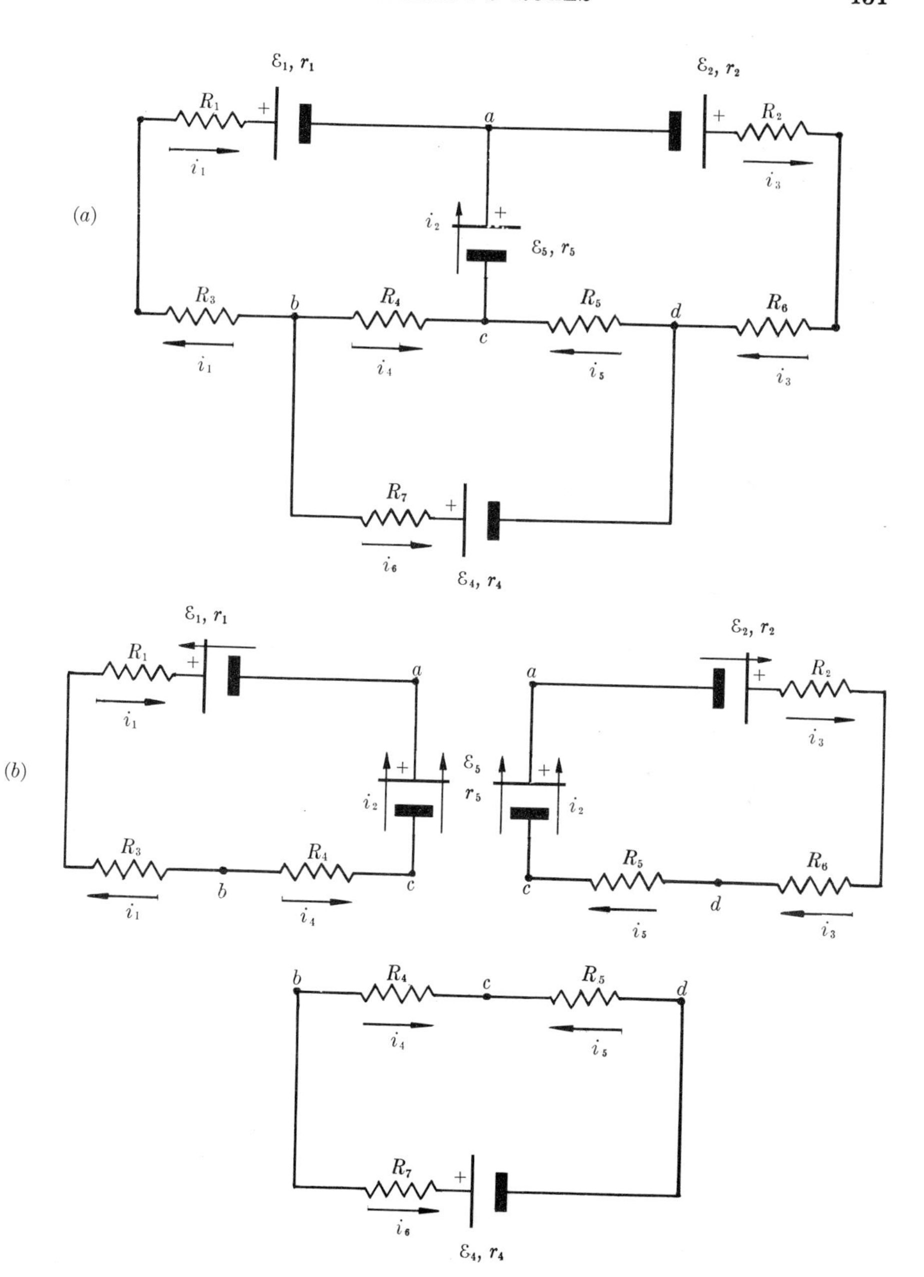

FIG. 5-15. (a) Diagram of a circuit arranged for the application of Kirchhoff's laws. (b) Same circuit cut up into loops.

may be clockwise and in other elements, counterclockwise. These and other points will be illustrated by examples.

In complicated networks, where a large number of unknown quantities may be involved, it is sometimes puzzling to know how to obtain a sufficient number of independent equations to solve for the unknowns. The following rules are useful:

(1) If there are n branch points in the network, apply the point rule at $n - 1$ of these points. Any points may be chosen. Application of the point rule at the nth point does not lead to an independent relation.

(2) Imagine the network to be separated into a number of simple loops, like the pieces of a jigsaw puzzle. Apply the loop rule to each of these loops.

Example. Let the magnitudes and directions of the emf's, and the magnitudes of the resistances in Fig. 5-15(a), be given. Solve for the current in each branch of the network.

Assign a direction and a letter to each unknown current. The assumed directions are entirely arbitrary. Note that the currents in R_3, R_1, and $\mathcal{E}_1$ are the same, and hence require only a single letter. The same is true for the currents in $\mathcal{E}_2$, R_2, and R_6.

The branch points are lettered a, b, c, and d.

At point a,

$$i_1 + i_2 - i_3 = 0.$$

At point b,

$$-i_1 - i_4 - i_6 = 0.$$

At point c,

$$i_4 + i_5 - i_2 = 0.$$

Since there are four branch points, there are only three independent "point" equations. If the point rule is applied at the fourth point, d, one finds

$$i_6 + i_3 - i_5 = 0.$$

But if the first three equations are added, the result is

$$-i_6 - i_3 + i_5 = 0$$

which is the same equation. Hence no new information is found by going to point d. We have, however, secured a check on the first three equations.

In Fig. 5-15(b) the circuit is shown cut up into its "jigsaw" sections. Let us consider the clockwise direction positive in each loop. The loop rule then furnishes the following equations:

$$\begin{aligned} -\mathcal{E}_1 - \mathcal{E}_5 &= i_1R_1 + i_1r_1 - i_2r_5 - i_4R_4 + i_1R_3 \\ +\mathcal{E}_2 + \mathcal{E}_5 &= i_3r_2 + i_3R_2 + i_3R_6 + i_5R_5 + i_2r_5 \\ +\mathcal{E}_4 &= i_4R_4 - i_5R_5 - i_6r_4 - i_6R_7. \end{aligned}$$

And one has six independent equations to be solved for the six unknown currents.

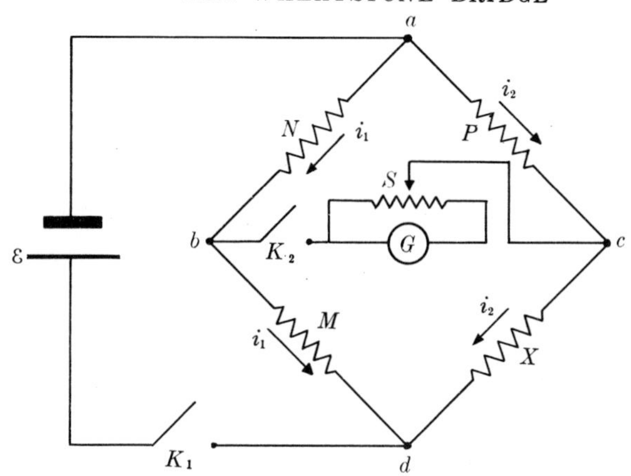

FIG. 5-16. Wheatstone bridge circuit.

5-10 The Wheatstone bridge. The Wheatstone bridge circuit, shown in Fig. 5-16, is widely used for the rapid and precise measurement of resistance. It was invented in 1843 by the English scientist, Charles Wheatstone. M, N, and P are adjustable resistors which have been previously calibrated, and X represents the unknown resistance. To use the bridge, switches K_1 and K_2 are closed and the resistance of P is adjusted until the galvanometer G shows no deflection. Points b and c must then be at the same potential or, in other words, the potential drop from a to b equals that from a to c. Also, the drop from b to d equals that from c to d. Since the galvanometer current is zero, the current in M equals that in N, say i_1, and the current in P equals that in X, say i_2. Then since $V_{ab} = V_{ac}$, it follows that

$$i_1N = i_2P$$

and since $V_{bd} = V_{cd}$,

$$i_1M = i_2X.$$

When the second equation is divided by the first, we find

$$X = \frac{M}{N}P.$$

Hence if M, N, and P are known, X can be computed. The ratio M/N is usually set at some integral power of 10 such as .01, 1, 100, etc., for simplicity in computation.

During preliminary adjustments when the bridge may be far from balance and V_{bc} large, the galvanometer must be protected by a shunt S

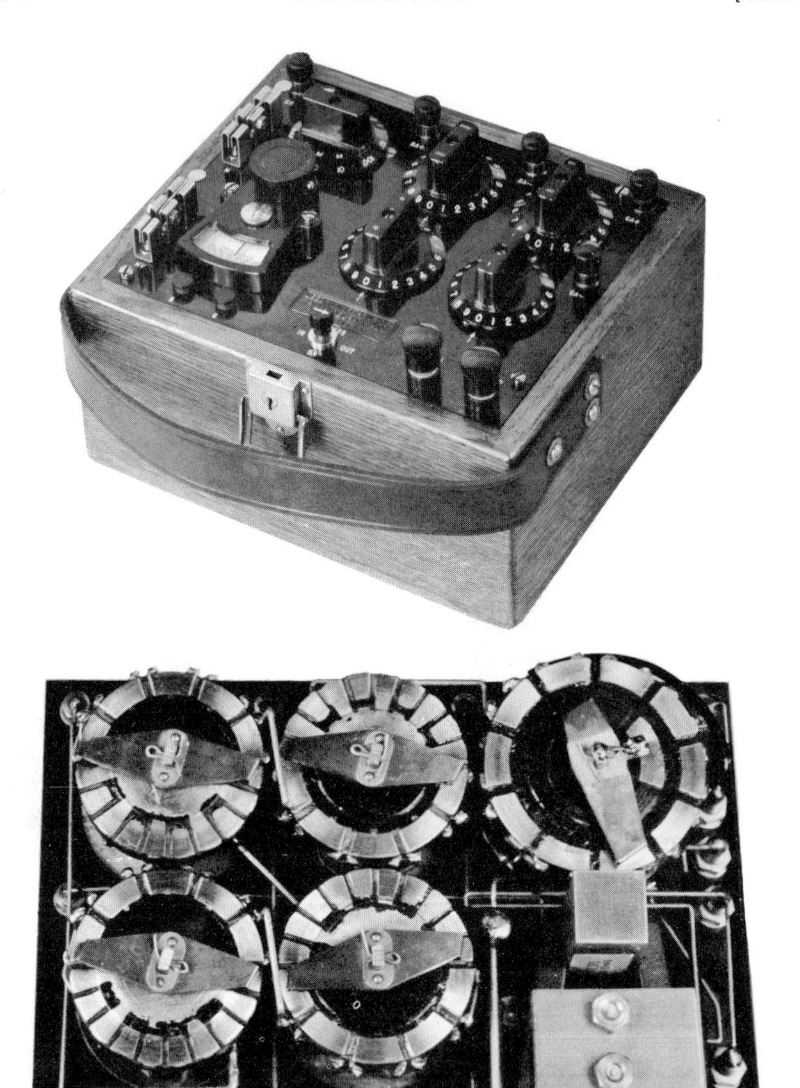

FIG. 5-17. Portable Wheatstone bridge.
(*Courtesy of Leeds & Northrup*)

to prevent damage to it. As balance is approached the shunt resistance is gradually increased and the final setting of R is made with the galvanometer unshunted to obtain maximum sensitivity.

If any of the resistances are inductive (see Sec. 13-3), the potentials V_b and V_c may attain their final values at different rates when K_1 is closed, and the galvanometer, if connected between b and c, would show an initial deflection even though the bridge were in balance. Hence K_1 and K_2 are frequently combined in a double key which closes the battery circuit first and the galvanometer circuit a moment later after the transient currents have died out.

Portable bridges are available having a self-contained galvanometer and dry cells. The ratio M/N can be set at any integral power of 10 between 0.001 and 1000 by a single dial switch, and the value of R adjusted by four dial switches. Fig. 5-17 is a photograph of such a bridge.

5-11 Power. Since our analysis of circuits has been developed from considerations of energy and power, little can be added under this heading except by way of summary.

The power input to any portion of a circuit between points a and b is given by

$$P = iV_{ab}. \tag{5-13}$$

The power is expressed in watts when i is in amperes and V_{ab} in volts. Current is considered positive if its direction is from a toward b in the portion of the circuit considered. If the power input as computed by this equation has a negative sign, which will be the case if the current is from b toward a or if $V_a < V_b$, the portion of the circuit considered is actually supplying power to the remainder of the circuit and the magnitude of P represents the *power output.* Eq. (5-13) is independent of the nature of the circuit between a and b, that is, the circuit may consist of resistors, motors, batteries, etc., arranged in any way.

If the circuit between a and b comprises resistance only, then $V_{ab} = Ri$ and

$$P = iV_{ab} = Ri^2 = dH/dt.$$

That is, the power input equals the rate of development of heat.

The rate of energy conversion in a seat of emf is equal to $\mathcal{E}i$. If $\mathcal{E}$ and i are in the same direction, electrical energy is developed at the expense of some other form. If $\mathcal{E}$ and i are in opposite directions (a seat of back emf) some other form of energy is developed at the expense of electrical energy.

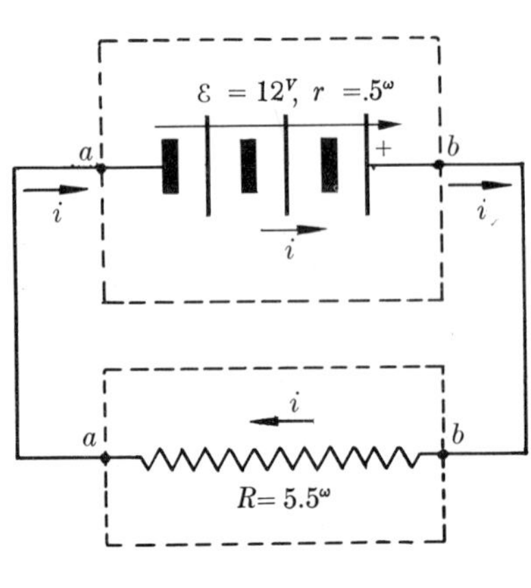

FIG. 5-18.

Examples. (1) Fig. 5-18 shows a battery and a resistor in series. The current in the circuit is

$$i = \frac{12}{5.5 + .5} = 2^a.$$

Point b is at a higher potential than point a, and

$$V_{ba} = iR = 11^V,$$

or

$$V_{ba} = \mathcal{E} - ir = 12 - 2 \times .5 = 11^V$$

and

$$V_{ab} = -V_{ba} = -11^V.$$

Consider first that part of the circuit within the upper dotted rectangle. The direction of the current through it is from a to b. Hence the power *input* to this portion is

$$P = iV_{ab} = 2 \times (-11) = -22 \text{ watts.}$$

The minus sign means that this is actually a power output.

In the lower rectangle, the direction of the current is from b to a. Hence,

$$\text{Power input} = iV_{ab} = (-2) \times (-11) = +22 \text{ watts}$$

and power is supplied to this part of the circuit.

As a check, since the energy supplied to the resistor is converted to heat,

$$\frac{dH}{dt} = i^2R = 4 \times 5.5 = 22 \text{ watts.}$$

The rate of conversion of nonelectrical to electrical energy in the seat of emf is

$$P = \mathcal{E}i = 12 \times 2 = 24 \text{ watts.}$$

The rate of development of heat in the seat of emf is

$$i^2r = 4 \times .5 = 2 \text{ watts.}$$

The power output is therefore 24 − 2 = 22 watts as previsiously computed.

To sum up:

The battery converts nonelectrical to electrical energy at the rate of 24 joules per second, or 24 watts.

The rate of heating in the battery is 2 watts.

The rate at which energy is supplied to the rest of the circuit by the battery is 24 − 2, or 22 watts.

This energy is transformed to heat in the resistor at the rate of 22 watts.

(2) Fig. 5-19 is a diagram of a shunt motor, the upper branch being the armature circuit and the lower branch the windings of the field coils. The armature, when the motor is running, develops a back emf. (See Sec. 5-2.) If the line current, i, is 5.5^a and V_{ab} is 100^v, find the mechanical power output and the efficiency of the motor.

The field winding is a pure resistance. Hence

$$i_f = \frac{100}{200} = 0.5^a.$$

From the point rule,

$$\begin{aligned} i &= i_a + i_f \\ 5.5 &= i_a + 0.5 \\ i_a &= 5.0^a. \end{aligned}$$

Since $\mathcal{E}$ is a back emf,

$$\begin{aligned} V_{ab} &= \mathcal{E} + i_a r = i_f \times 200 \\ 100 &= \mathcal{E} + 5 \times 2 \\ \mathcal{E} &= 90^v. \end{aligned}$$

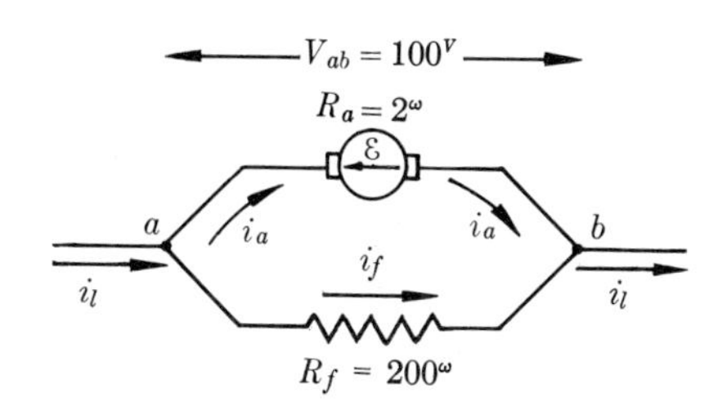

FIG. 5-19. Diagram of a shunt motor.

$\mathcal{E}$ may also be found by the loop rule. Assume its direction to be from a to b and consider the clockwise direction positive

$$\begin{aligned} \Sigma\mathcal{E} &= \Sigma iR \\ \mathcal{E} &= 5 \times 2 - .5 \times 200 \\ \mathcal{E} &= -90^v. \end{aligned}$$

Hence the correct direction is from b to a.

The power input to the entire motor is

$$P = iV_{ab} = 5.5 \times 100 = 550 \text{ watts.}$$

The rate of heating in the field is

$$i_f^2 R_f = .25 \times 200 = 50 \text{ watts.}$$

The rate of heating in the armature is

$$i_a^2 R_a = 5^2 \times 2 = 50 \text{ watts.}$$

The rate of conversion of electrical to mechanical energy is

$$P = \mathcal{E}i = 90 \times 5 = 450 \text{ watts.}$$

As a check,

$$50 + 50 + 450 = 550 \text{ watts.}$$

In the absence of mechanical friction in the bearings, and of windage losses, 450 watts of mechanical power are delivered by the motor. With friction present, a part of the 450 watts are lost as heat. If friction is neglected, the efficiency of the motor is

$$\text{Eff} = \frac{\text{output}}{\text{input}} = \frac{450}{550} = 82\%.$$

5-12 Measurement of power and energy. The power supplied to (or by) the device X in Fig. 5-20(a) may be measured by the ammeter and voltmeter connected as shown, correction being made, if necessary, for the fact that the ammeter reads the sum of the current in X and the voltmeter current. If meter corrections are neglected, the power in watts supplied to X is given by the product of the ammeter reading in amperes and the voltmeter reading in volts.

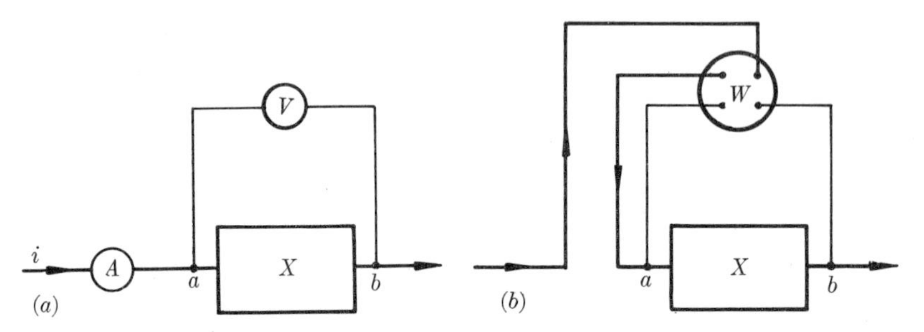

FIG. 5-20. (a) Voltmeter-ammeter method of measuring power. (b) A wattmeter.

The *wattmeter* is a single instrument which performs the combined functions of ammeter and voltmeter in Fig. 5-20(a) and indicates the power directly. It makes use of the dynamometer principle described in Sec. 10-5. Briefly, it consists of a coil with a pointer attached, pivoted so that it can swing in the magnetic field of a second fixed coil, and restrained by a hairspring whose restoring torque is proportional to the angle of deflection. The deflecting torque on the pivoted coil is jointly proportional to the current in it and to the magnetic field set up by the fixed coil. The latter is proportional to the current in the fixed coil. Hence if the fixed coil (with a suitable shunt) is connected as is the ammeter in Fig. 5-20(a), the current in it is proportional to the line current and its magnetic field is proportional to this current. If the pivoted coil (with suitable series resistance) is connected as is the voltmeter in Fig. 5-20(a) the current in it is proportional to the potential difference between the terminals of X. The deflecting torque is therefore proportional to the product of the current and the potential difference, or to the power supplied to X. Under the influence of this torque, the coil rotates to a position where the restoring torque equals the deflecting torque and, since the restoring torque is proportional to the angle of deflection, the latter is proportional to the power.

A wattmeter is provided with four terminals, two corresponding to an ammeter and two to a voltmeter. Connections are made as in Fig. 5-20(b).

The *watt-hour meter* indicates the total *energy* supplied to (or by) a device, rather than the *rate* of supply of energy or power. Essentially it is a small motor whose angular velocity is proportional to the power consumed. That is,

$$\omega = kp$$

where ω is the instantaneous angular velocity, p the instantaneous power and k is a constant of proportionality. The angular displacement θ in a time interval t_1 to t_2 is therefore

$$\theta = \int_{t_1}^{t_2} \omega\, dt = k \int_{t_1}^{t_2} p\, dt.$$

But $\int_{t_1}^{t_2} p\, dt$ is simply the energy supplied in the time interval, so the angular displacement is proportional to this energy.

The arrangement of dials in the familiar watt-hour meter found in most homes is simply a revolution counter for recording the total angular displacement of the meter. The dials are commonly calibrated to include the proportionality constant and read the energy directly in kilowatt-hours.

Problems—Chapter 5

(1) A 5-ohm resistor is connected across the terminals of a battery of emf 12 volts, internal resistance 1 ohm, as in Fig. 5-21. Point *c* is grounded. (a) What is the direction of the (conventional) current in the battery? in the external circuit? (b) How many coulombs per second flow out of the positive terminal of the battery? How many coulombs per second flow into the negative terminal? How many coulombs per second flow into the resistor at point *b*? How many coulombs per second flow out of the resistor at point *c*? (c) Find the potentials at points *a*, *b*, *c*, and *d*. (d) Find the terminal voltage of the battery, V_{ad}. (e) Find the potential difference across the resistor, V_{bc}. (f) What is the current in the conductor connecting point *c* to ground?

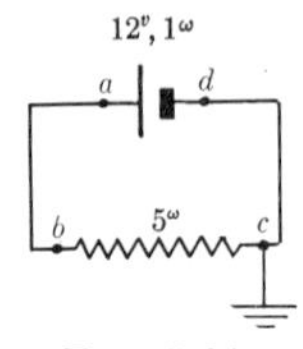

FIG. 5-21.

(2) In the circuit in Fig. 5-21, find (a) the rate of conversion of chemical energy to electrical energy within the battery, (b) the rate of development of heat in the battery, (c) the rate of development of heat in the external resistor.

(3) Fig. 5-22 represents a battery of emf 24 volts, internal resistance 1 ohm, across whose terminals is a rheostat of resistance R. (a) What should be the resistance R to obtain the following currents: 0, 4^a, 8^a, 12^a, 16^a, 20^a, 24^a? (b) What is the maximum current the battery can deliver in the circuit in Fig. 5-22? (c) Find the terminal voltage of the battery, V_{ab}, for each of the currents in part (a). Show the results in a graph of terminal voltage vs current. (d) The resistance R is set at 3^ω and a second battery of emf $\mathcal{E}$ and negligible internal resistance is inserted in series in the circuit. Find the magnitude of $\mathcal{E}$ if the current is 28^a, in the same direction as in part (a). Draw a diagram with the polarity of the battery clearly shown. (e) Same as (d) except the current is 4^a, opposite in direction to that in part (a). (f) Compute the terminal voltages of the first battery in parts (d) and (e) and include the values in the graph of part (c). Answer the following questions: Is it possible for the current in a battery to exceed the short-circuit current? Is the + terminal of a battery necessarily at a higher potential than the − terminal?

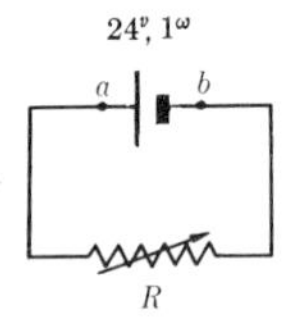

FIG. 5-22.

(4) In Fig. 5-23, $\mathcal{E}_1 = 24^v$, $r_1 = 2^\omega$, $\mathcal{E}_2 = 6^v$, $r_2 = 1^\omega$, $R_1 = 2^\omega$, $R_2 = 1^\omega$, $R_3 = 3^\omega$. Find (a) the current in the circuit, (b) the potentials at points *a*, *b*, *c*, and *d*, (c) the terminal voltages of the batteries, V_{ab} and V_{dc}.

(5) Same as problem 4 except that the 6-volt battery is reversed.

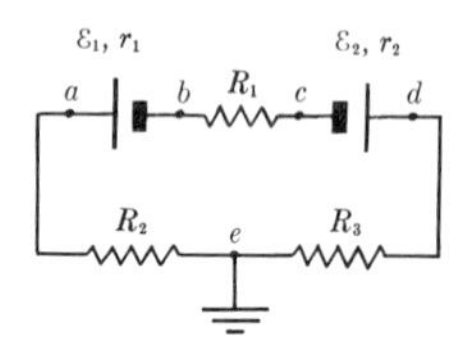

FIG. 5-23.

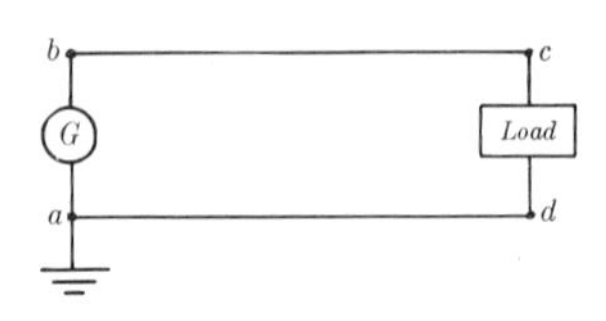

FIG. 5-24.

(6) Fig. 5-24 represents a DC generating plant G supplying electrical power to a distant factory along a transmission line. The terminal voltage of the generator, V_{ba}, is 230 volts, and the line current is 50 amperes. The resistances of the wires bc and ad are 0.1 ohm each. Point a is grounded. (a) What is the potential at point b? (b) What is the iR drop in each wire? (c) What is the potential at point c? (d) What is the potential at point d? (e) What is the line voltage at the load, i.e., V_{cd}? (f) Show in a diagram how voltmeters should be connected to measure the potential differences in parts (b) and (e).

(7) Fig. 5-25 shows a circuit containing a vacuum triode. The direction of the electron stream within the triode is from the cathode K to the plate or anode A, so the direction of the conventional current is from A to K within the triode. The current to the grid G is negligible. The current delivered by the battery, of emf 300 volts and negligible internal resistance, is 10 ma. The resistance of R_3 is 10,000 ohms. (a) What is the plate potential, V_p, above ground? (b) What is the potential of the grid above ground? (c) For proper operation of the triode the potential of the grid should be 15 volts below that of the cathode. What is the correct resistance of the so-called "biasing resistor," R_2? (d) What is the potential difference between anode and cathode, V_{pk}?

(8) The internal resistance of a dry cell gradually increases with age, even though the cell is not used. The emf, however, remains fairly constant at about 1.5 volts. Dry cells are often tested for age at the time of purchase by connecting an ammeter directly across the terminals of the cell and reading the current. The resistance of the ammeter is so small that the cell is practically short-circuited. (a) The short-circuit current of a fresh No. 6 dry cell is about 30 amp. Approximately what is the internal resistance? (b) What is the internal resistance if the short-circuit current is only 10 amp? (c) The short-circuit current of a 6-volt storage battery may be as great as 1000 amp. What is its internal resistance?

(9) When switch S is open, the voltmeter V, connected across the terminals of the dry cell in Fig. 5-26, reads 1.52 volts. When the switch is closed the voltmeter reading drops to 1.37 volts and the ammeter reads 1.5 amp. Find the emf and internal resistance of the cell. Neglect meter corrections.

(10) A battery of emf 24 volts and negligible internal resistance is connected in series with a fixed 6-ohm resistor and a variable resistor R as in Fig. 5-27. Consider the battery and the 6-ohm resistor as the power supply, and the resistor R as the external circuit. Let R have the following values: 0, 2^{ω}, 6^{ω}, 18^{ω}, 42^{ω}, ∞. For each value of R compute: (a) the power output of the battery, $\mathcal{E}i$. (b) the power wasted as heat in the 6-ohm resistor, (c) the power supplied to the external circuit. Arrange results in tabular form. (d) Construct graphs of the results of parts (a), (b), and (c), plotting power vertically and resistance R horizontally. A range of R from 0 to 18 ohms is sufficient. (e) A power supply of emf $\mathcal{E}$ and internal resistance r is connected to an

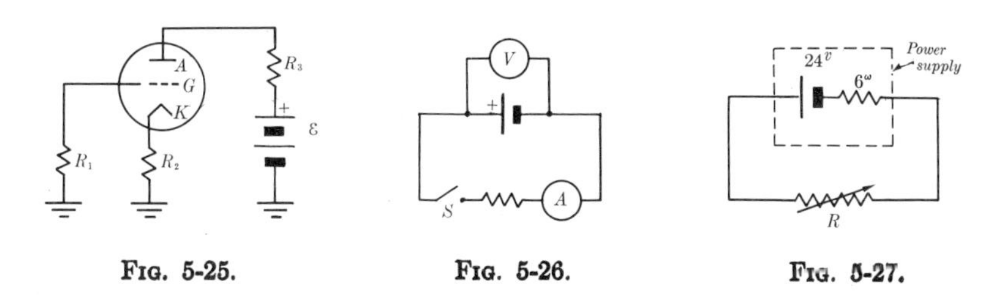

FIG. 5-25. FIG. 5-26. FIG. 5-27.

external circuit of adjustable resistance R. Set up the general expression for the power supplied to R in terms of $\mathcal{E}$, r, and R. Differentiate this expression with respect to R to find the external resistance that will make the power supplied a maximum. Does the result agree with the graph in part (d)?

This problem is an illustration of the more general principle of "impedance matching," which states that the power transferred from one part of a circuit to another is a maximum when the impedances (in this case, the resistances) are equal.

(11) (a) Prove that the power output (i.e., the power supplied to an external circuit) of a battery or generator of emf $\mathcal{E}$, internal resistance r, is a maximum when the current equals one-half the short-circuit current. (b) When the current has this value, compare the power output with the power wasted as heat within the seat of emf.

(12) A storage battery whose emf is 12 volts and whose internal resistance is 0.1 ohm (internal resistances of commercial storage batteries are actually only a few thousandths of an ohm) is to be charged from a 112-volt DC supply. (a) Should the + or the − terminal of the battery be connected to the + side of the line? (b) What will be the charging current if the battery is connected directly across the line? (c) Compute the resistance of the series resistor required to limit the current to 10 amp. With this resistor in the circuit, compute: (d) the potential difference between the terminals of the battery, (e) the power taken from the line, (f) the power wasted as heat in the series resistor, (g) the *useful* power input to the battery.

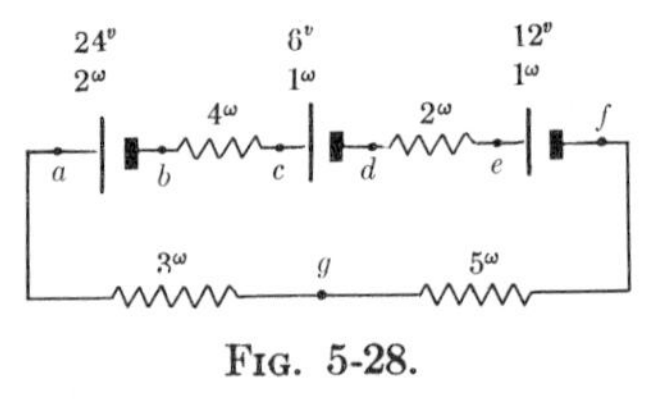

FIG. 5-28.

(13) In the series circuit in Fig. 5-28, compute V_{ea}, V_{fc}, and V_{gd}. State in each case which of the two points is at the higher potential.

(14) Refer to Fig. 5-11. Let the resistance of each resistor be 12 ohms. (a) Find the equivalent resistance of each network. (b) Find the rate of development of heat in each network, in watts, when the terminals of the network are connected across a 120-volt line.

(15) Prove that when two resistors are connected in parallel the equivalent resistance of the combination is always smaller than that of either resistor.

(16) (a) The power rating of a 10,000 ohm resistor is 2 watts. (The power rating is the maximum power the resistor can safely dissipate as heat without too great a rise in temperature.) What is the maximum allowable potential difference across the terminals of the resistor? (b) A 20,000-ohm resistor is to be connected across a potential difference of 300 volts. What power rating is required? (c) It is desired to connect a resistance of 1000 ohms across a potential difference of 200 volts. A number of 10-watt, 1000-ohm resistors are available. How should they be connected?

(17) (a) Find the equivalent resistance of the external circuit in Fig. 5-29. (b) Find the readings of the ammeter A and the voltmeter V.

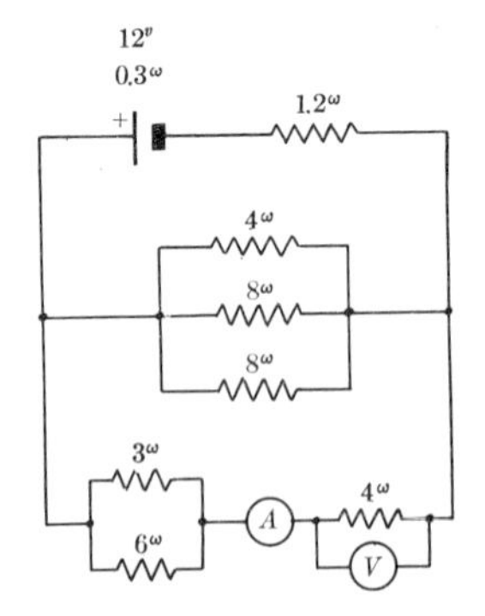

FIG. 5-29.

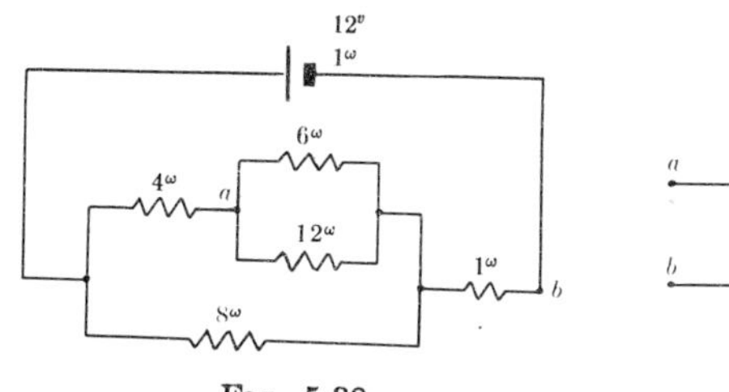

FIG. 5-30.

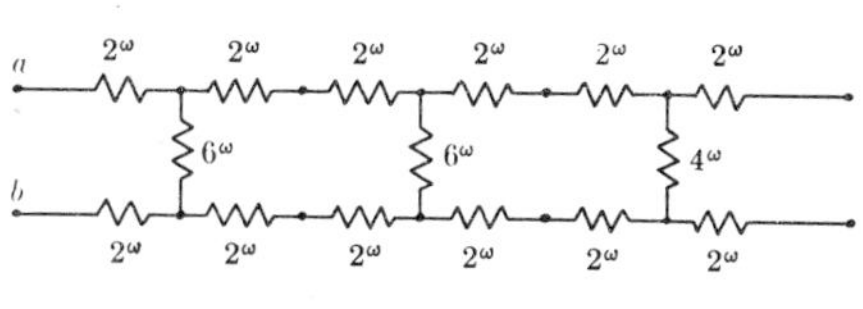

FIG. 5-31.

(18) The resistance of a certain galvanometer is 30 ohms. (a) What resistance in shunt (i.e., in parallel) with the galvanometer will reduce the equivalent resistance to 10 ohms? (b) If the potential difference between the galvanometer terminals is 100 mv, find the current in the galvanometer and in the shunt.

(19) Find the resistance of 1000 ft of a conductor consisting of a central iron core 0.2 inch in radius, surrounded by a copper sheath whose outer radius is 0.3 inch.

(20) In Fig. 5-30, find: (a) the current in the battery, (b) the current in each resistor, (c) the potential difference between points a and b.

(21) (a) Find the resistance of the network in Fig. 5-31, between the terminals a and b. (b) What potential difference between a and b will result in a current of 1 amp in the 4-ohm resistor?

(22) Two resistors A and B are in parallel with one another and in series with a 200-ohm resistor and a dry cell whose emf is 1.5 volts and whose internal resistance is negligible. The resistance of A is 100 ohms. When B is disconnected from A, an additional resistance of 50 ohms must be inserted in the circuit in order to keep the current through A unchanged. Find the resistance of B.

(23) The galvanometer current in part (b) of Fig. 5-32 is $1/n$th as great as in part (a), and the equivalent resistance of the network in part (b) equals the galvanometer resistance R_G. A constant potential difference is maintained between points a and b. Find the resistances of the series and shunt resistors, R_1 and R_2, in terms of n and R_G. For a numerical check, let $R_G = 90$ ohms, $n = 10$, $V_{ab} = 90$ mv. Find R_1, R_2, and the galvanometer current in both circuits.

(24) Cell A in Fig. 5-33 has an emf of 12 volts and an internal resistance of 2 ohms. Cell B has an emf of 6 volts and an internal resistance of 1 ohm. (a) Find the potential difference V_{ab} when switch S is open. (b) When switch S is closed, the current in R is 3 amp and its direction is from a to b. Use Kirchhoff's rules to compute the currents in the other branches of the circuit. (c) Find the resistance R. (d) Find the emf and internal resistance of a single cell which would be equivalent to the combination of A and B in parallel.

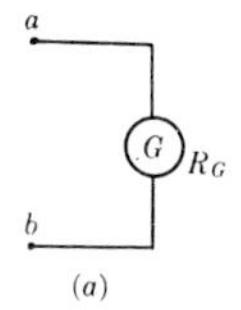

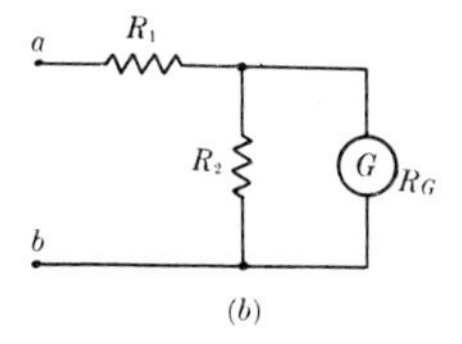

FIG. 5-32.

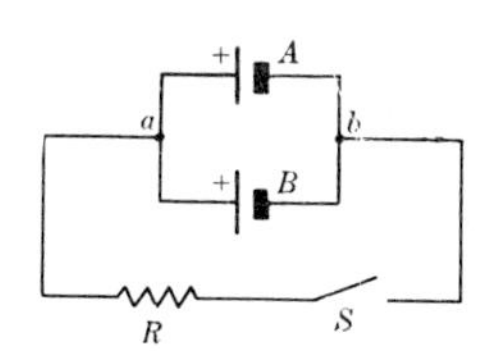

FIG. 5-33.

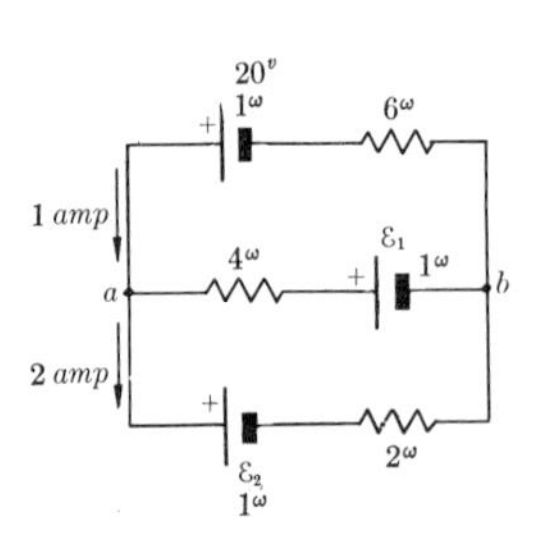

FIG. 5-34.

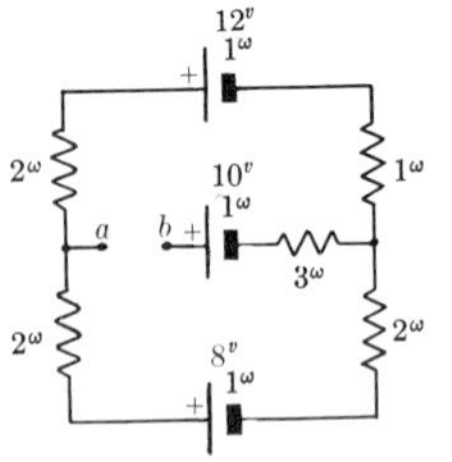

FIG. 5-35.

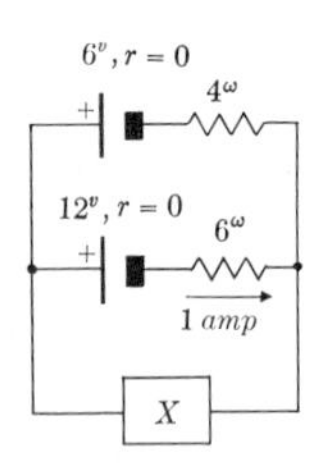

FIG. 5-36.

(25) Find the emf's $\mathcal{E}_1$ and $\mathcal{E}_2$ in the circuit of Fig. 5-34, and the potential difference between points a and b.

(26) (a) Find the potential difference between points a and b in Fig. 5-35. (b) If a and b are connected, find the current in the 12-volt cell.

(27) (a) Find the magnitude and direction of the current in the circuit element X in Fig. 5-36. (b) What information can be deduced as to the nature of this element?

(28) Three unlike cells of emf's $\mathcal{E}_1$, $\mathcal{E}_2$, and $\mathcal{E}_3$, and internal resistances r_1, r_2, and r_3, all have their positive terminals connected to a common point a and their negative terminals to a common point b. Find the expression for V_{ab}.

(29) (a) Find the equivalent resistance of the network in Fig. 5-37, between points a and b, in terms of R_1, R_2, and R. (b) Compute the equivalent resistance if $R_1 = 4^\omega$, $R_2 = 2^\omega$, $R = 1^\omega$. Compare with the equivalent resistance when resistor R is removed.

(30) Twelve metal rods, each of resistance R, are joined so as to form the edges of a cube as in Fig. 5-38. Find the equivalent resistance of the network, between a pair of diagonally opposite corners such as a and b.

(31) Fig. 5-39 is a diagram of a shunt wound DC motor, operating from 120-volt DC mains. The resistance of the field windings, R_f, is 240 ohms. The resistance of the armature, R_a, is 3 ohms. When the motor is running the armature develops a back emf $\mathcal{E}$ in the direction of the arrow. The motor draws a current of 4.5 amp from the line. Compute: (a) the field current, (b) the armature current, (c) the back emf $\mathcal{E}$, (d) the rate of development of heat in the field windings, (e) the rate of development of heat in the armature, (f) the power input to the motor, (g) the efficiency of the motor, if friction and windage losses amount to 50 watts.

(32) Refer to the shunt motor in problem 5-31. Assume that the given data apply when the motor is rotating at 1800 rpm, and that the back emf is proportional to the angular velocity. Assume also that friction and windage losses are constant. (a) As the result of an increased load on the motor its angular velocity decreases to 1700 rpm. What power does it now draw from the line? (b) What power does it deliver to the load? (c) At what angular velocity, in rpm, will the motor run when the external load is removed completely?

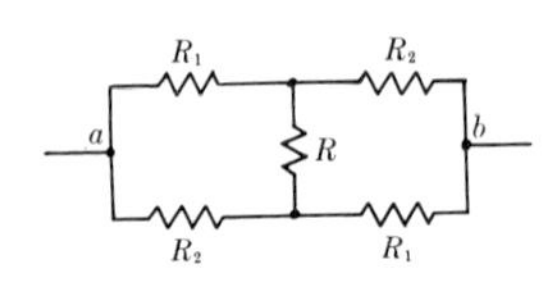

FIG. 5-37.

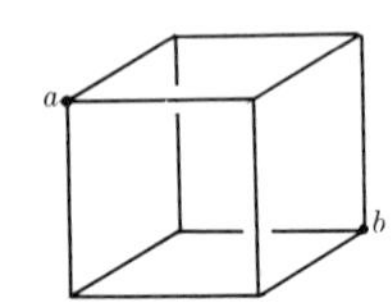

FIG. 5-38.

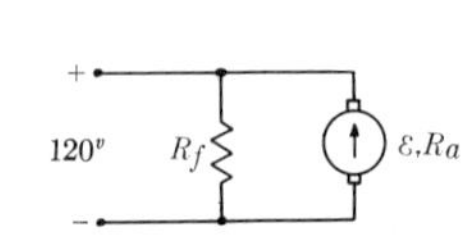

FIG. 5-39.

(33) Refer to problem 6 in Chap. 5. (a) Compute the power output of the generator, the power loss in the line, and the power delivered to the factory. (b) Suppose the terminal voltage at the generator were doubled and the power delivered to the factory kept constant. Find the power output of the generator and the power loss in the line.

(34) Show clearly with the aid of a diagram how the four terminals of a wattmeter should be connected to measure the power supplied to the motor in Fig. 5-39.

35. The table below lists corresponding values of terminal voltage V and current i for three 110-volt incandescent lamps.

	100-watt tungsten	40-watt tungsten	48-watt carbon
V (volts)	i (amp)	i (amp)	i (amp)
5	.20	.10	.009
10	.25	.14	.021
20	.35	.17	.060
40	.50	.22	.15
60	.62	.27	.23
80	.73	.31	.31
100	.83	.35	.39
120	.91	.37	.47

(a) Construct voltage-current graphs for each lamp.

(b) Compute the resistance of each lamp for terminal voltages of 10 volts and of 110 volts. Which lamp has the greatest and which the least resistance, at 110 volts?

(c) Which lamps have a positive and which a negative temperature coefficient of resistance?

(d) Compute the actual power taken by each lamp at a terminal voltage of 110 volts.

(e) If the 100-watt and the 40-watt lamps were connected in series across a 110-volt line, what would be the current in each and the voltage across each? Express the current through each lamp as a percentage of its rated current. (The rated current is the current drawn by the lamp at its rated terminal voltage of 110 volts.) Which lamp would appear brighter?

CHAPTER 6

CHEMICAL AND THERMAL EMF'S

6-1 Chemical energy and emf's. When an electrical cell or a battery of cells becomes part of a closed electrical circuit, a current results. This current may be made to do mechanical work or to develop heat, the energy being derived from chemical energy originally present inside the battery. In the language of Chap. 5, we say that a chemical emf arises in the electrical circuit by the conversion of chemical energy to electrical energy.

Chemical energy is the result of forces between the constituent atoms of molecules. Because of these forces, the product molecules of a chemical reaction may have a lower potential energy than the initial reactants. In such a reaction, energy is liberated partly as heat and partly as useful work. Conversely, this reaction can proceed in the reverse direction only if energy is introduced from some external source. In the reactions which occur in electrical cells one must consider not only forces between individual atoms but also between atoms and their orbital electrons. These reactions are called reactions of oxidation and reduction. In an oxidation reaction an atom or group of atoms loses one or more electrons. Reactions of electron gain are reduction reactions.

6-2 Electrode potentials. An example of the process of electron loss or oxidation occurs when a copper plate is dipped in pure water. Copper ions immediately go into solution. The chemical equation for the process may be written as follows:

$$Cu \longrightarrow Cu^{++} + 2e^{-}.$$

From a somewhat oversimplified point of view one may imagine that the attractive forces of the water molecules exceed the cohesive forces which hold the copper ions in their crystalline lattice. In the absence of other effects, a steady flow of ions into solution would occur, but since every positive ion which goes into solution leaves one electron behind for each valence which the ion possesses (for copper two altogether) the plate immediately acquires a negative potential with respect to the solution. As a consequence, the ions in solution are attracted by the copper plate, and a certain fraction is captured by the plate and neutralized. The greater the negative charge on the plate, the greater the tendency of the positive ions to return to the plate and the fewer the number of ions which escape neutralization. At equilibrium, the rates of flow and of return are equal and the potential of the plate is just sufficient to produce a reverse reaction equal to the forward reaction. The presence of more copper ions in the solution increases the reverse reaction rate and the potential decreases to a lower equilibrium value. It is therefore evident that the equilibrium potential of the copper with respect to the solution depends on the concentration of the copper ions around it. Since the two reaction rates are influenced to a different degree by the temperature of the system, the electrode potential varies with temperature. The process of electron loss occurs most easily for elements which contain only a few electrons in their outer valence shell. This is true in particular for the metals and for hydrogen.

Electron gain occurs in the case of a large group of elements whose outer atomic shells possess one or more vacancies into which free electrons may fit. When these vacancies are filled, the shell resembles that of the nearest noble gas in the periodic table and a particularly stable configuration results. The atoms now have an excess of negative electrons and thus are negative ions. As a typical example, a chlorine atom may pick up a single negative electron. The presence of the electron gives a single negative electronic charge to the chlorine. The resulting chlorine ion has a full outer shell which resembles that of its neighboring noble gas argon. The process is not restricted to single atoms, but can occur for groups of atoms. In the sulfate ion a single sulfur atom is bound chemically to four oxygen atoms and the group has two negative electronic charges.

The preparation of a typical gaseous electrode (that of hydrogen) will be discussed in a later section. In the case of a gaseous chlorine electrode, negative ions are formed by the reaction

$$\mathrm{Cl} + e^- \longrightarrow \mathrm{Cl}^-,$$

in which chlorine is reduced. As a result of the reduction, the electrode becomes positively charged and its potential is positive with respect to the solution. The same effect of temperature and ionic concentration on the electrode potential is observed for reduction as for oxidation. An additional factor is important when the electrode is a gas. The pressure of the gas also affects the electrode potential.

6-3 Electrical cells. When a single electrode is placed in a solution, it evidently forms a single seat of emf. This emf could be measured provided some means were available to make contact with the solution through an electrode having the same potential as the solution. In practice this is not feasible. The absolute measurement of single electrode potentials involve special methods of doubtful accuracy. However, no actual difficulty arises, since potential differences between pairs of electrodes are usually measured.

The introduction of a second electrode completes an electrical cell. The emf of the cell is determined in part by the difference between the emf's arising at the electrodes. These emf's depend upon the nature of the electrodes, upon the ionic concentrations around them, upon the temperature and, in special cases, upon the pressure. There is an additional source of emf in some cells. Just as a difference of potential arises between an electrode and a solution, potential differences may also occur at the junction between one solution and another. This effect is particularly marked when a concentrated solution is in contact with a dilute solution and the positive and negative ions are of such a nature that they diffuse with a considerable difference in velocities.

Since the emf of a cell without such a liquid junction is the difference of the emf's at the electrodes, one may form a particular kind of cell with a net emf by maintaining all but one of the factors which affect the two electrode potentials identical. One of the most interesting of such cells is that in which both electrodes are of the same material and only the concentrations of the ions about the electrodes are different. Such cells are important in the study of corrosion. By the exercise of a little ingenuity, the reader may mentally construct a large number of different types of cells, varying singly or simultaneously the electrode materials, the concentrations about the electrodes, the temperatures at the electrodes, and other factors which affect the individual electrode potentials.

6-4 The hydrogen electrode. In a very important class of cells, one of the electrodes consists of hydrogen gas at one atmosphere pressure. The solution has a 1 molal concentration of hydrogen ions. The hydrogen is allowed to bubble over a platinum electrode which is covered with finely-divided metallic platinum. Thus effective contact between the gaseous hydrogen and an external circuit can be made. The additional effect of the finely-divided platinum is to maintain the hydrogen in close equilibrium with its ions by holding it in an adsorbed form in which it has a large surface area. If this electrode is combined at 25° C with a series of other electrodes in contact with solutions 1 molal in their ions, a relative scale of electrode potentials may be obtained, using hydrogen as a reference potential. This scale is given in Table 6-1. From this table the potential difference between any two electrodes in contact with 1 molal solutions of their ions at 25° C may be obtained by taking the difference between their relative potential differences with respect to hydrogen.[1]

TABLE 6-1

ELECTRODE POTENTIALS*

Electrode	Ion	Potential (volts)
Li	Li^+	−3.02
K	K^+	−2.922
Ca	Ca^{++}	−2.87
Na	Na^+	−2.712
Mg	Mg^{++}	−2.34
Mn	Mn^{++}	−1.05
Zn	Zn^{++}	−0.7620
Cr	Cr^{++}	−0.71
Sn	Sn^{++}	−0.136
Pb	Pb^{++}	−0.126
H_2	H^+	0.00
Cu	Cu^{++}	+0.3448
Cu	Cu^+	+0.522
Ag	Ag^+	+0.7995
Cl_2	Cl^-	+1.3583
F_2	F^-	+2.85

* For other potentials consult Latimer and Hildebrand, "Reference book of Inorganic Chemistry." The sign convention used by these authors is opposite to that adopted here.

6-5 Calculation of emf's. By the application of the methods of thermodynamics it is possible to obtain an equation for the potential difference of an electrical cell under equilibrium conditions. In the present discussion a simplified form will be given which applies only to cells without internal liquid junctions and in which the electrodes are pure metals a and b immersed in very dilute solutions containing only the ions of metals a and b respectively. Let V_{ab} be the potential difference between the terminals of the cell under open circuit equilibrium conditions. Then

$$V_{ab} = V_a^o - V_b^o + \left(\frac{RT}{F}\right) \ln \frac{C_a^{\alpha}}{C_b^{\beta}}, \tag{6-1}$$

where V_a^o and V_b^o are the molal electrode potentials of metals a and b relative to hydrogen and are obtained from Table 6-1. R is the gas constant

[1] For the strict definition of a 1 molal solution in the present case, the reader should consult a more advanced text.

(8.314 joules/degree-mole), F is the Faraday (96,500 coulombs), T is the Kelvin temperature, C_a and C_b are the molal concentrations of the ions of metals a and b, α and β are the fractional numbers of ions which pass into or out of solution at a and b respectively when one electronic charge passes through the cell. The signs are chosen so that when V_{ab} is positive, terminal a has a higher potential than b.

Example. A cell consists of a lead and a tin electrode in a uniform solution containing bivalent tin ions at a concentration of 0.072 moles per liter, and bivalent lead ions at a concentration of 0.024 moles per liter. What is the potential difference at 25° C?

Call the lead electrode a, the tin b. In each case, two electronic charges flow through the circuit for each ion dissolving and therefore $\alpha = \beta = \frac{1}{2}$.

$$V_{ab} = -\,0.126 - (-0.136) + \frac{8.314 \times 298}{96{,}500} \ln \frac{(.024)^{1/2}}{(.072)^{1/2}}$$
$$= -0.004 \text{ volt.}$$

What is the potential difference at 100° C?

$$V_{ab} = .010 - \frac{8.314 \times 373}{96{,}500} \ln \frac{(.024)^{1/2}}{(.072)^{1/2}}$$
$$= .010 - .0175$$
$$= -.0075 \text{ volt.}$$

Which electrode is +?

At 100° C the lead electrode is at a lower potential than the tin. The tin is therefore the positive electrode.

6-6 Hydrogen ion concentration. The hydrogen electrode may be used to measure hydrogen ion concentrations. In its simplest application a cell is constructed which has two hydrogen electrodes. One electrode is dipped in a solution 1 normal (for hydrogen 1 normal = 1 molal) in hydrogen ions and the other in the solution of unknown concentration. In order to eliminate one of the hydrogen electrodes and thus simplify the arrangement, the normal hydrogen electrode is usually replaced by some other electrode of known potential. One of the most commonly used reference electrodes contains mercury and saturated mercurous chloride (calomel) in contact with a 1 normal (1 molal) solution of potassium chloride. It is called the *normal calomel electrode* and at 25° C it is positive with respect to the normal hydrogen electrode by 0.2801 volt. Since the hydrogen electrode is limited in its use for routine work, a number of other electrodes have been devised for measuring hydrogen ion concentrations. The most popular instrument is the "glass electrode."

It consists of a very thin glass membrane separating a solution of known hydrogen ion concentration from one of unknown concentration. The mechanism of its action is only partially understood. Commercial instruments are frequently calibrated in "pH" units. The pH of a dilute solution is the negative of the logarithm to the base of 10 of its hydrogen ion concentration.

As an example of pH calculations, one may compute the pH of normal human blood. The hydrogen ion concentration is 4×10^{-8} normal and the pH is therefore 7.4.

6-7 The Daniell cell. One of the simplest practical sources of electrical energy is the Daniell cell. It contains a zinc electrode immersed in a solution of zinc ions (Zn^{++}) and a copper electrode in a solution of bivalent copper ions(Cu^{++}). Negative ions are also present but need not be considered for the moment. In some cases the two solutions are separated by a porous cup and in others by gravity. The latter type of cell is called a "gravity cell" and is illustrated in Fig. 6-1. The maximum voltage is obtained by preparing the cell with a minimum concentration of zinc ions in the upper layer and a saturated solution of copper ions in the lower. The saturation of the copper solution is maintained by an excess of solid crystals of copper sulfate in the bottom. As current is drawn, zinc goes into solution and the concentration of zinc ions in the upper part of the cell increases. Thus the emf gradually declines with age.

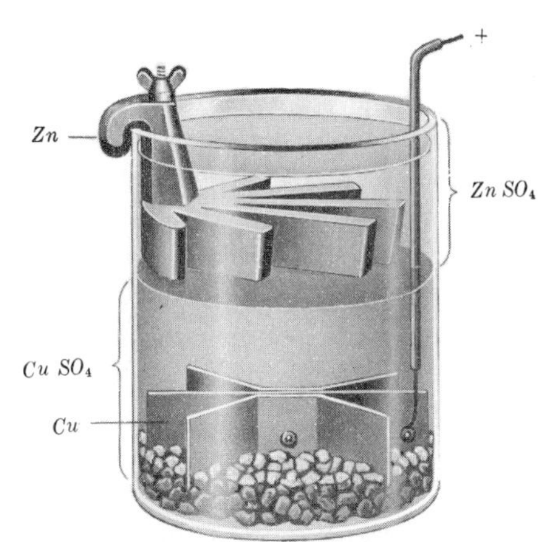

FIG. 6-1. Daniell cell.

Some insight will be gained concerning the behavior of ions in solution by considering the electrical and chemical behavior of a freshly-prepared Daniell cell. Under open circuit equilibrium conditions the potential differences of the electrodes with respect to the liquid will be determined by the temperature and concentration of the ions about them. Although experimental difficulties prevent the exact determination of these potentials, it will be assumed for purposes of discussion that they are both negative.

Since the zinc is much more negative with respect to the solution than the copper, the copper is at a higher potential than the zinc. This condition is illustrated in Fig. 6-2. When the zinc electrode is introduced, a few zinc ions go into solution. Thus a small positive charge accumulates around the zinc electrode. The same effect also occurs to a lesser extent around the copper so that after the copper electrode is introduced a somewhat smaller positive charge accumulates around the copper. When equilibrium is achieved the potential difference within the solution which was produced by the insertion of the electrodes is neutralized by a slight migration of ions. After the migration is complete the solution is at a uniform potential. Since the bottom of the cell is saturated with copper sulfate, the total concentration of dissolved salts is greater at the copper electrode than at the zinc. This concentration difference causes ions to diffuse from the region of high concentration to the region of low concentration. Both positive and negative ions move, thus maintaining electrical neutrality and equipotential conditions in the electrolyte. In this migration process copper ions diffuse upward and soon small local areas of copper deposition occur on the zinc. Small short-circuited cells are formed and zinc rapidly dissolves.

The cell may be short-circuited by establishing a very low resistance connection between the electrodes. Electrons then flow in the external circuit from the zinc to the copper until the copper and zinc are at the same potential. The concentrations of the ions about their respective electrodes now readjust themselves. Copper ions move to the copper electrode, since it is now more negative with respect to the solution. Zinc goes into solution from the zinc electrode, since it is now less negative. As a result of the concentration change, a gradient of potential is established in the solution, causing ions to move in the body of the solution, the positive ions moving toward the copper, the negative toward the zinc. The situation is that shown in Fig. 6-3. The fraction of the cur-

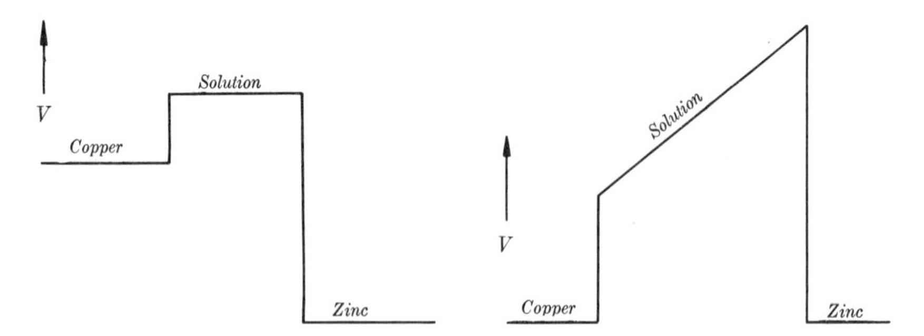

FIG. 6-2. Potentials of the copper and zinc electrodes of a Daniell cell with respect to that of the solution.

FIG. 6-3. Potentials of the copper, zinc, and solution of a short-circuited Daniell cell.

rent carried by different ions depends on their charge and on the ease with which they move. The negative sulfate ions of the original electrolyte move partly under the influence of the field and partly by diffusion from the region of high to the region of low concentration. As zinc ions dissolve at the zinc electrode and sulfate ions migrate upward, the concentration of zinc sulfate gradually increases in the upper part of the cell. The positive copper ions which under open circuit conditions diffused upward are now influenced by the field. If it is sufficiently strong, the upward diffusion tendency is completely overcome and copper ions cannot find their way to the zinc electrode and deposit themselves. For this reason it is advantageous to keep the external circuit of a Daniell cell closed at all times. Since a Daniell cell has a high internal resistance, the short circuit current is small. The internal resistance evidently arises in part from the forces which oppose the movement of the ions through the electrolyte. Foreign ions are often added to decrease the internal resistance.

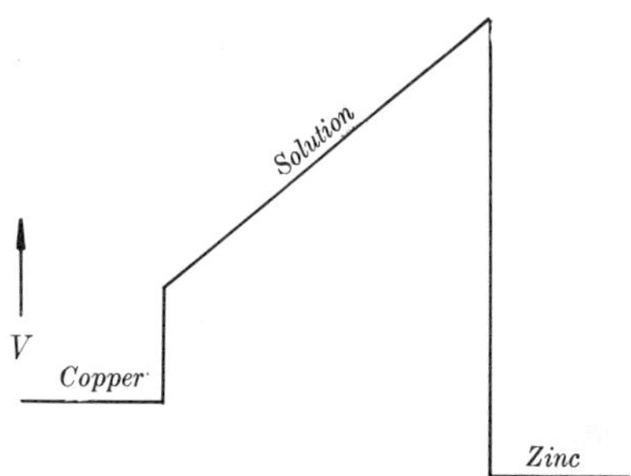

FIG. 6-4. Potentials of copper, zinc, and solution of a Daniell cell when it is connected to a moderately high resistance.

The Daniell cell works best when operating continuously in a closed circuit of moderately high resistance. The conditions in the cell are then as shown in Fig. 6-4. The current in the circuit is equal to the emf divided by the total resistance (internal and external) of the circuit. The copper plate is at a higher potential than the zinc by an amount equal to the voltage drop in the external circuit. The remainder of the emf is distributed through the solution and the resulting field is smaller than in the short-circuited cell.

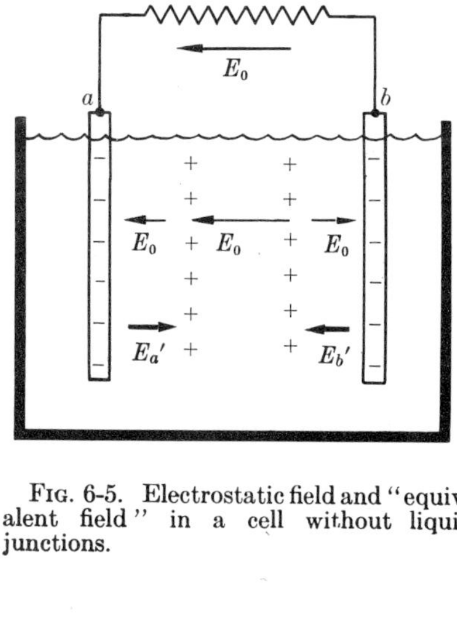

FIG. 6-5. Electrostatic field and "equivalent field" in a cell without liquid junctions.

Seats of emf in cells. It will be instructive to apply the ideas developed in Sec. 5-3 to the seats of emf in a typical cell without liquid junctions. Fig. 6-5 indicates schematically the situation within the cell. For ease of representation, the separation between the ions and the electrodes has been exaggerated. In practice it will be quite small. E_0 represents the electrostatic part of the field shown by the light arrows. E' represents the "equivalent field" which does work on the charge passing through the seat of emf. Starting at a and taking the line integral of the total field $E_0 + E'$ around the circuit through the electrolyte and returning to a through the external resistance, it is found that the total emf arises from the contribution of E' alone to the line integral. Thus

$$\mathcal{E} = \oint E' \cos \theta \, ds = \int E_a' \, ds_a - \int E_b' \, ds_b.$$

The effects of the two opposing seats of emf are the same as though the charges were under the action of the equivalent fields at electrodes a and b. At electrode a positive work is done on the charges by E_a' and at electrode b negative work is done by E_b'.

6-8 Reversibility. The principle of reversibility of electric cells is of great importance in thermodynamics since chemical energy may be converted to electrical energy reversibly (i.e., quasistatically) in an electric cell. In order for the conversion to be reversible, equilibrium must be maintained. This is achieved by inserting in the external circuit some opposing source of emf which is only infinitesimally smaller than the emf of the cell. If the opposing emf is increased so that it is infinitesimally greater than the emf of the cell, energy may be reconverted in the cell, appearing once more in the chemical form. Although a cell may be operated reversibly through infinitesimal changes, many practical cases occur where the passage of a finite amount of electrical charge produces, through side reactions, permanent changes in the cell materials which cannot be corrected by reverse operation of the cell. In such cases the cell is thermodynamically reversible but it is nonreversible in the practical sense, since it cannot be recharged.

6-9 Polarization. A number of effects occur in which the chemical conditions around the electrodes of a cell are modified temporarily during the operation. These effects are responsible for the phenomenon of *polarization*. The simplest kind of polarization occurs when the ionic concentration around an electrode is altered by the passage of current. Thus if the current in a Daniell cell were suddenly raised to a momentary high value, the concentration of copper ions immediately around the copper plate would suddenly drop to a very low value and, in a similar manner, the concentration around the zinc electrode would rise sharply. The result would be a temporary drop in the emf of the cell. This is quickly restored by diffusion of the ions, but for the moment the cell is polarized. There are many cases where polarization seriously limits the operation of a cell. One of the most common of these is encountered when hydrogen ions are discharged, forming gaseous hydrogen. If in the Daniell cell the copper sulfate is replaced by sulfuric acid, instead of copper ions being discharged at the copper electrode, hydrogen ions will discharge, forming bubbles of gaseous hydrogen. Although probably not the whole story, one effect of the hydrogen gas is to increase the resistance between the electrode and the solution, increasing the internal resistance of the cell. In some types of cells, a substance called a depolarizer is added to the electrolyte around the positive electrode. It reacts with the hydrogen to prevent the formation of free gas.

6-10 The dry cell. Although not reversible in the practical sense, the dry cell continues to be the most convenient small portable source of electrical energy. The positive pole is a rod of carbon and the negative pole is of metallic zinc. The ordinary fluid is replaced by a paste containing zinc chloride and ammonium chloride. As a depolarizer, manganese dioxide is added. Although many high voltage dry batteries are built in layers, single cells are usually cylindrical, the zinc being in the form of a closed can with the carbon electrode in the center. The paste occupies the space between and the top is sealed with a layer of pitch or wax. As the cell discharges, metallic zinc changes to bivalent zinc ions at the negative pole. At the positive pole ammonium ions are believed to interact with the manganese dioxide by the reaction

$$2NH_4^{++} + 2e^- + MnO_2 \longrightarrow 2NH_3 + MnO + H_2O.$$

The open circuit emf of a freshly-prepared dry cell lies between 1.5 and 1.6 volts, depending on conditions at the electrodes. As current passes, the emf falls with considerable rapidity, since the depolarization is slow. If the cell is allowed to stand for a short time, the depolarization reaction "catches up" and the emf increases, reaching nearly its original value again. After extensive use, the emf falls to a low value and the internal resistance increases until the cell becomes useless. This effect is believed to be due to depletion of the manganese dioxide. Freshly-exhausted cells may often be temporarily regenerated by the careful application for a short time of a moderate reverse current.

6-11 The lead storage battery. The cells of a lead storage battery are reversible in the practical sense. In the fully-charged state they consist of sets of positive plates of lead perioxide and sets of negative plates of metallic lead. The materials are coated on grids of metallic lead and are prepared in a highly porous form so that the electrolyte gains full access to the plates over a maximum surface area. Separation is maintained by the use of treated wood or glass fiber spacers. The electrolyte is sulfuric acid of specific gravity 1.280 at 80° F. The emf is 2.1 volts per cell. As charge passes through a cell, reactions occur at the plates. At the positive plate lead is reduced from the quadrivalent state to the divalent state by the reaction

$$PbO_2 + SO_4^{--} + 4H^+ + 2e^- \longrightarrow PbSO_4 + 2H_2O.$$

Since lead sulfate is poorly soluble, it adheres to the plate as it is formed. At the negative plate sulfate ions are neutralized by the reaction

$$Pb + SO_4^{++} \longrightarrow PbSO_4.$$

Again the sulfate adheres to the plate. When the battery is discharged, both plates have been almost entirely converted to lead sulfate. The loss of sulfate ions from the electrolyte causes the specific gravity to be lowered so that in the discharged condition it reaches 1.13. Under these conditions the open circuit emf is approximately 1.75 volts per cell.

At low temperatures the open circuit emf of the lead storage battery falls somewhat, as it does for other electrochemical reactions. The most important effect, however, is the decreasing ion mobility within the battery at low temperatures. This appears as an increased internal resistance. Thus the terminal voltage drops abnormally when current is drawn. As a result of the combined low temperature effects, the cranking ability of an automobile storage battery at 0° C is only about 40% of its value at 80° C.

(a) A Weston standard cell. (*Courtesy of Weston Electrical Instrument Corp.*)

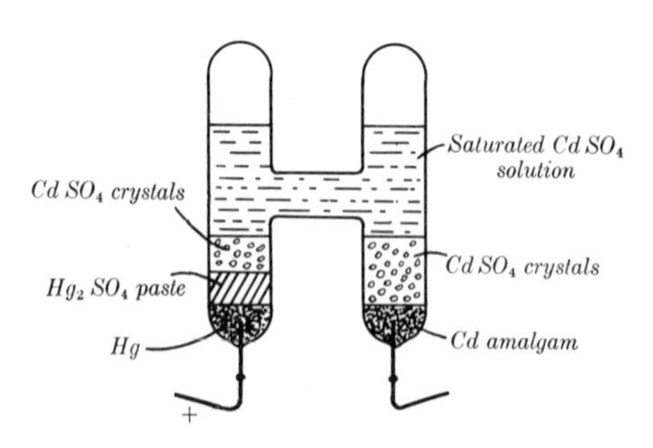

(b) Diagram of a Weston cell.

FIG. 6-6.

6-12 Standard cells. A number of cells have been devised which are extremely stable, maintaining a fixed emf over long periods of time. Probably the most common of these is the saturated Weston cell. It contains an electrode of mercury in contact with a paste of mercurous sulfate, and an electrode of cadmium amalgam in contact with a saturated solution of cadmium sulfate, as shown in Fig. 6-6. The construction is such that the passage of small amounts of charge in either direction has a negligible effect on the internal condition of the cell. The emf of the saturated Weston cell is affected by the temperature, not only through the temperature variation of the electrode emf's, but, in addition, the concentration of saturated cadmium sulfate changes slightly with temperature.

6-13 Electrolysis. By inserting in the circuit of an electrical cell an external source of counter emf which is infinitesimally greater than the original emf, it is possible to reverse the action of the cell and to convert electrical energy into chemical energy. This process is called *electrolysis*. In electrolysis the electrode at which positive ions are liberated or negative ions are neutralized is called the anode. The other electrode is called the cathode. In order to form a complete picture of electrolytic processes, it is first necessary to consider certain effects which occur in an electrolytic cell.

Decomposition potential. In all electrolytic reactions it is evident that a minimum external emf is necessary to maintain the system in equilibrium. This emf must be exceeded by an infinitesimal amount in order that a current may result. It is called the decomposition potential. It depends, as usual, on the concentration of electrolyte around the electrodes, on the electrode materials, and on the type of electrolytic reaction involved. Although the potential which can actually be measured is the difference of two individual potentials, one between each electrode and the solution, it is the individual potentials which are important.

Many decomposition potentials are measured in the following manner. Platinum electrodes are immersed in an electrolyte and a variable external emf is applied. The use of an inert material such as platinum permits the reaction to be studied without the complication of extraneous reactions between the products of decomposition and the electrodes. Under the influence of the applied field, positive ions in the electrolyte flow toward the negative electrode and negative ions toward the positive electrode. This separation of ions causes the cell to become polarized until the field of the electrodes is neutralized. Each electrode then has a certain potential relative to the electrolyte. If the external potential difference is increased, a point is reached where the attraction at one electrode for one of the groups of ions overcomes its solution tendency and neutral molecules deposit on the electrode. By further increasing the applied potential difference the same process occurs with another group of ions at the other electrode. Finally the decomposition potential is reached and a small current is observed. An electrical cell has now been formed in which the products of electrolytic decomposition coat the electrodes and form an equilibrium cell with their ions which are still in solution. Although platinum electrodes are employed, the situation is identical for any other electrode which is not attacked in the process. The addition of several ions of different types to the solution does not materially alter the situation. The cell which is formed is that between the pairs of positive and negative ions which have the lowest net decomposition potential between them. The potential for a given pair of ions depends, as usual, on their concentration and on the temperature.

Occasionally other substances in solution may react with the freshly-formed products of the electrolytic reaction. The electrodes themselves may also participate in this reaction. Thus, if chloride ion is decomposed at a silver electrode, silver chloride is formed. The most important electrode reaction is that which occurs in ordinary electroplating. Here both electrodes are of the same material and as fast as the ions of the metal are depleted in the electrolyte by precipitation at the cathode, new ions are formed by the solution of metal at the anode.

Overvoltage. Once the electrolytic cell is established the external emf just necessary to maintain it in equilibrium may still be considerably short of that required for the electrolysis to proceed at a satisfactory rate. The necessary additional voltage is called the overvoltage. Although overvoltage is not completely understood in all cases, it arises in part from the slowness of the reaction rate at the electrodes. Often included under the term overvoltage are the polarization effects which occur when ions are locally used up at the electrodes and not renewed sufficiently rapidly by diffusion from the body of the liquid. This effect may be reduced in practical electrolysis by rapid stirring of the solution or by rotation of the electrodes. One important consequence of overvoltage arises when the potential of the cathode is sufficiently large that the decomposition potential of hydrogen is exceeded. In an acid solution if overvoltage effects were absent, hydrogen would normally be evolved with an unfortunate effect on the electroplating of metals. Usually the overvoltage necessary for an appreciable production of hydrogen is not reached and the effect does not occur.

6-14 The electrolysis of water. When platinum electrodes are dipped in an electrolyte, the electrolytic reactions which occur depend upon the ions which are present in solution, the potential which exists between the platinum and the solution, and the decomposition potentials of the ions. If a platinum electrode could be made sufficiently negative with respect to the solution, platinum ions would dissolve. Under ordinary conditions, however, if the electrolyte is pure water, two other reactions of much lower decomposition potential take precedence. One of these is the conversion of hydrogen ions to hydrogen gas on the cathode by the reaction

$$2H^{+} + 2e^{-} \longrightarrow H_2.$$

The other reaction occurs at the anode and is

$$2OH^{-} \longrightarrow \tfrac{1}{2}O_2 + H_2O + 2e^{-}.$$

By these two reactions water is decomposed and two moles of hydrogen are produced per mole of oxygen. Since the resistance of such a cell is extremely high, it is customary to add a small amount of acid to the water. If hydrochloric acid is added, the process now becomes the electrolysis of hydrochloric acid solution. It is therefore pertinent to consider what modification has been introduced by the presence of chloride ions and of additional hydrogen ions. The reaction at the cathode is evidently the same as before, since only hydrogen ions are involved. Both hydroxyl ions (OH^{-}) and chloride ions are present at the anode. If the concentration of chloride ions is small compared to the hydroxyl ions, their decomposition potential is sufficiently greater than that of hydroxyl that only oxygen is liberated. However, even when this is the case for a short time at the beginning of the reaction, the gradual migration of chloride ions to the anode causes an increase in chloride concentration at this point and soon chlorine begins to appear. For this reason sulfuric rather than hydrochloric acid usually is added in the electrolysis of water, since sulfate ions are much less readily decomposed.

6-15 Chemical free energy. It is often of great importance in chemistry to determine whether or not a chemical reaction will occur when certain reactants are placed together. A very convenient method exists for determining the reacting tendency by measuring the amount of electrical energy released when the reaction is conducted in an electrical cell under reversible equilibrium conditions. In the language of thermodynamics this energy is termed the "free energy change" accompanying the reaction. By this method one may determine whether a system of a 1 molal solution of copper sulfate, metallic copper, metallic zinc, and a 1 molal solution of zinc sulfate, will react, causing zinc to be converted into zinc sulfate and copper sulfate to be converted to copper. This can be investigated by constructing a large "Daniell cell" containing 1 molal solutions of zinc and copper sulfate. The cell should be so large that the passage into solution at one electrode of one mole of ions and of a corresponding amount out of solution at the other will not materially affect the concentrations. One may then allow one mole of ions to pass under reversible conditions and determine the amount of electrical work which has been done. This will determine the free energy per mole which is available in the reaction. If zinc is allowed to go into solution and copper allowed to precipitate as metallic copper, a positive amount of free energy is released. The magnitude of this free energy will be a measure of the tendency of the reaction to occur.

An additional problem is to determine the concentrations of copper sulfate and zinc sulfate for which the reaction is in equilibrium. These may be found by preparing a cell whose voltage is zero and determining the concentration of copper sulfate and zinc sulfate. It should be noted that although a reaction may have a large reacting tendency and a large amount of free energy may be released, it may often proceed at an infinitesimally slow rate. The present discussion may be regarded by the reader as a mere glimpse into a large and very fruitful field of chemical science.

6-16 Thermal emf's. The conversion of electrical energy to heat in a resistor is an irreversible process—there is no way of recovering the heat and converting it back to electrical energy. Circuit elements do exist, however, in which heat is developed at the expense of electrical energy when the current is in one direction, while electrical energy is developed at the expense of heat when the current is in the other direction. In other words, if this phenomenon alone were present, it would be possible to cool the circuit by passing a current through it. A circuit or portion of a circuit in which heat is converted to electrical energy or vice versa (exclusive of i^2R heating) is called a seat of *thermal electromotive force.*

6-17 Thomson emf. A complete explanation of the theory of thermal emf's is beyond the scope of this text. About all that can be done is to make it appear reasonable that the effects should exist. Suppose that a conducting rod is supported by insulators and that one end of the rod is heated as in Fig. 6-7(a) so that there is a temperature gradient along the rod. If we picture the free electrons in a rod as analogous to a gas, we might expect that the higher temperature at one end would result in an increased pressure and a diffusion of the electron gas away from the

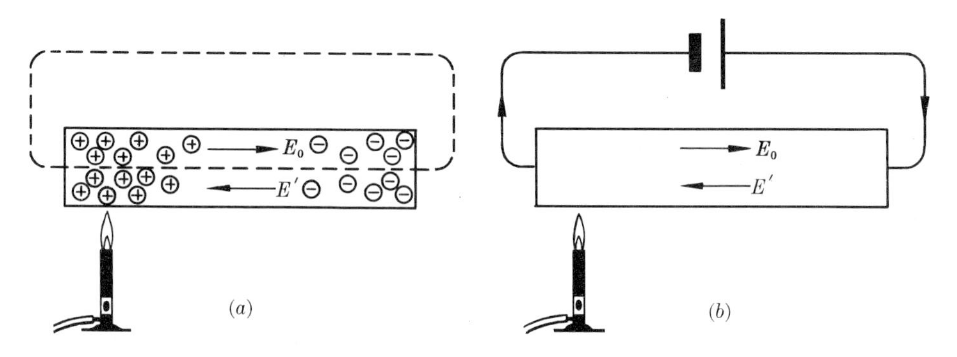

FIG. 6-7. The Thomson effect.

heated end. The accumulation of negative charge at the cooler end would set up an electric field within the rod in the direction shown by the vector E_0, and diffusion would cease when the force on the electrons, due to this field, was just great enough to balance their tendency to diffuse because of the temperature difference. It is an experimental fact that an electric field is set up in a conducting rod when there is a temperature difference between its ends, but the unsatisfactory nature of the simple theory outlined above becomes evident when one discovers that in some conductors the electric intensity is directed from the hot end toward the cold end, while in others it is in the opposite direction.

When the steady state has been established in the rod the electrons within it are in equilibrium. The electrostatic field exerts a force on the electrons in one direction and we may consider that the temperature differences give rise to an equivalent field which urges them in the opposite direction. The emf of a circuit, it will be recalled, can be defined as the line integral of the resultant field around the circuit. Since the line integral of the electrostatic field around any closed path is zero, the line integral of the resultant field around a closed path such as the dotted line in Fig. 6-7(a) reduces to the integral of the equivalent field along the rod, since this equivalent field exists only within the rod and is zero outside. Let us represent the equivalent field by E'. The magnitude of E' at any point of the rod depends on the temperature gradient at the point, dT/dl, (T is the Kelvin temperature) and may be written

$$E' = \sigma \frac{dT}{dl}. \tag{6-2}$$

The coefficient σ, known as the Thomson coefficient after Sir William Thomson (later Lord Kelvin) who was the first to investigate this type of emf, is characteristic of the material of the conductor. It is a function of temperature. The emf of the rod, $\mathcal{E}$, is then

$$\begin{aligned}\mathcal{E} &= \int E'dl \\ &= \int \sigma \frac{dT}{dl}\,dl \\ &= \int_{T_1}^{T_2} \sigma\,dT, \qquad (6\text{-}3)\end{aligned}$$

where T_1 and T_2 are the temperatures at the ends of the rod. The emf is called a *Thomson electromotive force.*

The existence of a Thomson emf may be demonstrated by sending a current through the rod as in Fig. 6-7(b). When the current and the emf are in opposite directions, electrical energy is converted to heat at a rate $\mathcal{E}i$. When the current is in the same direction as the emf, heat is converted to electrical energy. The Ri^2 rate of development of heat is therefore slightly augmented in the first case and slightly reduced in the second. Thomson emf's are never large, but are of the order of a few millivolts.

Obviously, a single Thomson emf can never be utilized to drive a current around a closed circuit. If a metal ring is heated at one point of its periphery and cooled at another point, equal and opposite emf's are set up in the two paths from the heated point to the cooled point. If different metals are used in the two paths from the heated to the cooled point, the Thomson emf's will, in general, not be equal but, as we shall show in the next section, other emf's develop at the junctions of the unlike metals.

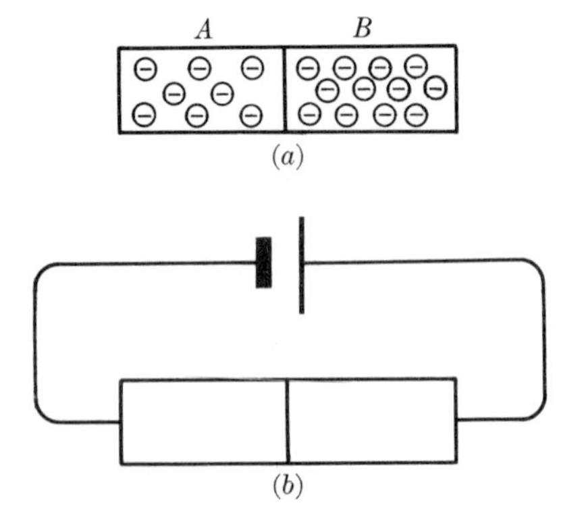

FIG. 6-8. The Peltier effect.

6-18 Peltier emf. When two unlike conductors at the same temperature are brought into contact as in Fig. 6-8(a), there will be a "diffusion" of electrons from one into the other unless the "electron gas" within each is at the same "pressure." (This theoretical explanation, like that of the Thomson emf, must not be taken too literally.)

Rearrangement of the electrons continues until at the junction an electric field is established of sufficient magnitude to bring about equilibrium. The junction of the metals then becomes a seat of emf called a *Peltier electromotive force* after Jean C. Peltier (1785–1845). The Peltier emf depends on the two metals and on the temperature of the junction between them. For two metals A and B, at a temperature T, we shall write the Peltier emf as

$$(\pi_{AB})_T.$$

Peltier emf's, like Thomson emf's, are of the order of millivolts. Their existence can also be shown by sending a current across a junction as in Fig. 6-8(b), either with or against the emf. With the metals antimony and bismuth, for which the Peltier emf is unusually large, it is not difficult to obtain sufficient cooling, when current and emf are in the same direction, to more than offset the Ri^2 heating and actually cause a lowering in temperature at the junction.

6-19 Seebeck emf. Consider next a closed circuit formed by two different metals A and B as in Fig. 6-9. The two junctions are at the temperatures T_1 and T_2. There is necessarily a temperature gradient in both metals, so that a Thomson emf is set up in each. Also, since the junctions are at different temperatures, the Peltier emf's at the junctions are unequal. The net emf in the circuit is then the algebraic sum of the two Thomson emf's and the two Peltier emf's. The net emf is, in general, not zero and hence a current will exist in the circuit as long as the junctions are kept at different temperatures. This phenomenon was first observed by Thomas J. Seebeck (1770–1831) in 1826, and the net emf is called a *Seebeck electromotive force.* If the individual emf's have the directions shown in Fig. 6-9 and the clockwise direction is considered positive, the Seebeck emf is

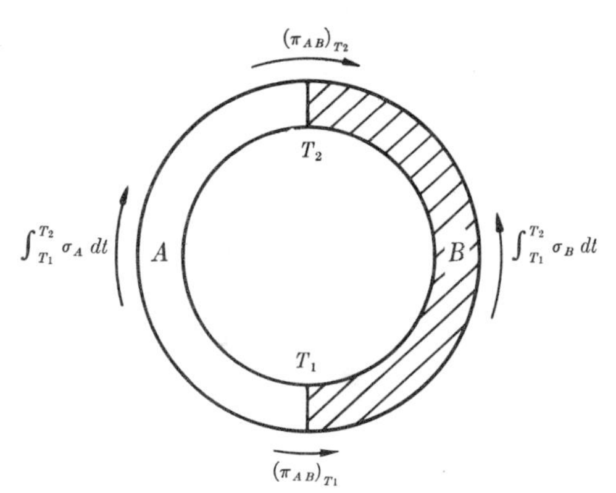

FIG. 6-9. Peltier and Thomson effects in a thermocouple.

$$\mathcal{E}_{AB} = (\pi_{AB})_{T_2} - (\pi_{AB})_{T_1} + \int_{T_1}^{T_2} (\sigma_A - \sigma_B)dT. \tag{6-4}$$

The circuit in Fig. 6-9 is called a *thermocouple.*

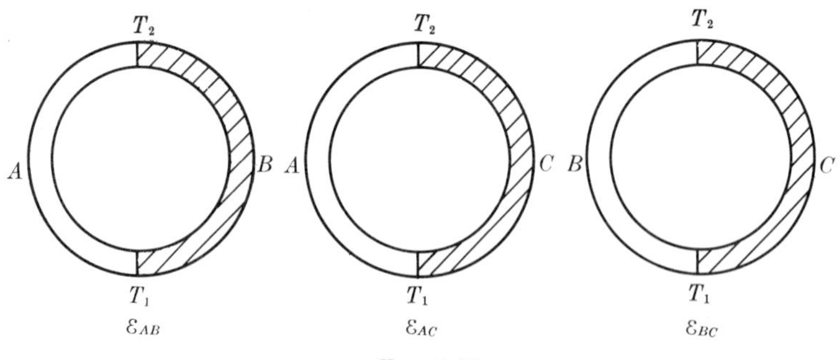

FIG. 6-10.

The emf of a thermocouple, even with large temperature differences between the junctions, is never very large and the effect is not a practical means of obtaining electrical energy directly from heat. Thermocouples are of great practical value, however, as a means of measuring temperature. The emf of the couple depends on the temperature of the junctions. Hence if the emf is measured with one junction at a known temperature, the temperature of the other junction can be found. The temperature range over which thermocouples can be used extends from the lowest attainable temperature to the melting point of the metals forming the couple.

For a given pair of junction temperatures, the emf of a circuit composed of metals A amd B ($\mathcal{E}_{AB}$) equals the difference between the emf of a circuit composed of metals A and C ($\mathcal{E}_{AC}$) and the emf of a circuit composed of metals B and C ($\mathcal{E}_{BC}$). (See Fig. 6-10.) That is,

$$\mathcal{E}_{AB} = \mathcal{E}_{AC} - \mathcal{E}_{BC}. \tag{6-5}$$

Hence if some reference metal C is agreed upon, the emf's of all others can be tabulated relative to this metal and the emf of any other pair can be found by subtraction. Lead is usually taken as the reference metal.

An *intermediate metal D* may be inserted in a thermocouple circuit as in Fig. 6-11(a) without affecting the

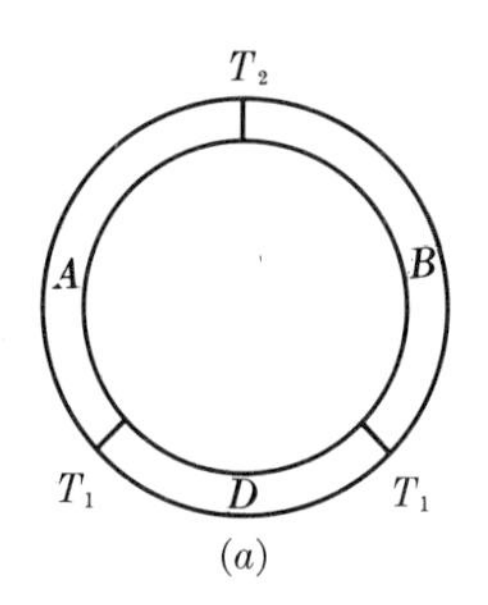

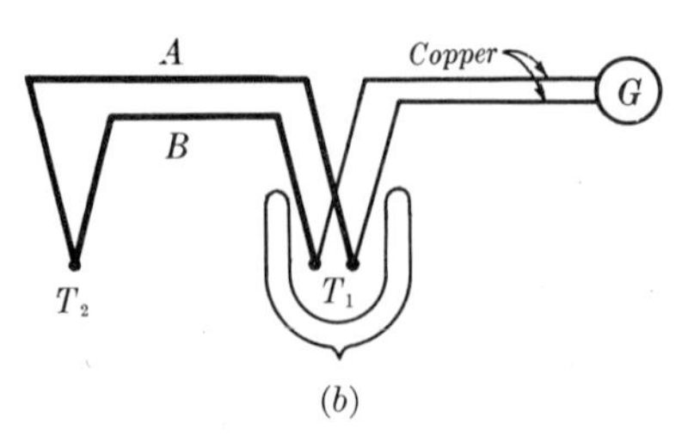

FIG. 6-11. Insertion of an intermediate metal in a thermocouple circuit.

emf, provided both ends of D are at the same temperature. That is, the emf of such a circuit is the same as that of one composed of metals A and B only, with junctions at temperatures T_1 and T_2.

The usual thermocouple circuit, shown in Fig. 6-11(b) makes use of this fact. Metals A and B might be nickel and iron, connected by copper wires to a galvanometer G at any convenient location. One junction is at the temperature T_2 to be measured, the other at temperature T_1, which might be maintained by melting ice in a Dewar flask. The emf is that of a nickel-iron couple with junctions at temperatures T_2 and T_1.

The emf of a thermocouple may be measured as in Fig. 6-11(b) by inserting a galvanometer or voltmeter at any convenient point. Such an instrument, of course, indicates only the potential difference between its own terminals, which is smaller than the emf of the couple by the "ir" drop in the circuit. Hence a high resistance instrument or, preferably, a potentiometer should be used in thermocouple circuits.

6-20 Dependence of emf on temperature. If the temperature dependence of the Thomson coefficients and the Peltier emf's is known, the net emf of a thermocouple can be computed from Eq. (6-4). In practice, the emf is usually measured over a range of junction temperatures and expressed by an empirical equation. Temperatures, for convenience, are expressed on the Centigrade scale and will be represented by t. If one junction of a thermocouple constructed of metals A and B is at a temperature t while the other is at a temperature t_0, the emf developed by the couple can be expressed to a good approximation by the second degree equation

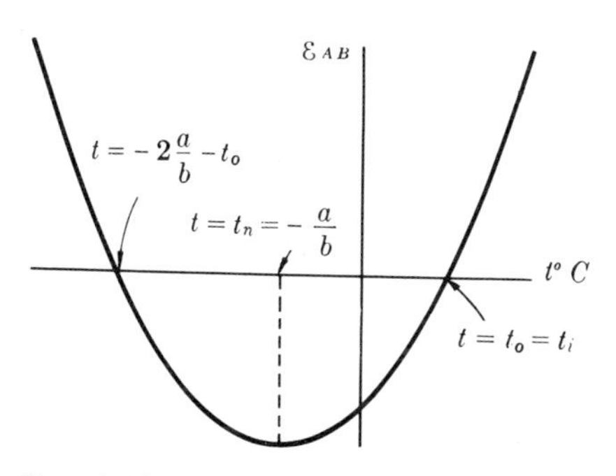

Fig. 6-12. Emf vs. hot junction temperature t with cold junction at temperature t_0.

$$\mathcal{E}_{AB} = a_{AB}(t - t_0) + \tfrac{1}{2}b_{AB}(t^2 - t_0^2), \tag{6-6}$$

where the coefficients a_{AB} and b_{AB} are characteristic of the metals A and B. (For more precise calculations a term in t^3 can be included.) Table 6-2 lists the values of a and b for a number of metals and alloys relative to lead. It follows from Eq. (6-5) that

$$a_{\text{AB}} = a_{\text{A-lead}} - a_{\text{B-lead}}, \quad b_{\text{AB}} = b_{\text{A-lead}} - b_{\text{B-lead}}.$$

A graph of $\mathcal{E}$ vs t, when both a_{AB} and b_{AB} are positive, and when t_0 is kept constant, is evidently a parabola as illustrated in Fig. 6-12. For no good reason, the rate of change of emf with temperature, $d\mathcal{E}/dt$, is commonly called the *thermoelectric power* of the couple. The temperature t_n at which $d\mathcal{E}/dt = 0$ is called the *neutral temperature*. From Eq. (6-6), when t_0 is constant,

$$\frac{d\mathcal{E}_{AB}}{dt} = a_{AB} + b_{AB}\,t,$$

and since

$$d\mathcal{E}/dt = 0 \text{ when } t = t_n,$$

$$t_n = -\frac{a_{AB}}{b_{AB}}.$$

TABLE 6-2

THERMOELECTRIC COEFFICIENTS

Reference metal—lead. $\mathcal{E}_{AB}$ in microvolts. When $\mathcal{E}_{AB}$ is positive, the direction of the (conventional) current is from metal A to metal B at the hot junction.

Substance	a	b
Antimony	+35.6	+0.145
Bismuth	−74.4	+0.032
Constantan (60% Cu, 40% Ni)	−38.1	−0.0888
Copper	+ 2.71	+0.0079
Iron	+16.7	−0.0297
Nickel	−19.1	−3.02
Platinum	− 3.03	−3.25

NOTE: These figures should be used with caution. Small amounts of impurities, or variations in heat treatment, have a pronounced effect on the thermoelectric coefficients and the values given should be considered as representative only.

The emf is zero at the temperatures

$$t = t_0 \quad \text{and} \quad t = -\frac{2a_{AB}}{b_{AB}} - t_0,$$

which are the two roots of Eq. (6-6) when $\mathcal{E}_{AB} = 0$. The second of these temperatures is called the *inversion temperature*, t_i.

For most of the metals or alloys commonly used in thermocouple circuits, the operating range is sufficiently far removed from the neutral point so that the relation between emf and hot junction temperature, for a constant temperature of the cold junction, is essentially linear.

Example. One junction of an iron-copper thermocouple is kept at 0° C. Find the neutral temperature, the inversion temperature, and the emf when the other junction is at 200° C. Consider the iron as metal A, the copper as metal B.

From Table 6-2,

$$a_{\text{iron-lead}} = 16.7, \quad b_{\text{iron-lead}} = -.0297$$
$$a_{\text{copper-lead}} = 2.71, \quad b_{\text{copper-lead}} = .0079$$

Hence

$$a_{\text{iron-copper}} = 16.7 - 2.71 = 14.0$$
$$b_{\text{iron-copper}} = -.0297 - .0079 = -.0376$$

The neutral temperature is

$$t_n = -\frac{a}{b} = -\frac{14.0}{-.0376} = 372° \text{ C}.$$

The inversion temperature is

$$t_i = -\frac{2a}{b} = 744° \text{ C}.$$

The general expression for the emf is

$$\mathcal{E}_{AB} = 14.0(t - t_0) - \tfrac{1}{2} \times .0376(t^2 - t_0^2),$$

and when $t = 200°$ C, $t_0 = 0°$ C,

$$\begin{aligned}\mathcal{E}_{AB} &= 14.0 \times 200 - \tfrac{1}{2} \times .0376 \times (200)^2 \\ &= 2050\ \mu\text{v} = 2.05 \text{ mv}.\end{aligned}$$

Since $\mathcal{E}_{AB}$ is positive, the direction of the (conventional) current is from iron (metal A) to copper (metal B) at the hot junction.

Theoretical reasoning based on the second law of thermodynamics shows that the Seebeck, Peltier, and Thomson emf's are related by the following equations:

$$(\pi_{AB})_T = T\frac{d\mathcal{E}_{AB}}{dT}, \tag{6-7}$$

$$(\sigma_A - \sigma_B) = -T\frac{d^2\mathcal{E}_{AB}}{dT^2}. \tag{6-8}$$

Example. Compute the Peltier and Thomson emf's of an iron-copper thermocouple whose junctions are at 200° C and 0° C.

In terms of Centigrade temperatures, Eq. (6-7) becomes

$$\begin{aligned}(\pi_{AB})_t &= (t + 273)\left(\frac{d\mathcal{E}_{AB}}{dt}\right) \\ &= (t + 273)(a_{AB} + b_{AB}t).\end{aligned}$$

Hence, using the values of a_{AB} and b_{AB} from the preceding example, we obtain for the Peltier emfs at the junctions,

$$(\pi_{\text{iron-copper}})_{100°\text{C}} = 473 \times (14.0 - 0.0376 \times 200) = 3070\ \mu\text{v},$$

$$(\pi_{\text{iron-copper}})_{0°\text{C}} = 273 \times 14.0 = 3820\ \mu\text{v}.$$

The Thomson emf may be found as follows. Since

$$\frac{d^2\mathcal{E}}{dT^2} = \frac{d^2\mathcal{E}}{dt^2} = \frac{d}{dt}(a + bt) = b,$$

we have

$$\text{Thomson emf} = \int_{T_1}^{T_2} (\sigma_A - \sigma_B)\, dT = \int_{T_1}^{T_2} -T\frac{d^2\mathcal{E}}{dT^2}\, dT$$

$$= \int_{T_1}^{T_2} -bT\, dT = \frac{b}{2}(T_1^2 - T_2^2)$$

$$= \frac{-.0376}{2}(273^2 - 473^2)$$

$$= 2800\ \mu\text{v}.$$

The net or Seebeck emf is therefore

$$\mathcal{E} = 3070 - 3820 + 2800 = 2050\ \mu\text{v},$$

which agrees with the result in the preceding example.

Problems—Chapter 6

(1) Refer to the example in Sec. 6-20. (a) Compute the emf of an iron-copper thermocouple when one junction is kept at 0°C and the other is at the following temperatures: −200°C, 0°C, 372°C, 744°C, 1000°C. (b) Repeat when one junction is kept at 200°C. (c) Construct a graph of the answers to parts (a) and (b). Could the junction actually be used to measure a temperature of 1000°C?

(2) One junction of a copper-constantan thermocouple is at 20°C and the other is at 300°C. (a) Compute the emf. (b) Is the emf the same if one junction is at 120°C and the other at 400°C?

(3) Compute the emf of a bismuth-antimony thermocouple when one junction is at 0°C and the other is at 100°C. Compare with the emf of an iron-copper couple whose junctions are at the same temperatures.

(4) (a) Compute the Peltier emf at an antimony-bismuth junction at a temperature of 20°C. (b) If a current of 20 amp is sent across the junction, find the rate at which thermal energy is converted to electrical energy, or vice versa. (c) What should be the direction of the current at the junction in order to produce a cooling effect?

CHAPTER 7

PROPERTIES OF DIELECTRICS

7-1 Induced Charges. When an uncharged body of any sort, a conductor or a dielectric, is brought into an electric field, a rearrangement of the charges in the body always results. If the body is a conductor, the free electrons within it move in such a way as to make the interior of the body a field-free, equipotential volume. If the body is a nonconductor, the electrons and the positive nuclei in each molecule are displaced by the field, but since they are not free to move indefinitely the interior of the body does not become an equipotential region. The net charge of the body in either case remains zero (the conductor is assumed to be insulated from other bodies from which it might acquire a charge) but certain regions of the body acquire excess positive or negative charges called *induced charges*. In this chapter we shall be concerned chiefly with the phenomena in a dielectric when it is in an external field, but by way of introduction let us consider first the charge distribution on an originally uncharged conductor in the form of a flat slab or sheet when it is inserted in the field between two plane parallel conductors having equal and opposite charges. If fringing effects are neglected, the field is uniform in the region between the charged plates, as in Fig. 7-1(a).

In Fig. 7-1(b) an uncharged conductor has been inserted in the field, without touching either of the charged plates. The free charges in the

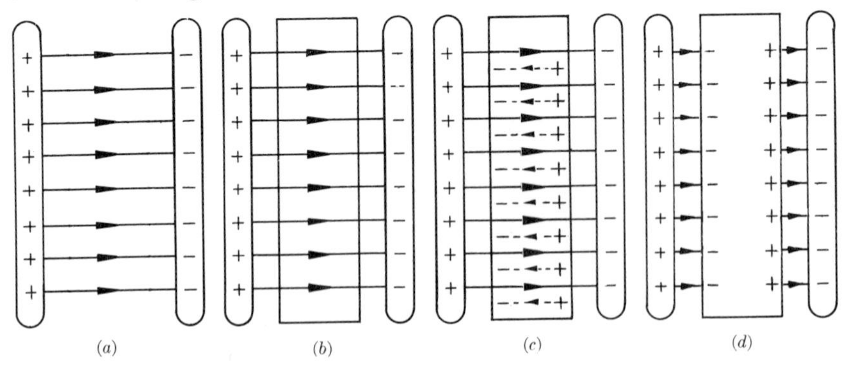

FIG. 7-1. (a) Electric field between two charged plates. (b) Introduction of a conductor. (c) Induced charges and their field. (d) Resultant field when a conductor is between two charged plates.

conductor, of course, immediately rearrange themselves as soon as the conductor is placed in the field, but let us assume for the moment that they do not. The field then penetrates the conductor. Under the influence of this field, the free electrons in the conductor move toward its left surface, leaving a positive charge on its right surface. This motion continues until at all points within the conductor the field set up by the layers of surface charge is equal and opposite to the original field. The motion of charge then ceases. The excess charges at the surfaces of the conductor are called induced charges. The net charge on the conductor remains zero.

The field set up by the induced charges is shown by dotted lines in Fig. 7-1(c). The resultant field is shown in Fig. 7-1(d). Inside the conductor the field is everywhere zero. In the gap between the conductor and the plates the field is the same as it was before the conductor was inserted. All of the lines of force that originate on the positive plate terminate on induced charges on the left face of the conductor. An equal number of lines originate at the induced positive charges on the right face of the conductor and terminate on the negative charges on the other plate. The induced charges at the faces of the conductor are equal and opposite in sign to the original charges on the plates, and, as far as the interior of the conductor is concerned, they effectively neutralize the charges on the plates. Hence the field in the conductor is zero.

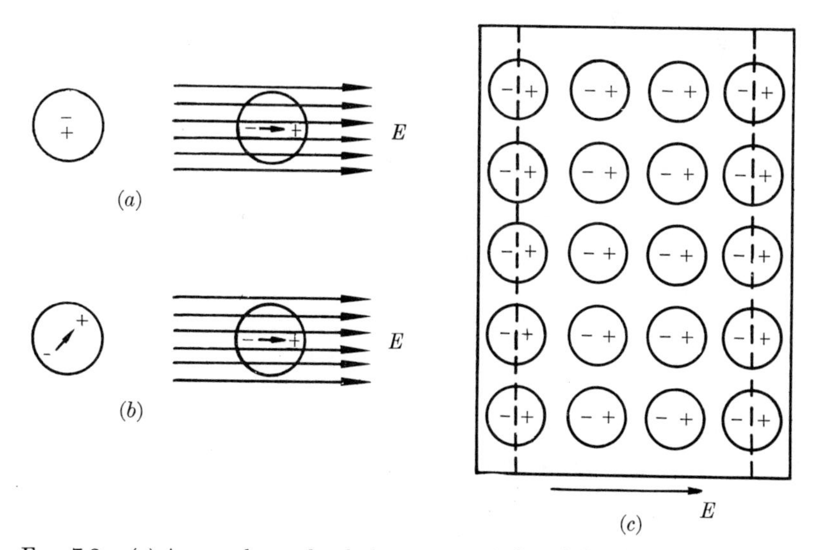

Fig. 7-2. (a) A nonpolar molecule becomes an induced dipole in an external field. (b) A polar molecule or permanent dipole is oriented in the direction of an external field. (c) Polarized molecules of a dielectric in an external field E directed from left to right.

Consider the behaviour of a dielectric in the same electric field. For our present purposes the molecules of a dielectric may be classified as either *polar* or *nonpolar*. A nonpolar molecule is one in which the "centers of gravity" of the protons and electrons normally coincide, while a polar molecule is one in which they do not. Under the influence of an electric field the charges of a nonpolar molecule become displaced as in Fig. 7-2(a). The molecule is said to become *polarized* by the field and is called an *induced dipole*, of dipole moment (see Sec. 2-3) equal to the product of either charge and the distance between them. The effect of an electric field on a polar molecule is to orient it in the direction of the field as in Fig. 7-2(b). The dipole moment may also be increased by the field. A polar molecule is called a *permanent dipole*.

When a nonpolar molecule becomes polarized, restoring forces come into play on the displaced charges. These are the interparticle binding forces which hold the molecule together. These forces are, in part at least, of electrical origin but, whatever their origin, we may think of them as elastic restoring forces pulling the displaced charges together much as if they were connected by a spring. Under the influence of a given external field the charges separate until the binding force is equal and opposite to the force exerted on the charges by the field. Naturally the binding forces vary in magnitude from one kind of molecule to another, with corresponding differences in the dipole moments developed by a given field.

Whether the polarization is induced or due to the alignment of permanent dipoles, the arrangement of charges within the molecules of a dielectric in an external field will be as shown in Fig. 7-2(c). The entire dielectric, as well as its individual molecules, is said to be polarized. Within the two extremely thin surface layers indicated by dotted lines there is an excess charge, negative in one layer and positive in the other. These layers constitute the induced surface charges. The charges are not free, however, but each is bound to an atom lying in or near the surface. Within the remainder of the dielectric the net charge remains zero. The internal state of a polarized dielectric is therefore characterized not by an excess charge but by the relative displacement of the charges within it.

The four parts of Fig. 7-3, which should be compared carefully with those of Fig. 7-1, illustrate the behaviour of a sheet of dielectric when inserted in the field between a pair of oppositely charged plane parallel plates. Fig. 7-3(a) shows the original field. Fig. 7-3(b) is the situation after the dielectric has been inserted but before any rearrangement of charges has occurred. Fig. 7-3(c) shows by dotted lines the field set up in the dielectric by its induced surface charges. As in Fig. 7-1(c), this field is opposite to the original field but, since the charges in the dielectric are

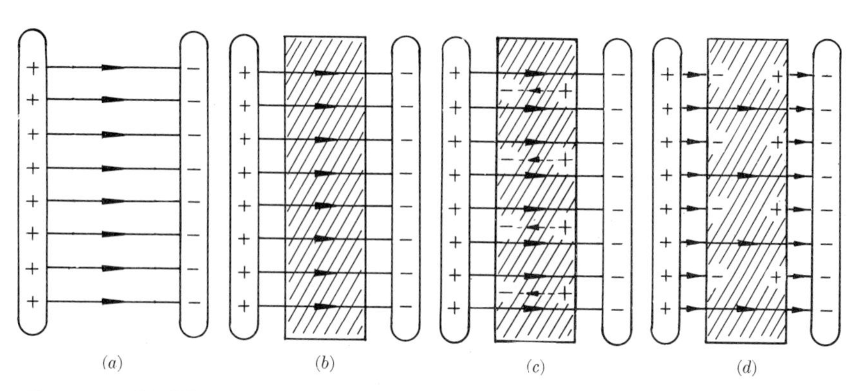

FIG. 7-3. (a) Electric field between two charged plates. (b) Introduction of a dielectric. (c) Induced surface charges and their field. (d) Resultant field when a dielectric is between charged plates.

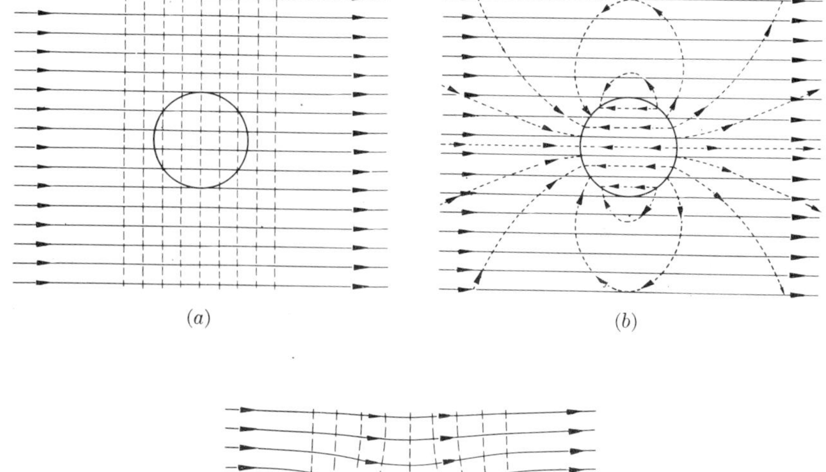

(c)

FIG. 7-4. (a) A conducting sphere in an external electric field. (b) The field due to induced charges is shown by dotted lines. (c) Resultant lines of force and equipotential surfaces of a conducting sphere in an electric field.

not free to move indefinitely, their displacement does not proceed to such an extent that the induced field is equal in magnitude to the original field. The field in the dielectric is therefore *weakened*, but not reduced to zero.

The resultant field is shown in Fig. 7-3(d). Some of the lines of force leaving the positive plate penetrate the dielectric; others terminate on the induced charges on the faces of the dielectric.

7-2 Induced charges on spheres. It is of some interest to consider the induced charges on a spherical conductor or insulator when inserted in an originally uniform field. The conducting sphere is illustrated in Fig. 7-4. As in Fig. 7-1, the free charges within the sphere rearrange themselves in such a way as to make the field zero at internal points. The field of the induced charges is shown by dotted lines in Fig. 7-4(b) and the resultant of this field and the original field in Fig. 7-4(c). Traces of a few equipotential surfaces, of which the surface of the sphere is one, are also shown. Since lines of force and equipotential surfaces are orthogonal, the lines of force intersect the surface of the sphere at right angles.

The same physical principles are involved whatever the shape of a conductor, but the mathematical expressions for the field and charge distribution are extremely complex except for spheres and ellipsoids. Thus if two conducting spheres in contact are placed in a field, one acquires an excess positive and the other an excess negative charge, the charge distribution being such as to bring both spheres to the same potential and reduce the field within them to zero. If slightly separated while still in the field and then removed from it, the induced charges become "trapped" on the spheres and may readily be detected by an electroscope. See Fig. 7-5.

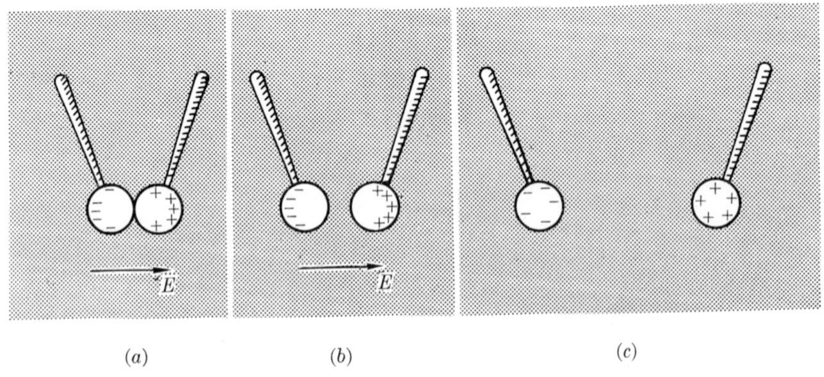

FIG. 7-5. (a) Conducting spheres in contact in a field. (b) Spheres slightly separated while still in the field. (c) Spheres are oppositely charged as field is removed.

A dielectric sphere in an originally uniform field is shown in Fig. 7-6. As in Fig. 7-3, the induced surface charges weaken the field in the sphere but do not reduce it to zero. The field extends inside as well as outside the dielectric sphere. Hence the surface of the sphere is not an equipotential and the lines of force do not intersect it at right angles.

The charges induced on the surface of a dielectric sphere in an external field afford an explanation of the attraction of an *uncharged* pith ball or bit of paper by a charged rod of rubber or glass. In Fig. 7-6, where the external field is uniform, the net *force* on the sphere is zero since the forces on the positive and negative induced charges are equal and opposite. However, if the field is *non-uniform* the induced charges are in regions where the electric intensity is different and the force in one direction is not equal to that in the other.

Fig. 7-7 shows an uncharged dielectric sphere B in the radial field of a positive charge A. The induced positive charges on B experience a force toward the right while the force on the negative charges is toward

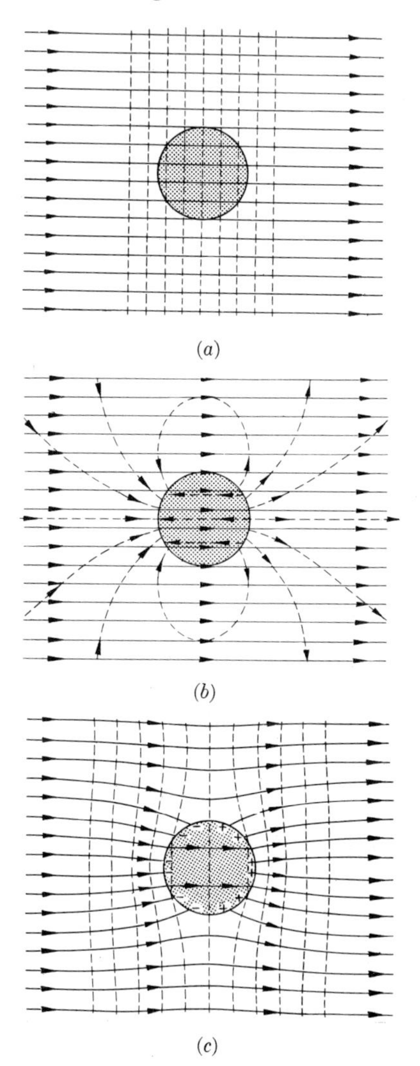
(*a*)

(*b*)

(*c*)

FIG. 7-6. A dielectric sphere in an originally uniform electric field.

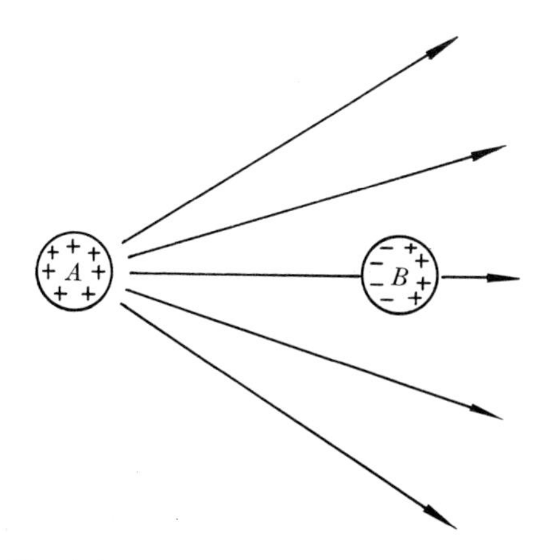

FIG. 7-7. An uncharged dielectric sphere B in the radial field of a positive charge A.

the left. Since the negative charges are closer to A and therefore in a stronger field than are the positive, the force toward the left exceeds that toward the right, and B, although its net charge is zero, experiences a resultant force toward A. The sign of A's charge does not affect the conclusion, as may readily be seen. Furthermore, the effect is not limited to dielectrics—a conducting sphere would be similarly attracted.

More general arguments based on energy considerations show that a dielectric body in a nonuniform field always experiences a force urging it from a region where the field is weak toward a region where it is stronger, provided the dielectric coefficient (see Sec. 7-3) of the body is greater than that of the medium in which it is immersed. If the dielectric coefficient is less, the reverse is true.

7-3 Susceptibility, dielectric coefficient, and permittivity. The field between a pair of oppositely charged plates and the induced charges on the surfaces of a dielectric adjacent to the plates are shown in Fig. 7-8. The dielectric is assumed to extend from one plate to the other, but in the diagram a small gap has been left between its surfaces and those of the plates for clarity. If edge effects are neglected, the surface density of the induced charges on the dielectric will be uniform by symmetry. Let σ_i represent the surface density of induced charge, and σ the surface density of charge on the plates. The former is often called the surface density of *bound* charge and the latter the surface density of *free* charge.

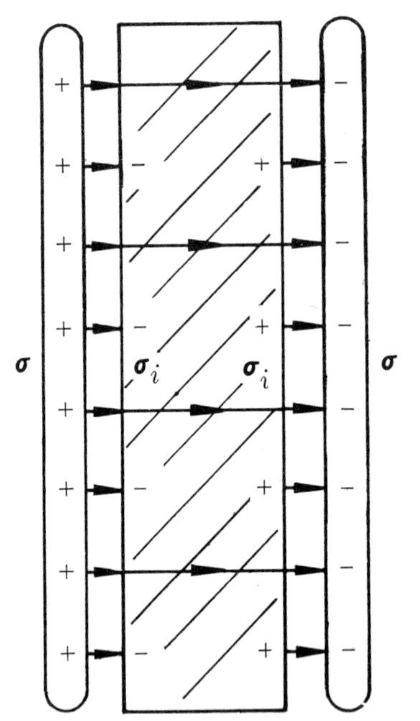

FIG. 7-8. Induced charges on the faces of a dielectric in an external field.

The induced charges neutralize a *part* of the free charges and reduce the effective surface density from σ to $\sigma - \sigma_i$. The resultant electric intensity within the dielectric is therefore

$$E = \frac{1}{\epsilon_0}(\sigma - \sigma_i) = \frac{1}{\epsilon_0}\sigma - \frac{1}{\epsilon_0}\sigma_i. \tag{7-1}$$

The term $\frac{1}{\epsilon_0}\sigma$ represents that component of the resultant field set up

TABLE 7-1

ELECTRIC SUSCEPTIBILITY η, DIELECTRIC COEFFICIENT K_e AND PERMITTIVITY ϵ
(From Handbook of Chemistry and Physics, 26th Ed.)

Material	η (coul²/newt-m.²)	$K_e = 1 + \frac{\eta}{\epsilon_0}$	$\epsilon = K_e\epsilon_0$ coul²/newt-m.²
Vacuum	0	1	$\frac{1}{36\pi \times 10^9} = 8.85 \times 10^{-12}$
Glass	$35 - 80 \times 10^{-12}$	5 – 10	$45 - 90 \times 10^{-12}$
Ice (−5° C)	17	2.9	26
Mica	18 – 45	3 – 6	27 – 54
Rubber	13 – 290	2.5 – 35	22 – 300
Sulfur	26	4	35
Wood	13 – 60	2.5 – 8	22 – 70
Alcohol, ethyl (0° C)	243×10^{-12}	28.4	252×10^{-12}
Alcohol, ethyl (−120° C)	474	54.6	483
Alcohol, ethyl (frozen)	14	2.7	23
Benzene (180° C)	11	2.3	20
Glycerine (15° C)	491	56	500
Petroleum	9	2	18
Water	708	81	717
Air (1 atm)		1.00059	
Air (100 atm)		1.0548	
CO_2 (1 atm)		1.000985	
CO_2 (40 atm)		1.060	
H_2 (1 atm)		1.000264	
Water vapor (4 atm)		1.00705	

by the free charges on the plates. The term $\frac{1}{\epsilon_0}\sigma_i$ is the reversed field set up by the induced charges.

Since the induced charges are brought about by the field E, their magnitude will depend on that of E and also on the material of the dielectric. The ratio of the induced charge density σ_i to the resultant electric intensity E is called the *electric susceptibility*[1] of the material and is represented by η.

$$\eta = \frac{\sigma_i}{E}, \quad \sigma_i = \eta E. \tag{7-2}$$

[1] But see Sec. 7-6 where a more general definition of electric susceptibility is given.

The greater the susceptibility of a material, the greater the induced charge in a given field. Experiment shows that at constant temperature and in fields that are not too great, the suceptibility of a given material is a constant, independent of E. That is, the surface density of induced charge is proportional to the resultant field. Some representative values of susceptibility are listed in the first column of Table 7-1. The "susceptibility of a vacuum" is a figure of speech. As we have defined susceptibility, this value must be zero, since in a vacuum there are no atoms whose charges can be displaced by an electric field. The mks units of suceptibility are those of surface density of charge divided by electric intensity, or

$$\frac{\text{coul}}{\text{m}^2} \div \frac{\text{newtons}}{\text{coul}} = \frac{\text{coul}^2}{\text{newton-m}^2}.$$

In terms of susceptibility, Eq. (7-1) becomes

$$E = \frac{1}{\epsilon_0}\sigma - \frac{1}{\epsilon_0}\eta E,$$

or

$$E = \frac{\sigma}{\left(1 + \frac{\eta}{\epsilon_0}\right)\epsilon_0}. \tag{7-3}$$

Let us represent the term $\left(1 + \frac{\eta}{\epsilon_0}\right)$ by the symbol K_e.

$$\boxed{K_e = 1 + \frac{\eta}{\epsilon_0}.} \tag{7-4}$$

Eq. (7-3) then becomes

$$E = \frac{1}{K_e\epsilon_0}\,\sigma. \tag{7-5}$$

The quantity K_e is called the *dielectric coefficient* of the material. Eq. (7-4) may be considered its definition. This quantity is also known by other names, and often is represented by other symbols. Some writers refer to it as the *specific inductive capacity* or the *dielectric constant*, and the symbol ϵ is often used for it. To speak of it as a "constant" is somewhat of a misnomer, since like η it may be a function of temperature and of electric intensity.

The dielectric coefficient of a material is a pure number, since η/ϵ_0 is a pure number. (The mks units of both η and ϵ_0 are coul^2/newton-m^2.) Some representative values are listed in the second column of Table 7–1. In empty space, where $\eta = 0$, $K_e = 1$.

Eqs. (7-1) and (7-5) are entirely equivalent in physical content. The former expresses the resultant field E in the dielectric as the *difference* between the field of the free charges and the field of the bound charges. The latter expresses the field as a *fraction* of the field of the free charges.

While in this special case (a slab of dielectric between a pair of parallel plates) the electric intensity within the dielectric is $\frac{1}{K_e}$ th of its value before the dielectric was inserted, the same proportional reduction in the field does not result in all instances. For example, if a dielectric sphere is inserted in a uniform field as in Fig. 7-6, the ratio of the resultant field within the sphere to the original field is

$$\frac{3}{K_e + 2}.$$

The product $K_e\epsilon_0$ which occurs in the denominator of Eq. (7-5) is called the *permittivity* of the dielectric and is represented by ϵ.

$$\boxed{\epsilon = K_e\epsilon_0.} \tag{7-6}$$

In empty space, where $K_e = 1$,

$$\epsilon = \epsilon_0.$$

The quantity ϵ_0 may therefore be described as "the permittivity of empty space," or "the permittivity of a vacuum." The term is unfortunate, because physically the quantities electric susceptibility η, dielectric coefficient $K_e = 1 + \frac{\eta}{\epsilon_0}$, and permittivity $\epsilon = K_e\epsilon_0$, are various ways of describing the relative displacement or orientation of the positive and negative charges within a substance when it is in an external electric field. Such effects cannot, of course, occur in empty space and the phrase "the permittivity of a vacuum" is at best a figure of speech. An improved terminology and notation in this part of the subject are both greatly to be desired.

The units of ϵ and ϵ_0 are evidently the same, coul^2/newton-m^2, since K_e is a pure number. Some representative values are listed in the third column of Table 7-1.

The dielectric properties of a material are completely specified if any one of the three quantities, electric susceptibility η, dielectric coefficient K_e, or permittivity ϵ, are known. The three are related by the equations

$$K_e = 1 + \frac{\eta}{\epsilon_0} = \frac{\epsilon}{\epsilon_0}$$

$$\epsilon = \epsilon_0 K_e = \epsilon_0 + \eta. \tag{7-7}$$

$$\eta = \epsilon_0(K_e - 1) = \epsilon - \epsilon_0$$

The only reason for introducing all three quantities is to simplify the form of certain common equations.

Examples: (1) The electric susceptibility of a material is 35.4×10^{-12} coul²/newton-m². What are the values of the dielectric coefficient and the permittivity of the material?

$$\eta = 35.4 \times 10^{-12} \text{ coul}^2/\text{newton-m}^2.$$

$$\text{Dielectric coefficient } K_e = 1 + \frac{\eta}{\epsilon_0}$$

$$= 1 + \frac{35.4 \times 10^{-12}}{8.85 \times 10^{-12}}$$

$$= 1 + 4$$

$$= 5 \text{ (no units).}$$

$$\text{Permittivity } \epsilon = K_e\epsilon_0$$

$$= 5 \times 8.85 \times 10^{-12}$$

$$= 44.3 \times 10^{-12} \text{ coul}^2/\text{newton-m}^2.$$

(2) Two parallel plates of area 1 m² are given equal and opposite charges of 30 microcoulombs each. A sheet of dielectric of permittivity 15×10^{-12} coul²/newton-m² occupies the space between the plates. Compute (a) the resultant electric intensity in the dielectric, (b) the surface density of induced charge on the faces of the dielectric, (c) the component of the resultant electric intensity in the dielectric due to the free charge, (d) the component due to the induced charge.

(a) The surface density of free charge is

$$\sigma = \frac{q}{A} = 30 \times 10^{-6} \text{ coul/m}^2.$$

The resultant electric intensity is

$$E = \frac{1}{K_e\epsilon_0}\sigma = \frac{1}{\epsilon}\sigma$$

$$= \frac{1}{15 \times 10^{-12}} \times 30 \times 10^{-6}$$

$$= 2 \times 10^6 \text{ volts/m.}$$

(b) The surface density of induced charge is

$$\begin{aligned}\sigma_i &= \eta E = (\epsilon - \epsilon_0)E \\ &= (15 \times 10^{-12} - 8.85 \times 10^{-12}) \times 2 \times 10^6 \\ &= 12.3 \times 10^{-6} \text{ coul/m}^2.\end{aligned}$$

(c) The component of the resultant field due to the free charge is

$$\begin{aligned}E_0 &= \frac{1}{\epsilon_0}\sigma \\ &= \frac{1}{8.85 \times 10^{-12}} \times 30 \times 10^{-6} \\ &= 3.39 \times 10^6 \text{ volts/m}.\end{aligned}$$

(d) The component due to the induced charge is

$$\begin{aligned}E_i &= \frac{1}{\epsilon_0}\sigma_i \\ &= \frac{1}{8.85 \times 10^{-12}} \times 12.3 \times 10^{-6} \\ &= 1.39 \times 10^6 \text{ volts/m}.\end{aligned}$$

The direction is opposite to that of E_0.
The resultant electric intensity is

$$\begin{aligned}E &= E_0 - E_i \\ &= 3.39 \times 10^6 - 1.39 \times 10^6 \\ &= 2 \times 10^6 \text{ volts/m},\end{aligned}$$

which agrees with the answer to part (a).

7-4 Extension of Gauss's law. Displacement. In Sec. 2-6 we showed that the number of lines of force passing outward through any closed surface in empty space equals $\frac{1}{\epsilon_0}$ times the quantity of charge within the surface (Gauss's law). In mathematical language,

$$\int E \cos \phi \, dA = \frac{1}{\epsilon_0} q. \quad (7\text{-}8)$$

We now wish to generalize this law when all or a portion of a closed surface cuts through the material of a dielectric. Fig. 7-9 shows a slab of

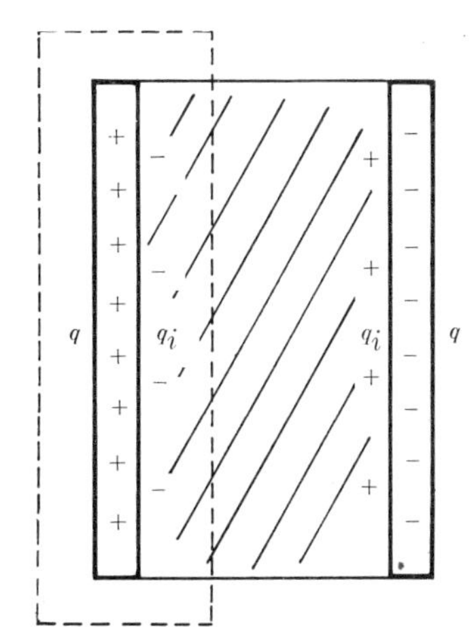

FIG. 7-9. A slab of dielectric between oppositely charged plates. The dotted line represents a gaussian surface.

dielectric between a pair of oppositely charged plates, and a gaussian surface, indicated by dotted lines, a portion of which cuts through the dielectric. Let σ represent the surface density of charge on the plates and σ_i the surface density of induced charge on the faces of the dielectric. The corresponding quantities of charge are $q = \sigma A$ and $q_i = \sigma_i A$, where A is the area of the plates. The field within the dielectric is

$$E = \frac{1}{\epsilon_0}(\sigma - \sigma_i) = \frac{1}{\epsilon_0}\frac{1}{A}(q - q_i).$$

Hence

$$EA = \frac{1}{\epsilon_0}(q - q_i). \tag{7-9}$$

But the product EA is the number of lines of force passing outward through the gaussian surface, since the field is zero at all points of the surface outside the dielectric. The term $(q - q_i)$ is the *net* charge enclosed by the surface, that is, the algebraic sum of the free and induced charges. Hence in Eq. (7-8), the charge q must be interpreted as *the algebraic sum of both the free and induced charges enclosed by the surface.* To bring out this meaning explicitly, we shall write Eq. (7-8) in the form

$$\boxed{\int E \cos\phi \, dA = \frac{1}{\epsilon_0}(q + q_i).} \tag{7-10}$$

(Both q and q_i are written as positive quantities in the general equation, but it is implied that we must take their *algebraic* sum. Hence if q_i is actually a *negative* charge, as in Fig. 7-9, the *algebraic sum* $(q + q_i)$ becomes the *difference* between the magnitudes of q and q_i as in Eq. (7-9).

On the other hand, from Eq. (7-5),

$$E = \frac{1}{K_e}\frac{1}{\epsilon_0}\sigma = \frac{1}{\epsilon}\sigma = \frac{1}{\epsilon}\frac{q}{A},$$

and hence

$$(\epsilon E) \times A = q. \tag{7-11}$$

The product of the resultant electric intensity E at any point within a dielectric, and the permittivity ϵ at the same point, is called the *displacement* at the point, and is represented by D.

$$\boxed{D = \epsilon E.} \tag{7-12}$$

Since the units of ϵ are coulomb^2/newton-m^2 and those of E are newtons/coulomb, the units of displacement are coulombs/m^2, the same as those of surface density of charge.

Displacement, like electric intensity, is a vector quantity whose direction at every point is the same as that of the electric intensity[1] E but whose magnitude is ϵ times that of E. Like electric intensity, displacement can be represented by lines called *lines of displacement*. The tangent to these lines at every point is in the direction of the displacement, and the number of lines per unit area perpendicular to their direction is numerically equal to the displacement. Hence the number of lines of displacement per unit area, at any point, is ϵ times the number of lines of force per unit area at the same point. In empty space, where $\epsilon = \epsilon_0$,

$$D = \epsilon_0 E.$$

Eq. (7-11) now takes the form

$$DA = q. \tag{7-12}$$

But the product DA equals the number of lines of displacement passing outward through the gaussian surface in Fig. 7-9, and q is the *free* charge within the surface. We may accordingly generalize this relation and state: *the surface integral of the normal component of D over a closed surface equals the free charge enclosed by the surface.*

$$\boxed{\int D \cos \phi \, dA = q.} \tag{7-13}$$

The concept of displacement, as we have defined it, does not have any direct physical interpretation like that of force-per-unit-charge for the electric intensity E. It may be looked on merely as an auxiliary quantity defined by the equation $D = \epsilon E$, and which has the useful property that its surface integral over any closed surface equals the enclosed *free* charge.

Examples. (1) Let us again compute the resultant electric intensity in the dielectric of Example (2) in Sec. 7-3, making use of the displacement concept. When Gauss's law (Eq. [7-13]) is applied to a closed surface surrounding either plate and cutting through the dielectric, as in Fig. 7-9, we obtain

$$\int D \cos \phi \, dA = DA = q.$$

[1] Except that in an anisotropic dielectric the directions of D and E may differ.

Hence

$$D = \frac{q}{A} = \sigma = 30 \times 10^{-6}\ \text{coul/m}^2,$$

and

$$E = \frac{D}{\epsilon} = \frac{30 \times 10^{-6}}{15 \times 10^{-12}} = 2 \times 10^6\ \text{volts/m}.$$

(2) Fig. 7-10 represents a pair of oppositely charged parallel plates between which are two dielectrics of thicknesses d_1 and d_2, and dielectric coefficients K_{e_1} and K_{e2}. If E_1 and E_2 are the electric intensities in the two dielectrics,

$$D_1 = \epsilon_1 E_1 = K_{e_1}\epsilon_0 E_1, \quad D_2 = \epsilon_2 E_2 = K_{e_2}\epsilon_0 E_2. \tag{7-14}$$

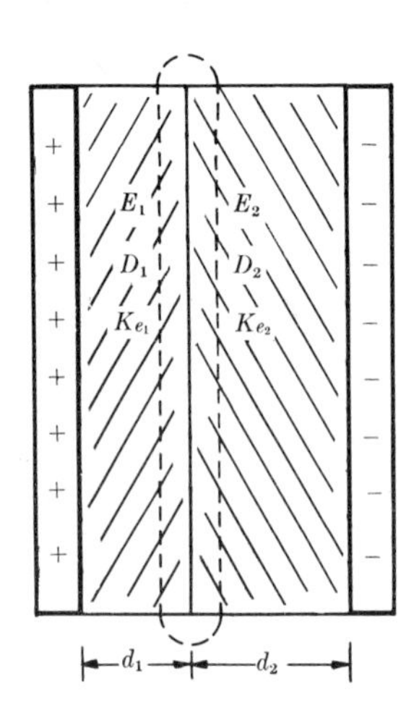

FIG. 7-10. Two dielectrics of different coefficients between oppositely charged plates.

Construct a gaussian surface as indicated by the dotted line. Since the *free* charge within the surface is zero and the direction of D is from left to right in each dielectric,

$$\int D \cos \phi \, dA = -D_1 A + D_2 A = 0,$$

or

$$D_1 = D_2 \tag{7-15}$$

The displacement is therefore the same in each dielectric.

From Eqs. (7-14) and (7-15) we find that

$$K_{e_1} E_1 = K_{e_2} E_2,$$

or

$$\frac{E_1}{E_2} = \frac{K_{e2}}{K_{e_1}},$$

and the electric intensities are inversely proportional to the corresponding dielectric coefficients.

7-5 Boundary conditions. It was shown in the preceding example that when two substances having different dielectric coefficients are "in series" between a pair of opposite charged parallel plates, the displacement D is the same in both dielectrics, and the electric intensities are related by the equation

$$\epsilon_1 E_1 = \epsilon_2 E_2,$$

or

$$K_{e_1} E_1 = K_{e_2} E_2.$$

In this special case the vectors D and E are both perpendicular to the plane separating the dielectrics. In the more general case, when a number of dielectrics of arbitrary shape are in an electric field, the directions of D and E may make any angle with the boundary surface between two dielectrics, and in general there is a discontinuous change in the magnitudes and directions of both D and E as the boundary is crossed.

Fig. 7-11 represents a small portion of the boundary surface between two dielectrics of dielectric coefficients K_{e_1} and K_{e_2} respectively. The field in each dielectric lies in the plane of the diagram. Let the X-axis be taken along the normal to the surface, and let ϕ_1 and ϕ_2 represent the angles between the normal and the respective directions of the fields.

Consider first the small closed surface of negligible length shaped like a pillbox, whose ends are of area A and whose curved wall is perpendicular to the boundary. Since there is no *free* charge within this box, the surface integral of the normal component of D over its surface is zero. The

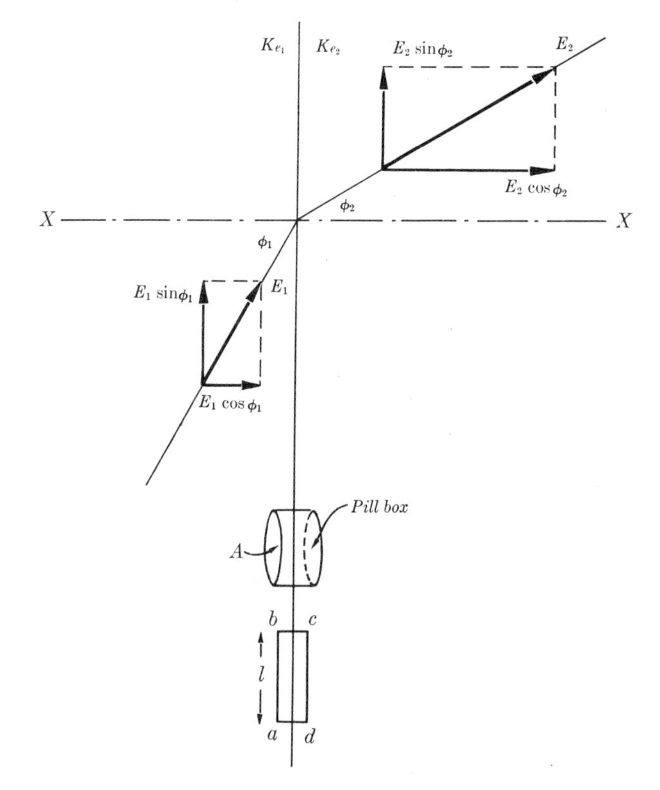

Fig. 7-11. The normal component of D and the tangential component of E are continuous across the boundary between two dielectrics.

walls of the box are very short and the surface integral over them is negligible. For the surface integral over the left face we have

$$-AD_1 \cos \phi_1,$$

and over the right face

$$AD_2 \cos \phi_2.$$

Hence

$$-AD_1 \cos \phi_1 + AD_2 \cos \phi_2 = 0,$$

and

$$D_1 \cos \phi_1 = D_2 \cos \phi_2. \tag{7-16}$$

It follows that when the direction of an electric field makes some arbitrary angle with a boundary surface between two dielectrics, the normal component of the displacement, $D \cos \phi$, has the same value on both sides of the surface. This is commonly expressed by stating that *the normal component of the displacement, $D \cos \phi$, is continuous across a boundary surface.*

Consider next the closed loop or path *abcd*, lying in the plane of the diagram. We have shown that the line integral of the electric intensity E around a closed loop in an electrostatic field, is zero. (See Sec. 3-3.) Sides *bc* and *ad* of the loop are so short that their contribution to the line integral is negligible. Along side *ab*, of length l, the line integral is

$$lE_1 \sin \phi_1,$$

while along side *cd* it is

$$-lE_2 \sin \phi_2.$$

Hence

$$lE_1 \sin \phi_1 - lE_2 \sin \phi_2 = 0,$$

or

$$E_1 \sin \phi_1 = E_2 \sin \phi_2. \tag{7-17}$$

Hence *the tangential component of the electric intensity, $E \sin \phi$, is continuous across the boundary surface.*

When Eq. (7-17) is divided by Eq. (7-16) we obtain

$$\frac{E_1 \sin \phi_1}{D_1 \cos \phi_1} = \frac{E_2 \sin \phi_2}{D_2 \cos \phi_2}$$

which, since $D_1 = K_{e_1}\epsilon_0 E_1$ and $D_2 = K_{e_2}\epsilon_0 E_2$, reduces to

$$\frac{1}{K_{e_1}} \tan \phi_1 = \frac{1}{K_{e_2}} \tan \phi_2,$$

or

$$\frac{\tan \phi_1}{\tan \phi_2} = \frac{K_{e_1}}{K_{e_2}}, \tag{7-18}$$

a relation which is similar to Snell's law for the refraction of a light ray at a boundary surface and is sometimes called the "law of refraction" of the lines of force of an electric field.

7-6 Polarization. The induced charges on the surface of a dielectric in an electric field are only one aspect of the influence of the field on the dielectric. We have shown that the surface charges come about as a result of the polarization or orientation of the molecules of the dielectric, so that we have to do in reality with a volume effect, not merely a surface effect.

The *dipole moment* of a polarized molecule is defined as the product of either of its charges, q, and their distance of separation, l. Let us assume for simplicity that all of the polarized molecules in a dielectric have the same dipole moment ql and that there are n such molecules per unit volume, all aligned in the same direction. The extent of polarization will then be measured by the product of the dipole moment of each molecule and the number of molecules per unit volume, or the *dipole moment per unit volume.* This product is called the *polarization* of the dielectric and is represented by P.

$$P = nql. \tag{7-19}$$

The dipole moment per unit volume can be expressed in another way. Let q_i be the induced charge at each surface of a polarized slab of area A and thickness d. Considering the entire slab as a large dipole, its dipole moment is $q_i d$ and its dipole moment per unit volume, or polarization P is

$$P = \frac{q_i d}{Ad} = \frac{q_i}{A} = \sigma_i. \tag{7-20}$$

That is, the *dipole moment per unit volume* (in this special case) *equals the surface density of induced charge.* If we wish to consider the influence of the field as a surface phenomenon, we describe it in terms of the surface density of induced charge, σ_i. If we wish to consider it as a volume phenomenon, we describe it in terms of the dipole moment per unit volume or the polarization P. The two are numerically equal, and it will be seen from their definitions that both are expressed as charge per unit area.

Electric susceptibility was defined in Sec. 7-3 as the ratio of surface density of induced charge to resultant electric intensity.

$$\eta = \frac{\sigma_i}{E}.$$

Since the polarization and induced charge density are equal, an alternate definition of susceptibility is the ratio of polarization to resultant electric intensity.

$$\eta = \frac{P}{E}, \quad P = \eta E. \tag{7-21}$$

Eq. (7-21) is actually the correct general definition of electric susceptibility, rather than Eq. (7-2). In the special case of a flat slab of dielectric inserted in a field at right angles to its faces, the polarization is constant throughout the dielectric and numerically equal to the surface density of induced charge. In more general cases the induced charge may not be wholly confined to the surfaces of the dielectric, and the polarization may vary from point to point. The electric susceptibility, at any point, is defined as the ratio of the polarization at that point to the electric intensity at the point.

The polarization of a dielectric having polar molecules is due in part to induced dipoles and in part to permanent dipoles which become aligned parallel to the field. It might appear at first that in any external field, however small, all of the permanent dipoles would line up in the direction of the field. This tendency toward alignment is opposed, however, by the thermal agitation of the molecules, which tends to destroy any regular orientation and replace it by a random one. The higher the temperature, the greater will be the tendency toward randomness due to thermal motion, and the smaller the susceptibility. Hence substances whose molecules are permanent dipoles show a dependence of susceptibility on temperature, while nonpolar substances do not.

It is useful to have an expression for the polarization of a dielectric in terms of the displacement D and the electric intensity E in the dielectric. From Eqs. (7-12) and (7-7),

$$D = \epsilon E = (\epsilon_0 + \eta)E = \epsilon_0 E + \eta E$$

and since $\eta E = P$,

$$D = \epsilon_0 E + P. \tag{7-22}$$

Eq. (7-22) is in fact a more general definition of displacement than is Eq. (7-12). If a dielectric is anisotropic its permittivity ϵ cannot be represented by a single number and Eq. (7-12) ceases to have a meaning. The displacement D and the electric intensity E in such a dielectric are, in general, in different directions. But Eq. (7-22) applies in all cases, with the extension that it is to be considered as a vector equation, that is,

$$D_x = \epsilon_0 E_x + P_x$$

$$D_y = \epsilon_0 E_y + P_y$$

$$D_z = \epsilon_0 E_z + P_z$$

7-7 Force between charges in a dielectric. We have built up the theory of electrostatics from Coulomb's law for the force between point charges,

$$F = \frac{1}{4\pi\epsilon_0}\frac{qq'}{r^2}.$$

When this law was introduced in Chapter 1 it was stated that although the purely electrostatic force exerted on one point charge by another is given by the expression above regardless of the medium in which the charges are immersed, the presence of a medium does affect the *apparent* force between the charges. We are now in a position to discuss this effect further.

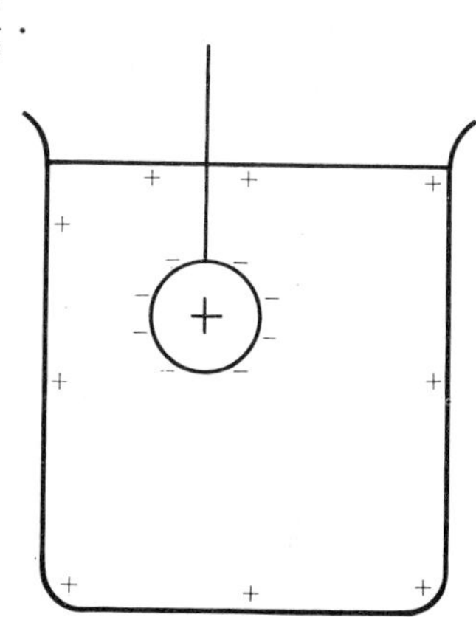

FIG. 7-12. Charged sphere suspended in a nonconducting liquid by an insulating thread.

Consider first a qualitative example. Let a charged sphere be suspended by an insulating thread in a beaker of nonconducting liquid as in Fig. 7-12. The field of the charge causes induced charges to appear at the surfaces of the liquid, including the surface that makes contact with the charged body. (The influence of the container is neglected for simplicity.) The field at any point is the resultant of the fields of the original charge and the induced charges. An exact mathematical treatment of the problem is obviously very complicated because of the lack of symmetry of the induced charges. Things would be even worse if the liquid were not homogeneous, or if the charged body were placed in an anisotropic solid, that is, one whose properties are different in different directions. However, if we assume the fluid around the charge to extend to infinity in all directions, and to be homogeneous and isotropic, the problem is considerably simplified.

In Fig. 7-13 a spherical conductor having a positive charge q is immersed in an infinite homogeneous isotropic fluid of dielectric coefficient K_e. No other charges are nearby. We wish to compute the magnitude of the electric intensity at a point P, at a distance r from the center of the charged body. The only induced charges that concern us are those at the surface of the medium just outside the conductor. Let q_i represent the magnitude of this induced charge.

Let us construct a gaussian surface in the form of a sphere of radius r, concentric with the charged sphere. Gauss's law states that

$$\int D \cos\phi \, dA = q.$$

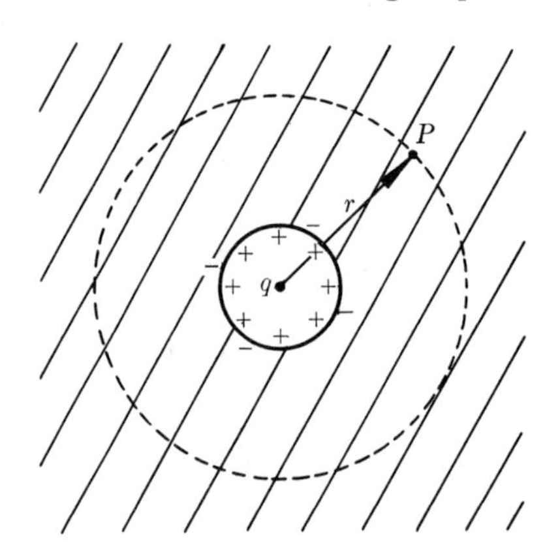

FIG. 7-13. A charged spherical conductor immersed in an infinite homogeneous isotropic fluid.

By symmetry, the magnitude of the displacement is constant at all points of this surface and its direction is radially outward. The surface integral then reduces to

$$\int D \cos\phi \, dA = D \times 4\pi r^2 = q$$

and hence

$$D = \frac{q}{4\pi r^2} = \epsilon E = K_e \epsilon_0 E,$$

and

$$E = \frac{1}{4\pi\epsilon_0} \frac{q}{K_e r^2}. \tag{7-23}$$

On the other hand, Gauss's law applied to the electric intensity gives

$$\int E \cos\phi \, dA = \frac{1}{\epsilon_0}(q - q_i) = 4\pi r^2 E,$$

from which

$$E = \frac{1}{4\pi\epsilon_0} \frac{q - q_i}{r^2}.$$

We may now equate the two expressions for E.

$$\frac{1}{4\pi\epsilon_0} \frac{q}{K_e r^2} = \frac{1}{4\pi\epsilon_0} \frac{q - q_i}{r^2},$$

or

$$q - q_i = \frac{q}{K_e}. \tag{7-24}$$

We see, therefore, that the *effective* charge of the sphere, $q - q_i$, is $\frac{1}{K_e}$th of its free charge q, and therefore the field at P is $\frac{1}{K_e}$th of that which would result at P if the dielectric were absent. That is,

$$E = \frac{1}{K_e} \frac{1}{4\pi\epsilon_0} \frac{q}{r^2}.$$

(We have, of course, already obtained this expression for E in the course of the argument [see Eq. (7-23)] without computing the magnitude of the induced charges, but it is instructive to carry through the analysis and see just why the field outside the charge is diminished by the presence of a dielectric.)

Consider next the force exerted on a second conducting sphere having a charge q' and placed at the point P. This sphere will also be surrounded by induced charges, say of magnitude q_i'. Furthermore, if the bodies are close together, the symmetry of both the original and induced charges is disturbed by the field of each charge at the point where the other is located. However, if the distance between the spheres is large compared with their dimensions, this effect is small and for sufficiently large separations we may still assume spherical symmetry.

The forces on the charge q' are those due to: (a) the charge q, (b) the induced charge q_i around q, and (c) the induced charge q_i' around q' itself. The charge q_i', if symmetrically distributed, produces no net force on the charge q', so the only outstanding effect is that of the field of the charges q and q_i whose value we have deduced. The force on q' may therefore be computed either directly from Coulomb's law, considering q' to be attracted by a net charge $q - q_i$, or it may be computed from the electric intensity at the point where q' is located. Either method gives

$$F = \frac{1}{4\pi\epsilon_0} \frac{qq'}{K_e r^2} = \frac{1}{4\pi\epsilon} \frac{qq'}{r^2}. \tag{7-25}$$

The force is therefore reduced below its value in vacuum, which would be

$$F_0 = \frac{1}{4\pi\epsilon_0} \frac{qq'}{r^2},$$

by the factor $1/K_e$.

A glance at the values of the dielectric coefficients of gases listed in Table 7-1 will show that the force between point charges in air is for all practical purposes the same as if the charges were in a vacuum.

Either Newton's third law or a consideration of the field set up by the charges q' and q_i', shows that the force on q is equal and opposite to that on q'.

It should not be inferred that the presence of a medium in any way diminishes the force exerted on q' *by the original charge* q. The force between q and q' is exactly the same whether the charges are in a vacuum or in a medium. The reduction in the net force between the charges arises from the induced charge q_i, opposite in sign to q, at the surface of the dielectric adjacent to the original charge.

Eq. (7-25) is an *approximate* formula, derived for the very special case of two charged bodies at a separation large compared with their dimensions and immersed in a homogeneous isotropic fluid filling all of space. It is not a good approximation, in general, when these conditions are not fulfilled and is not a general law of nature like Coulomb's law. Thus, if the dielectric is a solid, or if the charges are on nonconductors, or if the field is set up by permanently polarized bodies, Eq. (8-25) does not hold.

SUMMARY

The method followed in this chapter has been to develop general formulas by reasoning based on a special case (a sheet of dielectric between oppositely charged, closely spaced parallel plates). The steps in the derivation are summarized below, where formulas of *general* applicability have been indicated by enclosing them in a box. Formulas not so enclosed are correct for the special case of a sheet of dielectric between parallel plates. Equations are numbered as in the text.

The electric intensity E within the dielectric is the resultant of the fields of the free charges and the induced charges.

$$E = \frac{1}{\epsilon_0}\sigma - \frac{1}{\epsilon_0}\sigma_i. \tag{7-1}$$

Electric susceptibility η is defined by the equation

$$\sigma_i = \eta E. \tag{7-2}$$

(See Eq. (7-21) for a more general definition.)

Combine Eqs. (7-1) and (7-2) to get

$$E = \frac{\sigma}{\left(1 + \frac{\eta}{\epsilon_0}\right)\epsilon_0}. \tag{7-3}$$

Define dielectric coefficient K_e by the equation

$$K_e = 1 + \frac{\eta}{\epsilon_0}. \tag{7-4}$$

Combine Eqs. (7-3) and (7-4) to get

$$E = \frac{1}{K_e \epsilon_0} \sigma. \tag{7-5}$$

Define permittivity ϵ by the equation

$$\epsilon = K_e \epsilon_0. \tag{7-6}$$

Combine Eqs. (7-4) and (7-6) to get

$$\epsilon = \epsilon_0 + \eta. \tag{7-7}$$

Define displacement D by the equation

$$D = \epsilon E. \tag{7-12}$$

(See Eq. (7-22) for a more general definition.)
Define polarization P as

$$P = \text{dipole moment per unit volume.}$$

For the special case of a sheet of dielectric,

$$P = \sigma_i. \tag{7-20}$$

Combine Eqs. (7-2) and (7-20) to get

$$P = \eta E. \tag{7-21}$$

Note that Eq. (7-21), rather than Eq. (7-2), is the general definition of η.

Combine Eqs. (7-12), (7-7), and (7-21) to get

$$D = \epsilon_0 E + P. \tag{7-22}$$

Eq. (7-22), considered as a vector equation, is the general definition of D.

Problems—Chapter 7

(1) The dielectric coefficient of a certain material is 3.5. Compute the permittivity and the susceptibility of the material.

(2) Two oppositely charged conducting plates, having numerically equal surface densities of charge, are separated by a dielectric 5 mm thick, of dielectric coefficient 3. The resultant electric intensity in the dielectric is 10^6 volts/m. Compute: (a) the displacement in the dielectric, (b) the surface density of free charge on the conducting plates, (c) the polarization in the dielectric, (d) the surface density of induced charge on the surfaces of the dielectric, (e) the component of electric intensity in the dielectric, due to the free charge, (f) the component of electric intensity due to the induced charge.

(3) Two parallel conducting plates, 5 mm apart, are given equal and opposite surface densities of charge of 20 microcoulombs/m^2. The region between the plates is occupied by two sheets of dielectric, one 2 mm thick, of dielectric coefficient 3, the other 3 mm thick and of dielectric coefficient 4. Compute: (a) the electric intensity in each dielectric, (b) the displacement in each dielectric, (c) the surface density of induced charge on each dielectric.

(4) The Thomson model of a hydrogen atom is a sphere of positive electricity with an electron (a point charge) at its center. The total positive charge equals the electronic charge e. (a) Prove that when the electron is at a distance r from the center of the sphere of positive charge, it is attracted toward the center with a force F given by

$$F = \frac{1}{4\pi\epsilon_0}\frac{e^2 r}{R^3},$$

where R is the radius of the sphere of positive charge. (See problem 14, Chap. 2.) (b) What is the induced dipole moment of this atom model in an external field of intensity E?

(5) A uniform electric field of intensity 2×10^6 volts/m is set up within a large block of material of dielectric coefficient 3. A cavity is cut out of the block in the form of a short cylinder whose ends are perpendicular to the field. Find the electric intensity within the cavity and the surface density of induced charge on its end surfaces. Show the induced charges and the lines of force in a diagram.

CHAPTER 8

CAPACITANCE AND CAPACITORS

8-1 Capacitance of an isolated conductor. Reference to Chapter 3 will recall to mind that the potential V of an isolated charged sphere in empty space is given by the relation

$$V = \frac{1}{4\pi\epsilon_0}\frac{q}{a} \quad \text{or} \quad q = 4\pi\epsilon_0 aV, \tag{8-1}$$

where q is the charge on the sphere and a is its radius. The charge of the sphere is therefore proportional to its potential. It can be shown that the charge on an isolated charged body of any shape is also directly proportional to its potential. One can therefore write for any isolated charged conductor,

$$q \propto V, \quad q = CV, \quad \text{or}$$

$$C = \frac{q}{V}, \tag{8-2}$$

where C is a proportionality constant which depends on the size and shape of the conductor and is called its *capacitance. The capacitance of a conductor is the ratio of its charge to its potential.* (The reference level of potential is assumed at infinity.)

From its definition, capacitance is expressed in *coulombs per volt.* A capacitance of one coulomb per volt is called one *farad* (in honor of Michael Faraday).

The capacitance of an isolated spherical conductor, from Eqs. (8-1) and (8-2) is

$$C = \frac{q}{V} = 4\pi\epsilon_0 a. \tag{8-3}$$

The capacitance of a sphere is thus directly proportional to its radius.

The term "capacity" is sometimes used for the proportionality constant C, but "capacitance" is to be preferred since it avoids the implication that this quantity means the same thing as does the "capacity" of a bucket. The latter is the maximum volume of material the bucket can hold, whereas the quantity of charge a conductor "holds" can be increased indefinitely by raising its potential, except for the limit set by insulation problems. As we have seen, the maximum surface density of charge on a conductor in air is limited by the maximum electric intensity the air can support without breakdown.

8-2 Capacitors. If a number of charged conductors are in the vicinity of one another, the potential of each is determined not only by its own charge but by the magnitude and sign of the charges on the other conductors and by their shape, size, and location. For example, the potential of a positively-charged sphere is lowered if a second negatively-charged sphere is brought near the first. Thus the capacitance of the first sphere (the ratio of its charge to its potential) is increased by the presence of the second. In the same way the capacitance of the second is increased by the presence of the first.

An important special case arises in practice when two conductors in the same vicinity are given equal amounts of charge of opposite sign. This is usually accomplished by connecting the conductors, both initially uncharged, to the terminals of a seat of emf, which results in a transfer of charge from one conductor to the other. Such an arrangement of two conductors is called a *capacitor*. The fact that each conductor is in the vicinity of another carrying a charge of opposite sign makes possible the transfer of relatively large quantities of charge from one conductor to the other, with relatively small differences of potential.

The capacitance of a capacitor is defined as the ratio of the charge on either conductor to the potential difference between the conductors.

$$C = \frac{q}{V_{ab}}. \tag{8-4}$$

The net charge on the capacitor as a whole is, of course, zero, and "the charge on a capacitor" is understood to mean the charge on *either* conductor without regard to sign. The capacitance is *one farad* if *one coulomb is transferred from one conductor to the other, per volt of potential difference* between the conductors.

A capacitor is represented by the symbol

—||—

Capacitors find many applications in electrical circuits. A capacitor is used to eliminate sparking when a circuit containing inductance (see Sec. 13-4) is suddenly opened. The ignition system of every automobile engine contains a capacitor for this purpose. Capacitors are used in radio circuits for tuning, and for "smoothing" the rectified current delivered by the power supply. The efficiency of alternating current power transmission can often be increased by the use of large capacitors to improve the power factor. (See Sec. 16-8.)

The term "condenser" has long been used for the piece of apparatus we have described as a "capacitor." But "capacitor" is to be preferred, both because nothing is actually "condensed" in a "condenser," and also because of the corresponding usage of the terms resistance and resistor. That is, a resistor is a device that has resistance, and a capacitor is a device that has capacitance.

8-3 The parallel plate capacitor. The most common type of capacitor consists of two conducting plates parallel to one another and separated by a distance which is small compared with the linear dimensions of the plates. See Fig. 8-1. Practically the entire field of such a capacitor is localized in the region between the plates as shown. There is a slight "fringing" of the field at its outer boundary, but the fringing becomes relatively less as the plates are brought closer together. If the plates are sufficiently close, the fringing can be neglected, the field between the plates is uniform, and the charges on the plates are uniformly distributed over their opposing surfaces. This arrangement is known as a *parallel-plate capacitor.*

Let us assume first that the plates are in vacuum. It has been shown that the electric intensity between a pair of closely spaced parallel plates in vacuum is

$$E = \frac{1}{\epsilon_0}\sigma = \frac{1}{\epsilon_0}\frac{q}{A},$$

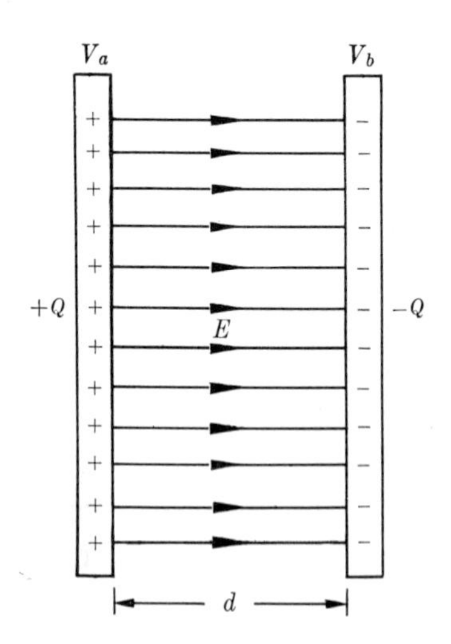

FIG. 8-1. Parallel plate capacitor.

where A is the area of the plates and q is the charge on *either* plate. Since the electric intensity or potential gradient between the plates is uniform, the potential difference between the plates is

$$V_{ab} = Ed = \frac{1}{\epsilon_0}\frac{qd}{A},$$

where d is the separation of the plates. Hence the capacitance of a parallel plate capacitor in vacuum is

$$C = \frac{q}{V_{ab}} = \epsilon_0 \frac{A}{d}. \qquad (8\text{-}5)$$

Since ϵ_0, A, and d are constants for a given capacitor, the capacitance is a constant independent of the charge on the capacitor, and is directly proportional to the area of the plates and inversely proportional to their separation. (The simple form of Eq. (8-5) results from the introduction of the factor 4π in the proportionality constant in Coulomb's law. Had this factor *not* been written explicitly in this law it would have appeared in Eq. (8-5). Since in practice the latter equation is used much more frequently than is Coulomb's law, there is an advantage in transferring the factor 4π to Coulomb's law.)

If mks units are used, A is to be expressed in square meters and d in meters. The capacitance C will then be in farads.

Eq. (8-5) indicates an alternate combination of units in which permittivity can be expressed. If we solve this equation for ϵ_0 we obtain

$$\epsilon_0 = \frac{Cd}{A},$$

and since C is in farads, d in meters, and A in square meters, ϵ_0 can be expressed in *farads per meter*. The permittivity ϵ of a material substance, which has the same units as ϵ_0, can also be expressed in farads per meter.

As an example, let us compute the area of the plates of a one farad parallel plate capacitor if the separation of the plates is one millimeter and the plates are in vacuum.

$$C = \epsilon_0 \frac{A}{d}$$

$$A = \frac{Cd}{\epsilon_0} = \frac{1 \times 10^{-3}}{8.85 \times 10^{-12}}$$

$$= 1.13 \times 10^{8}\ \text{m}^2.$$

This corresponds to a square 10,600 meters, or 34,600 ft, or about 6½ miles on a side!

Since the farad is such a large unit of capacitance, units of more convenient size are the *microfarad* ($1\mu\text{f} = 10^{-6}$ farads), and the *micro-microfarad* ($1\mu\mu\text{f} = 10^{-12}$ farads). For example, a common radio set contains in its power supply several capacitors whose capacitance is of the order of one to ten microfarads, while the capacitance of the tuning capacitors is of the order of a few hundred micro-microfarads.

We next compute the capacitance of a parallel plate capacitor when a substance of dielectric coefficient K_e occupies the space between its plates.

The electric intensity in the dielectric is

$$E = \frac{\sigma}{K_e \epsilon_0} = \frac{1}{K_e \epsilon_0} \frac{q}{A}. \tag{8-6}$$

The potential difference between the plates is therefore

$$V_{ab} = Ed = \frac{1}{K_e \epsilon_0} \frac{qd}{A},$$

and the capacitance is

$$C = \frac{q}{V_{ab}} = K_e \epsilon_0 \frac{A}{d}. \tag{8-7}$$

Let us represent by C_0 the capacitance of the same capacitor in vacuum Then

$$C_0 = \epsilon_0 \frac{A}{d}, \tag{8-8}$$

and we see that the capacitance is *increased* by a factor of K_e when a dielectric is inserted between the plates.

When Eq. (8-7) is divided by Eq. (8-8), we obtain

$$K_e = \frac{C}{C_0}. \tag{8-9}$$

This relation is often made the defining equation of the dielectric coefficient. That is, the dielectric coefficient of a material may be defined as the ratio of the capacitance of a given capacitor with the material between its plates, to the capacitance of the same capacitor in vacuum.

For the sake of emphasizing the atomic and electronic viewpoint, we began the theory of dielectrics in the preceding chapter with the concept of electric susceptibility η, defined as the ratio of induced charge density to electric intensity.

$$\eta = \frac{\sigma_i}{E}.$$

The dielectric coefficient, K_e, was then defined by the equation

$$K_e = 1 + \frac{\eta}{\epsilon_0}. \tag{8-10}$$

From the experimental viewpoint, however, the induced charge density σ_i is not susceptible of direct measurement so that the *defining* equation of η is of little value in determining the magnitude of this property of a dielectric. Eq. (8-9), however, affords a direct and relatively simple and accurate method of determining dielectric coefficients, since capacitances can be measured with high precision. Hence in practice it is usually K_e that is measured and Eq. (8-10) is then used to compute η.

Examples. (1) The plates of a parallel plate capacitor are 5 mm apart and 2 m² in area. The plates are in vacuum. A potential difference of 10,000 volts is applied across the capacitor. Compute (a) the capacitance, (b) the charge on each plate, (c) the surface density of charge, (d) the electric intensity, and (e) the displacement in the space between plates.

(a)
$$\begin{aligned} C_0 &= \epsilon_0 \frac{A}{d} \\ &= 8.85 \times 10^{-12} \times \frac{2}{5 \times 10^{-3}} \\ &= 3.54 \times 10^{-9} \text{ farads} \\ &= 3.54 \times 10^{-3}\ \mu\text{f} \\ &= 3540\ \mu\mu\text{f} \end{aligned}$$

(b) The charge on the capacitor is

$$\begin{aligned} q &= CV_{ab} \\ &= 3.54 \times 10^{-9} \times 10^4 \\ &= 3.54 \times 10^{-5} \text{ coulombs.} \end{aligned}$$

(c) The surface density of charge is

$$\begin{aligned} \sigma &= \frac{q}{A} \\ &= \frac{3.54 \times 10^{-5}}{2} \\ &= 1.77 \times 10^{-5} \text{ coulomb/meter}^2. \end{aligned}$$

(d) The electric intensity is

$$\begin{aligned} E &= \frac{1}{\epsilon_0} \sigma \\ &= 36\pi \times 10^9 \times 1.77 \times 10^{-5} \\ &= 2 \times 10^6 \text{ volts/meter.} \end{aligned}$$

The electric intensity may also be computed from the potential gradient.

$$\begin{aligned} E &= \frac{V_{ab}}{d} \\ &= \frac{10^4}{5 \times 10^{-3}} = 2 \times 10^6 \text{ volts/meter.} \end{aligned}$$

(e) The displacement is

$$D = \epsilon_0 E$$
$$= 8.85 \times 10^{-12} \times 2 \times 10^6$$
$$= 1.77 \times 10^{-5} \text{ coulomb/meter}^2.$$

The displacement is of course equal to the surface density of charge.

(2) The charged capacitor in Example (1) is disconnected from the charging voltage and insulated so that the charges on its plates remain constant. A sheet of dielectric 5 mm thick, of dielectric coefficient 5, is inserted between the plates. Compute (a) the displacement in the dielectric, (b) the electric intensity in the dielectric, (c) the potential difference across the capacitor, (d) its capacitance.

The insertion of a dielectric between the capacitor plates alters the electric intensity but not the displacement, since the free charge is kept constant. Hence

(a) $D = \sigma = 1.77 \times 10^{-5}$ coul/m².

(b) The electric intensity is

$$E = \frac{D}{\epsilon} = \frac{D}{K_e \epsilon_0}$$
$$= \frac{1.77 \times 10^{-5}}{5 \times 8.85 \times 10^{-12}}$$
$$= 4 \times 10^5 \text{ volts/m}.$$

(c) The potential difference across the capacitor is

$$V_{ab} = Ed$$
$$= 4 \times 10^5 \times 5 \times 10^{-3}$$
$$= 2000 \text{ volts}.$$

(d) The capacitance is

$$C = \frac{q}{V_{ab}}$$
$$= \frac{3.54 \times 10^{-5}}{2000}$$
$$= 1.77 \times 10^{-8} \text{ farads}$$
$$= 17{,}700\ \mu\mu\text{f}.$$

The capacitance may also be computed from

$$C = \epsilon \frac{A}{d} = K_e \epsilon_0 \frac{A}{d}$$
$$= 5 \times 8.85 \times 10^{-12} \frac{2}{5 \times 10^{-3}}$$
$$= 17{,}700\ \mu\mu\text{f}.$$

The capacitance is, of course, five times as great as its vacuum value of 3540 $\mu\mu$f.

(3) The sheet of dielectric in Example (2) is removed and replaced by two sheets, one 2 mm thick, of dielectric coefficient 5, and the other 3 mm thick, of dielectric coefficient 2. See Fig. 8-2. Compute (a) the displacement in each dielectric, (b) the electric intensity in each, (c) the potential difference across the capacitor, (d) its capacitance.

(a) The insertion of the dielectric between the capacitor plates does not alter the displacement, since the free charge is kept constant. Hence

$$D = \sigma = 1.77 \times 10^{-5} \text{ coulomb/meter}^2.$$

(b) In the first sheet,

$$E_1 = \frac{D}{K_{e_1}\epsilon_0} = \frac{1.77 \times 10^{-5}}{5 \times 8.85 \times 10^{-12}} = 4 \times 10^5 \text{ volts/meter.}$$

In the second sheet,

$$E_2 = \frac{D}{K_{e_2}\epsilon_0} = \frac{1.77 \times 10^{-5}}{2 \times 8.85 \times 10^{-12}} = 10 \times 10^5 \text{ volts/meter.}$$

(c) The potential difference across the first sheet is

$$V_{ax} = E_1 d_1 = 4 \times 10^5 \times 2 \times 10^{-3} = 800 \text{ volts.}$$

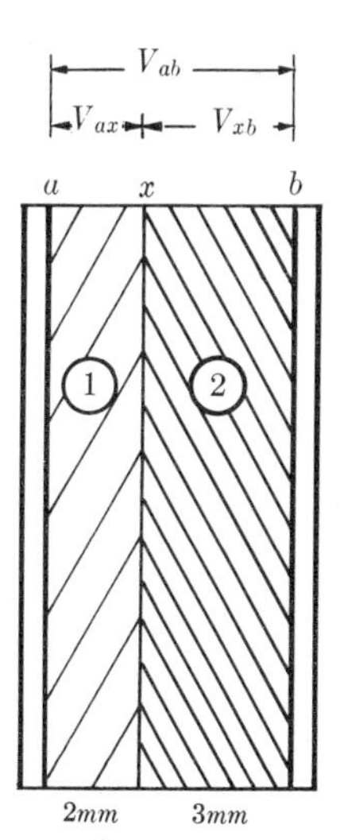

FIG. 8-2. Parallel plate capacitor with two sheets of different thicknesses and different dielectric coefficients.

The potential difference across the second is

$$V_{xb} = E_2 d_2 = 10^6 \times 3 \times 10^{-3} = 3000 \text{ volts.}$$

The potential difference across the capacitor is therefore

$$V_{ab} = V_{ax} + V_{xb} = 800 + 3000 = 3800 \text{ volts.}$$

(d) The capacitance is

$$C = \frac{q}{V_{ab}} = \frac{3.54 \times 10^{-5}}{3800} = 9.33 \times 10^{-9} \text{ farad} = 9330\ \mu\mu\text{f}.$$

The capacitance in vacuum was 3540 $\mu\mu$f.

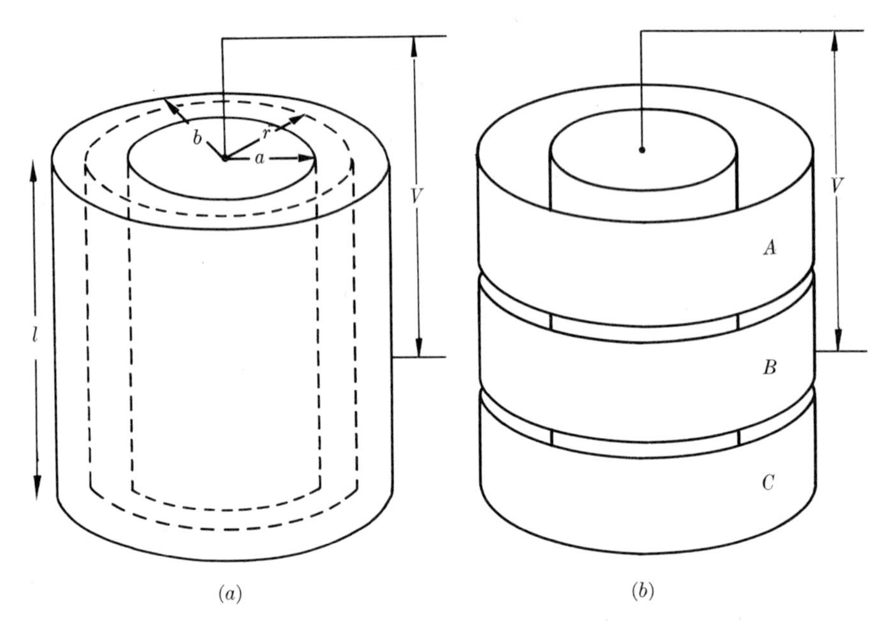

FIG. 8-3. (a) Cylindrical capacitor. (b) Standard cylindrical capacitor with guard rings.

8-4 Other types of capacitor. Cylindrical and spherical capacitors are sometimes used as standard capacitors, since their capacitance can be computed accurately from their dimensions. Suppose two coaxial cylinders of radii a and b and length l (Fig. 8-3) are given equal and opposite charges $+q$ and $-q$. Take as a gaussian surface a cylinder of radius r, intermediate between a and b, and of length l. If end effects are neglected, lines of displacement cut this surface only over the curved surface of area $2\pi rl$. Let D represent the displacement at radius r. Then

$$\int D \cos \phi \, dA = 2\pi rlD = q,$$

$$D = \frac{q}{2\pi rl}.$$

If the region between the cylinders contains a dielectric of permittivity ϵ,

$$E = \frac{D}{\epsilon} = \frac{1}{2\pi\epsilon} \frac{q}{rl}.$$

The magnitude of the potential difference between the cylinders is

$$V_{ab} = \int_a^b E\,dr = \frac{1}{2\pi\epsilon}\frac{q}{l}\int_a^b \frac{dr}{r}$$
$$= \frac{1}{2\pi\epsilon}\frac{q}{l}\ln\frac{b}{a}.$$

The capacitance is therefore

$$C = \frac{q}{V_{ab}} = 2\pi\epsilon\frac{l}{\ln\frac{b}{a}}. \tag{8-11}$$

To minimize end effects, standard cylindrical capacitors are constructed as shown in Fig. 8-3(b). The end sections A and C, known as *guard rings*, are maintained at the same potential as the central section B but only the latter is used in a measurement. Fringing effects are thus transferred to the end sections A and C, which are not used.

It will be left as an exercise to show that the capacitance of two concentric spheres is

$$C = 4\pi\epsilon\frac{ab}{b - a}. \tag{8-12}$$

Variable capacitors whose capacitance may be varied at will (between limits) are widely used in the tuning circuits of radio receivers. These are usually air capacitors of relatively small capacitance and are constructed of a number of fixed parallel metal plates connected together and constituting one "plate" of the capacitor, while a second set of movable plates also connected together forms the other "plate" (Fig. 8-4). By rotating a shaft on which the movable plates are mounted, the second set may be caused to interleave the first to a greater or lesser extent. The effective area of the capacitor is that of the interleaved portion of the plates only. A variable capacitor is represented by the symbol

FIG. 8-4. Variable air capacitor. (*Courtesy of A. D. Cardwell Mfg. Corp.*)

Most capacitors utilize a solid dielectric between their plates. A common type is the paper and foil capacitor, in which strips of metal foil form the plates and a sheet of paper impregnated with wax is the dielectric. By rolling up such a capacitor, a capacitance of several microfarads can be obtained in a relatively small volume. The "Leyden jar," constructed by cementing metal foil over a portion of the inside and outside surfaces of a glass jar, is essentially a parallel plate capacitor with the glass forming the dielectric.

The function of a solid dielectric between the plates of a capacitor is threefold. First, it solves the mechanical problem of maintaining two large metal sheets at an extremely small separation but without actual contact. Second, since its dielectric strength is larger than that of air, the maximum potential difference which the capacitor can withstand without breakdown is increased. Third, because its permittivity is greater than that of empty space or of air, the capacitance of a capacitor of given dimensions is several times larger than if the plates were in vacuum or in air.

Electrolytic capacitors utilize as their dielectric an extremely thin layer of nonconducting oxide between a metal plate and a conducting solution. Because of the small thickness of the dielectric, electrolytic capacitors of relatively small dimensions may have a capacitance of the order of 50 microfarads.

8-5 Charge and discharge currents of a capacitor. When an uncharged capacitor is connected to two points at different potentials, the capacitor does not become charged instantaneously, but at a rate which depends on its capacitance and on the resistance of the circuit.[1] Fig. 8-5 shows a capacitor and resistor in series connected to two points maintained at a potential difference V_{ab}. Let i represent the current in the circuit, and q the charge on the capacitor, at some instant after the switch S is closed. We then have the following relations:

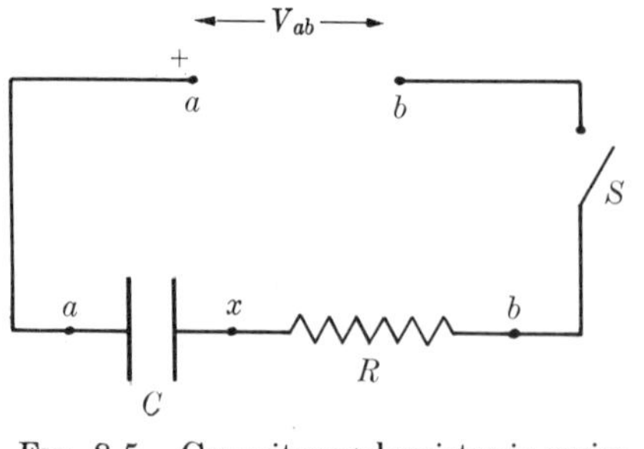

FIG. 8-5. Capacitor and resistor in series.

$$i = dq/dt$$
$$V_{ax} = q/C$$
$$V_{xb} = iR$$
$$V_{ab} = V_{ax} + V_{xb}$$

These equations may be combined to give

$$\frac{dq}{dt} + \frac{1}{RC}q - \frac{V_{ab}}{R} = 0, \quad (8\text{-}13)$$

[1] The effect of inductance, which is neglected here, is discussed in Sec. 13-3.

and hence

$$q = CV_{ab}(1 - \epsilon^{-t/RC}). \tag{8-14}$$

(ϵ, of course, represents here the base of natural logarithms and not the permittivity.)

Since CV_{ab} equals the final charge on the capacitor, say Q, one may also write

$$q = Q(1 - \epsilon^{-t/RC}). \tag{8-15}$$

A graph of Eq. (8-15) is given in Fig. 8-6. The charge on the capacitor is seen to approach its final value asymptotically, and hence an infinite time is required for the capacitor to attain its final charge. The time for the charge to increase to any stated fraction of its final amount, however, is perfectly definite, and with any values of R and C encountered in practice a very short time suffices for the charge to grow to essentially its final value.

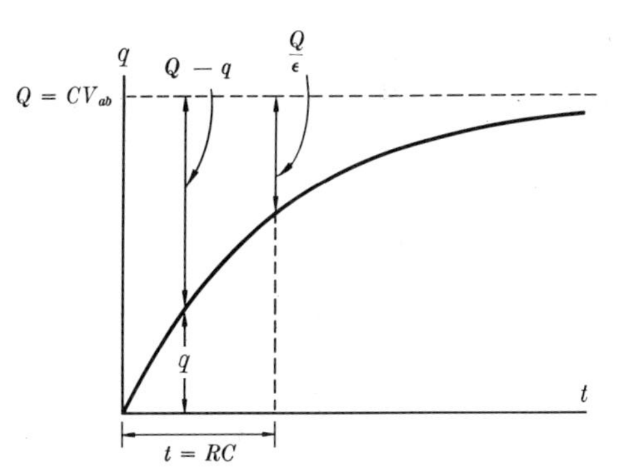

FIG. 8-6. The charge on a capacitor as a function of the time in the case of a circuit with capacitance and resistance.

At any time t, the difference between the final charge Q and the charge q already acquired is, from Eq. (8-15),

$$Q - q = Q\epsilon^{-t/RC}.$$

If we set this difference equal to Q/ϵ and solve for t, we find

$$t = RC.$$

That is, the product RC equals the time required for the charge on the capacitor to increase to within $\frac{1}{\epsilon}$ th $\left(\frac{1}{2.71} \text{ or } 0.369\right)$ of its final value, and it is called the *time constant* of the circuit. For a given capacitor this time is longer the larger the resistance and vice versa. Thus, although the general shape of the graph of q vs t is the same whatever the resistance, the curve rises rapidly to its final value if R is small, and slowly if R is large.

As an example, if $C = 1\ \mu f = 10^{-6}$ farads, and $R = 1$ megohm $= 10^6$ ohms, the time constant is

$$RC = 10^6 \times 10^{-6} = 1 \text{ second}$$

and the capacitor acquires $(1 - .369)$ or about 63% of its final charge in one second.

If $R = 1000$ ohms,

$$RC = 0.001 \text{ second}$$

and if $R = 10$ megohms,

$$RC = 10 \text{ seconds.}$$

Since $i = dq/dt$, the equation of the charging current can be obtained from Eq. (8-15) by differentiation. One finds

$$i = \frac{V_{ab}}{R} \epsilon^{-t/RC}. \tag{8-16}$$

The initial charging current (when $t = 0$) is therefore the same as if the circuit contained resistance R only, and the current decreases exponentially in the same way that the charge increases, falling to $\frac{1}{\epsilon}$th of its original value after a time equal to the time constant.

If the charged capacitor is disconnected from the line and its terminals connected to one another through a resistance R, it is easy to show that the charge remaining on either plate after a time t is

$$q = Q\epsilon^{-t/RC}$$

where Q is the initial charge, and that the discharge current is

$$i = \frac{Q}{RC} \epsilon^{-t/RC}$$

$$= \frac{V_{ab}}{R} \epsilon^{-t/RC},$$

where V_{ab} is the initial potential difference across the capacitor.

8-6 Capacitors in series and parallel. The capacitance of a capacitor is defined as the ratio of the charge Q on either plate to the potential difference between the capacitor terminals. The charge Q may be described as the charge displaced past any point of the external circuit in the process of charging the capacitor.

The *equivalent* capacitance of a capacitor network is similarly defined as the ratio of the displaced charge to the potential difference between the terminals of the network. Hence the method of computing the equivalent capacitance of a network is to assume a potential difference between the terminals of the network, compute the corresponding charge, and take the ratio of the charge to the potential difference.

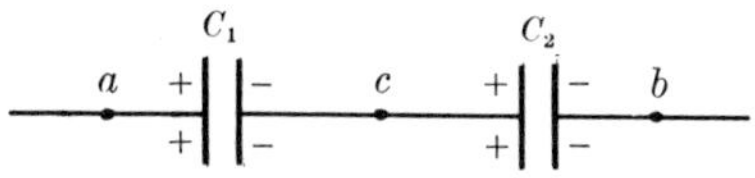

FIG. 8-7. Two capacitors in series.

If two capacitors of capacitances C_1 and C_2 respectively are connected in series as in Fig. 8-7 and the terminals a and b are maintained at a potential difference V_{ab}, one has

$$Q_1 = C_1 V_{ac}, \quad Q_2 = C_2 V_{cb}$$

$$V_{ab} = V_{ac} + V_{cb}.$$

But Q_1 must equal Q_2, since the charging process (in terms of the motion of positive charge) consists of the transfer of charge from the right plate of capacitor 2 around the circuit to the left plate of capacitor 1, and from the right plate of capacitor 1 to the left plate of capacitor 2. Hence, if Q stands for the charge on either capacitor,

$$V_{ac} = \frac{Q}{C_1}, \quad V_{cb} = \frac{Q}{C_2}, \text{ and}$$

$$V_{ab} = V_{ac} + V_{cb} = \frac{Q}{C_1} + \frac{Q}{C_2} = Q\left(\frac{1}{C_1} + \frac{1}{C_2}\right).$$

Finally, since Q/V_{ab} is by definition the equivalent capacitance C,

$$\frac{V_{ab}}{Q} = \frac{1}{C} = \frac{1}{C_1} + \frac{1}{C_2}.$$

Evidently the reciprocal of the equivalent capacitance of any number of capacitors in series equals the sum of the reciprocals of the individual capacitances. Also, since

$$Q = C_1 V_1 = C_2 V_2 = C_3 V_3, \text{ etc.},$$

when a number of capacitors are connected in series the potential differences across the individual capacitors are inversely proportional to their capacitances, although the charges on all of the capacitors are equal.

If two capacitors are connected in parallel as in Fig. 8-8, the potential difference across each is the same, and the total displaced charge is the sum of the individual charges. Hence

$$Q = Q_1 + Q_2 = C_1V_{ab} + C_2V_{ab} = V_{ab}(C_1 + C_2)$$

$$\frac{Q}{V_{ab}} = C = C_1 + C_2.$$

Evidently the equivalent capacitance of any number of capacitors in parallel equals the sum of the individual capacitances.

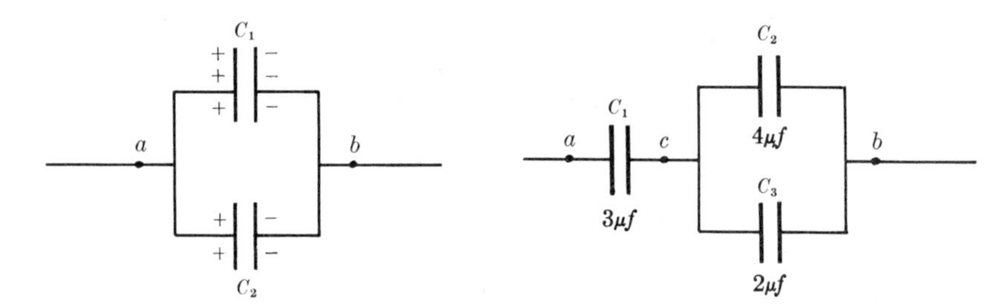

FIG. 8-8. Two capacitors in parallel. FIG. 8-9. Simple network of capacitors.

Example. Refer to Fig. 8-9. If point b is grounded and point a maintained at a potential of $+1200$ volts, find the charge on each capacitor and the potential of point c.

The 4 μf and the 2 μf capacitor are in parallel, and are hence equivalent to a single capacitor of capacitance

$$C = 4 + 2 = 6\ \mu\text{f}.$$

This equivalent capacitor is in series with the 3 μf capacitor. Hence the equivalent capacitance C_0 of the network is

$$\frac{1}{C_0} = \frac{1}{6} + \frac{1}{3} = \frac{1}{2}, \quad \text{or}$$

$$C_0 = 2\ \mu\text{f}.$$

The charge on the equivalent capacitor is

$$Q = C_0V = 2 \times 10^{-6} \times 1200 = 2.4 \times 10^{-3} \text{ coulombs.}$$

This must equal the charge on the 3 μf capacitor, and the sum of the charges on the 4 μf and 2 μf capacitors.

Now

$$V_{ac} = \frac{Q_1}{C_1} = \frac{2.4 \times 10^{-3}}{3 \times 10^{-6}} = 800^{\text{v}}$$

$$V_{ac} \equiv V_a - V_c = 800^{\text{v}}, \quad V_a = 1200^{\text{v}}, \quad \therefore V_c = 400^{\text{v}}$$

and

$$V_{cb} \equiv V_c - V_b = 400 - 0 = 400^{\text{v}}.$$

Hence

$$Q_2 = C_2V_{cb} = 4 \times 10^{-6} \times 400 = 1.6 \times 10^{-3} \text{ coulombs}$$
$$Q_3 = C_3V_{cb} = 2 \times 10^{-6} \times 400 = 0.8 \times 10^{-3} \text{ coulombs}$$
$$Q_2 + Q_3 = (1.6 + 0.8) \times 10^{-3} = 2.4 \times 10^{-3} \text{ coulombs.}$$

If the network contains cross-connections as in Fig. 8-10 it is not possible to consider it as made up of series and parallel groupings. Such networks are solved by an obvious extension of Kirchhoff's rules. For example, the algebraic sum of the potential differences across C_1, C_3, and C_4 in Fig. 8-10 must be zero, and similarly for C_2, C_5, and C_3. Also, the algebraic sum of the charges on the capacitors connected to branch points x and y must be zero.

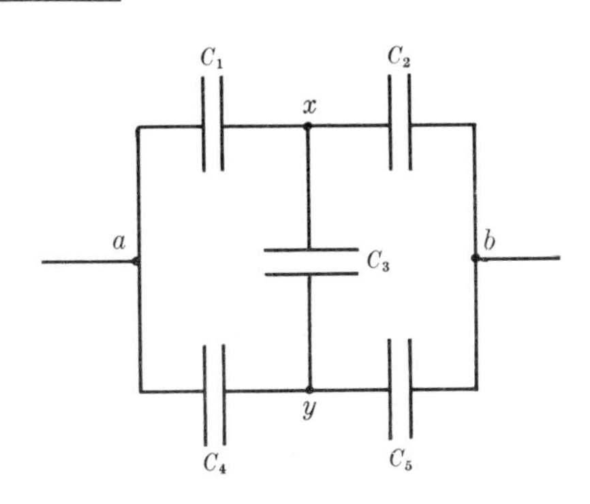

FIG. 8-10. Network of capacitors requiring extension of Kirchhoff's rules.

8-7 Energy of a charged capacitor. The process of charging a capacitor consists of transferring charge from the plate at lower potential to the plate at higher potential. The charging process therefore requires the expenditure of energy. Imagine the charging process to be carried out by starting with both plates completely uncharged, and then repeatedly removing small positive charges dq from one plate and transferring them to the other plate. At a stage of this process when the total quantity of charge transferred has reached an amount q, the potential difference between the plates is

$$V_{ab} = \frac{q}{C}$$

and the work dW necessary to transfer the next charge dq is

$$dW = V_{ab}\, dq = \frac{q}{C}\, dq.$$

The total work done in the charging process can be found by integrating the expression above from $q = 0$ to $q = Q$, where Q is the final charge.

$$W = \int dW = \frac{1}{C}\int_0^Q q\, dq = \frac{1}{2}\frac{Q^2}{C}. \tag{8-17}$$

The energy is expressed in joules when Q is in coulombs and C in farads.

Since $V_{ab} = \dfrac{Q}{C}$, Eq. (8-17) is equivalent to

$$W = \tfrac{1}{2}CV_{ab}^2 = \tfrac{1}{2}QV_{ab} \tag{8-18}$$

where V_{ab} now represents the potential difference across the capacitor when its charge is Q.

8-8 Energy density in an electric field. The principles of electricity and magnetism as they stand today may be said to have a dual aspect. On the one hand they deal with charged *particles*, the forces on them, their kinetic and potential energies, and the manner in which they move. On the other hand they are concerned with the electric and magnetic *fields* in the space surrounding the particles. In many problems, particularly those having to do with electromagnetic waves, it is on the fields rather than the particles that emphasis is placed. One of the most useful concepts of the field theory is that of the energy associated with an electric or magnetic field.

From this point of view, the energy of a charged capacitor is considered to be associated with the electric field of the capacitor rather than the charges on its plates, and to be distributed throughout the field with a certain energy per unit volume or *energy density*. Since the field of a parallel plate capacitor is uniform, the energy density is presumably uniform also, and equal to the total energy divided by the volume of the field. The latter is given by the product of the area of the capacitor plates, A, and their separation, d. Hence

$$\text{Energy density} = \frac{\tfrac{1}{2}QV_{ab}}{Ad}.$$

But

$$\frac{Q}{A} = \sigma = \epsilon E$$

and

$$\frac{V_{ab}}{d} = E.$$

Hence

$$\text{Energy density} = \frac{\epsilon E^2}{2}. \tag{8-19}$$

Since $D = \epsilon E$, this can also be written

$$\text{Energy density} = \frac{ED}{2} = \frac{D^2}{2\epsilon}. \tag{8-20}$$

In the mks system, energy density is expressed in joules per cubic meter.

Although derived for the special case of a uniform field, Eqs. (8-19) and (8-20) may be used for the energy density at any point of a nonuniform field, D and E referring to the particular point in question.

8-9. Force between the plates of a capacitor. There is, of course, a force of attraction between the positive and negative charges on the plates of a capacitor. This force can be computed by applying Coulomb's law to a pair of infinitesimal charges on the plates and performing a double integration over both plates. If the plates are to be in mechanical equilibrium, the electrical force must be balanced by a mechanical force. The latter is provided by the mountings of the plates of an air capacitor, or by the dielectric itself in a paper and foil capacitor. The *electrical* force between the charges on the plates is the same whether the region between the plates is empty or is occupied by a dielectric. But if there is a dielectric between the plates, the dielectric becomes stressed by the electric field, a phenomenon known as *electrostriction*. As a consequence of the stresses in the dielectric the latter will, in general, exert mechanical forces on the plates. The latter are then in equilibrium under a set of forces, partly electrical and partly mechanical, and the general problem is a complex one involving both electrostatics and elasticity. The elastic aspect is simplest if the dielectric is a fluid (liquid or gas), since the only type of stress that can exist in a fluid at rest is a hydrostatic pressure. This is the only case we shall consider and a more advanced text should be consulted for further details.

To understand the origin of the stresses set up in a dielectric in an electric field, consider the special case of two circular plates immersed in a large tank of liquid as in Fig 8-11(a). Each plate is in equilibrium under (a) the force due to the attraction of the charges on the other plate, (b) the force exerted by the liquid between the plates, (c) the force exerted by the liquid outside the plates, and (d) the mechanical force exerted by the supports. If the plates are in vacuum, the mechanical force (d) must be equal and opposite to the electrical force (a). If the plates are in a fluid, the mechanical force necessary to maintain equilibrium is diminished over its value in vacuum by the resultant effect of forces (b) and (c).

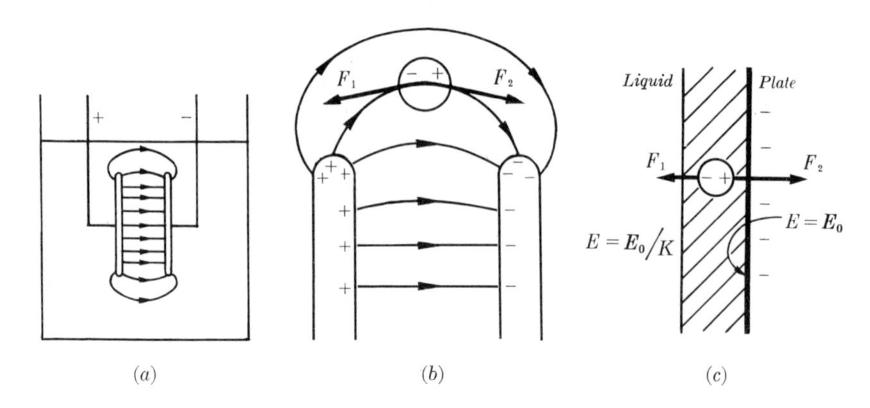

FIG. 8-11. Parallel plate capacitor with liquid dielectric.

We have to examine the situation in the liquid in two regions: the region of fringing field around the rims of the plates and the boundary of the liquid where it makes contact with the plates. The region of fringing field is shown in Fig 8-11(b). A molecule of the liquid, greatly exaggerated in size, is represented by the circle above the plates. The general direction of the field at the molecule is from left to right, and the molecule becomes polarized as indicated by the + and − signs. But because the lines of force are concave downward, the forces F_1 and F_2 on the induced charges are both inclined slightly below the horizontal. Hence there is a net downward force on this particular molecule. It is evident that all molecules lying in the region of fringing field around the rims of the plates will experience a force drawing them in toward the center of the plates. They accordingly move in a small distance and compress the liquid in the region between the plates until sufficient hydrostatic pressure is established to balance the inward electrical force. The excess pressure between the plates, over that outside, causes the liquid to push the plates apart. The mechanical force which the supports have to exert to maintain equilibrium is therefore less than the electrical force of attraction between the plates But this is not the only effect.

The boundary of the liquid adjacent to the right plate is shown in Fig. 8-11(c). In the interior of the liquid, as we have shown, the electric intensity is E_0/K_e, where E_0 is the field of the free charges and K_e is the dielectric coefficient of the liquid. *Outside* the liquid, i.e., at the surface of the plate, the intensity is E_0. The rise in electric intensity from E_0/K_e to E_0 is not discontinuous, but takes place in a small region (a few molecular diameters thick) indicated by shading. A molecule in this region, represented by the circle, becomes polarized as shown. But since the positive induced charge is in a region of greater intensity than is the negative charge, the force toward the right is greater than that toward the left. This has the effect of further increasing the pressure exerted by the liquid against the charged plate, and hence of reducing further the force the supports must exert to maintain equilibrium.

As in so many instances, a detailed calculation, both of the forces on the induced dipoles and of the increased pressure, can be circumvented with the help of energy considerations. Let the plates be charged and then disconnected from the line so that the charge Q on each plate remains constant. We then imagine the plates to be pulled apart slightly, and compute the force by equating the work done to the increase in energy of the system. Let A represent the areas of the plates, x the distance between them, and F the force exerted on each by its supports. The work dW done in a small increase in separation dx is

$$dW = F\,dx.$$

The energy density in the fluid, from Eq. (8-20), is

$$\text{Energy density} = \frac{D^2}{2\epsilon},$$

where ϵ is the permittivity of the fluid. The volume of the fluid between the plates is Ax, so the energy W in the field is

$$W = \frac{D^2}{2\epsilon} Ax.$$

The change in energy when x is increased by dx is

$$dW = \frac{D^2}{2\epsilon} A\,dx.$$

Equating the work done to the increase in energy, we obtain

$$F\,dx = \frac{D^2}{2\epsilon} A\,dx,$$

and hence

$$F = \frac{D^2}{2\epsilon} A.$$

Since $D = \sigma = Q/A$, and $\epsilon = K_e\epsilon_0$, the force can also be written as

$$F = \frac{Q^2}{2\epsilon A} = \frac{1}{K_e}\frac{Q^2}{2\epsilon_0 A}. \quad (8\text{-}21)$$

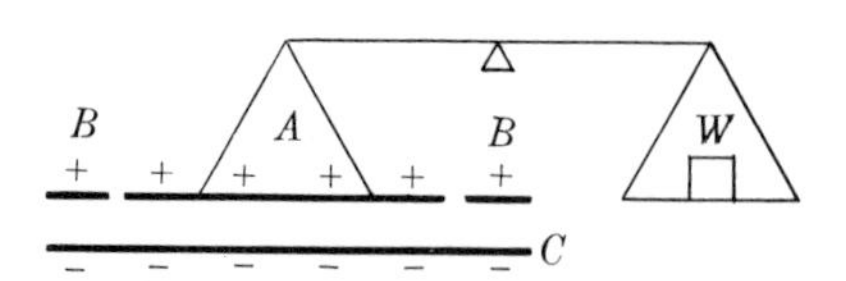

FIG. 8-12. Schematic diagram of an attracted-disk electrometer.

If the plates are in vacuum, $K_e = 1$ and the force is

$$F = \frac{Q^2}{2\epsilon_0 A}.$$

The latter equation is the result obtained by applying Coulomb's law and performing the integration mentioned at the beginning of the section. The force needed to hold the plates in equilibrium, given by Eq. (8-21), is therefore reduced by a factor of $1/K_e$ as a consequence of electrostriction. This is another illustration of the general law that the net force between charged conductors in a fluid dielectric is reduced by a factor of $1/K_e$ over its value in vacuum.

The mechanical force between a pair of oppositely-charged plates is utilized in one form of electrostatic voltmeter, the *attracted-disk electrometer*, illustrated in Fig. 8-12. A circular disk A surrounded by a closely-fitting guard ring B-B forms one plate of a parallel plate capacitor. Both A and B are kept at the same potential, while plate C is at a different potential. Plate A is suspended from one arm of an equal arm balance. The force of attraction between A and C can be measured directly by the counterweight W.

Since the charge on A is proportional to the potential difference between A and C, the force of attraction, from Eq. (8-21), is proportional to the square of this potential difference. The instrument is of no great practical value but it is an absolute instrument in the sense that it requires no calibration, but only an accurate knowledge of its dimensions.

8-10 Displacement current. Fig. 8-13 shows a series circuit consisting of a seat of emf, a resistor, a capacitor, and a switch. When the switch is closed the capacitor charges and the current in the circuit is given by Eq. (8-16). We have stated that a current consists of a flow of charge, and that the current is the same at all cross sections of a series circuit.

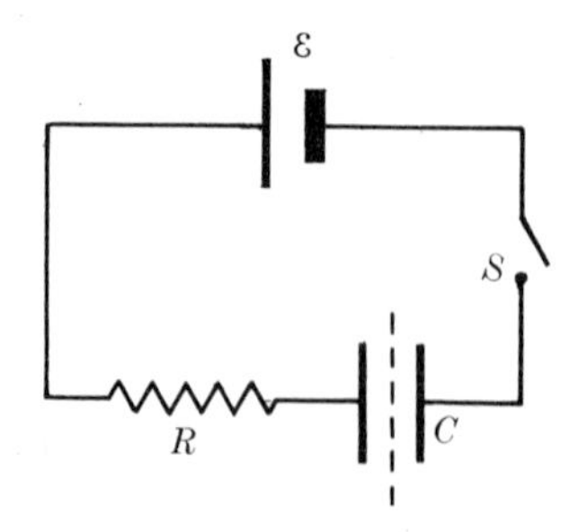

FIG. 8-13. Capacitor, resistor, and seat of emf in series.

It is evident, however, that this statement must be modified if the circuit contains a capacitor as in Fig. 8-13. The dielectric between the capacitor plates is, presumably, a nonconductor, and hence the *conduction* current at any cross section passing through the capacitor (such as the one indicated by the dotted line) is necessarily zero. The conduction current is the same at all cross sections of the circuit *except* those passing through the dielectric; as a result of the current, free electrons flow into one plate of the capacitor and force other free electrons out of the other plate, but no free electrons flow *through* the dielectric.

It was pointed out by the Scotch scientist James Clerk Maxwell (1831–1879) that by revising or extending our definition of current one could still say that the current, in the revised sense of the word, is the same at all cross sections of the circuit *including those through the dielectric.*

Let q represent the charge on the capacitor at some instant. The conduction current in the line, say i, is the rate at which charge is delivered to or removed from either plate of the capacitor, and is therefore equal to the rate of change of the charge q.

$$i = \frac{dq}{dt}\cdot$$

The displacement D in the dielectric is

$$D = \sigma = \frac{q}{A},$$

or

$$q = DA,$$

where A is the area of the capacitor plates. Then

$$\frac{dq}{dt} = A\frac{dD}{dt},$$

and

$$i = A\frac{dD}{dt}\cdot \qquad \textbf{(8-20)}$$

Maxwell's proposal was that the term $A\dfrac{dD}{dt}$ should be considered *a current in the dielectric,* which he called a *displacement current.*

$$i_D = \text{Displacement current} = A\frac{dD}{dt}\cdot \qquad \textbf{(8-21)}$$

It follows that the term $\frac{dD}{dt}$ represents the *displacement current density.*

$$J_D = \text{Displacement current density} = \frac{dD}{dt}. \qquad (8\text{-}22)$$

Displacement current obviously has the same units as conduction current, since D can be expressed in coul/m^2. Then, from Eq. (8-22), the (conduction) current in the line equals the (displacement) current in the dielectric, and the current is the same at all cross sections of the circuit, including those through the dielectric, even though the circuit is not "closed" in the sense of consisting of a closed conducting path. In other words, all circuits can be considered as closed circuits even if they consist in part of an insulator such as the dielectric of a capacitor.

Of course, no real dielectric is a perfect insulator and hence there will, in general, be a conduction current through the dielectric of any charged capacitor in addition to the displacement current. Also, if the current in a conductor is varying, the electric intensity and hence the displacement in the conductor will vary also and there will be a displacement current in the conductor in addition to the conduction current. The actual current at any cross section of a circuit is considered to be the sum of the conduction and displacement currents at that section. In most instances the conduction currents in dielectrics are negligible in comparison with displacement currents, and the displacement currents in conductors are negligible in comparison with conduction currents. At ultra-high frequencies, however, the displacement current in conductors may become an important factor.

We have shown in Sec. 7-6 that

$$D = \epsilon_0 E + P.$$

It follows that the displacement current density, dD/dt, may be written as the sum of two terms

$$J_D = \frac{dD}{dt} = \epsilon_0 \frac{dE}{dt} + \frac{dP}{dt}.$$

The second term on the right, dP/dt, does represent a "conduction" current density in the sense that while the polarization in a dielectric is changing, there is an actual shift of displaced (bound) charges across a section of the dielectric. The first term, $\epsilon_0 \frac{dE}{dt}$, does not represent any motion of charged particles but is the term which must be added to make the current the same at all points. If the plates of a capacitor are in vacuum, the dP/dt term is necessarily zero and the entire displacement current is due to the rate of change of the electric intensity.

Example. In Fig. 8-14 there is shown a "leaky" capacitor whose capacitance is 100 μf and whose resistance is 10^6 ohms, connected between points a and b, between which the potential difference may be varied at will. Consider the following situations:

(1) The capacitor has been connected for a long time to a constant pd of 100 volts. The current through the dielectric and the charge on the capacitor have therefore reached steady values. Since the charge on the capacitor is not changing, the displacement in the dielectric is constant and the displacement *current* is zero. There is a conduction current of

$$i = \frac{V_{ab}}{R} = \frac{100}{10^6} = 10^{-4} \text{ amp.}$$

through the dielectric, and an equal current in the line.

(2) The potential difference between a and b is increased from 100 volts at a uniform rate of 10 volts/sec, i.e.,

$$V_{ab} = 100 + 10t.$$

Find the conduction and displacement currents after 20 seconds.

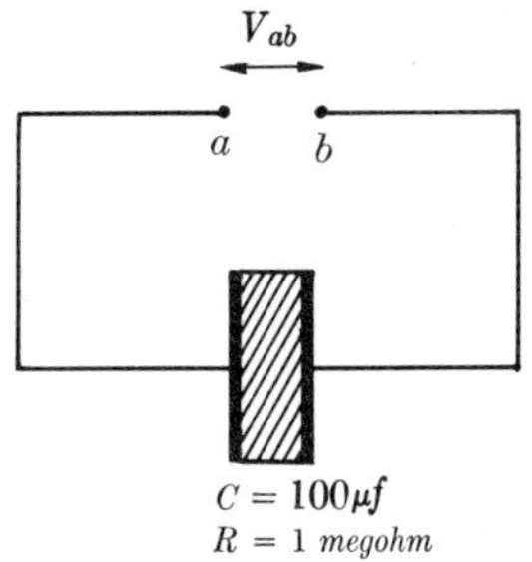

FIG. 8-14.

The conduction current in the dielectric at time t is

$$i_{\text{cond}} = \frac{V_{ab}}{R} = \frac{100 + 10t}{10^6}$$

$$= 10^{-4} + 10^{-5}t,$$

and when $t = 20$ sec,

$$i_{\text{cond}} = 3 \times 10^{-4} \text{ amp.}$$

The displacement in the dielectric at any time t is

$$D = \frac{q}{A} = \frac{CV_{ab}}{A} = \frac{10^{-4}(100 + 10t)}{A} = \frac{1}{A}(10^{-2} + 10^{-3}t),$$

and the displacement current is

$$i_{\text{displ}} = A\frac{dD}{dt} = A\frac{d}{dt}\left[\frac{1}{A}(10^{-2} + 10^{-3}t)\right]$$

$$= 10 \times 10^{-4} \text{ amp.}$$

The displacement current is constant, since the displacement increases at a constant rate.

The line current at 20 sec is therefore

$$\begin{aligned} i_{\text{line}} &= i_{\text{capacitor}} = i_{\text{cond}} + i_{\text{displ}} \\ &= 3 \times 10^{-4} + 10 \times 10^{-4} \\ &= 13 \times 10^{-4} \text{ amp.} \end{aligned}$$

That is, there is in the line a conduction current of 13×10^{-4} amperes, while in the dielectric there is a conduction current of 3×10^{-4} amperes and a displacement current of 10×10^{-4} amperes. The total current is the same at all cross sections of the circuit.

Problems—Chapter 8

(1) Considering the earth as a spherical conductor in empty space, compute its capacitance in microfarads. The earth's radius may be taken as 6730 km.

(2) The paper dielectric, in a paper and tinfoil capacitor, is 0.005 cm thick. Its dielectric coefficient is 2.5 and its dielectric strength is 50×10^6 volts/m. (a) What area of paper, and of tinfoil, is required for a 0.1 μf capacitor? (b) If the electric intensity in the paper is not to exceed one-half the dielectric strength, what is the maximum potential difference that can be applied across the capacitor? (c) Find the resistance of the dielectric, if its resistivity is 10^{14} ohm-meters.

(3) Derive Eq. (8-12) by the method used in Sec. 8-4.

(4) Complete the derivation of Eq. (8-14) from Eq. (8-13).

(5) A 10 μf capacitor is connected through a 1 megohm resistor to a constant potential difference of 100 volts. (a) Compute the charge on the capacitor at the following times after the connections are made: 0, 5 sec, 10 sec, 20 sec, 100 sec. (b) Compute the charging current at the same instants. (c) How long a time would be required for the capacitor to acquire its final charge, if the charging current remained constant at its initial value? (d) Find the time required for the charge to increase from zero to 5×10^{-4} coulombs. (e) Construct graphs of the results of parts (a) and (b) for a time interval of 20 sec.

(6) A capacitor of capacitance C_1 is charged to a potential difference V_1. Its terminals are connected through a resistance R to the terminals of an uncharged capacitor of capacitance C_2. Find the expression for the potential difference across the first capacitor at any time t after it is connected to the second.

(7) The circuit in Fig. 8-15 is used to obtain a "saw-tooth" voltage wave. The 884 tube is a small gas-filled thyratron. The tube is originally nonconducting. When switch S is closed the capacitor C charges through the resistor R. When the voltage V_{ab} reaches a critical value the tube suddenly starts to conduct. The capacitor discharges rapidly through the tube and V_{ab} decreases nearly instantaneously to a small value called the maintaining voltage. When this voltage is reached the tube again becomes non-conducting and the cycle is repeated.

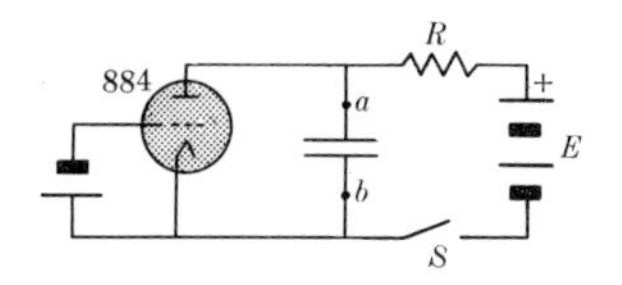

FIG. 8-15.

Assume that $\mathcal{E} = 300$ volts, $R = 200{,}000$ ohms, $C = 0.05$ μf. Assume also that the critical voltage at which the tube starts to conduct is 210 volts, and that the maintaining voltage is zero. (a) At what time after switch S is closed will the tube start to conduct? (b) Compute the potential difference V_{ab} across the capacitor at times equal to $\frac{1}{4}$, $\frac{1}{2}$, and $\frac{3}{4}$ of that in part (a). (c) Construct a graph of V_{ab} as a function of time.

(8) A capacitor of capacitance C is charged by connecting it through a resistance R to the terminals of a battery of emf $\mathcal{E}$ and of negligible internal resistance. (a) How much energy is supplied by the battery in the charging process? (b) What fraction of this energy appears as heat in the resistor?

(9) The capacitance of a variable radio capacitor can be changed from 50 $\mu\mu$f to 950 $\mu\mu$f by turning the dial from 0° to 180°. With the dial set at 180° the capacitor is connected to a 400-volt battery. After charging, the capacitor is disconnected from the battery and the dial is turned to 0°. (a) What is the charge on the capacitor? (b) What is the potential difference across the capacitor when the dial reads 0°? (c) What is the energy of the capacitor in this position? (d) How much work is required to turn the dial, if friction is neglected?

(10) An air capacitor, consisting of two closely spaced parallel plates, has a capacitance of 1000 $\mu\mu$f. The charge on each plate is 1 microcoulomb. (a) What is the potential difference between the plates? (b) If the charge is kept constant, what will be the potential difference between the plates if the separation is doubled? (c) How much work is required to double the separation?

(11) A 20 μf capacitor is charged to a potential difference of 1000 volts. The terminals of the charged capacitor are then connected to those of an uncharged 5 μf capacitor. Compute: (a) the original charge of the system; (b) the final potential difference across each capacitor; (c) the final energy of the system; (d) the decrease in energy when the capacitors are connected.

(12) A 1 μf capacitor and a 2 μf capacitor are connected in series across a 1000-volt supply line. (a) Find the charge on each capacitor and the voltage across each. (b) The charged capacitors are disconnected from the line and from each other, and re-connected with terminals of like sign together. Find the final charge on each and the voltage across each.

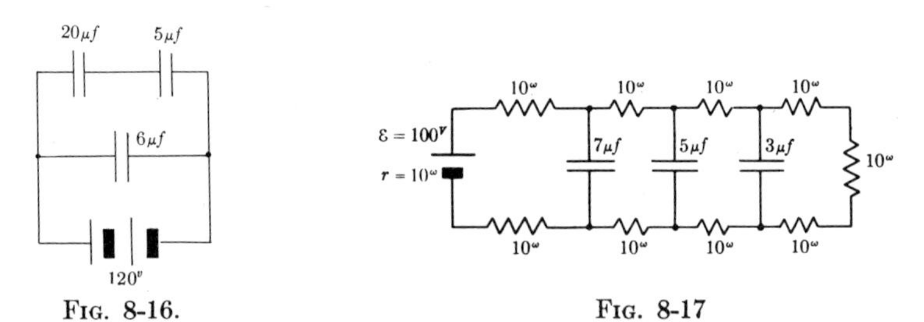

FIG. 8-16. FIG. 8-17

(13) A 1 μf capacitor and a 2 μf capacitor are connected in parallel across a 1000-volt supply line. (a) Find the charge on each capacitor and the voltage across each. (b) The charged capacitors are then disconnected from the line and from each other, and re-connected with terminals of unlike sign together. Find the final charge on each and the voltage across each.

(14) In the network in Fig. 8-16, find: (a) the charge on each capacitor; (b) the voltage across each; (c) the total energy stored in the three capacitors.

(15) In the network in Fig. 8-17, find the charge on each capacitor and the current in the battery.

(16) In Fig. 8-18, each capacitance $C_3 = 3$ μf and each capacitance $C_2 = 2$ μf. (a) Compute the equivalent capacitance of the network between points a and b. (b) Compute the charge on each of the capacitors nearest a and b, when $V_{ab} = 900$ volts. (c) With 900 volts across a and b, compute V_{cd}.

(17) In the network in Fig. 8-19, compute the equivalent capacitance between points a and b.

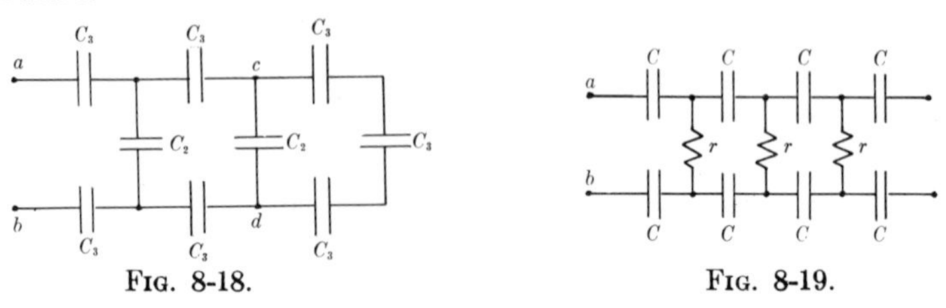

FIG. 8-18. FIG. 8-19.

(18) Refer to example 2 at the end of Sec. 8-3. Suppose the capacitor plates had been kept connected to the charging voltage of 10,000 volts while the dielectric was inserted. What would then have been the displacement and electric intensity in the dielectric, and the surface density of charge on the capacitor plates?

(19) Refer to example 3 at the end of Sec. 8-3. (a) Find the polarization P in each dielectric. (b) Find the surface density of induced charge at the surfaces of each dielectric. (c) What is the net or resultant surface density of charge at the surface of contact between the dielectrics? (d) What is the potential at this surface, if the negative plate of the capacitor is grounded?

(20) A conducting sphere of radius a, in empty space, has a charge Q. (a) What is the energy density in the electric field of the sphere at points within a thin spherical shell of radius r ($r > a$) and thickness dr? (b) What is the total energy in the shell? (c) Integrate the preceding expression to find the total energy in the field of the sphere. (d) Equate this energy to that of a charged capacitor, $\frac{1}{2}\frac{Q^2}{C}$, and derive the expression for the capacitance of a sphere. Compare with Eq. (8-3).

(21) Two coaxial cylinders of length l and radii a and b are given equal and opposite charges Q, as in Fig. 8-3. The space between the cylinders contains a dielectric of permittivity ϵ. (a) What is the energy density in the dielectric, at points within a thin cylindrical shell of radius r ($a < r < b$) and thickness dr? (b) What is the total energy in the shell? (c) Integrate the preceding expression to find the total energy in the dielectric. (d) Equate this energy to that of a charged capacitor, $\frac{1}{2}\frac{Q^2}{C}$, and derive the expression for the capacitance of a cylindrical capacitor. Compare with Eq. (8-11).

(22) The resistance of the dielectric of an 0.1 μf capacitor is 100 megohms. The dielectric coefficient of the dielectric is 3. Suppose the potential difference across the capacitor at a certain instant is 1000 volts, and is increasing at the rate of 10^6 volts/sec. Find the conduction current in the dielectric, the displacement current in the dielectric, and the current in the leads to the capacitor terminals.

(23) The market price of fifty 600-volt, 1 μf capacitors, is roughly the same as that of a 6-volt, 120 ampere-hour storage battery. Compare the energy obtainable from the fully charged battery with that stored in the fifty capacitors, each charged to a potential difference of 600 volts.

CHAPTER 9

THE MAGNETIC FIELD

9-1 Magnetism. The first magnetic phenomena to be observed were undoubtedly those associated with so-called "natural" magnets, rough fragments of an ore of iron found near the ancient city of Magnesia (whence the term "magnet"). These natural magnets have the property of attracting to themselves unmagnetized iron, the effect being most pronounced at certain regions of the magnet known as its *poles*. It was known to the Chinese as early as A.D. 121 that an iron rod, after being brought near a natural magnet, would acquire and retain this property of the natural magnet, and that such a rod when freely suspended about a vertical axis would set itself approximately in the north-south direction. The use of magnets as aids to navigation can be traced back at least to the eleventh century.

The study of magnetic phenomena was confined for many years to magnets made in this way. Not until 1819 was there shown to be any connection between electrical and magnetic phenomena. In that year the Danish scientist Hans Christian Oersted (1770–1851) observed that a pivoted magnet (a compass needle) was deflected when in the neighborhood of a wire carrying a current. Twelve years later, after attempts extending over a period of several years, Faraday found that a momentary current existed in a circuit while the current in a nearby circuit was being started or stopped. Shortly afterward followed the discovery that the motion of a magnet toward or away from the circuit would produce the same effect. Joseph Henry (1797–1878), an American scientist who later became the first director of the Smithsonian Institution, had anticipated Faraday's discoveries by about twelve months, but since Faraday was the first to publish his results he is usually assigned the credit for them. The work of Oersted thus demonstrated that magnetic effects could be produced by moving electric charges, and that of Faraday and Henry that currents could be produced by moving magnets.

It is believed at the present time that all so-called magnetic phenomena arise from forces between electric charges in motion. That is, moving charges exert "magnetic" forces on one another, over and above the purely "electrical" or "electrostatic" forces given by Coulomb's law. We shall therefore begin the subject of magnetism with a study of the forces between moving charges and postpone the question of magnetic poles until a later chapter, where they will be discussed in connection with the magnetic properties of matter in general.

Since the electrons in atoms are in motion about the atomic nuclei, and since each electron appears to be in continuous rotation about an axis passing through it, all atoms can be expected to exhibit magnetic effects and, in fact, such is found to be the case. The possibility that the magnetic properties of matter were the result of tiny atomic currents was first suggested by Ampere in 1820. Not until recent years has the verification of these ideas been possible.

As with electrostatic forces, the medium in which the charges are moving may have a pronounced effect on the observed magnetic forces between them. In the present chapter we shall assume the charges or conductors to be in otherwise empty space. For all practical purposes the results will apply equally well to charges and conductors in air.

9-2 The magnetic field. Induction. Instead of dealing directly with the forces exerted on one moving charge by another, it is found more convenient to adopt the point of view that a moving charge sets up in the space around it a *magnetic field,* and that it is this field which exerts a force on another charge moving through it. The magnetic field around a moving charge exists in addition to the electrostatic field which surrounds the charge whether it is in motion or not.[1] A second charged particle in these combined fields experiences a force due to the electric field whether it is in motion or at rest. The magnetic field exerts a force on it only if it is in motion.

A magnetic field is said to exist at a point if a force (over and above any electrostatic force) *is exerted on a moving charge at the point.*

The electric field set up by moving charges or by currents is, in many instances, so small that the electrostatic force on a moving charge can be neglected in comparison with the magnetic force.

There are two aspects to the problem of computing the magnetic force between moving charges. The first is that of finding the magnitude and direction of the magnetic field at a point, given the data on the moving charges that set up the field. The second is to find the magnitude and direction of the force on a charge moving in a given field. We shall take

[1] In order to specify a motion some reference frame is necessary. The precise formulation of the expression for the fields associated with moving charges and the forces experienced by them, requires a careful analysis of the assumptions regarding the reference frame and the role of the observer. The interested reader is referred to a text such as J. A. Stratton, *Electromagnetic Theory.*

up the latter aspect of the problem first. That is, let us accept for the present the fact that moving charges and currents do set up magnetic fields, and study the laws that determine the force on a charge moving through the field.

There are many similarities between electric and magnetic fields, although the two are inherently different in nature and relate to entirely distinct sets of phenomena. Like an electric field, a magnetic field is a vector quantity having at every point a certain magnitude and a certain direction. Furthermore, just as it was found useful to introduce two electric vectors E and D to describe various aspects of an electric field, so it is useful to introduce two similarly related magnetic field vectors B and H. We shall begin by discussing the *magnetic induction* vector, B.

A magnetic field, like an electric field, can be represented by lines called *lines of induction*, whose direction at every point is that of the magnetic induction vector. By convention, the number of these lines per unit area normal to their direction is made equal to the magnitude of the induction. Hence the induction at a point can be expressed in "lines per unit area." In the mks system, a line of induction is called a *weber*, and in this system the magnetic induction B is expressed in *webers per square meter* (w/m^2). A submultiple of this unit, one one-thousandth as great and called a *milliweber per square meter* (mw/m^2) is often of more convenient size.

Although we shall not use it in this book, there is another system of units known as the *electromagnetic system* in terms of which magnetic quantities are often expressed. The unit of charge in this system is *one abcoulomb*, equal to 10 coulombs. The current unit is the *abampere*, equal to one abcoulomb per second or 10 amperes. The units of mass, length, time, etc., are those of the cgs system.

In the electromagnetic system a line of induction is called a *maxwell*, and magnetic induction is expressed in *maxwells per square centimeter*. One maxwell per square centimeter is called one *gauss*. There is no similar single term in the mks system for one weber per square meter. One gauss, or one maxwell per square centimeter, corresponds to a magnetic induction one ten-thousandth as great as a weber per square meter.

$$1 \text{ w/m}^2 = 10^4 \text{ maxwells/cm}^2 = 10^4 \text{ gauss}.$$

Since $1 \text{ m}^2 = 10^4 \text{ cm}^2$, it follows that

$$1 \text{ weber} = 10^8 \text{ maxwells}.$$

The largest values of magnetic induction that can be produced in the laboratory are of the order of 10 w/m² or 100,000 gauss, while in the magnetic field of the earth the induction is only a few hundred-thousandths of a weber per square meter or a few tenths of a gauss.

The total number of lines of induction threading through a surface is called the *magnetic flux* through the surface and is denoted by Φ. From the discussion in Sec. 2-5 it is evident that in general

$$\Phi = \int B \cos \phi \, dA, \tag{9-1}$$

and in the special case where B is uniform and normal to a finite area A,

$$\Phi = BA. \tag{9-2}$$

In the mks system, magnetic flux is expressed in webers. Since the induction at a point equals the flux per unit area, it is often referred to as the *flux density*.

9-3 Force on a moving charge. Fig. 9-1(a) represents a region throughout which the magnetic flux density is uniform and perpendicular to the Y-Z plane. That is, the lines of induction are straight, parallel to the X-axis, and uniformly spaced. A positive charge q, moving with velocity v perpendicular to the direction of the induction, is found to ex-

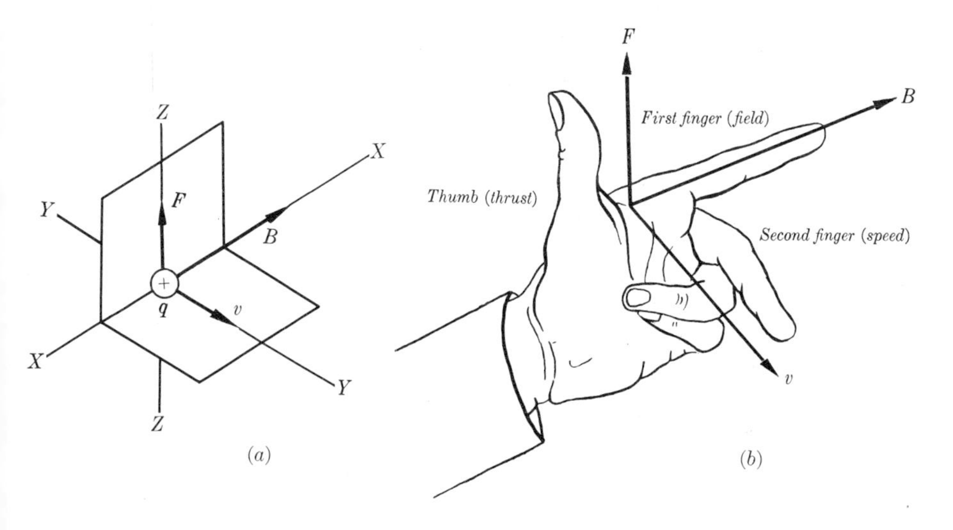

FIG. 9-1. (a) Charge moving perpendicular to a uniform magnetic field. (b) The left-hand rule.

perience a force F in the direction shown, perpendicular both to its velocity v and to the induction B. The magnitude of this force is given by

$$F = qvB \qquad (9\text{-}3)$$

In the mks system, the force F is in newtons when q is in coulombs, v is in m/sec, and B is in w/m^2.

The vectors B, v, and F form a mutually perpendicular set. The relation between their directions may be kept in mind by the *left-hand rule,* which utilizes the thumb and the first two fingers of the left hand to represent these mutually perpendicular directions (Fig. 9-1(b)). The First finger points in the direction of the Field, the Second finger in the direction of the velocity (or Speed), and the THumb points in the direction of the force or (THrust). (The direction of the force on a negative charge is opposite to that on a positive charge.)

Fig. 9-1 represents a special case in which the velocity of the moving charge is perpendicular to the direction of the field. More generally, if the velocity vector makes an angle ϕ with the induction vector as in Fig. 9-2, the magnitude of the force exerted on the charge is given by

$$\boxed{F = qvB \sin \phi} \qquad (\phi \text{ is the angle between } v \text{ and } B) \quad (9\text{-}4)$$

Hence the force is proportional to the *component* of the velocity ($v \sin \phi$) perpendicular to the direction of the magnetic induction. The *direction* of the force is always perpendicular to the plane determined by v and B. Obviously, from Fig. 9-2, when $\phi = 90°$ the situation reduces to that of Fig. 9-1 and when $\phi = 0$, *no* force is experienced by the moving charge.

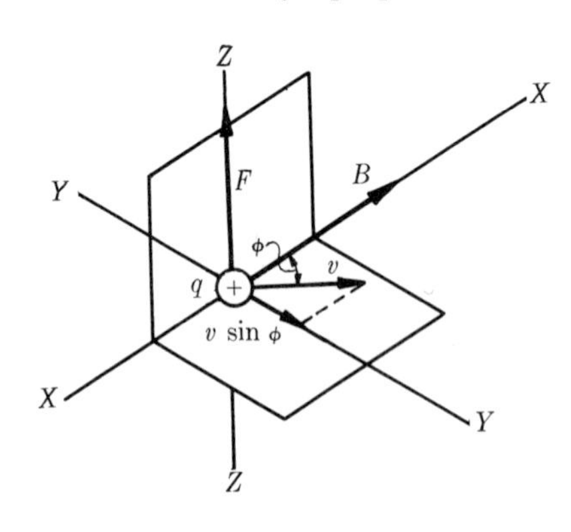

FIG. 9-2. Charge moving at an angle to a uniform magnetic field.

The vector product, or the cross product of two vectors is defined as a vector whose magnitude is the product of the magnitudes of the two vectors and the sine of the angle between them. The cross product is at right angles to the plane determined by the two vectors, and its sense is that of the advance of a right-hand screw when rotated so as to turn the first vector into the second. (See Phillips, Analytical Geometry and Calculus, page 376.) Evidently, then, the force $\vec{F}$ on a charge q moving with velocity $\vec{v}$ in a magnetic field of flux density $\vec{B}$ is, in vector notation,

$$\vec{F} = q\vec{v} \times \vec{B} \tag{9-5}$$

Eq. (9-4) may be used to *define* the magnetic induction. That is,

$$B = \frac{F}{qv \sin \phi} \quad \text{(Definition of } B\text{)}.$$

In words, *the magnitude of the magnetic induction at a point is the quotient obtained when the force on a moving charge at the point is divided by the product of the charge and its component of velocity normal to the induction.* (Compare with the corresponding definition of electric intensity E in Sec. 2-1.) Hence, in the mks system, magnetic induction may also be expressed in newtons per (coulomb-meter per second). Since one ampere is one coulomb per second, these units are equivalent to newtons per ampere-meter. In practice the term webers per square meter is much more commonly used.

When v and B are parallel, the component $v \sin \phi$ is zero and the force on the moving charge is zero. The *direction* of the induction vector at a point may therefore be defined as *the direction of the velocity of a moving charge such that the force on the charge is zero.* Later on we shall describe more practical experimental methods, consistent with the above definitions, for determining both the magnitude and direction of the magnetic induction vector.

The direction of an electric field is, by definition, the same as that of the force on a positive charge. It is puzzling at first sight to understand why the direction of a magnetic field is not defined in the same way, namely, as the direction of the force on a *moving* positive charge, but a little thought shows that such a definition is not possible. Let the direction of a magnetic field at the point P, Fig. 9-3, be away from the reader perpendicular to the plane of the diagram. The B-vectors and the lines of induction are then seen end-on. A vector directed away from the reader is represented by a cross (×) which may be thought of as representing the tail feathers of an arrow. A vector directed toward the reader is represented by a dot (•), corresponding to the point of the arrow. The same convention is used to show the direction of a current when it is at right angles to the plane of a diagram.

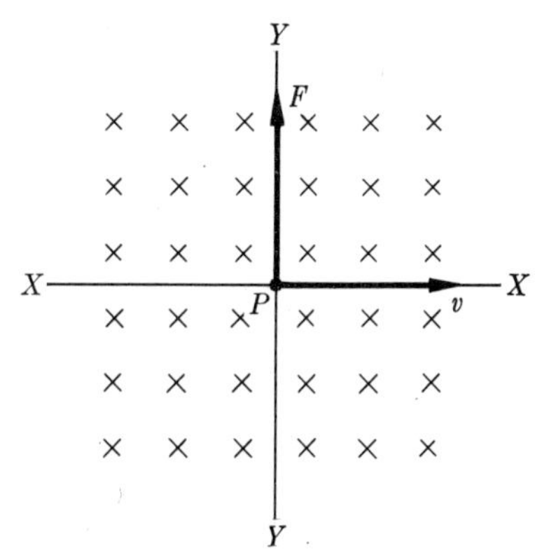

FIG. 9-3. The force on a moving charge and the velocity of the charge are always perpendicular to each other.

A positive charge at P, moving in the direction indicated by the arrow v, experiences a force F. If the velocity were in the direction of the positive Y-axis the force would be along the negative X-axis and so on, F and v always being at right angles. Hence *there is no one direction* in which force is exerted on a moving charge at a point in a magnetic field, and the direction of the force cannot be used to define the direction of the field.

One can say this, however, that regardless of the direction in which the charge is moving *the force vector must lie somewhere in the plane of the diagram* of Fig. 9-3. Hence when describing magnetic fields it is a *plane* which must be defined and not a line. But the direction of the magnetic field does define a plane perpendicular to the direction of this field, in the same way that an angular velocity vector defines a plane of rotation perpendicular to this vector. The particular orientation of the force vector in this plane is fixed by the requirement that it must be perpendicular to the velocity.

Example. An electron is projected into a magnetic field of flux density $B = 10\ \text{w/m}^2$ with a velocity of 3×10^7 m/sec in a direction at right angles to the field. Compute the magnetic force on the electron and compare with the weight of the electron.

The magnetic force is

$$F = qvB = 1.6 \times 10^{-19} \times 3 \times 10^7 \times 10 = 4.8 \times 10^{-11} \text{ newtons.}$$

The gravitational force, or the weight of the electron, is

$$F = mg = 9 \times 10^{-31} \times 9.8 = 8.8 \times 10^{-30} \text{ newtons.}$$

The gravitational force is therefore negligible in comparison with the magnetic force.

9-4 Orbits of charged particles in magnetic fields.

Let a positively-charged particle at point O in a uniform magnetic field of flux density B be given a velocity v in a direction at right angles to the field. (Fig. 9-4.) The left-hand rule shows that an upward force F, equal to qvB, is exerted on the particle at this point. Since the force is at right angles to the velocity, it will not affect the magnitude of this velocity but will merely alter its direction. At points such as P and Q the directions of force and velocity will have altered as shown, the magnitude of the force remaining constant since the magnitudes of q, v, and B are constant. The particle therefore moves under the influence of a force whose magnitude is constant but whose direction is always at right angles to the velocity of the particle. The orbit of the particle is therefore a circle described with

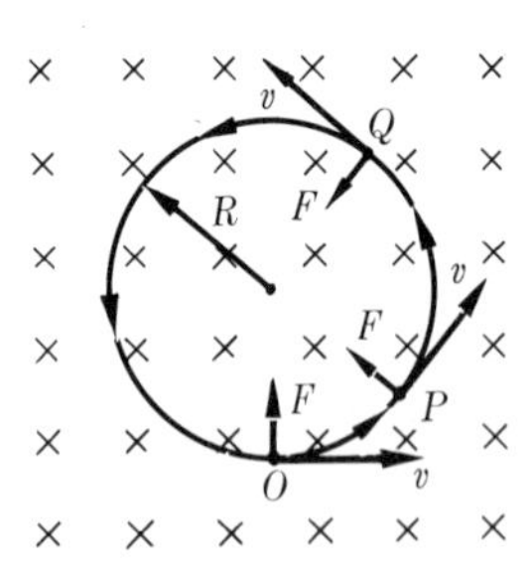

FIG. 9-4. The orbit of a charged particle in a uniform magnetic field is a circle when the initial velocity is perpendicular to the field.

constant tangential speed v, the force F being the centripetal force. Since

$$\text{centripetal force} = \frac{mv^2}{R},$$

we have

$$qvB = \frac{mv^2}{R},$$

and the radius of the circular orbit is

$$R = \frac{mv}{Bq}.$$

If the direction of the initial velocity is not perpendicular to the field, the particle moves in a helix. The cross section of the helix is a circle of radius

$$\frac{mv \sin \phi}{Bq}.$$

The axial velocity along the helix is constant and equal to

$$v \cos \phi,$$

where ϕ is the angle between v and B.

9-5 The cyclotron. The cyclotron is an instrument developed in 1931 by Drs. Ernest O. Lawrence and M. Stanley Livingston at the University of California at Berkeley, for the purpose of securing a beam of charged atomic particles traveling at high speed. Despite its size and complexity, the basic theory of its operation is quite simple.

The heart of the cyclotron is a pair of metal chambers shaped like the halves of a pillbox that has been cut along one of its diameters. (See Fig. 9-5.) These hollow chambers, referred to as "dees" or "D's" because of their shape, have their diametric edges parallel and slightly separated from one another. A source of ions—the positively charged nuclei of heavy hydrogen (deuterons) are commonly used—is located near the midpoint of the gap between the dees. The latter are connected to the terminals of an electric circuit of the same sort as that used in a radio transmitter. The potential between the dees is thus caused to alternate rapidly, some millions of times per second, so the electric field in the gap between the dees is directed first toward one and then toward the other. But because of the electrical shielding effect of the dees the space within each is a region of zero electric field.

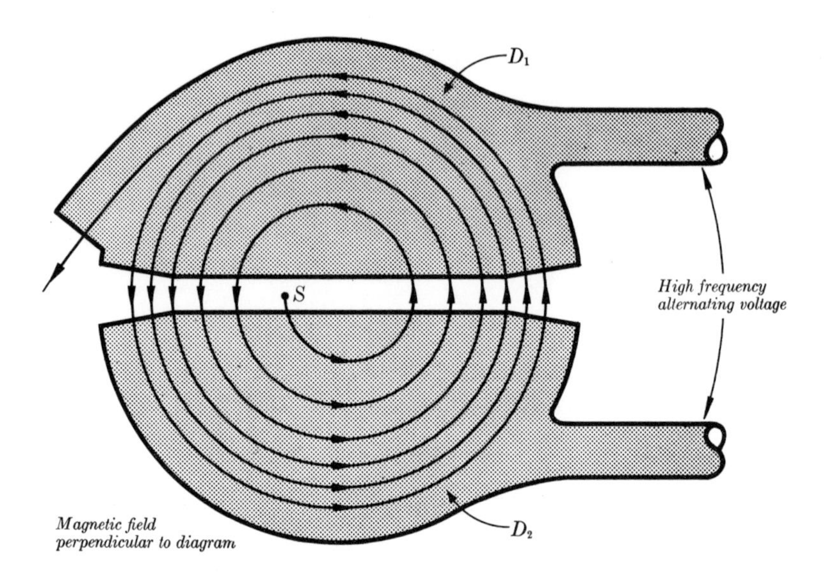

FIG. 9-5. Schematic diagram of a cyclotron.

The two dees are enclosed within, but insulated from a somewhat larger cylindrical metal container from which the air is exhausted, and the whole apparatus is placed between the poles of a powerful electromagnet which provides a magnetic field whose direction is perpendicular to the ends of the cylindrical container.

Consider an ion of charge $+q$ and mass m, emitted from the ion sources at an instant when D_1 in Fig. 9-5 is positive. The ion is accelerated by the electric field in the gap between the dees and enters the (electric) field-free region within D_2 with a speed, say, of v_1. Since its motion is at right angles to the magnetic field, it will travel in a circular path of radius

$$r_1 = \frac{mv_1}{Bq}.$$

If now, in the time required for the ion to complete a half-circle, the *electric* field has reversed so that its direction is toward D_1, the ion will again be accelerated as it crosses the gap between the dees and will enter D_1 with a larger velocity v_2. It therefore moves in a half circle of larger radius within D_1 to emerge again into the gap.

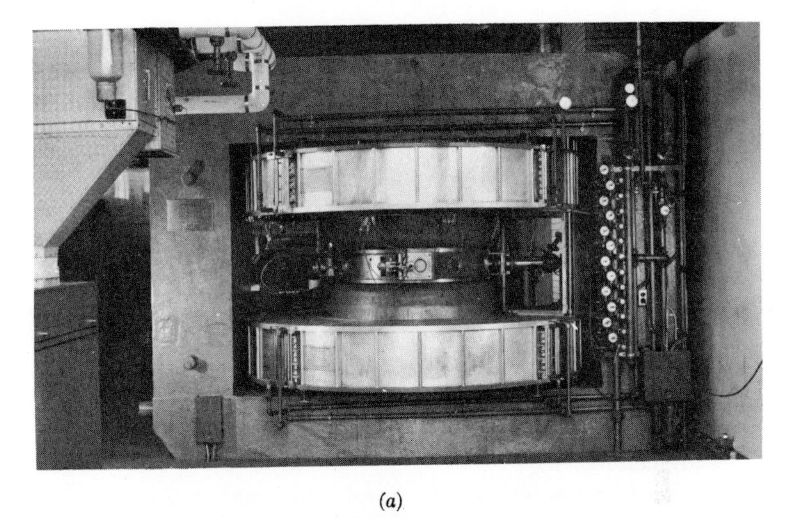

(a)

(b)

FIG. 9-6. The M.I.T. cyclotron.

The angular velocity ω of the ion is

$$\omega = \frac{v}{r} = B\frac{q}{m}.$$

Hence the angular velocity is *independent of the speed of the ion and of the radius of the circle* in which it travels, depending only on the magnetic induction and the charge-to-mass ratio (q/m) of the ion. If, therefore, the electric field reverses at regular intervals, each equal to the time required for the ion to make a half revolution, the field in the gap will always be in the proper direction to accelerate an ion each time the gap is crossed. It is this feature of the motion, the independence of time of rotation on radius, which makes the cyclotron feasible, since the regularly timed reversals are accomplished automatically by the "radio" circuit to which the dees are connected.

The path of an ion is a sort of spiral, composed of semicircular arcs of progressively larger radius, connected by short segments along which the radius is increasing. If R represents the outside radius of the dees and $v_{\max}$ the speed of the ion when traveling in a path of this radius,

$$v_{\max} = BR\frac{q}{m},$$

and the corresponding kinetic energy of the ion is

$$\frac{1}{2}mv_{\max}^2 = \frac{1}{2}m\left(\frac{q}{m}\right)^2 B^2R^2.$$

The potential difference V which would be required to produce the same kinetic energy in a single step, as in the Van de Graaff generator, can be found from

$$\frac{1}{2}mv_{\max}^2 = qV,$$

or

$$V = \frac{1}{2}\frac{q}{m}\cdot B^2R^2. \tag{9-7}$$

If the ions are protons,

$$\frac{q}{m} = 9.6 \times 10^7 \text{ coul/kgm}.$$

In the M.I.T. cyclotron, B is about 1.3 w/m^2 and $R = 0.48$ m. Hence

$$\begin{aligned} V &= \tfrac{1}{2} \times 9.6 \times 10^7 \times (1.3)^2 \times (.48)^2 \\ &= 19 \times 10^6 \text{ volts, or 19 million volts,} \end{aligned}$$

and the deuterons have the same speed as if they had been accelerated through a potential difference of 19 million volts.

Fig. 9-6(a) is a photograph of the M.I.T. cyclotron, and Fig. 9-6(b) shows the dees removed from the gap between the poles of the electromagnet.

A new cyclotron under construction at Berkeley will have dees 4.5 m in diameter (about 15 ft) and is expected to accelerate particles to a speed equivalent to a potential difference of 100 million volts.

The cyclotron operates successfully only with relatively massive particles such as protons or deuterons. It cannot be used to accelerate electrons for the following reason. In order that a particle shall remain in phase with the alternating electric field its angular velocity, Bq/m, must remain constant. Now although q and B are constants, the same is not true of the mass m. The latter increases with increasing velocity because of relativistic effects. For a given energy, the velocity of an electron is much greater than that of a more massive proton or deuteron and the relativistic increase of mass is correspondingly more pronounced. For example, the mass of a 2-million volt electron is about five times its rest mass, while the mass of a 2-million volt deuteron differs from its rest mass by only about .01 percent. Hence electrons very quickly get out of phase with the electric field and do not arrive at the gap between the dees at the proper times to be accelerated.

This difficulty does not arise in a generator of the Van de Graaff type, and it can also be surmounted by a new device called a *betatron*, as described in Sec. 12-4.

9-6 Measurement of e/m. The charge to mass ratio of an electron, e/m, was first measured by Sir J. J. Thomson in 1897 at the Cavendish Laboratory in Cambridge, England. The apparatus used by Thomson was a glass tube having the form shown in Fig. 9-7, from which most of the air had been evacuated. A potential difference of a few thousand volts is maintained between anode A and cathode C. The few positive ions which are always present in a gas, caused by radioactivity or cosmic radiation, are accelerated toward the cathode by the electric field in the anode-cathode space. When these rapidly moving ions strike the cathode, they liberate electrons from its surface and these electrons accelerate in the opposite direction toward the anode A. Occasional collisions between them and the residual gas atoms maintain the supply of positive ions. This mechanism of releasing electrons from a surface is called "secondary emission."

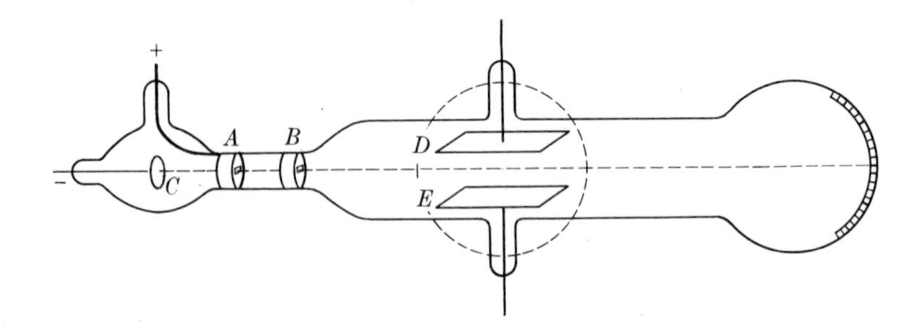

FIG. 9-7. Apparatus of J. J. Thomson to measure the ratio e/m of electrons.

At the time these experiments were being performed the nature of the electron stream was not known, and the particles composing it (if they were particles at all) were simply called *cathode rays.*

Most of the electrons emitted at the cathode are stopped by the anode, but a narrow beam passes through the slit in the anode and through a second slit in the metal plug B. The end of the tube is coated on the inside with a fluorescent material and the point of impact of the electron stream appears as a bright luminous spot. Except for the mechanism of electron emission, the tube is essentially the same as that now used in cathode ray oscillographs and television receivers.

The tube is provided with two metal plates D and E between which a vertical electric field can be set up, and by means of an external electromagnet a magnetic field can be established, perpendicular to the plane of the diagram, within the region indicated by the dotted circle. If the electric field is upward, and if there is no magnetic field, the electron stream will be deflected downward as it passes between the plates and its deflection can be observed on the fluorescent screen. If a magnetic field only is present, directed out from the plane of the diagram, the electron stream will be deflected upward. By the proper adjustment of the combined electric and magnetic fields, the deflection can be made zero. Under these conditions the electric force eE on an electron in the region of "crossed fields" is equal and opposite to the magnetic force evB.

$$eE = evB, \quad \text{or}$$

$$v = \frac{E}{B}, \tag{9-8}$$

from which the velocity of the electrons can be found.

Now let the electric field be cut off. The electron stream then moves through the magnetic field in a circular arc, the radius of this arc being, as we have seen,

$$R = \frac{mv}{eB}.$$

Hence

$$\frac{e}{m} = \frac{v}{RB},$$

and combining with Eq. (9-8) we obtain

$$\frac{e}{m} = \frac{E}{RB^2}.$$

The radius R can be deduced from the displacement of the fluorescent spot, the extent of the magnetic field, and the distance from the field to the screen, and hence e/m can be found.

Many modifications of Thomson's original method have been used to determine e/m. The most precise value to date is

$$\frac{e}{m} = 1.7592 \times 10^{11} \text{ coulomb/kgm.} \tag{9-9}$$

In one sense, these experiments of Thomson constituted the "discovery" of the electron, although they measured only the ratio of the electronic charge to its mass and the charge on an individual electron was not determined until 12 years later by Millikan. If the assumption is made that the electronic charge is numerically equal to the charge on a hydrogen ion, the mass of an electron can be computed from this charge and the charge-to-mass ratio. On the basis of this assumption Thomson concluded that the mass of an electron was only about 1/1860 as great as that of a hydrogen atom, the lightest particle which had hitherto been known. The way was thus opened for the study of subatomic particles and the structure of the "indivisible" atoms, a subject which has dominated the field of experimental physics during the last forty years.

It should be mentioned that Thomson and one of his pupils, H. A. Wilson, pointed the way for Millikan's later measurements of the electronic charge when they attempted to measure this quantity by determining the *total* charge on a cloud of water droplets formed in the Wilson cloud chamber. Millikan's great contribution was his success in making measurements on a single isolated droplet.

If we insert in Eq. (9-9) the value of e based on Millikan's measurements, we find for the mass of an electron,

$$m = 9.1066 \times 10^{-31} \text{ kgm} = 9.1066 \times 10^{-28} \text{ gm}.$$

Now the mass of a hydrogen atom is

$$\frac{1.00813}{6.0228 \times 10^{23}} = 1.67339 \times 10^{-24} \text{ gm}.$$

Hence the electronic mass is only 1/1824 as great as the mass of a hydrogen atom.

9-7 The mass spectrograph. As another illustration of the motion of charged particles in electric and magnetic fields we shall describe briefly one type of *mass spectrograph.* The mass spectrograph is an instrument, similar in principle to Thomson's apparatus for measuring the charge-to-mass ratio of an electron, but used for determining the masses of positively charged ions. The first apparatus was designed by Thomson, but many modifications and improvements have been made by other investigators, notably Aston in England, and Dempster and Bainbridge in this country.

A mass spectrograph designed by Bainbridge is illustrated schematically in Fig. 9-8. S_1 and S_2 are metal plates in each of which is a narrow slit whose long dimension is perpendicular to the plane of the diagram. Plate S_1 is maintained at a potential of a few thousand volts above S_2. Positive ions produced in the space above S_1 pass through the slit in this plate and are accelerated toward S_2 by the electric field. We shall consider only singly-charged ions, that is, those that have lost one electron and hence have a positive charge of magnitude e. Not all ions, however, have the same velocity when they go through the slit S_1, and hence their velocities will differ when they emerge from S_2.

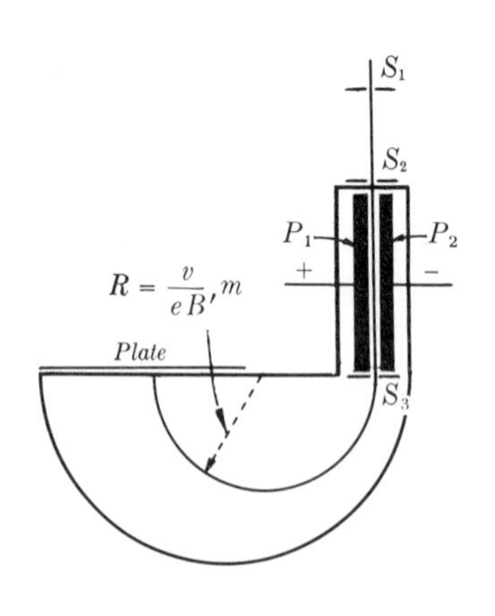

FIG. 9-8. Bainbridge's mass spectrograph utilizing a velocity selector.

The next stage in the apparatus is a "velocity filter" which selects only those ions having a predetermined velocity. An electric field, directed from left to right, is maintained between the plates P_1 and P_2, together with a magnetic field perpendicular to the plane of the diagram. A third slit S_3 is placed below the plates P_1 and P_2. Only those ions can pass through the slit S_3 which are undeviated by the crossed electric and magnetic fields between S_2 and S_3, that is, only those whose velocity is such that the magnetic force on them is just balanced by the electric force, or

$$Ee = evB$$

and

$$v = \frac{E}{B}.$$

In the region below S_3 there is also a magnetic field B' perpendicular to the plane of the figure, but no electric field. In this region an ion moves in a circular path of radius

$$R = \frac{mv}{eB'} = \left(\frac{v}{eB'}\right) m.$$

The velocity filter ensures that all ions in this region have the same velocity. Hence the ratio v/eB' is the same for all ions and the radius R is directly proportional to the mass of the ion, m. Ions of different mass travel in different semicircular paths and strike a photographic plate after having made one-half a revolution. The emulsion on the plate is rendered developable when struck by the ions, just as it is by exposure to light. Since the long dimensions of the slits are at right angles to the plane of the figure, the semicircular paths are like curved ribbons or tapes, and each path produces a line on the plate at the point where it strikes. The distance of any line from the slit S_3 is twice the radius in which that particular ion moves and, since the radii are proportional to the masses, equal mass differences appear as equal separations of lines on the developed photographic plate. The apparatus spreads the beam of ions into a "mass spectrum" much as a prism spreads a beam of light into a spectrum; hence the name "mass spectrograph."

The mass spectrograph is used chiefly for the study of *isotopes*. The term refers to atoms that have the same atomic number but differ in mass. Since the atomic number equals the number of electrons in the atom, and the *chemical* properties of an atom depend only on the number and arrangement of its electrons, isotopes cannot be separated by chemical means, but of course they are separated by the mass spectrograph. The differences in mass arise from different numbers of neutrons in the nucleus.

Most elements as they occur in Nature are mixtures of isotopes. Oxygen, for example, always occurs as a mixture of three isotopes. The most abundant (relative abundance = 99.76%) is arbitrarily assigned an atomic weight of exactly 16.0000. The other two, relative to the first, have atomic weights of 17.0045 (relative abundance = 0.04%) and 18.005 (relative abundance = 0.20%). The average atomic weight is therefore not exactly 16.0000, but slightly larger. (The chemists, however, assign the value of exactly 16.0000 to this average, so the physical and chemical scales of atomic weights are not exactly the same.) As another example, chlorine, whose atomic weight is 35.46, is a mixture of 75.4% of an isotope of atomic weight 34.980 and 24.6% of an isotope of atomic weight 36.978, all on the physical scale.

The nearest integer to the atomic weight of an isotope is called its *mass number* and is written as a superscript. The number of extranuclear electrons, or the number of protons in the nucleus, is written below and to the left of the chemical symbol. Thus oxygen is a mixture of ${}_8O^{16}$, ${}_8O^{17}$, and ${}_8O^{18}$, while chlorine is a mixture of ${}_{17}Cl^{35}$ and ${}_{17}Cl^{37}$.

The masses of both the proton and the neutron, on the physical scale, are very nearly unity. Since an oxygen atom contains eight electrons, its nucleus must contain eight protons. Therefore all of the isotopes of oxygen have eight protons in the nucleus, but ${}_8O^{16}$ has eight neutrons, ${}_8O^{17}$ nine neutrons, and ${}_8O^{18}$ ten neutrons.

Fig. 9-9 is a reproduction of a plate showing the isotopes of germanium as they are separated by a mass spectrograph. The numbers are the mass numbers of the isotopes.

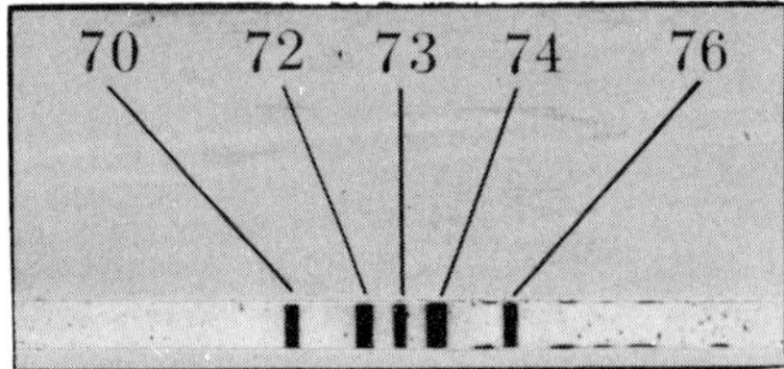

FIG. 9-9. The mass spectrum of germanium showing the isotopes of mass numbers 70, 72, 73, 74, 76.

(*Courtesy of Dr. K. T. Bainbridge*)

9-8 Force on a current-carrying conductor. When a current-carrying conductor lies in a magnetic field, magnetic forces are exerted on the moving electrons within the conductor. These forces are transmitted to the material of the conductor and hence the conductor as a whole experiences a force or torque, or both. The electric motor and the moving coil galvanometer both depend for their operation on the torque exerted on a current-carrying loop in a magnetic field.

Fig. 9-10 represents a portion of a straight conductor of length l and cross section A, within which there is a current i. A magnetic field of flux density B is perpendicular to the conductor. Since the force on the conductor as a whole is the resultant of the forces on the moving charges within it, let us express the current in the conductor in terms of the number of moving charges per unit volume, n, the charge q on each, and their velocity v. This relation (see Sec. 4-1) has been shown to be

$$i = nqvA.$$

The force on each charge, say f, is

$$f = qvB.$$

The number of charges in the length l is

$$N = nlA.$$

The resultant force F is therefore

$$F = Nf = nlAqBv,$$

and since $i = nqvA$, this can be written,

$$F = ilB. \tag{9-9}$$

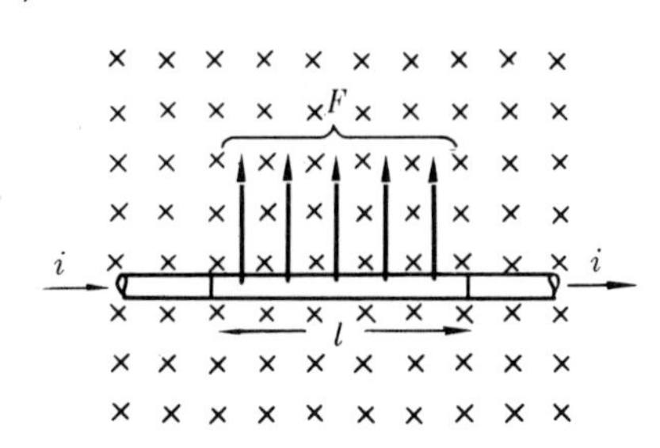

FIG. 9-10. Force on a straight conductor of length l at right angles to a magnetic field of flux density B directed away from the reader.

If the direction of the conductor makes an angle ϕ with the direction of the field, it is evident from Eq. (9-4) that one has in general

$$\boxed{F = iBl \sin \phi.} \tag{9-10}$$

If the conductor is not straight, and/or if the magnetic field is not uniform, the relation

$$\boxed{dF = iBdl \sin \phi} \tag{9-11}$$

gives the force dF on an element of the conductor of length dl at a point where the induction is B and the angle between dl and B is ϕ. In general, both B and ϕ will vary from point to point along a conductor.

The direction of the force dF is at right angles to the plane determined by dl and B, and its direction can be found from the left-hand rule, the second (or Center) finger now pointing in the direction of the conventional Current in the conductor.

Eq. (9-11), in vector notation, is

$$\vec{dF} = i\vec{dl} \times \vec{B}$$

The vector $\vec{dl}$ is in the direction of the current, and the direction of the vector $\vec{dF}$ is the same as that of the cross product of $\vec{dl}$ and $\vec{B}$.

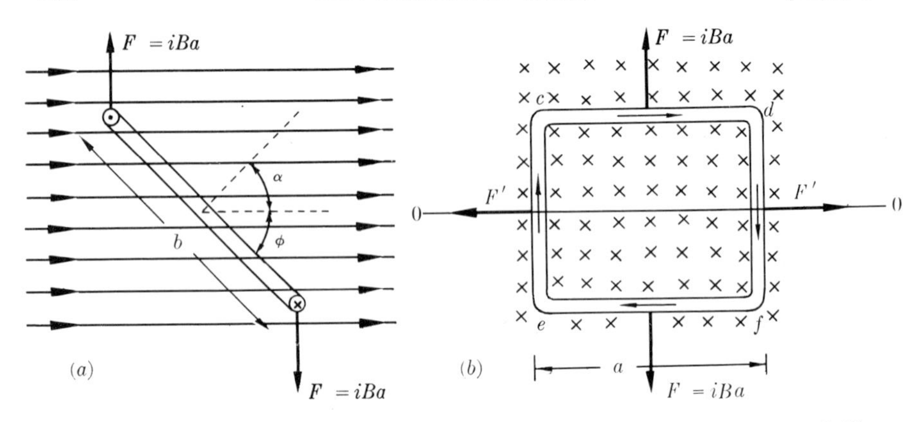

FIG. 9-11. Rectangular loop of wire carrying a current in a uniform magnetic field.

9-9 Force and torque on a complete circuit. The net force and torque on a complete circuit in a magnetic field can be found from Eq. (9-11) by summation or integration over all elements comprising the circuit. Three simple cases will be analyzed.

Rectangular loop. In Fig. 9-11 there is shown a rectangular loop of wire the lengths of whose sides are a and b. The normal to the plane of the loop makes an angle α with the direction of a uniform magnetic field. The loop is pivoted about an axis OO, and it carries a current i. (Provision must be made for leading the current into and out of the loop, or for inserting a seat of emf. This is omitted from the diagram for simplicity.)

Sides *cd* and *ef* of the loop are perpendicular to the field. Hence equal and opposite forces of magnitude

$$F = iBa$$

are exerted on them, vertically upward on *cd*, vertically downward on *ef*.

Sides *ce* and *df* make an angle ϕ with the field. Equal and opposite forces of magnitude

$$F' = iBb \sin \phi$$

are exerted on them, to the right on *df* and to the left on *ce*. (These forces are in reality distributed along each side of the loop. They are shown as single forces in Fig. 9-11.)

The resultant *force* on the loop is evidently zero, since the forces on opposite sides are equal and opposite. The resultant *torque*, however, is not zero, since the forces on sides *cd* and *ef* constitute a couple of moment

$$\tau = iBa \times b \sin \alpha. \tag{9-12}$$

The couple is a maximum when $\alpha = 90°$ or when the plane of the coil is

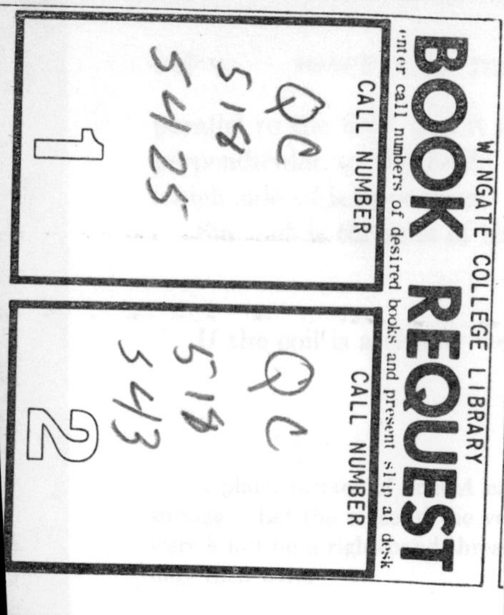

s zero when $\alpha = 0$ and the plane of the coil is
The position of stable equilibrium is that in

e coil, A, Eq. (9-12) may be written

$$\tau = iAB \sin \alpha.$$

und one having N turns, then evidently

$$\tau = NiAB \sin \alpha.$$

n be represented by a vector $\vec{A}$ perpendicular to the
ctor be that in which the surface would advance if it
ded rod, turned in the direction of the current i. We

$$\vec{\tau} = i\vec{A} \times \vec{B}$$

where $\vec{\tau}$ is the vector torque on the loop.

Circular loop. In Fig. 9-12(a) there is shown a circular loop of radius a in which there is a current i. The loop lies in the X-Z plane and is in a uniform magnetic field of flux density B, parallel to the X-axis. The force dF on an element dl $(= a d\phi)$ of the loop is directed toward the reader in Fig. 9-12(a). Its magnitude is

$$dF = iBdl \sin \phi = iBa \sin \phi \, d\phi.$$

The small arrows in Fig. 9-12(b) show how the forces on other elements of length dl vary from point to point of the loop. The magnetic flux lines are not shown in the figure, to avoid confusion.

The torque about the Z-axis due to the force dF is

$$d\tau = dF \times a \sin \phi = iBa^2 \sin^2 \phi \, d\phi,$$

and the total torque is

$$\tau = \int d\tau = iBa^2 \int_0^{2\pi} \sin^2 \phi \, d\phi = iB\pi a^2.$$

But πa^2 is simply the area of the loop, so

$$\tau = iAB. \tag{9-12}$$

This is the same as the expression for the torque on a rectangular loop parallel to a magnetic field, and it can be shown to hold for a plane loop of any shape whatever.

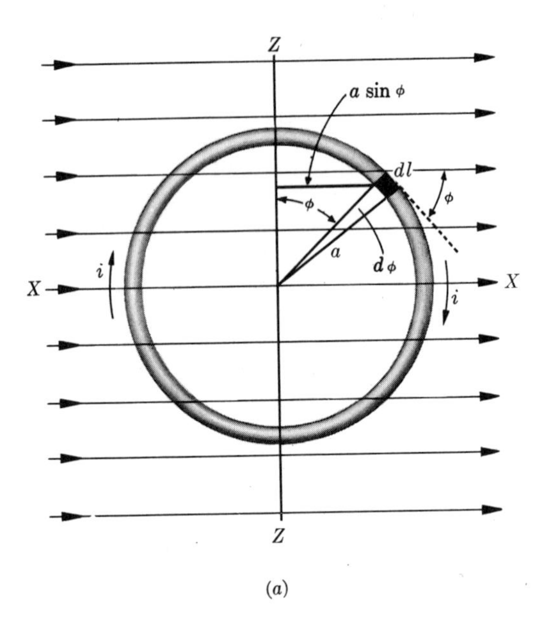

(a)

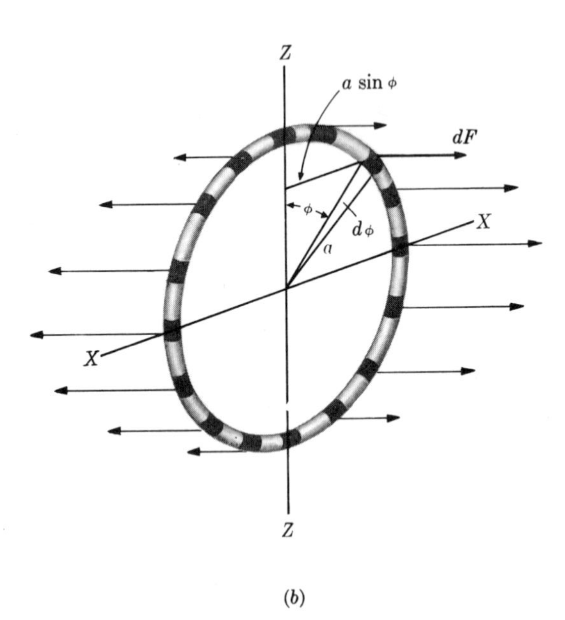

(b)

Fig. 9-12. Forces acting on a circular loop of wire carrying a current in a uniform magnetic field.

If the normal to the plane of the loop makes an angle α with the field, one has

$$\tau = iAB \sin \alpha. \tag{9-13}$$

Solenoid. A helical winding of wire is called a *solenoid* If the solenoid is closely wound, it can be approximated by a number of circular turns lying in planes at right angles to its long axis. The total torque acting on a solenoid in a magnetic field is simply the sum of the torques on the individual turns. Hence for a solenoid of N turns in a uniform field of flux density B,

$$\tau = NiAB \sin \alpha,$$

where α is the angle between the axis of the solenoid and the direction of the field.

The torque is a maximum when the induction is parallel to the planes of the individual turns or perpendicular to the long axis of the solenoid. The effect of this torque, if the solenoid is free to turn, is to rotate it into a position in which each turn is perpendicular to the field and the axis of the solenoid is parallel to the field.

Although little has been said thus far regarding permanent magnets, everyone will probably recognize that the behaviour of the solenoid as described above is the same as that of a bar magnet or compass needle, in that both the solenoid and the magnet will, if free to turn, set themselves with their axes parallel to a magnetic field. The behaviour of a bar magnet or compass is usually explained by ascribing the torque on it to magnetic forces exerted on "poles" at its ends. We see, however, that no such interpretation is demanded in the case of the solenoid. May it not be, therefore, that the whirling electrons in a bar of magnetized iron are equivalent to the current in the windings of a solenoid, and that the observed torque arises from the same cause in both instances? We shall return to this question later on.

Problems—Chapter 9

(1) The magnetic induction or flux density B in a certain region is 2w/m^2 and its direction is that of the positive X-axis in Fig. 9–13. (a) What is the magnetic flux across the surface *abcd* in Fig. 9–13. Express the answer in webers and in maxwells. (b) What is the magnetic flux across the surface *becf*? (c) What is the magnetic flux across the surface *aefd*?

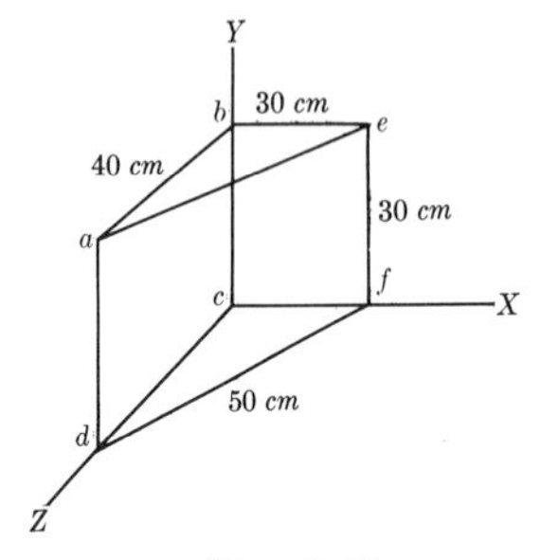

FIG. 9-13.

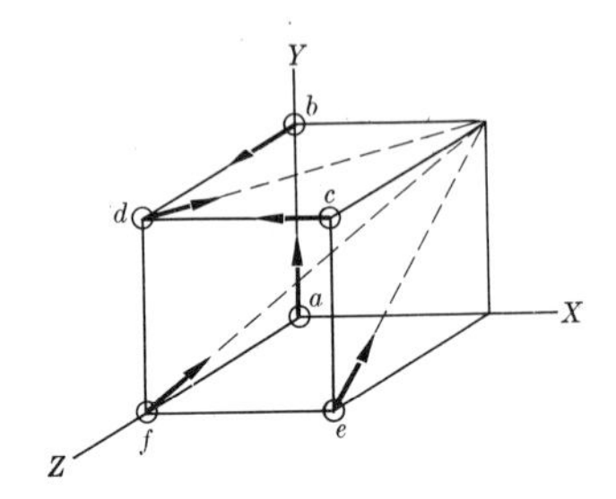

FIG. 9-14.

(2) Each of the lettered circles, at the corners of the cube in Fig. 9-14, represents a positive charge q moving with a velocity of magnitude v in the directions indicated. The region in the figure is a uniform magnetic field of flux density B, parallel to the X-axis and directed toward the right. Copy the figure, find the magnitude and direction of the force on each charge, and show the force in your diagram.

(3) A deuteron travels in a circular path of radius 40 cm in a magnetic field of flux density $1.5\ \text{w/m}^2$. (a) Find the speed of the deuteron. (b) Find the time required for it to make one-half a revolution. (c) Through what potential difference would the deuteron have to be accelerated to acquire this velocity?

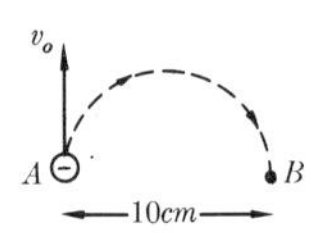

FIG. 9-15.

(4) An electron at point A in Fig. 9-15 has a velocity v_0 of 10^7 m/sec. Find (a) the magnitude and direction of the magnetic induction that will cause the electron to follow the semicircular path from A to B; (b) the time required for the electron to move from A to B.

(5) (a) What is the velocity of a beam of electrons when the simultaneous influence of an electric field of intensity 34×10^4 volts/m and a magnetic field of flux density $2 \times 10^{-3}\ \text{w/m}^2$, both fields being normal to the beam and to each other, produces no deflection of the electrons? (b) Show in a diagram the relative orientation of the vectors v, E, and B. (c) What is the radius of the electron orbit when the electric field is removed?

(6) Suppose the electric intensity between the plates P_1 and P_2 in Fig. 9-8 is 150 volts/cm and the magnetic induction in both magnetic fields is $0.5\ \text{w/m}^2$. If the source contains the three isotopes of magnesium, ${}_{12}\text{Mg}^{24}$, ${}_{12}\text{Mg}^{25}$, and ${}_{12}\text{Mg}^{26}$, and the ions are singly charged, find the distance between the lines formed by the three isotopes on the photographic plate. Assume the atomic weights of the isotopes are equal to their mass numbers.

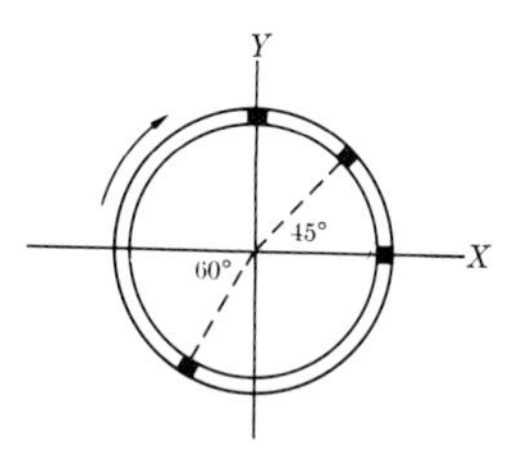

FIG. 9-16.

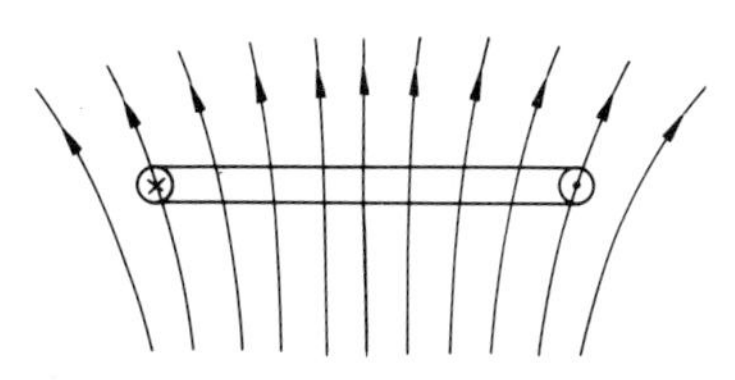
FIG. 9-17.

(7) Fig. 9-16 represents a circular loop of wire 20 cm in radius and carrying a current of 20 amp in a clockwise direction. The loop lies in a uniform magnetic field of flux density 0.8 w/m^2, perpendicular to the plane of the diagram and directed away from the reader. (a) Compute and show in a diagram the magnitude and direction of the force exerted by the field on the three shaded elements, each 1 cm in length. (b) Repeat, if the direction of the field is parallel to the X-axis, toward the right.

(8) A wire ring whose radius is 4 cm is at right angles to the general direction of a radially symmetrical diverging magnetic field as shown in Fig. 9-17. The flux density in the region occupied by the wire itself is 0.1 w/m^2 and the direction of the field is everywhere at an angle of 60° with the plane of the ring. Find the magnitude and direction of the force on the ring when the current in it is 15.9 amperes.

(9) Compute the tension in a circular loop of flexible wire of radius a, carrying a current I, and lying in a uniform magnetic field of flux density B perpendicular to the plane of the loop.

(10) The rectangular loop in Fig. 9-18 is pivoted about the Y-axis and carries a current of 10 amp in the direction indicated. (a) If the loop is in a uniform magnetic field of flux density 0.2 w/m^2, parallel to the X-axis, find the force on each side of the loop, in dynes, and the torque in dyne-cm required to hold the loop in the position shown. (b) Same as (a) except the field is parallel to the Z-axis. (c) What torque would be required if the loop were pivoted about an axis through its center, parallel to the Y-axis?

(11) The rectangular loop of wire in Fig. 9-19 has a mass of 0.1 gm per centimeter of length, and is pivoted about side ab as a frictionless axis. The current in the wire is 10 amp in the direction shown. (a) Find the magnitude and sense of the magnetic field, parallel to the Y-axis, that will cause the coil to swing up until its plane makes an angle of 30° with the Y-Z plane. (b) Discuss the case where the field is parallel to the X-axis.

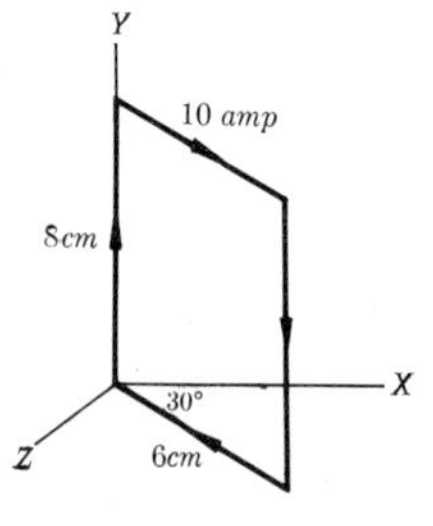

FIG. 9-18.

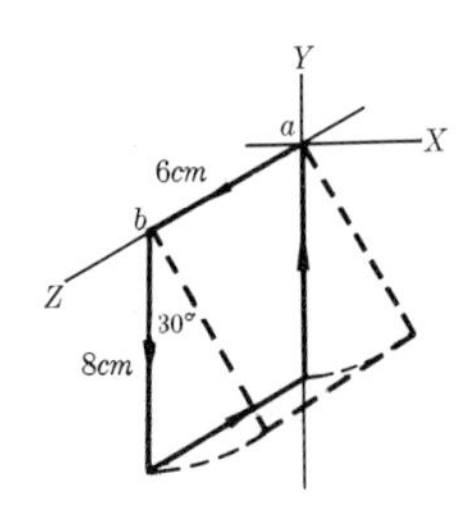

FIG. 9-19.

CHAPTER 10

GALVANOMETERS, AMMETERS, AND VOLTMETERS. THE D.C. MOTOR

10-1 The galvanometer. Any device used for the detection or measurement of current is called a galvanometer, and the majority of such instruments depend for their action on the torque exerted on a coil in a magnetic field. The earliest form of galvanometer was simply the apparatus of Oersted, namely, a compass needle placed below the wire in which the current was to be measured. Wire and needle were both aligned in the north-south direction with no current in the wire. The deflection of the needle when a current was sent through the wire was then a measure of the current. The sensitivity of this form of galvanometer was increased by winding the wire into a coil in a vertical plane with the compass needle at its center, and instruments of this type were developed by Lord Kelvin in the 1890's to a point where their sensitivity is scarcely exceeded by any available at the present time.

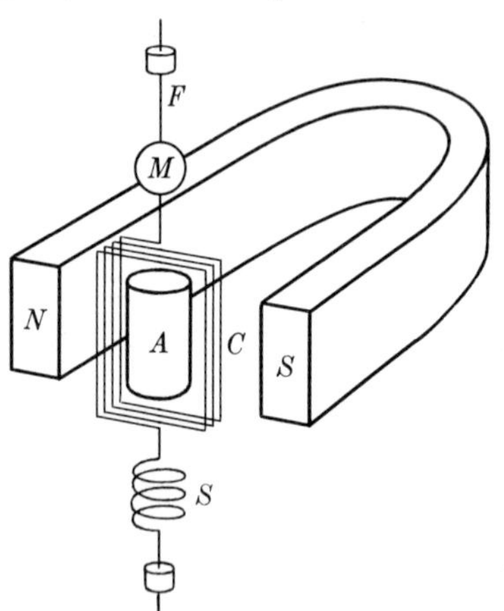

Fig. 10-1. Principle of D'Arsonval galvanometer.

Practically all galvanometers used today, however, are of the D'Arsonval moving coil or pivoted coil type, in which the roles of magnet and coil are interchanged. The magnet is made much larger and is stationary, while the moving element is a light coil swinging in the field of the magnet.[1] The construction of a moving coil galvanometer is illustrated in Fig. 10-1. The magnetic field of a horseshoe magnet whose poles are designated by N and S is concentrated in the vicinity of the

[1] The magnetic field surrounding a permanent magnet is discussed more fully in Chapter 15. For our present purposes, we may take it for granted that in the region between the magnet poles of Fig. 10-1 there does exist a field whose general direction is from N to S.

coil C by the soft iron cylinder A. The coil consists of from 10 to 20 turns, more or less, of insulated copper wire wound on a rectangular frame and suspended by a fine conducting wire or thin flat strip F which provides a restoring torque when the coil is deflected from its normal position, and which also serves as one current lead to the coil. The other terminal of the coil is connected to the loosely wound spiral S which serves as the second lead, but which exerts a negligible control on the coil.

When a current is sent through the coil, horizontal and oppositely directed side-thrusts are exerted on its vertical sides, producing a couple about a vertical axis through its center. The coil rotates in the direction of this couple and eventually comes to rest in such a position that the restoring torque exerted by the upper suspension equals the deflecting torque due to the side-thrust. The angle of deflection is observed with the aid of a beam of light reflected from a small mirror M cemented to the upper suspension, the light beam serving as a weightless pointer. Since light incident on the mirror is reflected at an angle of reflection equal to the angle of incidence, rotation of the mirror through an angle θ deflects the light beam through an angle 2θ. It is standard practice to observe the reflected beam on a scale at a distance of one meter from the galvanometer.

The *current sensitivity* of a galvanometer is defined as the current in the galvanometer coil required to produce a displacement of the reflected light beam through one millimeter, on a scale one meter distant from the galvanometer. Typical current sensitivities range from 0.01 microampere (10^{-8} amp.) to 0.0001 microampere (10^{-10} amp.) per mm. division at one meter. Fig. 10-2 is a photograph of a high-sensitivity galvanometer.

FIG. 10-2. D'Arsonval galvanometer. (*Courtesy of Leeds & Northrup*)

With a given current in the galvanometer coil the deflection is proportional to the flux density, to the number of turns in the coil and to its breadth, and is inversely proportional to the torque constant k ($\tau = -k\theta$) of the upper suspension.

On the other hand, increasing the number of turns and the breadth of the coil, and decreasing the torque constant of the suspension all increase the period of swing and hence make the instrument slow in coming to its final equilibrium position. In practice a compromise must be made between sensitivity and time of swing.

Because of the geometry of the field in which the moving coil swings, the deflections of a D'Arsonval galvanometer are not directly proportional to the current in the galvanometer coil except for relatively small angles. Hence these instruments are used chiefly as *null* instruments, that is, in connection with circuits such as those of a Wheatstone bridge or a potentiometer, in which other circuit elements are adjusted so that the galvanometer current is zero.

10-2 The pivoted coil galvanometer. The pivoted coil galvanometer, while essentially the same in principle as the D'Arsonval instrument, differs from the latter in two respects. One is that the moving coil, instead of being suspended by a fine fiber, is pivoted between two jewel bearings. The instrument may hence be used in any position and is much more rugged and conveniently portable. The second difference is that the permanent magnetic field is modified by the use of soft iron pole pieces attached to the permanent magnet as shown in Fig. 10-3 so that the coil swings in a field which is everywhere radial. The side-thrusts on the coil are therefore always perpendicular to the plane of the coil, and the angular deflection of the coil is directly proportional to the current in it. The restoring torque is provided by two hairsprings, which serve also as current leads. A length of aluminum tubing, flattened at its tip in a vertical plane, serves as a pointer.

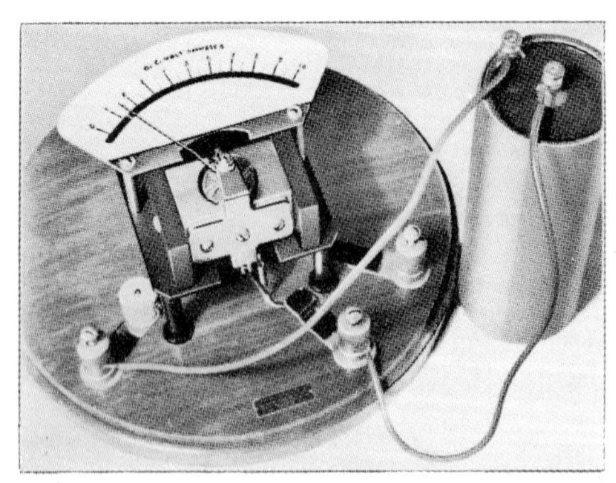

FIG. 10-3. Pivoted coil galvanometer, modified for use as an ammeter or a voltmeter. Series resistor may be seen at left. (*Courtesy of Houghton Mifflin Co.*)

The frictional torque of the jewel bearings, while small, is greater than that of a supporting fiber. Since the deflecting and restoring torques must both be considerably larger than the friction torque, pivoted coil instruments cannot be made as sensitive as the D'Arsonval type. The smallest currents which can be read on such an instrument are of the order of magnitude of 0.1 microampere.

In addition to the double pivot type, many other modifications are in use. For example, the coil may rest on a single pivot with some gain in sensitivity but with a sacrifice in ruggedness.

10-3 Ammeters and voltmeters. The coil of a pivoted-coil galvanometer is deflected because of the interaction between the field of the permanent magnet and the current in the coil. We shall assume that the deflection is proportional to this current. Since the coil and associated leads are metallic conductors obeying Ohm's law, the current in the coil is directly proportional to the potential difference between the terminals of the instrument. Hence the deflection of the instrument is proportional to the potential difference between its terminals as well as to the current through it, and it may be calibrated and used to measure either this potential difference or the current.

For example, suppose that the resistance of the coil and leads of a pivoted-coil galvanometer is 20 ohms, and that the galvanometer deflects full-scale with a current in the coil of 10 milliamperes or 0.010 ampere. The potential difference between the terminals, with a current of 0.010 ampere, is

$$V = iR = 0.010 \times 20 = 0.20 \text{ volts}$$

The scale of the instrument could therefore be calibrated to read either from zero to 0.010 ampere, or from zero to 0.20 volt. Then if in some particular circuit the pointer were deflected by one-half the full-scale amount, one could conclude that the current through the instrument was 0.005 ampere, and also that the potential difference between the terminals of the instrument was 0.10 volt.

It has been explained earlier (see Sec. 4-8) that an ammeter must be inserted in series in a circuit and hence must be a low resistance instrument, while a voltmeter, which is connected in parallel between the points whose potential difference is to be measured, needs to have a relatively high resistance. The physical limitations imposed on the size of the pivoted coil prevent the use of either very large wire, to obtain a low resistance, or a very large number of turns to obtain a high resistance. However, by inserting a low resistance shunt in parallel with the pivoted coil, or a resistor of high resistance in series with it, any galvanometer may be modified to serve as an ammeter or a voltmeter.

As an illustration, consider the pivoted-coil galvanometer described in the preceding example, whose resistance was 20 ohms and which deflected full-scale with a current of 0.010 amp. through it. Suppose it is desired to convert this instrument to an ammeter which will be deflected full-scale by a current of 10 amperes. The coil and its shunt are shown schematically in Fig. 10-4. The line current is assumed to be 10.0 amperes. Since the instrument is to deflect full scale with this line current, the current through the coil must be 0.010 amperes. Hence by the point rule, the current in the shunt is

$$10 - 0.010 = 9.99 \text{ amp.}$$

Since the currents are inversely proportional to the resistances,

$$0.01/9.99 = R_{sh}/20$$

and R_{sh}, the required shunt resistance, is 0.0200 ohm (to three significant figures).

The resistance of the ammeter as a whole is

$$R = \frac{20 \times .02}{20.02} = .0200 \text{ ohm}$$

so that one has a low resistance instrument which at the same time requires 10 amperes for full-scale deflection.

Since it is difficult to adjust precisely the value of a low resistance shunt, the shunt resistance is usually made slightly too large and the final adjustment is made by inserting a resistor in series with the moving coil. The shunt and series resistors are ordinarily enclosed within the case of the instrument, although in some instances external shunts are provided.

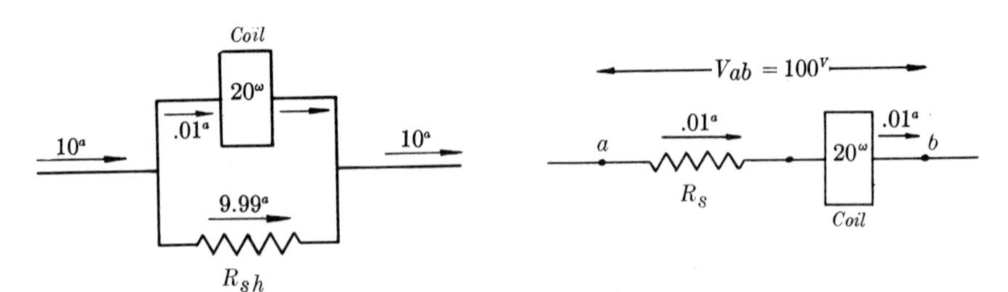

FIG. 10-4. Conversion of a galvanometer to an ammeter with the aid of a shunt.

FIG. 10-5. Conversion of a galvanometer to a voltmeter with the aid of a series resistor.

Suppose next that it is desired to modify the same galvanometer for use as a voltmeter which will deflect full-scale with 100 volts across its terminals. The coil and its series resistor are shown in Fig. 10-5. Assume that $V_{ab} = 100$ volts. The current in the moving coil must then be 0.010 amp. Hence

$$.010 = 100/(R_s + 20),$$

and R_s, the required series resistance, is 9980 ohms.

The resistance of the voltmeter as a whole is 10,000 ohms and one has a high resistance instrument which draws only 0.010 amperes with a potential difference of 100 volts between its terminals. Paradoxically, an ammeter can be considered a low range voltmeter which measures the potential difference across a low resistance shunt, while a voltmeter is a sensitive ammeter which measures the current in a large resistance.

10-4 The ballistic galvanometer. A ballistic galvanometer is used for measuring the *quantity of charge* displaced by a current of short duration, as for example in the charging or discharging of a capacitor. That is, it measures $\int i\,dt$ rather than i. While any moving coil galvanometer can be used ballistically, instruments designed specifically for the purpose have coils with somewhat larger moments of inertia, and suspensions with somewhat smaller torque constants, than are found in instruments designed primarily for current measurement.

The angular impulse exerted on a galvanometer coil during the passage of a transient current produces an equal angular momentum of the coil. The coil then proceeds to swing until its initial kinetic energy has been converted to potential energy of the suspension. The maximum angle of throw is observed, and this can be shown to be proportional to the quantity of charge which passed through the coil. Notice that the process is entirely analogous to that which takes place when a ballistic pendulum is struck by a bullet.

To derive the relation between the charge and the maximum angle of throw, let J be the moment of inertia of the galvanometer coil, A its area, N the number of turns and B the flux density of the field in which it swings. The torque acting on the coil at an instant when the current through it is i is (see Sec. 9-8).

$$\tau = NiBA.$$

The angular impulse imparted to the coil during the time of duration of the current is

$$\int \tau\,dt = NBA \int i\,dt = NBAq,$$

and since angular impulse equals change in angular momentum,

$$J\omega_0 = NBAq,$$

where ω_0 is the angular velocity imparted to the coil.

The initial kinetic energy of the coil is

$$\text{KE} = \tfrac{1}{2}J\omega_0^2.$$

If θ_m is the maximum angle of swing, and k is the torque constant of the suspension (restoring torque $= -k\theta$), the potential energy of the suspension in its position of maximum deflection is

$$\text{PE} = \tfrac{1}{2}k\theta_m^2.$$

Equating the kinetic and potential energies, one obtains

$$q = \left(\frac{\sqrt{Jk}}{NBA}\right)\theta_m = K\theta_m, \tag{9-14}$$

where K is an abbreviation for the expression in parentheses. The quantity of charge displaced through the galvanometer is thus proportional to the maximum deflection, which was to be shown. While Eq. (9-14) indicates the manner in which the galvanometer constants affect its sensitivity and hence serves as a basis for design, the proportionality constant K is ordinarily found experimentally by observing the deflection due to the passage of a known charge through the instrument. For example, a capacitor of known capacitance, charged to a known potential difference, may be discharged through it.

The damping of the galvanometer has been neglected in the preceding derivation, but is discussed in Sec. 12-9.

10-5 The dynamometer. The D'Arsonval galvanometer and its modifications are suitable for the measurement of direct currents and potential differences only. If alternating current is sent through the coil of such an instrument, the direction of the torque reverses with each reversal of current, and since the moment of inertia of the moving element is too large for it to follow alternations at the commercial frequency of 60 cycles per second or higher, the instrument shows no deflection.

One method of obtaining a steady deflection with an alternating current is to employ a *dynamometer* type of movement, in which the moving coil is deflected, not by the field of a permanent magnet, but by the field of a second fixed coil. If the current in the second coil alternates at the same frequency as that in the moving coil, the magnetic field reverses whenever the current in the moving coil reverses and the deflecting torque, although pulsating, is always in the same direction. The moving element then assumes a position in which the restoring torque equals the *average* deflecting torque. The same type of movement is used in the *wattmeter*. (See Sec. 5-12.)

The instantaneous torque on the moving coil of a dynamometer is proportional to the product of the instantaneous current in the moving coil and the instantaneous flux density of the field set up by the fixed coil. The latter is proportional to the current in the fixed coil. If the two coils are connected in series the current in each is the same. Hence if i is the instantaneous current,

$$\tau = k_1 i B, \quad B = k_2 i,$$

and

$$\tau = k_1 k_2 i^2$$

where k_1 and k_2 are proportionality constants and τ is the instantaneous torque.

The *average* torque is proportional to the average squared or *mean squared current*, with the result that the scale of a dynamometer is not uniform but is compressed toward the zero end.

Because of the smaller fields available, dynamometer movements cannot be made as sensitive as those employing permanent magnets.

10-6 The direct current motor. The direct current motor is illustrated schematically in Fig. 10-6. The armature, A, is a cylinder of soft steel laminated to minimize eddy current losses (see Sec. 12-10) and mounted on a shaft so that it can rotate about its axis. Embedded in longitudinal slots in the surface of the armature are a number of copper conductors C. Current is led into and out of these conductors through graphite brushes making contact with a segmented cylinder on the shaft called the commutator (not shown in Fig. 10-6). The commutator is an automatic switching arrangement which maintains the currents in the conductors in the directions shown in the figure whatever the position of the armature.

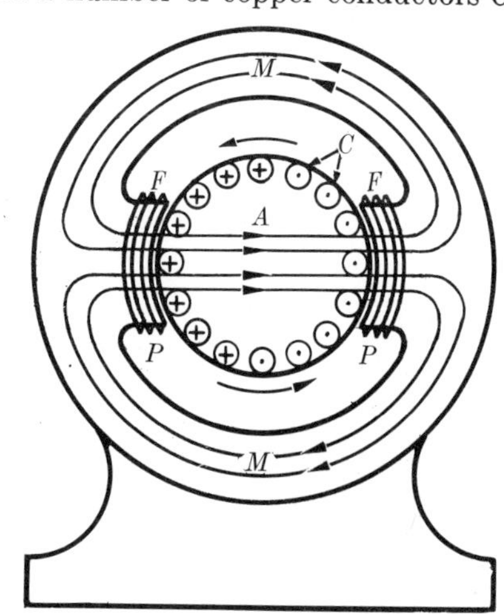

FIG. 10-6. Schematic diagram of a direct current motor.

The current in the field coils F,F, sets up a magnetic field which because of the shape of the pole pieces P,P is essentially radial in the gap between them and the armature. The motor frame M,M

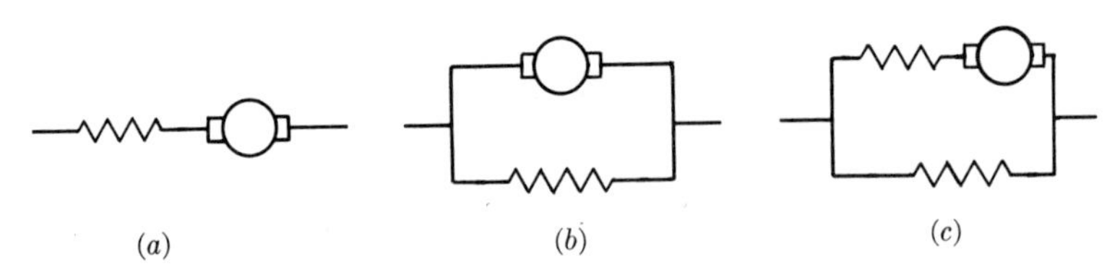

FIG. 10-7. Three ways of connecting the armature and field coils of a DC motor. (a) Series. (b) Shunt. (c) Compound.

provides a low reluctance path (see Sec. 15-11) for the magnetic field. Some of the lines of induction are indicated by the light lines in the figure.

Application of the left-hand rule shows that with the relative directions of field and armature currents as shown, the side thrust on each conductor is such as to produce a counterclockwise torque on the armature. When the motor is running, the armature develops mechanical energy at the expense of electrical energy. It must therefore be a seat of back emf. This is an "induced" emf and is discussed further in Chap. 12. The field windings, however, are static and behave like a pure resistance.

If the armature and the field windings are connected in series we have a *series* motor; if they are connected in parallel, a *shunt* motor. In some motors the field windings are in two parts, one in series with the armature and the other in parallel with it; the motor is then *compound*. The three corresponding electrical circuits are shown in Fig. 10-7. For further details a text on DC machinery should be consulted.

The energy interchanges in the DC shunt motor have already been discussed in Example 2 at the end of Sec. 5-11, which should be re-read at this point.

Problems—Chapter 10

(1) The coil of a pivoted coil galvanometer has 50 turns and encloses an area of 6 cm². The magnetic induction in the region in which the coil swings is 100 gauss and is radial. The torsional constant of the hairsprings is $k = 0.1$ dyne-cm/degree. Find the angular deflection of the coil for a current of 1 milliampere.

(2) The resistance of a galvanometer coil is 25 ohms and the current required for full scale deflection is 0.02 amp. (a) Show in a diagram how to convert the galvanometer to an ammeter reading 5 amp full scale, and compute the shunt resistance. (b) Show how to convert the galvanometer to a voltmeter reading 150 volts full scale, and compute the series resistance.

(3) Fig. 10-8 shows the internal wiring of a "three-scale" voltmeter whose binding posts are marked +, 3^v, 15^v, 150^v. The resistance of the moving coil, R_G, is 15 ohms, and a current of 1 milliampere in the coil causes it to deflect full scale. Find the resistances R_1, R_2, R_3, and the overall resistance of the meter on each of its ranges.

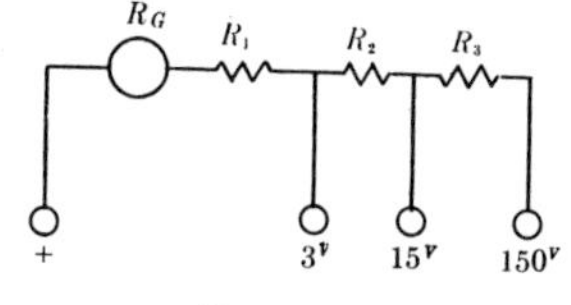

FIG. 10-8.

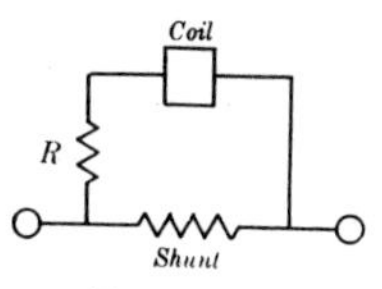

FIG. 10-9.

(4) A certain DC voltmeter is said to have a resistance of "one thousand ohms per volt." What current, in milliamperes, is required for full scale deflection?

(5) The resistance of the coil of a pivoted coil galvanometer is 10 ohms, and a current of 0.02 amp causes it to deflect full scale. It is desired to convert this galvanometer to an ammeter reading 10 amp full scale. The only shunt available has a resistance of 0.03 ohm. What resistance R must be connected in series with the coil? (See Fig. 10-9.)

(6) A thermocouple has a total resistance of 1 ohm and is connected directly to the terminals of a galvanometer of resistance 15 ohms. The emf of the couple is 4 millivolts. The resulting galvanometer current is inconveniently large, so a shunt is desired which will reduce the galvanometer current to one-half its original value. Find the required shunt resistance.

(7) A 150-volt voltmeter has a resistance of 20,000 ohms. When connected in series with a large resistance R across a 110-volt line the meter reads 5 volts. Find the resistance R. (This problem illustrates one method of measuring large resistances.)

(8) A 600-ohm resistor and a 400-ohm resistor are connected in series across a 90-volt line. A voltmeter across the 600-ohm resistor reads 45 volts. (a) Find the voltmeter resistance. (b) Find the reading of the same voltmeter if connected across the 400-ohm resistor.

(9) Two identical resistors of 60,000 ohms each are connected in series across an 800-volt D.C. line. A voltmeter with a resistance of 30,000 ohms is connected across one resistor. (a) What is the voltmeter reading? (b) What must be the power rating of the resistors if one of them is not to burn out when the voltmeter is connected?

(10) In Fig. 10-10, point a is maintained at a constant potential above ground. R_1 = 20,000 ohms, $R_2 = R_3 = 10,000$ ohms. A 150-volt voltmeter whose resistance is 15,000 ohms reads 45 volts when connected between point c and ground. (a) What was the potential of point c above ground before the voltmeter was connected? (b) What is the potential of point a above ground?

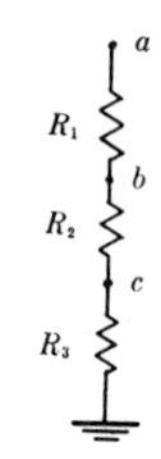

FIG. 10-10.

(11) The ranges of a "three-scale" voltmeter (see problem 3, Chap. 10) are 1.5^v, 3^v, and 15^v. The resistance of the meter between its 15-volt terminals (i.e. between the + terminal and the terminal marked 15^v) is 1500 ohms. (a) What is the resistance between the 1.5^v terminals and between the 3^v terminals? (b) When a cell is connected across the 1.5^v terminals the meter reads 1.42 volts, when the same cell is connected across the 3^v terminals the meter reads 1.48 volts. Compute the emf and internal resistance of the cell.

(12) Two 150-volt voltmeters, one of resistance 15,000 ohms and the other of resistance 150,000 ohms, are connected in series across a 120-volt D.C. line. Find the reading of each voltmeter.

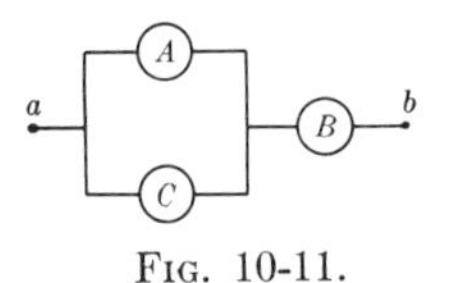

FIG. 10-11.

(13) Three voltmeters A, B, and C, in Fig. 10-11, have the same range but different resistances. $R_A = 1200$ ohms, $R_B = 1000$ ohms, $R_C = 800$ ohms. When connected as shown across a constant difference of potential V_{ab}, meter A reads 4.8 volts and meter B reads 10 volts. What will be the reading of each meter when all are reconnected in series across the same potential difference V_{ab}?

(14) The voltmeter V in Fig. 10-12 reads 117 volts and the ammeter A reads 0.130 amp. The resistance of the voltmeter is 9000 ohms and the resistance of the ammeter is 0.015 ohm. Compute (a) the resistance R; (b) the power input to R.

(15) Let A and V represent the readings of the ammeter and voltmeter in Fig. 10-13. The meter resistances are R_A and R_V. (a) In each circuit, find the error made in computing the true resistance R from the equation $R = V/A$, i.e., find the difference between R and V/A. If meter corrections are *not* to be made, and if R is known approximately, how would you decide which circuit to use?

(16) A dynamometer movement whose resistance (both coils in series) is 50 ohms, has a scale of 100 equally spaced divisions either side of zero, and deflects 36 divisions clockwise when carrying a current of 9 milliamperes. How much additional resistance is required to convert this instrument into a voltmeter deflecting full scale with 120 volts across its terminals?

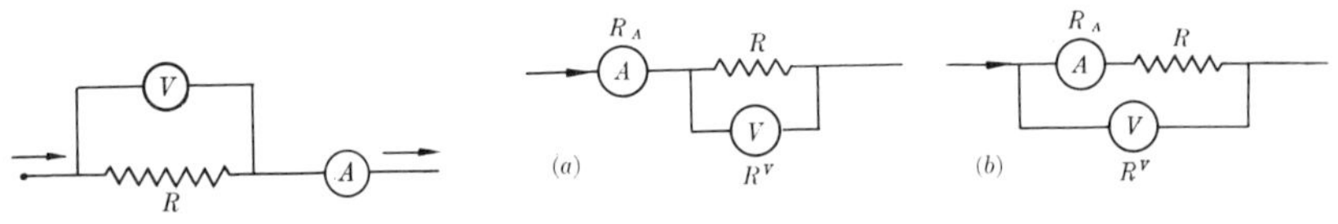

FIG. 10-12.

FIG. 10-13.

(17) The resistance of the moving coil of the galvanometer G in Fig. 10-14 is 25 ohms and it deflects full scale with a current of 0.010 amp. Find the magnitudes of the resistances R_1, R_2, and R_3, to convert the galvanometer to a multi-range ammeter deflecting full scale with currents of 10 amp, 1 amp, and 0.1 amp.

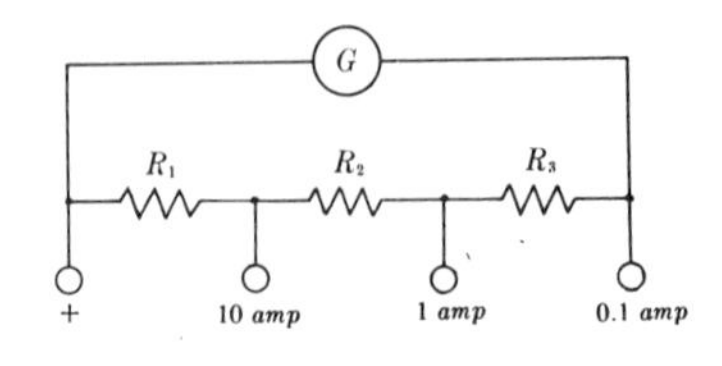

FIG. 10-14.

CHAPTER 11

MAGNETIC FIELD OF A CURRENT AND OF A MOVING CHARGE

11-1 Magnetic field of a current element. An electric charge in motion sets up, in the space around it, a magnetic field. A second charge, moving in a magnetic field, experiences a force. The two preceding chapters were devoted to the second aspect of the problem, that is, to the forces on moving charges or current-carrying conductors in magnetic fields. The existence of the field was taken for granted. In this chapter we return to the first part of the problem and correlate the magnitude and direction of a magnetic field with the motion of the charged particles responsible for it. In practice, the moving charges setting up a magnetic field are most commonly those constituting the current in a conductor. We shall therefore begin by discussing the magnetic field around a conductor in which there is a current.

The first recorded observations of magnetic fields set up by currents were those of Oersted, who discovered that a pivoted compass needle, beneath a wire in which there was a current, set itself with its long axis perpendicular to the wire. Later experiments by Biot and Savart, and by Ampere, led to a relation by means of which one can compute the flux density at any point of space around a circuit in which there is a current (provided the integrals can be evaluated).

The circuit is to be divided, in imagination, into short elements of length dl, one of which is shown in Fig. 11-1. The moving charges in each element set up a field at all points of space, and the field of the entire circuit, at any point, is the resultant of the infinitesimal fields of all the elements of the circuit. The direction of the infinitesimal field dB set up at point P by the element of length dl is shown in Fig. 11-1. The vector dB lies *in* a plane perpendicular to the axis of dl, and is itself *perpendicular* to the plane determined by dl and the line joining P and dl. It follows that the lines of induction, to which the vectors are tangent, are circles lying in planes perpendicular to the axis of the element. The direction of these lines is clockwise when viewed along the direction of the conventional current in dl. The direction may also be described as that in which a right-hand screw would have to be rotated in order that it should advance in the direction of the current. This relation between the lines of induction and the current is known as the "right-hand screw rule."

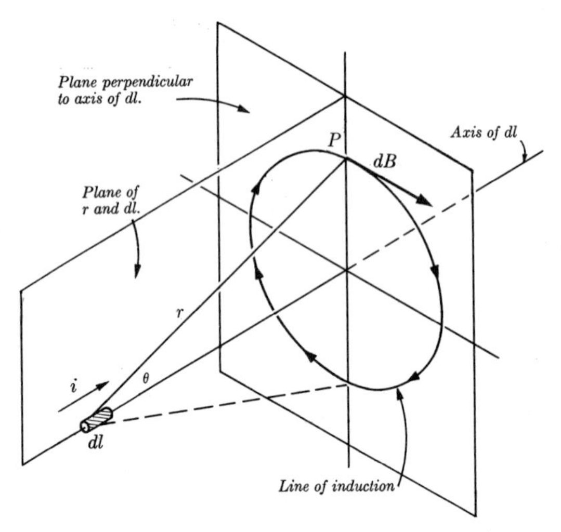

FIG. 11-1. Magnetic field due to a current element.

Another useful rule is to grasp the element (in imagination) in the right hand with the extended thumb pointing in the direction of the current. The fingers then encircle the element in the direction of the flux lines.

The magnitude of dB is given by the following relation,

$$dB = k' \frac{i \, dl \sin \theta}{r^2}, \tag{11-1}$$

where r is the distance between dl and the point P, and θ is the angle between r and dl. This equation is often called *Ampere's formula*, or *Ampere's law*, but it should be credited to Biot, who first proposed it in 1820.

The factor k' in Eq. (11-1) is a proportionality constant whose magnitude, like that of the constant k in Coulomb's law, depends on the choice of units. If the units of flux density, current, and length are independently defined, then k' must be found by experiment. On the other hand, some arbitrary value may be assigned to k' and to the units of any two of the quantities in Eq. (11-1) and this equation may then be used to define the unit of the third quantity. The latter procedure is the one that is actually adopted. In the mks system, k' is set equal to exactly 10^{-7} webers per ampere-meter, and Eq. (11-1) (or more precisely, an equation derived from it) is used to define the unit of current, the ampere. In the electromagnetic system, k' is set equal to unity, and Eq. (11-1) then defines the unit current in this system, the abampere.

For the sake of eliminating the factor 4π from other equations derived from the Biot law, and which are used more frequently than is the law itself, it is convenient to define a new proportionality constant μ_0 by the equation

$$\frac{\mu_0}{4\pi} = k', \quad \mu_0 = 4\pi k'.$$

This procedure is analogous to replacing the proportionality constant k in Coulomb's law by the term $\frac{1}{4\pi\epsilon_0}$. To four significant figures,

$$\mu_0 = 4\pi k' = 12.57 \times 10^{-7} \text{ w/amp-m},$$

$$k' = \frac{\mu_0}{4\pi} = 10^{-7} \text{ w/amp-m}.$$

The Biot law then becomes

$$\boxed{dB = \frac{\mu_0}{4\pi} \frac{i\,dl \sin\theta}{r^2}.} \tag{11-2}$$

It follows from this equation that the flux density dB due to a current element is zero at all points on the axis of the element, since $\sin\theta = 0$ at all such points. At a given distance r from the element the flux density is a maximum in a plane passing through the element perpendicular to its axis, since $\theta = 90°$ and $\sin\theta = 1$ at all points in such a plane.

The expression for the resultant flux density at any point of space, due to a complete circuit, is

$$B = \int dB = \frac{\mu_0}{4\pi} \int \frac{i\,dl \sin\theta}{r^2} \quad \text{(vector sum)}, \tag{11-3}$$

where the integration is to be extended over the entire circuit (or circuits) setting up the field, and it is understood that the integral refers to the *vector sum* of the infinitesimal flux density vectors. Except in relatively simple cases where all of the infinitesimal vectors at a given point are in the same direction, it is necessary to resolve them along a set of axes and integrate their components separately, as we have done previously in evaluating electric and gravitational fields.

Biot's law, in its differential form, can not, of course, be verified directly, since one can never obtain an isolated element of a current-carrying circuit. The integrated expression for the field set up by a complete circuit can be compared with experiment, however, and in every case the two are found to agree.

Eq. (11-3) can be considered an alternate definition of the magnetic induction at a point, analogous to the definition of electric intensity by the relation

$$E = \frac{1}{4\pi\epsilon_0} \int \frac{dq}{r^2} \text{ (vector sum).}$$

If there is matter in the space around a circuit, the flux density at a point will be due not entirely to currents in conductors, but in part to the magnetization of this matter. The magnetic properties of matter are discussed more fully in Chaps. 14 and 15. However, unless iron or some of the ferromagnetic materials are present, this effect is so small that while strictly speaking Eq. (11-2) holds only for conductors in vacuum, as a practical matter it can be used without correction for conductors in air, or in the vicinity of any nonferromagnetic material.

In vector notation, Eq. (11-2) becomes

$$d\vec{B} = \frac{\mu_0}{4\pi} \frac{i\, d\vec{l} \times \vec{r}}{r^3}. \tag{11-4}$$

The current i is a scalar, $\vec{r}$ is the vector from the current element $d\vec{l}$ to the point at which $d\vec{B}$ is to be evaluated, and r is the scalar magnitude of this vector. The relative directions of $d\vec{B}$, $d\vec{l}$, and $\vec{r}$, as expressed, for example, by the right-hand screw rule, follow at once from the fact that the vector $d\vec{B}$ has the same direction as the cross product of $d\vec{l}$ and $\vec{r}$.

11-2 Field of a straight conductor. Let us use Biot's law to compute the magnetic flux density at a point P outside a straight wire in which there is a current. Let the X-axis in Fig. 11-2 coincide with the wire and let the point be taken on the Z-axis. The direction of the field dB set up at point P by an element of length dx lies in the Y-Z plane, and is perpendicular to the X-Z plane, since both r and dx lie in this plane. The resultant field at P is therefore the *arithmetic* sum of the dB's and the integration may be carried out directly. Simplification will result if θ, rather than x, is chosen as the independent variable. From inspection of the diagram, we see that $r = a \csc\theta$, $x = a \cot\theta$, and $dx = a\, d(\cot\theta) = -a \csc^2\theta\, d\theta$.

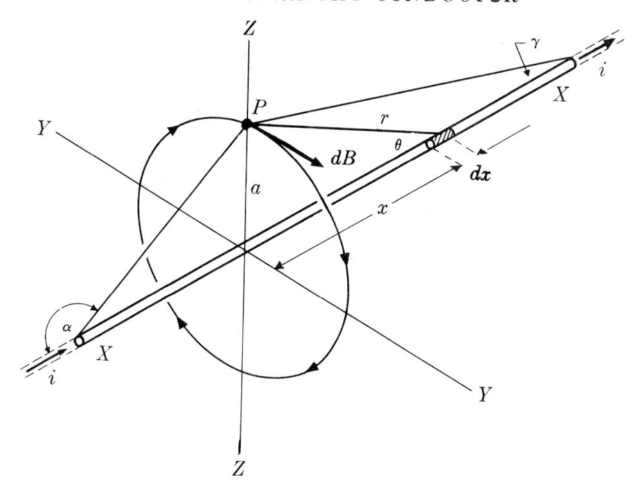

FIG. 11-2. Field of a straight conductor.

Hence

$$B = \frac{\mu_0}{4\pi} \int \frac{i\,dx \sin\theta}{r^2} = -\frac{\mu_0}{4\pi}\frac{i}{a} \int_{\alpha}^{\gamma} \sin\theta\, d\theta$$

$$= \frac{\mu_0}{4\pi}\frac{i}{a}(\cos\gamma - \cos\alpha). \tag{11-5}$$

If the wire is very long compared with the distance a, and the point P is not too near either end, then $\gamma = 0$, $\alpha = \pi$, and

$$B = \frac{\mu_0}{2\pi}\frac{i}{a}. \tag{11-6}$$

This relation was deduced from experimental observations by Biot and Savart before the differential form (Eq. 11-2) had been discovered. It is called the Biot-Savart law.

Like the electric intensity around a charged wire or cylinder, the magnetic induction around a current-carrying wire is inversely proportional to the *first* power of the radial distance from the wire. Unlike the electric field around a charged wire, which is radial, the lines of magnetic induction are *circles* concentric with the wire and lying in planes perpendicular to it. It will also be noted that each line of induction is a *closed* line, and that in this respect lines of induction differ from the lines of force in an electric field which terminate on positive or negative charges. This property of lines of induction is true whatever the geometry of the circuit setting up the field—every line of induction closes on itself.

11-3 Surface and line integrals of magnetic induction. Since lines of induction are always closed curves, it follows that if a closed *surface* is constructed in a magnetic field, every line of induction which enters the surface must leave it also, since no lines can terminate or originate within the space bounded by the surface. Hence *the surface integral of the normal component of B over a closed surface is always zero,* in contrast with the surface integral of the normal component of the electric intensity E, which is $\frac{1}{\epsilon_0}$ times the charge enclosed by the surface.

$$\int B \cos \phi \, dA = 0. \tag{11-7}$$

Another useful generalization, analogous to the above, can be obtained by considering the *line integral* of the induction around a closed curve. The line integral, it will be recalled, is formed by taking the product, at every point of a curve, of an element of length of the curve and the component of induction in the direction of this element, and integrating these products. See Fig. 11-3. The line integral around a closed curve is written

$$\oint B \cos \theta \, dl, \tag{11-8}$$

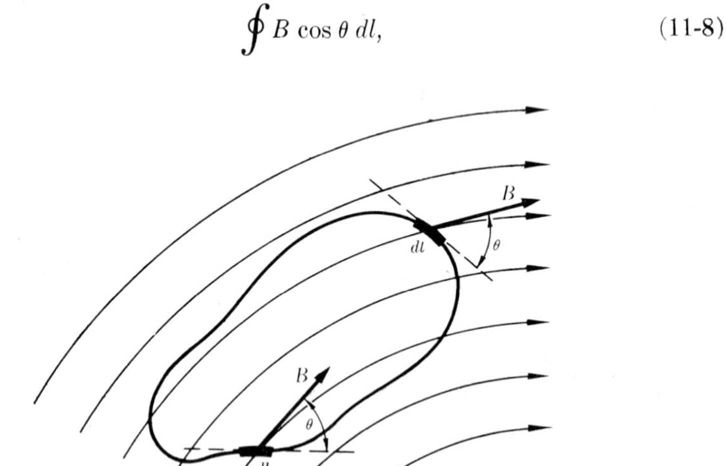

FIG. 11-3. The line integral of B around the closed curve is the sum of the products $B \cos \theta \, dl$ taken at every point on the curve.

where θ is the angle between any element dl and the direction of the induction B at the element. A sign convention of traverse around the curve must be adopted—let us take the clockwise direction as positive.

We shall deduce the general expression for the line integral by considering the special case of the field around a long straight conductor. Consider first the closed curve shown by the full circle in Fig. 11-4(a), a circle of radius r concentric with the conductor. The current i is directed away from the reader. The lines of induction are shown by dotted circles. The induction at every point is tangent to the circle, hence in Eq. (11-8) $\theta = 0$ and $\cos\theta = 1$. The magnitude of B, at every point of the circle, is

$$B = \frac{\mu_0}{2\pi}\frac{i}{r},$$

and the line integral becomes

$$\oint B\cos\theta\, dl = \oint \frac{\mu_0}{2\pi}\frac{i}{r}\,dl = \frac{\mu_0}{2\pi}\frac{i}{r}\,2\pi r = \mu_0 i.$$

Consider next the loop *abcd* in Fig. 11-4(b), whose sides *ab* and *cd* are radial and whose sides *bc* and *da* are circular arcs subtending an angle α and of radii r_2 and r_1. Along *ab* and *cd* the angle θ is $\pi/2$, $\cos\theta = 0$, and these sides contribute nothing to the integral. An element of length along *ad* or *bc* can be written

$$dl = r_1\,d\alpha, \quad dl = r_2\,d\alpha.$$

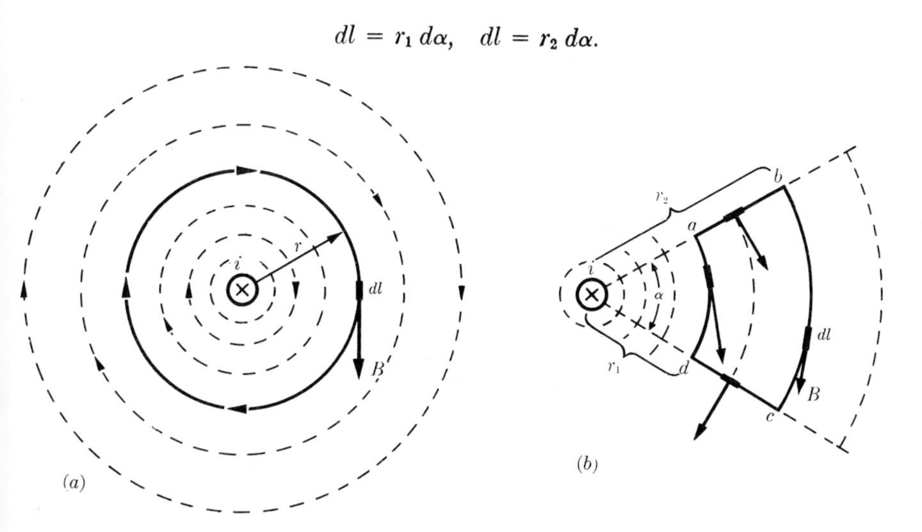

FIG. 11-4. The line integral of the magnetic induction around a closed curve. (a) Enclosing a current. (b) Not enclosing a current.

The contribution of side bc to the line integral is

$$\int \frac{\mu_0}{2\pi} \frac{i}{r_2} r_2 \, d\alpha = \frac{\mu_0}{2\pi} i\alpha,$$

since $\theta = 0$ along this side.

Side da contributes

$$-\int \frac{\mu_0}{2\pi} \frac{i}{r_1} r_1 \, d\alpha = -\frac{\mu_0}{2\pi} i\alpha,$$

since $\theta = \pi$ along this side (dl and B are in opposite directions). Hence the line integral around the loop $abcd$ is zero.

These two special cases can be generalized for a closed loop of any shape as follows: *The line integral of the magnetic induction around any closed curve equals the product of* μ_0 *and the current through the surface bounded by the curve.* If there is no current through the surface bounded by the curve, the line integral is zero. In accord with the convention that the direction of traverse around the loop is clockwise when one faces the loop, the current is considered positive if its direction is away from the observer as in Fig. 11-4.

$$\boxed{\oint B \cos\theta \, dl = \mu_0 i.} \tag{11-9}$$

The further generalization of Eq. 11-8 when magnetized matter is present is given in Sec. 15-12.

In vector notation, these surface and line integrals are written as follows. Eq. (11-7) becomes

$$\int \vec{B} \cdot d\vec{A} = 0 \tag{11-10}$$

where $d\vec{A}$ is a vector directed outward from the surface element dA.

Eq. (11-9) takes the form

$$\oint \vec{B} \cdot d\vec{l} = \mu_0 i. \tag{11-11}$$

11-4 Force between parallel straight conductors. The ampere. Fig. 11-5(a) shows a portion of two long straight parallel conductors separated by a distance a and carrying currents i and i' respectively in the same direction. Since each conductor lies in the magnetic field of the other, each will experience a force, exerted on it by the magnetic field set up by the current in the other conductor.

Fig. 11-5(b) shows some of the lines of induction of the field of the lower conductor. The induction B at the upper conductor is

$$B = \frac{\mu_0}{2\pi}\frac{i}{a}.$$

The left-hand rule shows that the direction of the force on the upper conductor is downward, and from Eq. (9-10),

$$\text{Force per unit length} = i'B = \frac{\mu_0}{2\pi}\frac{i\,i'}{a}.$$

There is an equal and opposite force per unit length on the lower conductor, as may be seen by considering the field around the upper conductor, or by Newton's third law. Hence the conductors attract one another.

If the direction of either current is reversed, the forces reverse also. Parallel conductors carrying currents in opposite directions repel one another.

The fact that two straight parallel conductors exert forces of attraction or repulsion on one another is made the basis of the definition of the ampere in the mks system. The ampere is defined as follows:

One ampere is that unvarying current which, if present in each of two parallel conductors of infinite length and one meter apart in empty space, causes each conductor to experience a force of exactly 2×10^{-7} *newtons per meter of length.*

It follows from this definition that the numerical value of μ_0 in the rationalized mks system, is ex-

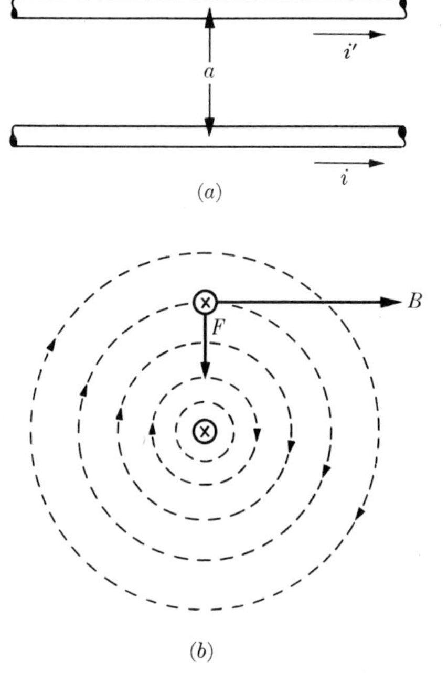

FIG. 11-5. Force between parallel straight conductors.

actly $4\pi \times 10^{-7}$ or, to four significant figures,

$$\mu_0 = 12.57 \times 10^{-7} \text{ webers/amp-m.}$$

Now the coulomb is defined in terms of the ampere, as the quantity of charge that in one second crosses a section of a circuit in which there is a constant current of one ampere. It follows that the unit of charge in the mks system is defined in terms of the force between *moving* charges, as contrasted with the electrostatic system in which the statcoulomb is defined in terms of the electrostatic force between charges.

From the definition above the ampere can be established, in principle, with the help of a meter stick and a spring balance. For the practical standardization of the ampere, coils of wire are used instead of straight wires and their separation is made only a few centimeters. The complete instrument, which is capable of measuring currents with a high degree of precision, is called a *current balance.*

The ampere as defined above, in terms of the force between current-carrying conductors, is called the *absolute ampere.* The *international ampere* was defined in Sec. 4-10 in terms of the rate of deposition of silver in a specified electrolytic cell. At the time of its definition the international ampere was believed to be equal to the absolute ampere, but later more precise measurements have shown a slight discrepancy. The best value at present is

$$1 \text{ international ampere} = 0.99986 \text{ absolute ampere.}$$

An *international volt* is defined as the potential difference between the terminals of an international ohm (see Sec. 4-6) when the current in it is one international ampere. The potential difference between two points is one *absolute volt* when one joule of work is required to take one absolute coulomb from one point to the other, the absolute coulomb being an absolute ampere-second. The absolute volt differs slightly from the international volt, and the best value is

$$1 \text{ international volt} = 1.00034 \text{ absolute volt.}$$

11-5 Field of a circular turn. Fig. 11-6 represents a circular turn of wire. The radius of the turn is a and the current in it is i. Of course a seat of emf is necessary to maintain the current, and one may either imagine the turn cut and a small seat of emf inserted, or consider that current is led into and out of the turn through two long wires side by side. The currents in the wires are in opposite directions and annul one another's magnetic effects.

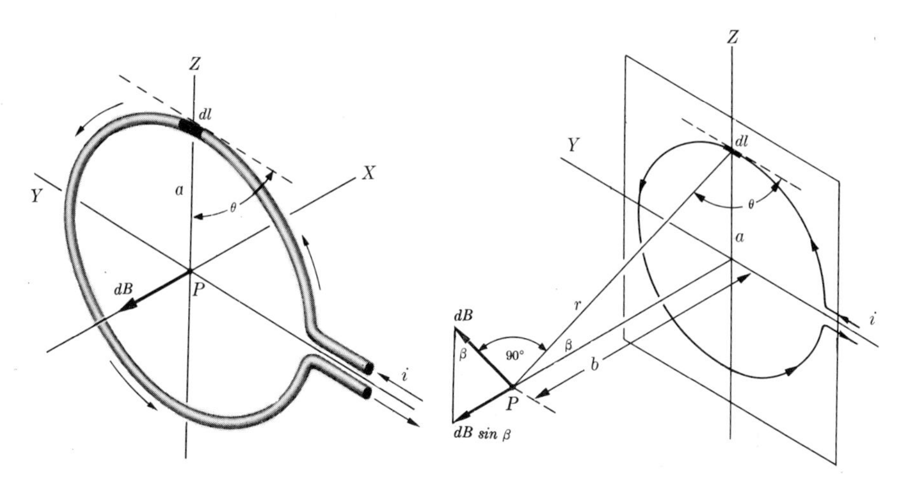

FIG. 11-6. Field at the center of a circular turn.

FIG. 11-7. Field on the axis of a circular turn.

Let us first calculate the induction at P, the center of the turn. Each element dl lies in the Y-Z plane as does each of the distances a from any dl to the point P. Hence the direction of each vector dB at P is perpendicular to the Y-Z plane or along the X-axis, and the resultant field can be found by direct integration of the dB's. The angle θ between dl and a is 90° and $\sin\theta = 1$. Hence

$$B = \int dB = \frac{\mu_0}{4\pi}\int \frac{i\,dl\sin\theta}{a^2} = \frac{\mu_0}{4\pi}\frac{i}{a^2}\int dl$$

But $\int dl$ is simply the circumference of the turn, $2\pi a$. Therefore

$$B = \frac{\mu_0}{2}\frac{i}{a}. \tag{11-12}$$

Consider next a point on the axis of the turn, not at its center, as in Fig. 11-7. The flux density dB at P, produced by the current in the element dl, is in the X-Z plane, perpendicular to r. It is evident that the dB's produced by the other elements are not all in the same direction, but lie on the surface of a cone with apex at P. Hence each must be resolved into components and the components integrated separately. However, one can see by symmetry that the components perpendicular to the X-axis annul one another, so that only the components along the X-axis, given by $dB \sin \beta$, need be summed. The angle θ between each dl and its corresponding r is 90°. Hence

$$B = \int dB \sin \beta = \frac{\mu_0}{4\pi} \int \frac{i\, dl \sin \theta \sin \beta}{r^2}$$

$$= \frac{\mu_0}{2} \frac{ia \sin \beta}{r^2} = \frac{\mu_0}{2} \frac{ia^2}{r^3} = \frac{\mu_0}{2} \frac{ia^2}{(a^2 + b^2)^{3/2}}. \qquad (11\text{-}13)$$

If $b = 0$, Eq. (11-13) reduces to Eq. (11-12).

The induction at points not on the axis of the turn can be computed by the same method but the calculations become rather lengthy and the results, in terms of familiar functions, are expressible only as an infinite series. However, at distances from the turn large compared with its radius, the equations reduce to a simple form when polar coordinates are used. At point P, Fig. 11-8, let the induction B be resolved into components B_r, in the direction of increasing r, and B_θ, perpendicular to r and in the direction of increasing θ. (Note that r and θ are polar coordinates and do not have the same meaning as in Figs. 11-6 and 11-7.)

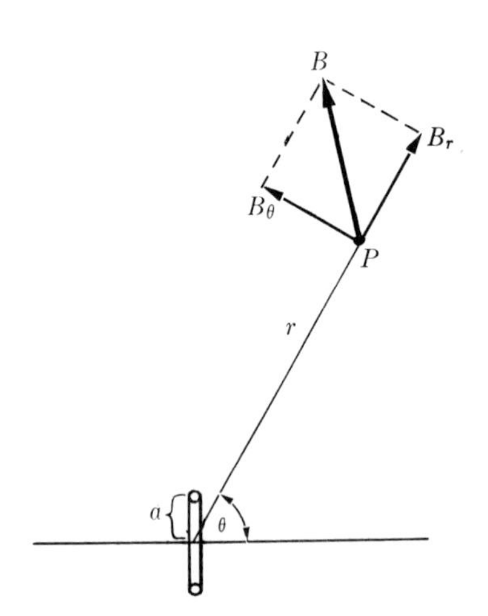

FIG. 11-8. Field at a point very far from a circular turn.

Then

$$\left.\begin{aligned} B_r &= \frac{\mu_0}{2}\, ia^2 \frac{\cos \theta}{r^3} \\ B_\theta &= \frac{\mu_0}{4}\, ia^2 \frac{\sin \theta}{r^3} \end{aligned}\right\} \text{ approximately, when } r \gg a. \qquad (11\text{-}14)$$

These equations can be put in the following useful form. Let us multiply numerator and denominator by π. Then

$$B_r = \frac{\mu_0}{2\pi}\, i\pi a^2 \frac{\cos \theta}{r^3},$$

$$B_\theta = \frac{\mu_0}{4\pi}\, i\pi a^2 \frac{\sin \theta}{r^3}.$$

The product πa^2 is the area A of the turn, and the factor $\mu_0 i \pi a^2$, or $\mu_0 iA$, is called the *magnetic moment* of the turn and is represented by M.

$$M = \mu_0 iA \tag{11-15}$$

The magnetic moment of a loop is a vector, defined in vector notation as

$$\vec{M} = \mu_0 i\vec{A},$$

where $\vec{A}$ is a vector perpendicular to the plane of the loop.

Although the preceding discussion has been in terms of a circular turn the magnetic moment of a turn of any shape is defined by Eq. (11-15). The components of flux density, at large distances from the turn, can then be written

$$\left.\begin{aligned} B_r &= \frac{1}{4\pi} 2M \frac{\cos\theta}{r^3}, \\ B_\theta &= \frac{1}{4\pi} M \frac{\sin\theta}{r^3}. \end{aligned}\right\} \tag{11-16}$$

Notice that these equations are of the same form as those given in Sec. 2-3 for the components E_r and E_θ of the electric intensity of an electric dipole of moment P, at distances large compared with the length of the dipole. (The factor μ_0, corresponding to ϵ_0, does not appear in these expressions because it was included in the definition of magnetic moment.) This means that the B-field around a current loop is geometrically identical with the E-field around an electric dipole, at distances large compared with the dimensions of the loop or dipole. Close to the loop or the dipole the similarity breaks down.

The torque on a current-carrying loop can also be expressed in terms of the magnetic moment of the loop. We have shown in Eq. (9-13) that the torque is

$$\tau = iBA \sin\alpha.$$

From Eq. (11-15)

$$iA = \frac{M}{\mu_0}.$$

Hence

$$\tau = \frac{1}{\mu_0} MB \sin\alpha.$$

If instead of a single turn, as in Fig. 11-6, one has a coil of N closely spaced turns all of essentially the same radius, each turn contributes equally to the field and Eqs. (11-12) and (11-13) become

$$B = \frac{\mu_0}{2}\frac{Ni}{a} \tag{11-17}$$

$$B = \frac{\mu_0}{2}\frac{Nia^2}{(a^2 + b^2)^{3/2}} \tag{11-18}$$

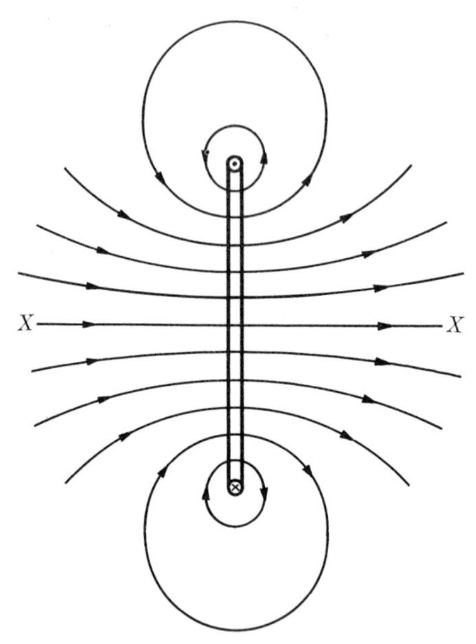

FIG. 11-9. Lines of induction surrounding a circular turn.

Some of the lines of induction surrounding a circular turn and lying in a plane through the axis are shown in Fig. 11-9. In reality, of course, the field is three-dimensional, and it may be visualized by imagining the plane to be rotated about the axis X-X. Note that each line of induction is a closed line.

It can now be seen that the positions of stable equilibrium of the loops shown in Figs. 9-11 and 9-12 are such that within the area enclosed by the loop the induction of the loop's own field is in the same direction as that of the external field. In other words, a loop, if free to turn, will set itself in such a plane that the flux passing through the area enclosed by it has its maximum possible value. This is found to be true in all instances and is a useful general principle. For example, if a current is sent through an irregular loop of flexible wire in an external magnetic field, the loop will assume a circular form with its plane perpendicular to the field and with its own flux adding to that of the field. The same conclusion can, of course, be drawn by analyzing the side thrusts on the elements of the conductor.

11-6 Field of a solenoid. The flux density produced at any point by a current in a solenoidal winding is simply the resultant of the flux densities set up at that point by each turn of the solenoid. To find the flux density at a point P on the axis of the solenoid (Fig. 11-10) consider an elementary length dx of the solenoid, at an axial distance x from P. Let l represent the length of the solenoid and N the number of turns in its

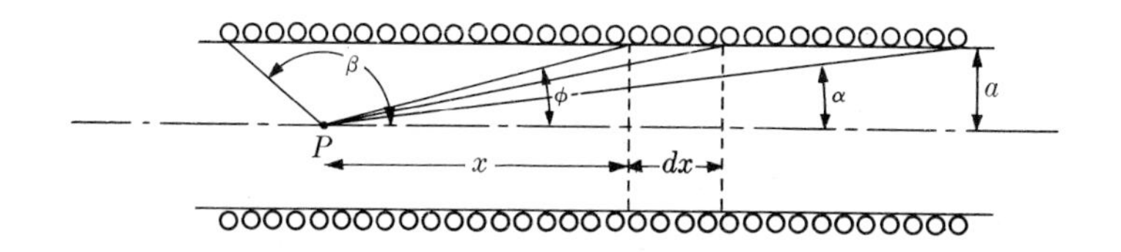

FIG. 11-10. Field of a solenoid.

winding. The number of turns per unit length is then N/l, and the number of turns in the length dx is $\frac{N}{l}\,dx$. (We are assuming that each turn can be considered to lie in a plane perpendicular to the axis of the solenoid.)

The flux density at P, set up by a current i in the element of length dx is, from Eq. (11-13),

$$dB = \frac{\mu_0}{2}\frac{Ni}{l}\frac{a^2}{(a^2 + x^2)^{3/2}}\,dx.$$

Let us use the angle ϕ instead of x as the independent variable. Then

$$x = a\cot\phi, \quad dx = -a\csc^2\phi\,d\phi$$

and

$$B = -\frac{\mu_0}{2}\frac{Ni}{l}\int_\beta^\alpha \sin\phi\,d\phi$$

$$= \frac{\mu_0}{2}\frac{Ni}{l}(\cos\alpha - \cos\beta).$$

Note that this equation applies to *any* axial point and is not restricted to points within the solenoid.

At any axial point within a long solenoid and not too near either end, $\alpha = 0$, $\beta = 180°$. Hence at such a point

$$B = \mu_0\frac{Ni}{l}\cdot \tag{11-19}$$

At an axial point at one end of a long solenoid, $\alpha = 0$, $\beta = 90°$ (or vice versa). Hence at either end,

$$B = \frac{\mu_0}{2}\frac{Ni}{l}, \tag{11-20}$$

and the flux density at either end is one-half its magnitude at the center. However, the field is very nearly constant except close to the ends and the

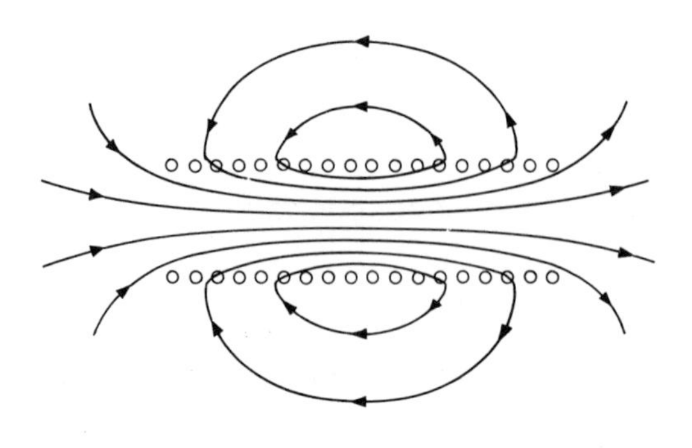

FIG. 11-11. Lines of induction surrounding a solenoid.

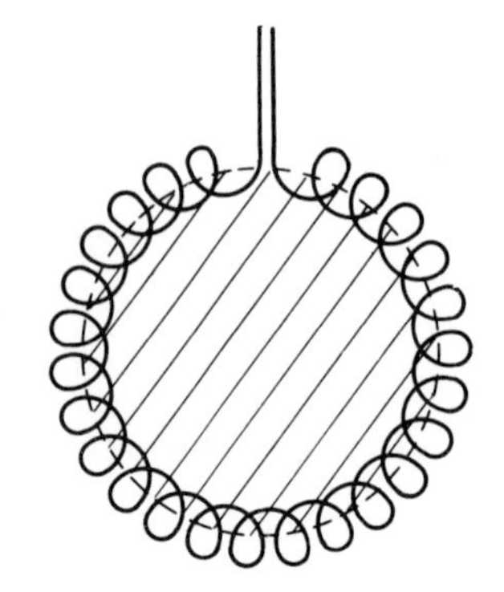

FIG. 11-12. A toroid.

lines of induction enter and leave the solenoid in relatively small regions near its ends. See Fig. 11-11 and problem 15 at the end of the chapter.

Although Eq. (11-19) was derived for a point on the axis of the solenoid, it is a good approximation for any point within the solenoid not too close to its ends.

A winding such as that sketched in Fig. 11-12 is called a *toroid*. It may be thought of as a solenoid which has been bent into a circular form so that its ends are joined together. Practically all of the magnetic flux is confined to the interior of the toroid, and the flux density at any point within the winding is given by Eq. (11-19), where l is to be interpreted as the *mean circumference* of the toroid.

As an example of the simplification made possible in some problems by the use of the line integral relation deduced in Sec. 11-3, let us use it as an alternate method of computing the flux density within a toroid. Let the closed path around which the line integral is to be evaluated be that indicated by the dotted line in Fig. 11-12. The surface enclosed by the path is shown shaded. Each of the N turns of the winding of the toroid passes once through this surface and the total current through it is Ni. The induction along the path is parallel to it at all points and by symmetry is constant. Let B represent the induction and l the length of the path. The line integral is then simply Bl, and from Eq. (11-9)

$$Bl = \mu_0 Ni,$$

$$B = \mu_0 \frac{Ni}{l},$$

which agrees with Eq. (11-19).

11-7 Field of a moving point charge. The magnetic induction at points around a current element and given by

$$dB = \frac{\mu_0}{4\pi} \frac{i\, dl \sin\theta}{r^2},$$

can be considered as the resultant of the induction due to all of the moving charges in the element, each charge contributing equally if all move with the same velocity. We have shown that the current in a conductor may be written

$$i = nqvA$$

where n is the number of moving charges per unit volume, q is the charge of each and v is the common velocity. It follows that

$$i\, dl = nqvA\, dl.$$

But $nqA\, dl$ is simply the quantity of charge in the element, say dq. The field dB can therefore be written

$$dB = \frac{\mu_0}{4\pi} \frac{v\, dq \sin\theta}{r^2}.$$

Hence the field of a finite charge q moving with velocity v is

$$B = \frac{\mu_0}{4\pi} \frac{vq \sin\theta}{r^2} \tag{11-21}$$

where θ is the angle between v and r, and B is perpendicular to the plane of v and r. If a number of charges contribute to the field,

$$B = \frac{\mu_0}{4\pi} \Sigma \frac{qv \sin\theta}{r^2} \text{ (vector sum).} \tag{11-22}$$

This equation, rather than Biot's rule, may equally well be taken as the starting point in an analysis of the magnetic fields around moving charges and currents. It may also be considered as the definition of the magnetic induction B, rather than Eq. (11-3).

Consider now two charges, q and q', moving with velocities v and v' respectively and separated by a distance r. The magnetic field set up by the charge q, at the point of space occupied by q', is from Eq. (11-22).

$$B = \frac{\mu_0}{4\pi} \frac{qv \sin\theta}{r^2},$$

and the force on the charge q' is

$$F = q'v'B \sin \phi$$

where ϕ is the angle between v' and B, and F is perpendicular to the plane of v' and B. Combining these equations, we find

$$F_{\text{mag}} = \frac{\mu_0}{4\pi} \frac{qv \sin \theta \cdot q'v' \sin \phi}{r^2}, \tag{11-23}$$

where the force is written F_{mag} to indicate that it arises from "magnetic" causes. This equation should be compared with Coulomb's law for the "electrical" force between charges,

$$F_{\text{elec}} = \frac{1}{4\pi\epsilon_0} \frac{qq'}{r^2}.$$

A perfectly logical way to develop the subjects of electricity and magnetism would be to start with both of these equations as experimental laws describing the force between electric charges. The right side of each equation may be split into two factors.

$$F_{\text{elec}} = \left[\frac{1}{4\pi\epsilon_0} \frac{q}{r^2}\right] \times [q'] \tag{11-24}$$

and

$$F_{\text{mag}} = \left[\frac{\mu_0}{4\pi} \frac{qv \sin \theta}{r^2}\right] \times [q'v' \sin \phi]. \tag{11-25}$$

The first bracketed term in both equations involves the charge q only, the second the charge q' only. The terms involving the charge q only are described as "fields" set up by this charge. The electric field is

$$E = \frac{1}{4\pi\epsilon_0} \frac{q}{r^2}, \tag{11-26}$$

and the magnetic field is

$$B = \frac{\mu_0}{4\pi} \frac{qv \sin \theta}{r^2}. \tag{11-27}$$

Eqs. (11-24) and (11–25) become

$$F_{\text{elec}} = Eq',$$

$$F_{\text{mag}} = Bq'v' \sin \phi.$$

In the most general case, both an electric and a magnetic field exist at a point, and the force on a moving charge at the point is

$$F = F_{\text{elec}} + F_{\text{mag}} \text{ (vector sum)}$$

$$= Eq' + Bq'v' \sin \phi,$$

where q' is the magnitude of the moving charge, v' the magnitude of its velocity, and ϕ the angle between B and v'.

Problems—Chapter 11

(1) A long straight wire, carrying a current of 200 amp, runs through a cubical wooden box, entering and leaving through holes in the centers of opposite faces as in Fig. 11-13. The length of each side of the box is 20 cm. Consider an element of the wire 1 cm long at the center of the box. Approximate ds by Δs, and compute the magnitude of the magnetic induction ΔB produced by this element at the points lettered a, b, c, d, and e in Fig. 11-13. Points a, c, and d are at the centers of the faces of the cube, point b is at the midpoint of one edge, and point e is at a corner. Copy the figure and show by vectors the directions and relative magnitudes of the field vectors.

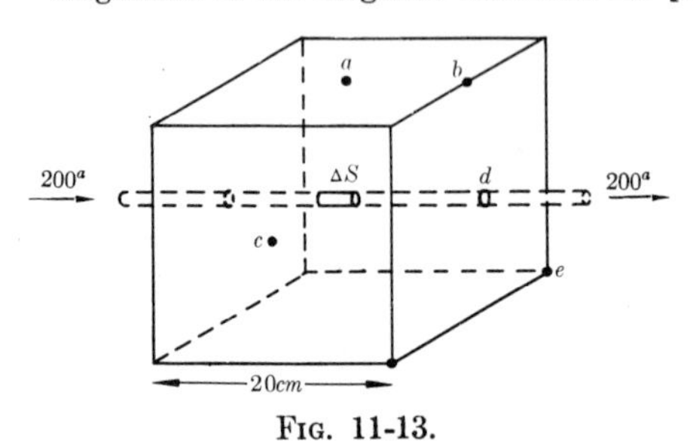

FIG. 11-13.

(2) A wire is bent to form a closed turn in the shape of a regular hexagon, each of whose sides is 20 cm long. The wire carries a current of 100 amp. Compute the magnetic induction at the center of the turn.

(3) A square loop of wire 6 m on a side carries a current of 5 amp. Compute the magnetic flux density at a point 4 m from the center of the square on a line through the center perpendicular to the plane of the square.

(4) A long horizontal wire AB rests on the surface of a table. (See Fig. 11-14.) Another wire CD vertically above the first is 100 cm long and is free to slide up and down on the two vertical metal guides C and D. The two wires are connected through the sliding contacts and carry a current of 50 amp. The mass of the wire CD is 0.05 gm/cm. To what equilibrium height will the wire CD rise?

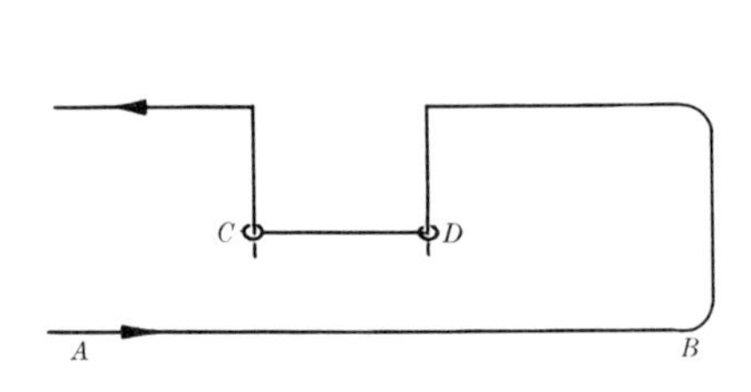

FIG. 11-14.

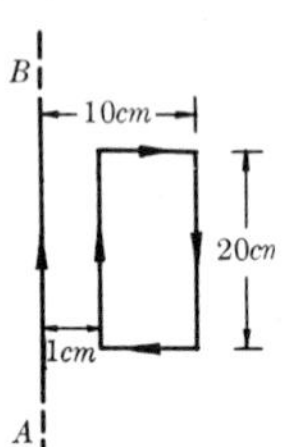

FIG. 11-15.

(5) The long straight wire AB in Fig. 11-15 carries a current of 20 amp. The rectangular loop whose long edges are parallel to the wire carries a current of 10 amp. Find the magnitude and direction of the resultant force exerted on the loop by the magnetic field of the wire.

(6) The two long straight wires in Fig. 11-16 are spaced 40 cm between centers. The current in each wire is 20 amp. (a) Compute the magnetic flux density at a point in the plane of the wires and midway between them. (b) Compute the total flux through the shaded area. Each long side of the rectangle is 10 cm from the axis of its neighboring wire, and is 25 cm long.

(7) A wooden ring shown in section in Fig. 11-17 has a long straight wire carrying a current i along its axis. Find the total number of flux lines (webers) within the material of the ring.

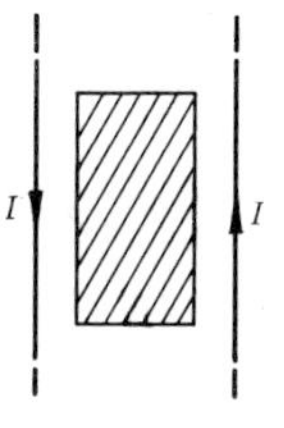

FIG. 11-16.

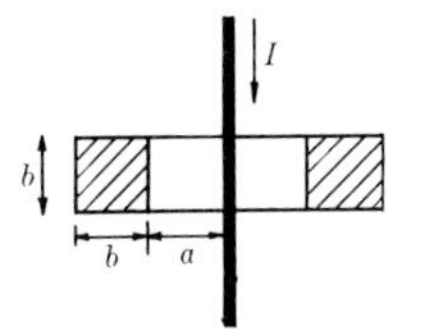

FIG. 11-17.

(8) A long thin straight metallic strip whose width is b carries a constant current i. The current density is uniform. Derive the expression for the magnetic flux density at a point in the plane of the strip and at a distance a from the nearer edge of the strip.

(9) A closely wound coil of 100 turns, 5 cm in radius, carries a current of 2 amp. Compute the magnetic flux density at points on the axis of the coil, at the following distances from its center: 0, 2 cm, 5 cm, 8 cm, 10 cm. Show the results in a graph. Compare with the answer to problem 11, Chap. 2.

(10) Two circular turns of wire, each of radius a, are placed with their planes parallel at a separation a between centers, in such a way that a line joining their centers is perpendicular to the planes of both turns. Deduce an expression for the magnetic induction at a point on this line midway between the planes of the turns. The currents in the turns are equal and are in the same direction in both turns. As a numerical example, let $a = 20$ cm, let each turn be a closely wound coil of 100 turns, and let the current be 5 amp.

(11) Two circular turns, each of radius a, and carrying equal currents in the same direction, are placed with their planes parallel at a separation b. Find the magnetic induction at the center of either turn.

(12) Compute the magnetic induction at the point P in Fig. 11-18. There are three circular turns arranged as shown, each carrying a current of 8 amp. The radius of the largest turn is 3 cm.

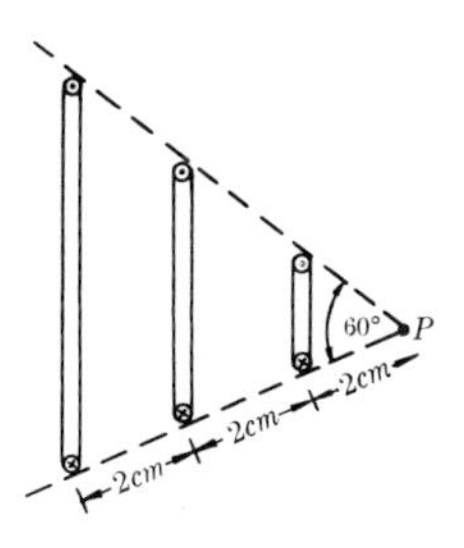

FIG. 11-18.

(13) A large number N of closely spaced turns of fine wire are wound in a single layer upon the surface of a wooden sphere with the planes of the turns perpendicular to the axis of the sphere and completely covering its surface. The current in the winding is i. Find the magnetic flux density at the center of the sphere.

(14) A thin disk of dielectric material, having a total charge $+Q$ distributed uniformly over its surface, rotates n times per second about an axis perpendicular to the surface of the disk and passing through its center. Find the magnetic induction at the center of the disk.

(15) A solenoid of length 20 cm and radius 2 cm is closely wound with 200 turns of wire. The current in the winding is 5 amp. Compute and show in a graph the magnetic induction at points on the axis of the solenoid at the following distances from its center: 0, 4 cm, 8 cm, 10 cm, 12 cm, 16 cm, 20 cm.

(16) A wooden ring whose mean diameter is 10 cm is wound with a closely spaced toroidal winding of 500 turns. Compute the flux density at a point on the mean circumference of the ring when the current in the windings is 0.3 amp.

(17) A brass ring of square cross section, 2 cm on a side, is bent into a ring of inner radius 7 cm, and the ends are welded together as in Fig. 11-19. Wire is wound toroidally around this ring to form a coil of 1600 turns and a current of 2 amp is sent through the winding. (a) Find the magnetic induction at a distance r from the center of the toroid. (b) What is the total flux across any section of the ring?

(18) A long straight cylindrical copper tube having an inside radius of 1 cm and an outside radius of 2 cm carries a current of 200 amp. Compute the magnetic induction at the following distances from the axis of the tube: 0.5 cm, 1.5 cm, 4 cm.

(19) The positive point charges q_1 and q_2 in Fig. 11-20 are moving with velocities v_1 and v_2 as shown. (a) What is the magnetic induction set up by the charge q_1 at the point occupied by q_2? (b) Find the magnitude and direction of the force acting on q_2. (c) What is the magnetic induction set up by the charge q_2 at the point occupied by q_1? (d) What is the force acting on q_1? (e) What about Newton's third law?

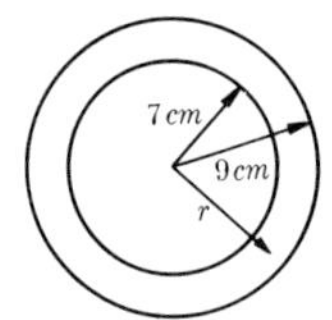

FIG. 11-19.

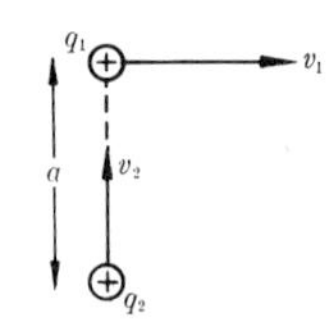

FIG. 11-20.

(20) A coaxial cable is constructed of a solid conductor of radius 1 cm, on the axis of a hollow cylindrical conductor of inner radius 2 cm and outer radius 2.5 cm. The inner conductor carries a steady current of 20 amp in one direction and the outer conductor a steady current of 20 amp in the opposite direction. Find the magnitude of the magnetic induction at the following distances from the axis of the cable: 0.5 cm, 1.5 cm, 2.2 cm, 3 cm.

CHAPTER 12

INDUCED ELECTROMOTIVE FORCE

12-1 Motional electromotive force. Our present day large scale production and distribution of electrical energy would not be economically feasible if the only seats of emf available were those of chemical nature such as dry cells. The development of electrical engineering as we now know it began with Faraday and Henry, who independently and at nearly the same time discovered the principles of induced emf's and the methods by which mechanical energy can be converted directly to electrical energy.

Fig. 12-1 represents a conductor of length l in a uniform magnetic field, perpendicular to the plane of the diagram and directed away from the reader. If the conductor is set in motion toward the right with a velocity v, perpendicular to both its own length and to the magnetic field, every charged particle within it experiences a force $F = qBv$ directed along the length of the conductor. The direction of the force on a negative charge is from a toward b in Fig. 12-1, while the force on a positive charge is from b toward a.

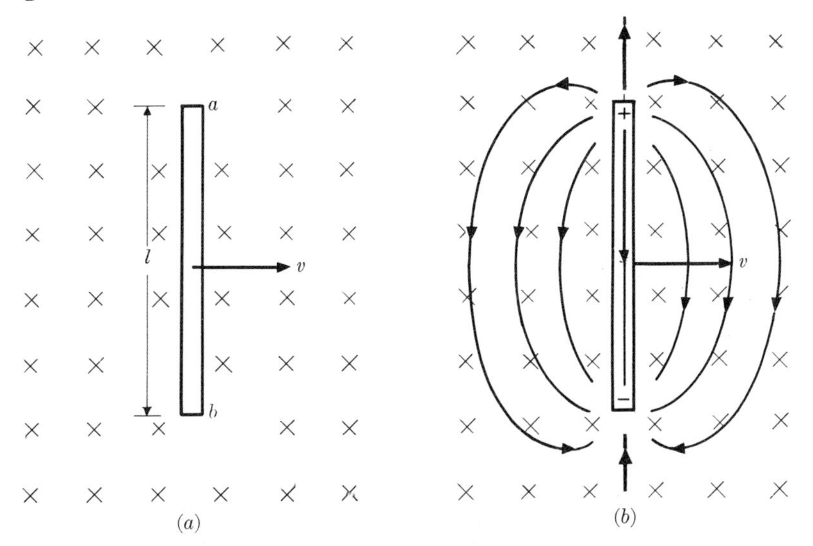

Fig. 12-1. Conductor moving in a uniform magnetic field.

The state of affairs within the conductor is therefore the same as if it had been inserted in an electric field of intensity Bv whose direction was from b toward a. The free electrons in the conductor will move in the direction of the force acting on them until the accumulation of excess charges at the ends of the conductor establishes an electrostatic field such that the resultant force on every charge within the conductor is zero. The general nature of this electrostatic field is indicated in Fig. 12-1(b). The upper end of the wire acquires an excess positive, and the lower end an excess negative charge.

That such a separation of charge actually takes place in a conductor moving in a magnetic field was shown by Barnet, who carried out experiments equivalent to cutting the rod at its center while it was still in motion, and then bringing it to rest. The upper half was found to be positively and the lower half negatively charged.

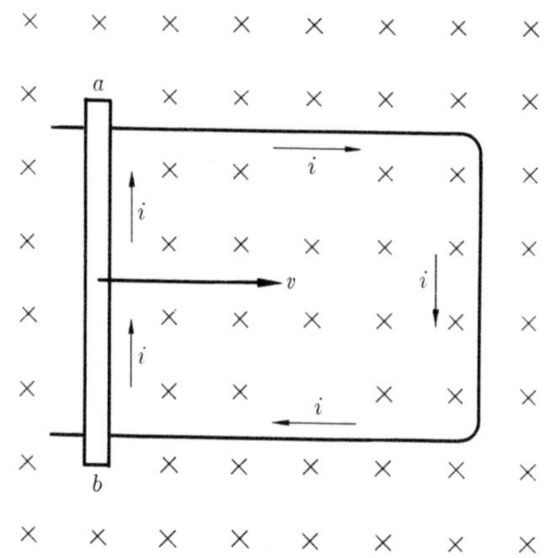

FIG. 12-2. Current produced by the motion of a conductor in a magnetic field.

Imagine now that the moving conductor slides along a stationary U-shaped conductor as in Fig. 12-2. There is no magnetic force on the charges within the stationary conductor, but since it lies in the electrostatic field surrounding the moving conductor, a current will be established within it, the direction of this current (in the conventional sense) being clockwise or from b toward a. As a result of this current the excess charges at the ends of the moving conductor are reduced, the electrostatic field within the moving conductor is weakened, and the magnetic forces cause a further displacement of the free electrons within it from a toward b. As long as the motion of the conductor is maintained there will, therefore, be a continual displacement of electrons counterclockwise around the circuit, or a conventional current in a clockwise direction. The moving conductor corresponds to a seat of electromotive force, and is said to have *induced* within it a *motional electromotive force* whose magnitude we now compute.

The definition of the emf of a seat is the ratio of the work done on the circulating charge to the quantity of charge displaced past a point of the circuit. Let i be the current in the circuit of Fig. 12-2. Because of the existence of this current, a side thrust toward the left is exerted on the moving conductor by the field, and therefore an external force provided by some working agent is needed to maintain the motion. The work done by this agent is the work done on the circulating charge—hence the direct conversion in this device of mechanical to electrical energy.

The side thrust on the moving conductor is

$$F = Bli.$$

The distance moved in time dt is

$$ds = v\,dt,$$

and the work done is

$$dW = F\,ds = Blvi\,dt.$$

But $i\,dt$ is the charge dq displaced in this time. Hence

$$dW = Blv\,dq$$

and the emf, dW/dq, is therefore

$$\boxed{\mathcal{E} = Blv.} \tag{12-1}$$

If B is expressed in webers/m^2, l in meters and v in meters/sec, the emf is in joules per coulomb or volts as may readily be verified.

An alternate definition of the emf in a circuit is (see Sec. 5-3)

$$\mathcal{E} = \oint E \cos\theta\,dl.$$

The significance of the "E" in this equation is the net force per unit charge. In the problem under discussion both electric and magnetic forces must be included. The net force on a charge q is

$$\begin{aligned} F &= F_{\text{elec}} + F_{\text{mag}} \\ &= Eq + Bqv \sin\phi, \end{aligned}$$

and since $\sin\phi = 1$, the force per unit charge is

$$\frac{F}{q} = E + Bv,$$

where E is the electrostatic field set up by the displaced charges and B is the external magnetic field. Then

$$\mathcal{E} = \oint (E + Bv)\cos\theta\,dl = \oint E\cos\theta\,dl + \oint Bv\cos\theta\,dl.$$

We have shown in Sec. 3-3 that the line integral of an *electrostatic* field around a closed curve is zero. Hence

$$\oint E \cos\theta \, dl = 0.$$

In the second term, v is zero in the stationary part of the circuit so that the integral extends only along the moving rod where $\theta = 0$ and $\cos\theta = 1$. Then

$$\oint Bv \cos\theta \, dl = Bv \int dl = Bvl,$$

and

$$\mathcal{E} = Bvl,$$

which is the same as Eq. (12-1).

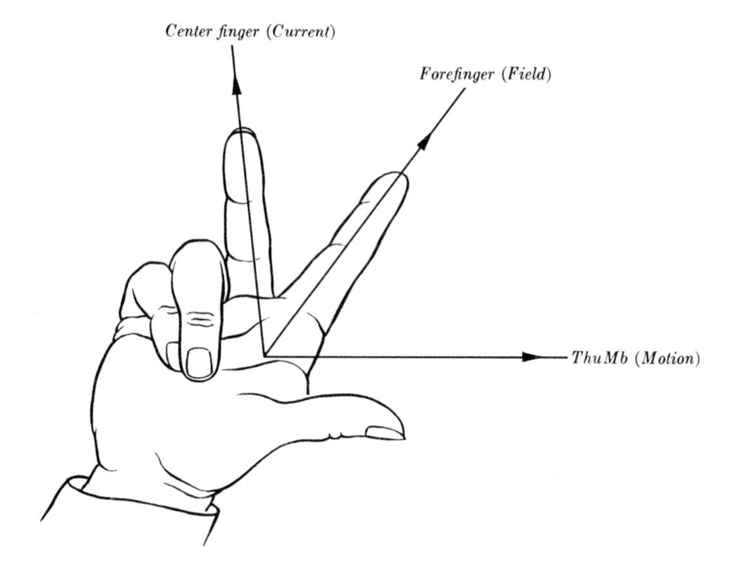

FIG. 12-3. The right-hand rule.

The relative directions of emf, field, and motion may be kept in mind by the *right hand rule*. (See Fig. 12-3.) Extend the thumb and first two fingers of the right hand in mutually perpendicular directions. Point the thuMb in the direction of Motion, the Forefinger in the direction of the Field. The Center finger then points in the direction of the emf, or the (conventional) Current if there is a closed circuit.

It is important to understand that two velocities (or velocity components) and two side thrusts are involved in this problem. The charges within the moving rod in Fig. 12-2 have a velocity component v toward the right because they are carried along by the moving rod. As a result of this velocity of transport they experience a thrust lengthwise along the rod (upward in Fig. 12-2) and if the rod is part of a closed circuit, this thrust causes a general drift of charge upward along the rod. (We are describing the process in terms of the motion of positive charge.) As soon as the charges acquire an upward velocity component, a second component of side thrust develops which, by the left-hand rule, is toward the left in Fig. 12-2. This force is perpendicular to the length of the rod and does not affect the circulation of charge, but it does give rise to a side thrust on the rod which opposes its motion toward the right. Consequently an external force must be applied to the rod to maintain its motion and the energy input to the circuit consists of the work done by this external force.

An emf is presumed to develop in the rod even in the absence of any closed circuit, although of course there can be no steady current unless such a circuit is provided. If the conductor is part of a closed circuit, the resulting current is called an *induced current.*

Eq. (12-1) was deduced for the special case where the magnetic field was uniform and the velocity, the length of the conductor, and the magnetic field were mutually perpendicular. The general equation is

$$d\mathcal{E} = Bv\, dl \sin\theta \cos\phi, \tag{12-2}$$

where dl is an element of length moving with velocity v, B is the flux density at the point where dl is located, θ is the angle between B and dl, and ϕ is the angle between v and the normal to the plane determined by dl and B. See Fig. 12-4. If dl is at right angles to B, then $\theta = 90°$ and $\sin\theta = 1$. If v is at right angles to dl and B, $\phi = 0$ and $\cos\phi = 1$. Eq. (12-2) then reduces to Eq. (12-1).

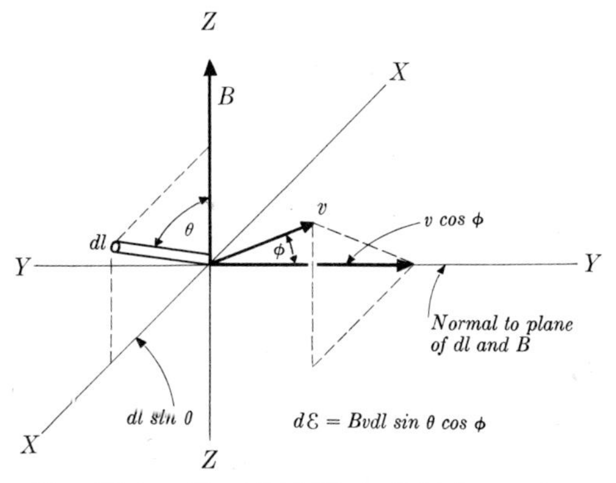

FIG. 12-4. $d\mathcal{E} = B \times (dl \sin\theta) \times (v \cos\phi)$.

In vector notation, Eq. (12-2) becomes

$$d\mathcal{E} = (\vec{B} \times \vec{dl}) \cdot \vec{v}.$$

Example: The induction B in the region between the pole faces of an electromagnet is 0.5 weber/meter^2. Find the induced emf in a straight conductor 10 cm long, perpendicular to B, and moving perpendicular both to B and its own length with a velocity of one meter/second.

The induced emf is

$$\mathcal{E} = Blv \sin\theta \cos\phi$$

$$\theta = 90°, \quad \phi = 0$$

$$B = .5 \text{ weber/m}^2, \quad l = 0.10 \text{ m}, \quad v = 1 \text{ m/sec.}$$

$$\mathcal{E} = .5 \times .1 \times 1 = 0.05 \text{ volt.}$$

12-2 The Faraday law. The induced emf in the circuit of Fig. 12-2 may be considered from another viewpoint. While the conductor moves toward the right a distance ds, the cross sectional area of the closed circuit *abcd*, Fig. 12-5, decreases by

$$dA = l\,ds,$$

and the change in flux through the circuit is

$$d\Phi = -B\,dA = -Bl\,ds.$$

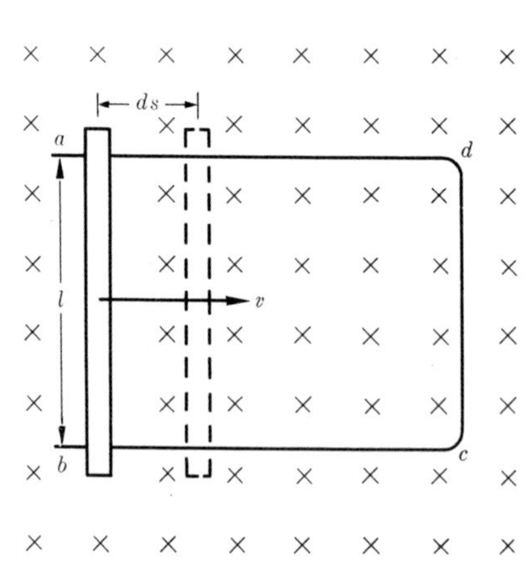

FIG. 12-5. Change of flux through the circuit *abcd*.

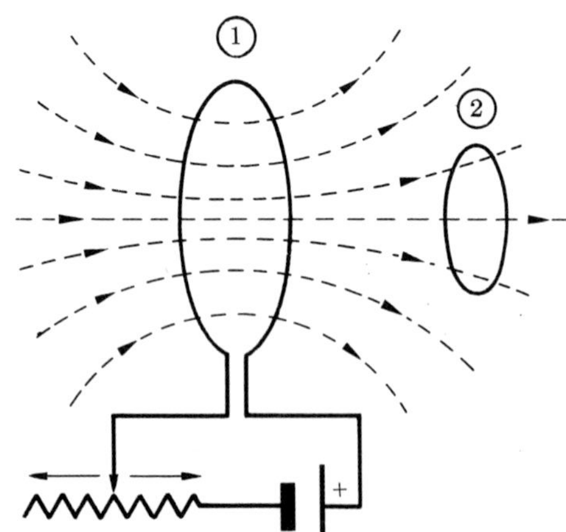

FIG. 12-6. As the current in circuit 1 is varied, the magnetic flux through circuit 2 changes.

When both sides are divided by dt, we obtain

$$-\frac{d\Phi}{dt} = Bl\frac{ds}{dt} = Blv. \tag{12-3}$$

But Blv equals the induced emf, and Eq. (12-3) states that *the induced emf in the circuit is numerically equal to the negative rate of change of the magnetic flux through the circuit.*

$$\boxed{\mathcal{E} = -\frac{d\Phi}{dt}.} \tag{12-4}$$

Eq. (12-4) as it stands appears to be merely an alternate form of Eq. (12-1) for the emf in a moving conductor. It turns out, however, that the relation has a much deeper significance than might be expected from its derivation. That is, it is found to apply to any circuit through which the flux is caused to vary by any means whatever, even though there is no motion of any part of the circuit and hence no emf directly attributable to a force on a moving charge.

Suppose, for example, that two loops of wire are located as in Fig. 12-6) A current in circuit 1 sets up a magnetic field whose magnitude at all points is proportional to this current. A part of this flux passes through circuit 2, and if the current in circuit 1 is increased or decreased the flux through circuit 2 will also vary. Circuit 2 is not moving in a magnetic field and hence no "motional" emf is induced in it, but there is a change in the flux through it and it is found experimentally that an emf appears in circuit 2 of magnitude

$$\mathcal{E} = -\frac{d\Phi}{dt}.$$

In a situation such as that shown in Fig. 12-6 it is evident that no one portion of circuit 2 can be considered the seat of emf; the entire circuit constitutes the seat.

The induced emf may also be expressed as follows. Since

$$\Phi = \int B\cos\phi\, dA,$$

we obtain by differentiating both sides

$$\mathcal{E} = -\frac{d\Phi}{dt} = -\int \frac{d(B\cos\phi)}{dt}\, dA. \tag{12-5}$$

This relation is known as the *Faraday law.*

Let us again make use of the definition of emf as

$$\mathcal{E} = \oint E \cos \theta \, dl.$$

When this is combined with the Faraday law, we obtain

$$\oint E \cos \theta \, dl = -\frac{d\Phi}{dt} = -\int \frac{d(B \cos \phi)}{dt} \, dA, \tag{12-6}$$

where the first integral is a line integral along a circuit, and the second is a surface integral over the area bounded by the circuit. We shall need Eq. (12-6) in Chap. 17 in the derivation of the velocity of electromagnetic waves.

To sum up, then, an emf is induced in a circuit whenever the flux through the circuit varies with time. The flux may be caused to vary in two ways, (1) by the motion of a conductor, as in Fig. 12-2 or (2) by a change in the magnitude or direction of the induction through a stationary circuit as in Fig. 12-5. For case (1), the emf may be computed either from

$$\mathcal{E} = Blv$$

(or the more general relation $d\mathcal{E} = Bv \, dl \sin \theta \cos \phi$) or from

$$\mathcal{E} = -\frac{d\Phi}{dt}.$$

For case (2), the emf may be computed only by

$$\mathcal{E} = -\frac{d\Phi}{dt} = -\int \frac{d(B \cos \phi)}{dt} \, dA. \tag{12-7}$$

Example. With a certain current in circuit 1 of Fig. 12-5, a flux of 5×10^{-4} webers links with circuit 2. When circuit 1 is opened the flux falls to zero in 0.001 sec. What average emf is induced in circuit 2?

The average rate of decrease of flux in circuit 2 is

$$\frac{\Delta\Phi}{\Delta t} = \frac{5 \times 10^{-4}}{.001} = 0.5 \text{ webers/sec}$$

The average induced emf is therefore 0.5 volt.

12-3 Lenz's Law. H. F. E. Lenz (1804–64) was a German scientist who without knowledge of the work of Faraday and Henry duplicated many of their discoveries nearly simultaneously. The law which goes by his name is a useful rule for predicting the direction of an induced emf It states:

The direction of an induced emf is such as to oppose the cause producing it.

The law is to be interpreted in somewhat different ways depending on the particular "cause" of the emf.

If the emf is "caused" by the motion of a conductor in a magnetic field, it "opposes" this cause by setting up a current (always provided there exists a closed circuit) in such a direction that the side thrust on this current is opposite to the direction of motion of the conductor. The *motion* of the conductor is therefore "opposed."

If the emf is "caused" by the change in flux through a closed circuit, the current resulting from the emf is in such direction as to set up a flux of its own which within the plane of the circuit is (a) *opposite* to the original flux if this flux is *increasing,* but (b) is in the *same* direction as the original flux if the latter is *decreasing.* Thus it is the *change in flux* (not the flux itself) which is "opposed."

12-4 The betatron. The magnetic induction accelerator, or betatron, is a recent addition to the family of instruments designed for the purpose of accelerating charged particles to high speeds. It was invented in 1941 by Donald W. Kerst of the University of Illinois. Kerst's original apparatus was capable of accelerating electrons to energies of 2 million electron-volts. A second machine constructed by the General Electric Company

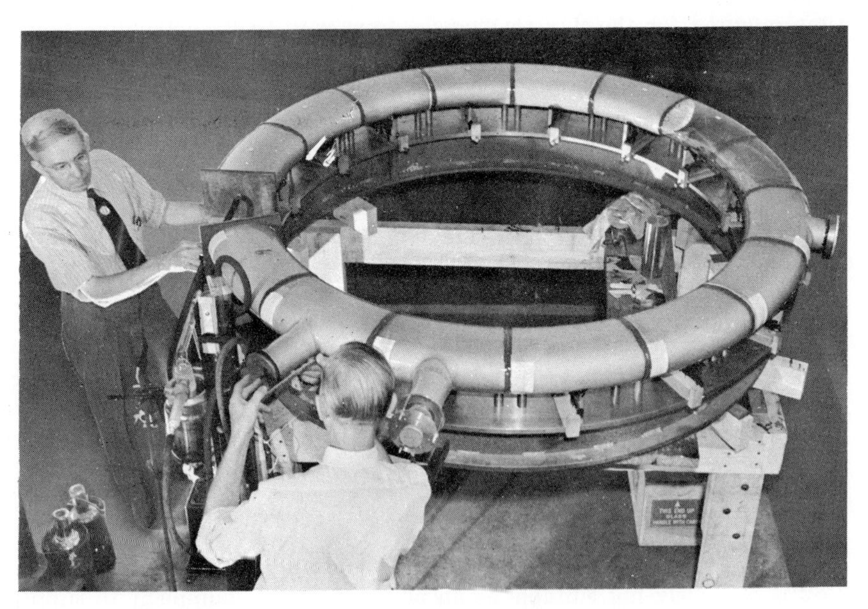

FIG. 12-7. Assembling the huge vacuum tube of the betatron. (*Courtesy of General Electric Co.*)

developed 20-million volt electrons, and in 1945 a third model was completed which can accelerate electrons to an energy of 100 million electron-volts. The following description refers to the 100-million volt machine.

An evacuated toroidal or doughnut-shaped glass tube of elliptical cross section, 74 inches in outside diameter and 58 inches in inside diameter, is placed horizontally in an air gap between the pole faces of an electromagnet. Alternating current at a frequency of 60 cycles/sec is sent through the windings of the electromagnet so that the magnetic flux through the plane of the toroid reverses from a maximum in one direction to a maximum in the opposite direction in 1/120 sec. Electrons accelerated through approximately 50,000 volts by an electron gun are shot tangentially into the tube and are caused by the magnetic field to circle around within the tube in an orbit 66 inches in diameter. In each revolution they are accelerated through the same voltage (about 400 volts) as would be induced in a single turn of wire through which the flux varied at the same rate. The accelerator may be compared with an ordinary transformer, with the usual high voltage secondary winding of many turns replaced by the electrons in the evacuated tube. The electrons are accelerated by the *changing* magnetic field and at the same time are forced to move in a circular orbit by the *existence* of the magnetic field. The velocity acquired is so great that an electron may make 250,000 revolutions in the time required for the flux to increase from zero to its maximum value. Since each revolution is equivalent to an acceleration through 400 volts, the final energy is $250{,}000 \times 400$ or 100 million electron volts.

At any desired stage in the accelerating process an additional surge or pulse of current may be sent through the magnet coils. This causes the electrons to spiral out of their circular path and strike a target, which then becomes a source of x-rays of extremely short wave length and great penetrating power.

Note that the relativistic increase of mass with velocity, which prevents the acceleration of electrons in a cyclotron, does not preclude their acceleration in this apparatus since they gain their final energy in less than one cycle of the varying field and do not have to remain in phase with the field for a large number of cycles.

Fig. 12-7 is a photograph showing the assembly of the vacuum tube.

12-5 The Faraday disk dynamo. The arrangement shown in Fig. 12-2 is obviously not suitable as an actual piece of equipment to supply current to a circuit. Mechanical power is usually available at the shaft of some rotating device such as a turbine or internal combustion engine, and hence some device in which an emf is developed in a rotating conductor is to be preferred. One of the first of such devices was invented by Faraday and is called a *disk dynamo*.

Fig. 12-8 represents a metal disk of radius R rotating about an axis through its center. The disk is in a uniform magnetic field perpendicular to the plane of the diagram. Brushes B-B make sliding contact with the shaft and rim of the disk, and are connected to the terminals of the external circuit. Consider a radial element of the disk lying between the two brushes. A segment of length dr is moving transverse to the field with a velocity

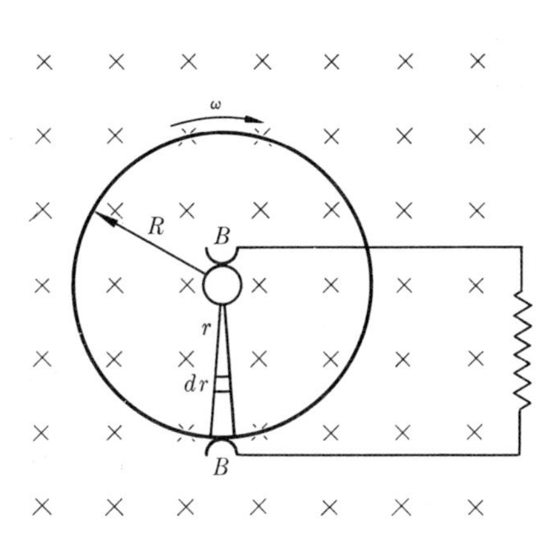

FIG. 12-8. The Faraday disk dynamo.

$$v = \omega r$$

and hence an emf is induced in it of magnitude

$$d\mathcal{E} = B\omega r\, dr.$$

Since the segments between shaft and rim are in series, the emf in the entire radial element is

$$\mathcal{E} = \int d\mathcal{E} = B\omega \int_0^R r\, dr = \frac{B\omega R^2}{2} \cdot \qquad (12\text{-}8)$$

The same emf is of course induced in every other radial element, and since all of these are connected in parallel, the entire disk may be considered a seat of emf of magnitude given by Eq. (12-8), with the rim constituting one terminal and the shaft the other.

Because of the relatively small emf's obtainable with this apparatus, it is not used commercially.

12-6 Induced emf in a rotating coil. The principle of the present-day form of dynamo is illustrated in Fig. 12-9. A closely wound rectangular coil *abcd* of N turns rotates about an axis OO which is perpendicular to a uniform magnetic field of flux density B. The terminals of the coil are connected to *slip-rings* S-S concentric with the axis of the coil and rotating with it, but insulated from one another. Brushes bearing against these rings connect the coil to the external circuit. The magnetic field of a commercial dynamo is provided by an electromagnet and the coil

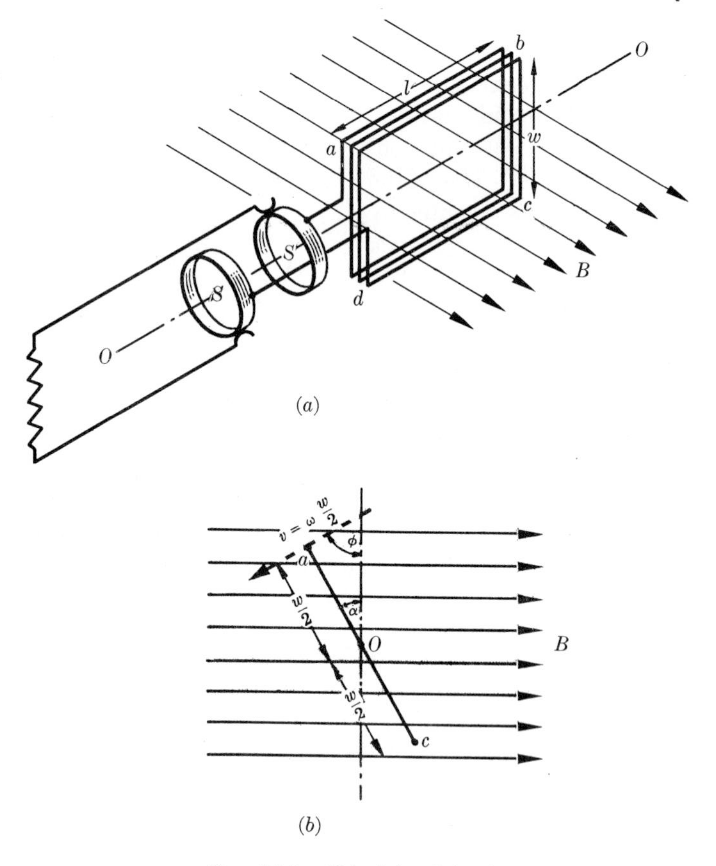

FIG. 12-9. Principle of the dynamo.

itself is wound on an iron cylinder or *armature*. See Chap. 16 for a further description. (The assembly of coil plus cylinder is also referred to as the armature.)

The magnitude of the emf induced in the coil of Fig. 12-9 may be computed either from the rate of change of flux through the coil, or from the velocities of its sides transverse to the magnetic field.

At an instant when the plane of the coil makes an angle α with the normal to the field as in Fig. 12-8(b), the flux through the coil is

$$\Phi = AB \cos \alpha,$$

where A is the area enclosed by the coil. The rate of change of flux is therefore

$$\frac{d\Phi}{dt} = -AB \sin \alpha \frac{d\alpha}{dt} \cdot$$

The induced emf in each turn of wire in the coil equals $-\frac{d\Phi}{dt}$, and if the coil has N turns the induced emf is

$$\mathcal{E} = -N\frac{d\Phi}{dt} = NAB\omega \sin \alpha \tag{12-9}$$

where $\omega = d\alpha/dt$ is the angular velocity of the coil.

Alternately, if w is the width and l the length of the coil, the tangential velocity of sides ab and cd is

$$v = \omega \frac{w}{2}$$

and the motional emf in each is

$$\begin{aligned} \mathcal{E} &= Blv \cos \phi \\ &= \tfrac{1}{2}Bl\omega w \sin \alpha. \end{aligned}$$

(There is no motional emf in the sides bc and da. Why not?) Since these emf's are in series aiding, the net emf in N turns is

$$\mathcal{E} = NBlw\omega \sin \alpha$$

and since

$$lw = A,$$

$$\mathcal{E} = NBA\omega \sin \alpha$$

which agrees with Eq. (12-9).

It is easy to show that Eq. (12-9) applies to a coil of any shape, rotating about an axis perpendicular to a uniform magnetic field.

From Eq. (12-9) the emf is a maximum when the plane of the coil is parallel to the field, and zero when it is perpendicular to the field. This is in agreement with the fact that in the parallel position the sides ab and cd are moving normally to the field, while in the perpendicular position their motion is parallel to the field. The maximum emf is

$$\mathcal{E}_{\max} = NBA\omega.$$

Hence Eq. (12-9) can be written,

$$\mathcal{E} = \mathcal{E}_{\max} \sin \alpha$$

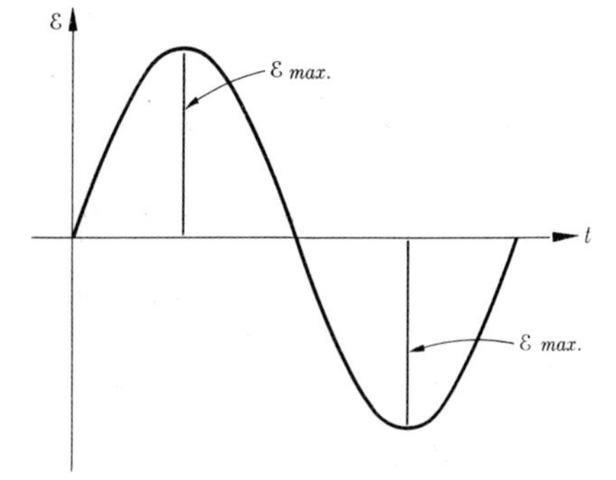

FIG. 12-10. Graph of the alternating emf induced in the coil of Fig. 12-9.

If ω is constant,

$$\alpha = \omega t \quad \text{or} \quad \alpha = 2\pi f t$$

and hence

$$\mathcal{E} = \mathcal{E}_{\text{max}} \sin \omega t$$

$$= \mathcal{E}_{\text{max}} \sin 2\pi f t, \tag{12-10}$$

which brings out explicitly the dependence of the instantaneous emf on the time t. A graph of Eq. (12-9), is given in Fig. 12-10. The rotating coil is the simplest form of *alternating current generator* or *alternator.*

12-7 The direct current generator. A coil of wire rotating in a magnetic field develops, as we have seen, a sinusoidal alternating emf. A unidirectional emf may be obtained by connecting each terminal of the coil to one side of a split ring or *commutator.* (See Fig. 12-10.) At the instant when the emf in the coil reverses, the connections to the external circuit are interchanged and the emf between the terminals, although pulsating, is always in the same direction.

In practice a more uniform emf is obtained by winding a large number of coils on the generator armature, each coil being brought out to its own pair of commutator segments. The brushes thus make connection to each coil during a short period when the emf in that coil is near its maximum value. Thus if there were six coils on the armature, equally spaced around its circumference, six emf's would develop, as shown by

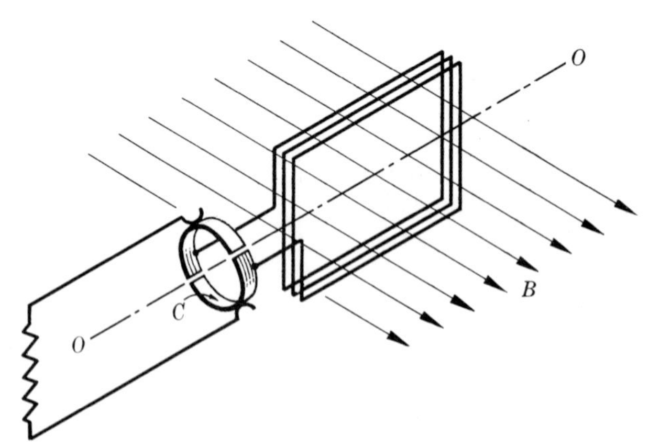

FIG. 12-11. Split ring commutator for unidirectional emf.

the dotted curves in Fig. 12-12. The potential difference at the terminals of the generator is shown by the heavy line.

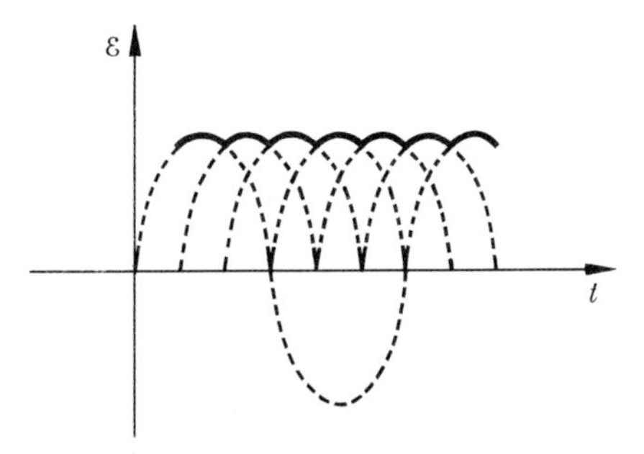

FIG. 12-12. Unidirectional emf obtained with an armature composed of 6 coils with 12 commutator segments.

We are now in a position to understand why the armature of a motor is a seat of back emf, a fact which was stated without proof in Sec. 5-2. A motor armature is simply a coil (or coils) in continuous rotation in a magnetic field. The armature is forced to rotate by the side thrust on the current sent through it by the external seat of emf to which its terminals are connected. The fact that it does rotate results in the generation of an emf within it, opposite to the external emf. Every motor is therefore necessarily a generator also.

12-8 Search coil method of measuring magnetic flux. A useful experimental method of measuring the flux density at a point in a magnetic field will now be described. The apparatus consists of a ballistic galvanometer connected by flexible leads to the terminals of a small, closely-wound coil called a *search coil* or a *snatch coil.* Assume first, for simplicity, that the search coil is placed with its plane perpendicular to a magnetic field of flux density B. If the area enclosed by the coil is A, the flux Φ through it is $\Phi = BA$. Now if the coil is quickly given a quarter-turn about one of its diameters so that its plane becomes parallel to the field, or if it is quickly snatched from its position to another where the field is known to be zero, the flux through it decreases rapidly from BA to zero. During the time that the flux is decreasing, an emf of short duration is induced in the coil and a "kick" is imparted to the ballistic galvanometer. The maximum deflection of the galvanometer is noted.

The galvanometer current at any instant is

$$i = e/R,$$

where R is the combined resistance of galvanometer and search coil, e is the instantaneous induced emf and i the instantaneous current.

Since
$$e = -N\, d\Phi/dt,$$

$$i = -\frac{N}{R}\, d\Phi/dt.$$

Hence

$$\int_0^t i dt = q = -\frac{N}{R}\int_\Phi^0 d\Phi = \frac{N\Phi}{R},$$

$$\Phi = \frac{R}{N} q, \tag{12-10}$$

$$B = \frac{\Phi}{A} = \frac{Rq}{NA}. \tag{12-11}$$

We have shown (see Sec. 10-4) that the maximum deflection of a ballistic galvanometer is proportional to the quantity of charge displaced through it. Hence, if this proportionality constant is known, q may be found, and from q one may obtain Φ and B.

Strictly speaking, while this method gives correctly the total flux through the coil, it is only the *average* flux density over the area of the coil which is measured. However, if the area is sufficiently small, this approximates closely the flux density at, say, the center of the coil.

Eq. (12-11) provides still another possible *definition* of the magnetic induction, in addition to Eqs. (9-4), (11-3), and (11-22).

The preceding discussion assumed the plane of the coil to be initially perpendicular to the direction of the field. If one is "exploring" a field whose direction is not known in advance, the same apparatus may be used to find the direction by performing a series of experiments in which the coil is placed at a given point in the field in various orientations, and snatched out of the field from each orientation. The deflection of the galvanometer will be a maximum for the particular orientation in which the plane of the coil was perpendicular to the field. Thus the magnitude and direction of an unknown field can both be found by this method.

Since the search coil is permanently connected to the galvanometer terminals, the galvanometer will be highly damped and must either be calibrated with the search coil connected, or the corrections mentioned in Sec. 12-9 must be applied.

12-9 Galvanometer damping. Suppose a ballistic galvanometer is connected as in Fig. 12-13(a) to measure the quantity of charge on a capacitor. Let the switch S be closed momentarily, allowing the capacitor to discharge through the galvanometer, and then immediately opened. The surge of charge starts the galvanometer coil swinging and, since it is rotating in a magnetic field, an emf is induced in it. The current through it is zero, however, since the switch has been opened and there is no closed

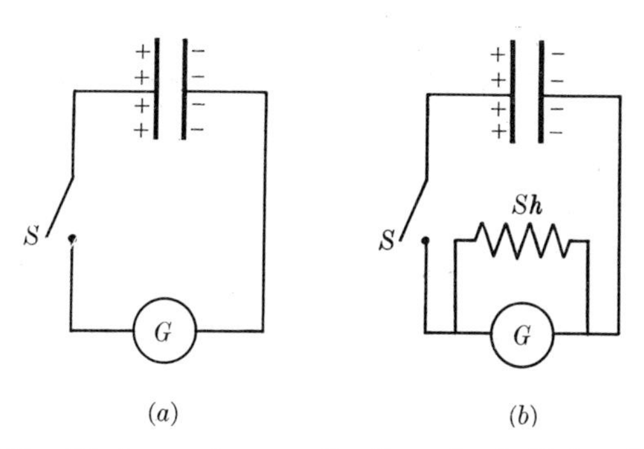

Fig. 12-13. Discharge of a capacitor through a ballistic galvanometer.

circuit. The motion of the coil is controlled solely by the suspension and friction. If the latter were entirely absent, the coil would oscillate indefinitely with angular harmonic motion.

Now let the shunt resistor Sh in Fig. 12-13(b) be connected across the galvanometer terminals and the experiment repeated. The motion of the galvanometer will be affected for two reasons. First, a part of the discharge current of the capacitor will be by-passed by the shunt and the impulse imparted to the coil will be correspondingly less. Second, the galvanometer and shunt now form a closed circuit even when switch S is opened, so that there will now be a current in the swinging coil. The side thrusts on this current give rise to a torque on the coil, and from Lenz's law the direction of this torque is such as to oppose the motion of the coil and aid in bringing it to rest. From the energy standpoint, a part of the kinetic energy of the swinging coil becomes converted to heat developed by the induced currents. The motion of the coil is accordingly *damped harmonic.*

It should be evident that the smaller the shunt resistance, the larger will be the induced current and the greater the damping. With a sufficiently small shunt resistance the motion ceases to be oscillatory, the galvanometer making but one swing and returning slowly to its zero position. The particular resistance for which the motion just ceases to be oscillatory is called the *critical external damping resistance* ($CXDR$) and when shunted by its $CXDR$ the galvanometer is said to be *critically damped.* With more resistance it is *underdamped* and with less it is *overdamped.*

Since the presence of damping reduces the maximum swing of a ballistic galvanometer, the simple theory in Sec. 10-4, which assumes that all of the initial kinetic energy of the coil is converted to potential energy of the suspension, must be extended if damping is present. The complete analysis shows that the quantity of charge displaced is still proportional to the maximum angle of swing, although with a modified proportionality constant. However, if the galvanometer is calibrated with the same external resistance as that with which it is to be used, the modified constant is automatically determined.

In many pivoted coil instruments, such as portable ammeters and voltmeters, the necessary amount of damping is "built in" to the moving coil, so to speak, by winding this coil on a light aluminum frame. The frame itself then forms a closed circuit and the currents induced in this circuit quickly bring the swinging coil to rest.

12-10 Eddy currents. Thus far we have considered only instances in which the currents resulting from induced emf's were confined to well-defined paths provided by the wires and apparatus of the external circuit. In many pieces of electrical equipment, however, one finds masses of metal moving in a magnetic field or located in a changing magnetic field, with the result that induced currents circulate throughout the volume of the metal. Because of their general circulatory nature these are referred to as *eddy currents.*

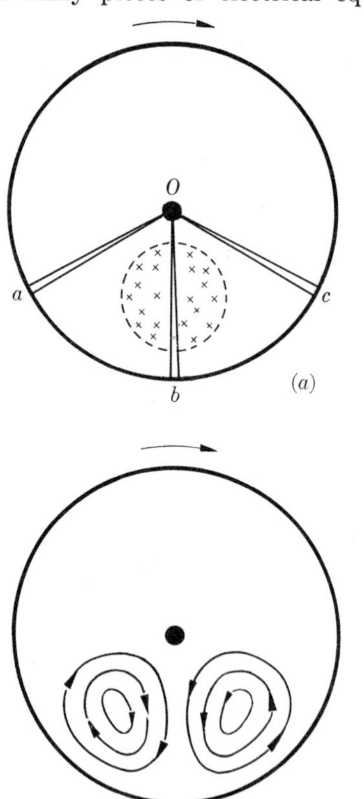

FIG. 12-14. Eddy currents in a rotating disk.

Consider a rotating disk such as that of the disk dynamo, in a magnetic field perpendicular to the plane of the disk but confined to a limited portion of its area as in Fig. 12-14(a). Element Ob is moving across the field and has an emf induced in it. Elements Oa and Oc are not in the field and hence are not seats of emf. They do, however, in common with all other elements located outside the field, provide return conducting paths along which positive charges displaced along Ob can return from b to O. A general eddy current circulation is therefore set up in the disk somewhat as sketched in Fig. 12-14(b).

Application of the left-hand rule or of Lenz's law shows that the currents in the neighborhood of radius Ob experience a side thrust which opposes the motion of the disk, while the return currents, since they lie outside the field, do not experience such a thrust. The interaction between the eddy currents and the field therefore results in a braking action on the disk. The apparatus finds some technical applications and is known as an "eddy current brake."

As a second example of eddy currents, consider the core of an alternating current transformer, shown in Fig. 12-15. The alternating current in the transformer windings sets up an alternating flux within the core, and an induced emf develops in the secondary windings because of the continual change in flux through them. The iron core, however, is also a conductor, and any section such as that at A-A can be thought of as a number of closed conducting circuits, one within the other. The flux through each of these circuits is continually changing, so that there is an eddy current circulation in the entire volume of the core, the lines of flow lying in planes perpendicular to the flux. These eddy currents are very undesirable both because of the large I^2R heating which they produce, and because of the flux which they themselves set up.

In all actual transformers the eddy currents are nearly, although not completely, eliminated by the use of a *laminated* core, that is, one built up of thin sheets or laminae. The electrical resistance between the surfaces of the laminations (due either to a natural coating of oxide or to an insulating varnish) effectively confines the eddy currents to individual laminae. The resulting length of path is greatly increased, with consequent increase in resistance. Hence although the induced emf is not altered, the currents and their heating effects are minimized.

In open core transformers or spark coils a bundle of iron wires is often used as a core. Powdered iron formed into a core under high pressure is also used in small transformers where eddy current loss must be kept at an absolute minimum.

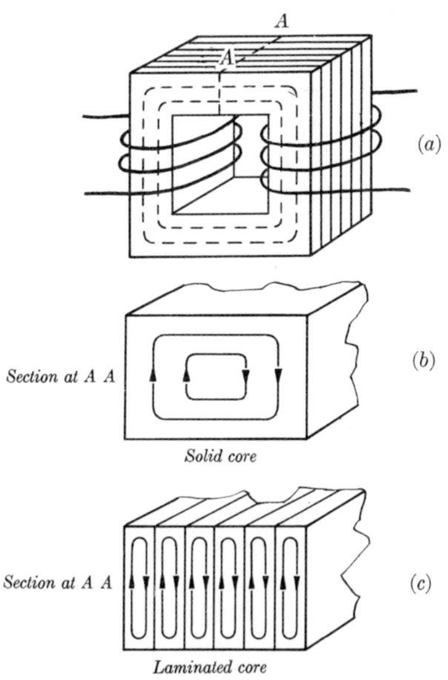

FIG. 12-15. Reduction of eddy currents by use of a laminated core.

Problems—Chapter 12

(1) A conducting rod AB in Fig. 12-16 makes contact with the metal rails CA and DB. The apparatus is in a uniform magnetic field of flux density 500 milliwebers/m², perpendicular to the plane of the diagram. (a) Find the magnitude and direction of the emf induced in the rod when it is moving toward the right with a velocity of 4 m/sec. (b) If the resistance of the circuit $ABCD$ is 0.2 ohm (assumed constant) find the force required to maintain the rod in motion. Neglect friction. (c) Compare the rate at which mechanical work is done by the force (Fv) with the rate of development of heat in the circuit (i^2R).

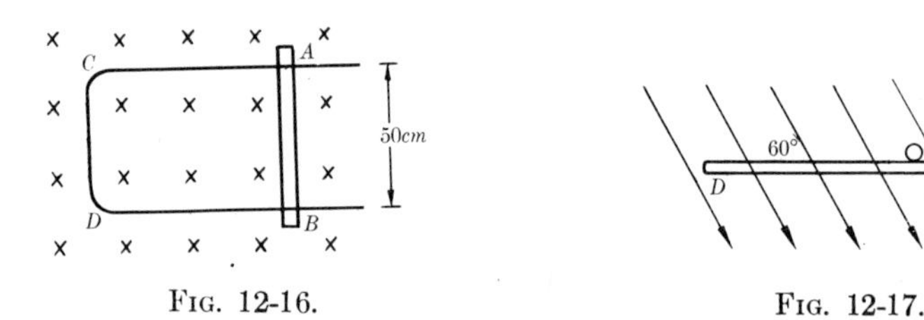

FIG. 12-16. FIG. 12-17.

(2) Fig. 12-17 is a side view of the same rod and metal rails as in Fig. 12-16, except that the magnetic induction makes an angle of 60° with the plane of the loop $ABCD$. Find the induced emf. The flux density is 500 milliwebers/m² and the velocity of the rod is 4 m/sec toward the right.

(3) In Fig. 12-18, AB is a metal rod moving with a constant velocity v of 2 m/sec parallel to a long straight wire in which the current i is 40 amp. Compute the induced emf in the rod. Which end of the rod is at the higher potential?

FIG. 12-18. FIG. 12-19.

(4) The long straight wire in Fig. 12-19 carries an alternating current given by

$$i = I_m \sin \omega t$$

where i is the instantaneous and I_m the maximum current. Find the expressions for (a) the flux passing through the rectangular loop at any instant, (b) the induced emf in the loop at any instant.

(5) A cardboard tube is wound with two windings of insulated wire as in Fig. 12-20. Terminals a and b of winding A may be connected to a seat of emf through a reversing switch.

State whether the induced current in the resistor R is from left to right, or from right to left, in the following circumstances: (a) the current in winding A is from a to b

and is increasing; (b) the current is from b to a and is decreasing; (c) the current is from b to a and is increasing.

(6) The orbit of an electron in a betatron is a circle of radius R. Suppose the electron is revolving tangentially in this orbit with a velocity v.

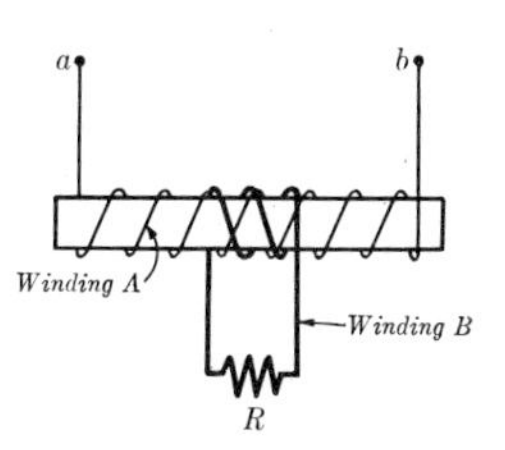

FIG. 12-20.

(a) What flux density is required to maintain the electron in this orbit if the magnitude of its velocity is constant?

(b) If the flux density is uniform over the plane of the orbit, and is increasing at a rate dB/dt, what is the equivalent voltage accelerating the electron in each revolution?

(7) A metal disk 10 cm in radius rotates with an angular velocity of 3600 rpm about an axis through its center perpendicular to its plane. The disk is in a uniform magnetic field of flux density 200 mw/m^2, parallel to the axis of the disc. (a) Find the reading of a millivoltmeter connected to brushes making contact with the rim and axis of the disk. (b) Show in a diagram the directions of rotation and magnetic field if the rim of the disk is positive with respect to the axis.

(8) A circular turn of wire 4 cm in radius rotates with an angular velocity of 1800 rpm about a diameter which is perpendicular to a uniform magnetic field of flux density 0.5 w/m^2. (a) What is the instantaneous induced emf in the turn when the plane of the turn makes an angle of 30° with the direction of the flux? (b) What is the angle between the plane of the turn and the flux, when the instantaneous emf has the same value as the average emf for a half cycle?

(9) The cross-sectional area of a closely wound search coil having 20 turns, is 1.5 cm^2 and its resistance is 4 ohms. The coil is connected through leads of negligible resistance to a ballistic galvanometer of resistance 16 ohms. Find the quantity of charge displaced through the galvanometer when the coil is pulled quickly out of a region where $B = 1.8$ w/m^2 to a point where the magnetic field is zero. The plane of the coil, when in the field, made an angle of 60° with the magnetic induction.

(10) A solenoid 50 cm long and 8 cm in diameter is wound with 500 turns. A closely wound coil of 20 turns of insulated wire surrounds the solenoid at its midpoint, and the terminals of the coil are connected to a ballistic galvanometer. The combined resistance of coil, galvanometer, and leads is 25 ohms.

(a) Find the quantity of charge displaced through the galvanometer when the current in the solenoid is quickly decreased from 3 amp to 1 amp.

(b) Draw a sketch of the apparatus, show clearly the directions of winding of the solenoid and coil, and of the current in the solenoid. What is the direction of the current in the coil when the solenoid current is decreased?

CHAPTER 13

INDUCTANCE

13-1 Mutual inductance. It has been shown that an emf is induced in a stationary circuit whenever the flux linking the circuit is increasing or decreasing. If the variation in flux is brought about by a varying current in a second circuit, it is found convenient to express the induced emf in terms of this varying current, rather than in terms of the varying flux.

Fig. 13-1 is a sectional view of two closely-wound coils of wire. A current in circuit 1 sets up a magnetic field as indicated, and a part of this flux passes through circuit 2. Since every line of induction is a closed line, each line passing through the plane of circuit 2 links with this circuit in the same way that two consecutive links of a chain are joined together. If the circuit comprises N_2 turns, and Φ flux lines link with each turn, the product $N_2\Phi$ is called the number of *flux-linkages* with circuit 2. The mks unit of flux-linkage is one *weber-turn*.

For a given current in circuit 1, the number of flux-linkages with circuit 2 depends on the geometry of the arrangement. Regardless of the geometry, however, the flux density at every point of the field is directly proportional to the current in circuit 1.[1] Hence the flux linking circuit 2 is also proportional to the current in circuit 1, and we can write

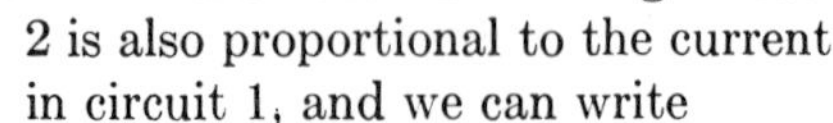

$$\Phi_{21} = Ki_1$$

where Φ_{21} means the flux linking circuit 2, due to a current i_1 in circuit 1, and K is a constant of proportionality.

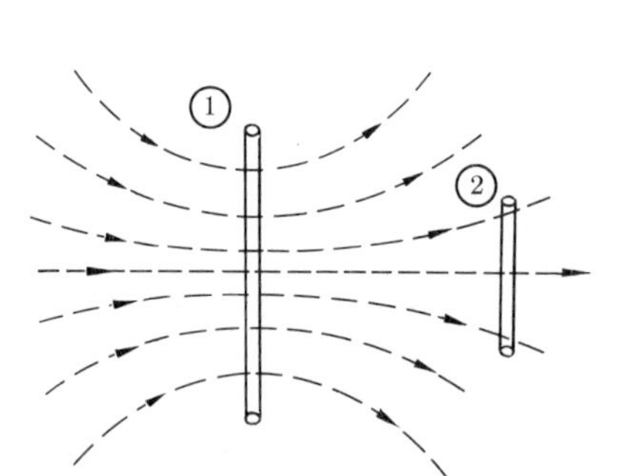

Fig. 13-1. The flux set up by a current in coil 1 links with coil 2.

If i_1 is varied, Φ_{21} will vary also, and an emf will appear in circuit 2, of magnitude

$$\mathcal{E}_2 = -N_2 \frac{d\Phi_{21}}{dt} = -N_2K \frac{di_1}{dt}$$

where N_2 is the number of turns of circuit 2. (We are assuming circuit 2 to be a closely-wound coil so that the same flux links with each turn.)

[1] Except if ferromagnetic materials are present. See Chap. 15.

Let us represent the product N_2K by a single constant M. Then

$$\boxed{\mathcal{E}_2 = -M\frac{di_1}{dt},} \quad \text{or} \tag{13-1}$$

$$M = -\frac{\mathcal{E}_2}{di_1/dt}. \tag{13-2}$$

The factor M is called the *coefficient of mutual inductance*, or, more briefly, the *mutual inductance* of the two circuits. The *mutual inductance* of two circuits may be defined as *the ratio of the induced emf in one circuit to the rate of change of current in the other.* The unit of mutual inductance, in the mks system, is one volt per ampere-per-second, which is called one *henry* in honor of Joseph Henry. The mutual inductance of two circuits is *one henry if an emf of one volt is induced in one of the circuits when the current in the other is changing at the rate of one ampere per second.*

It can be shown that the same mutual inductance is obtained whichever circuit is taken as the starting point. That is, the same emf is induced in either of two circuits between which there is mutual inductance, when the current in the other circuit changes at a given rate.

An alternate expression for mutual inductance can be derived as follows.

$$\mathcal{E}_2 = -N_2\frac{d\Phi_{21}}{dt} = -M\frac{di_1}{dt}.$$

Cancelling dt's and integrating leads to the result

$$N_2\Phi_{21} = Mi_1 + \text{a constant}.$$

Since

$$\Phi_{21} = 0 \text{ when } i_1 = 0,$$

$$N_2\Phi_{21} = Mi_1,$$

and

$$\boxed{M = \frac{N_2\Phi_{21}}{i_1}.} \tag{13-3}$$

From what was said in the preceding paragraph, this can also be written

$$\boxed{M = \frac{N_1\Phi_{12}}{i_2}.} \tag{13-4}$$

That is, *the mutual inductance of two circuits is the ratio of the flux-linkages set up in either circuit to the current in the other circuit.* The mutual inductance will therefore be large when the circuits are arranged so that a large part of the flux set up by a current in either circuit links with the other circuit, as for example when both are wound on the same iron core. From Eq. (13-3) or (13-4), mutual inductance may also be expressed in *weber-turns per ampere.* The mutual inductance between two circuits is *one henry if a current of one ampere in either circuit results in a flux-linkage of one weber-turn in the other.*

Examples: (1) A long solenoid of length l, cross section A, having N_1 turns, has wound about its center a small coil of N_2 turns as in Fig. 13-2. Compute the mutual inductance of the two circuits.

Consider the long solenoid as circuit 1. A current i_1 in it sets up a field at its center

$$B = \mu_0 \frac{N_1 i_1}{l}.$$

The flux through the central section is

$$\Phi = BA = \mu_0 \frac{A N_1 i_1}{l}.$$

Since all of this flux links with coil 2,

$$M = \frac{N_2 \Phi_{21}}{i_1} = \mu_0 \frac{A N_1 N_2}{l}. \quad \left\{ \begin{array}{c} \text{volts per (amp/sec)} \\ \text{weber-turns/amp} \\ \text{henrys} \end{array} \right\}$$

If $l = 1$ meter, $A = 10 \text{ cm}^2 = 10^{-3} \text{ m}^2$, $N_1 = 1000$ turns, $N_2 = 20$ turns,

$$M = \frac{12.57 \times 10^{-7} \times 10^{-3} \times 1000 \times 20}{1}$$

$$= 25.1 \times 10^{-6} \text{ henrys}$$

$$= 25.1 \times 10^{-3} \text{ millihenrys}$$

$$= 25.1 \text{ microhenrys.}$$

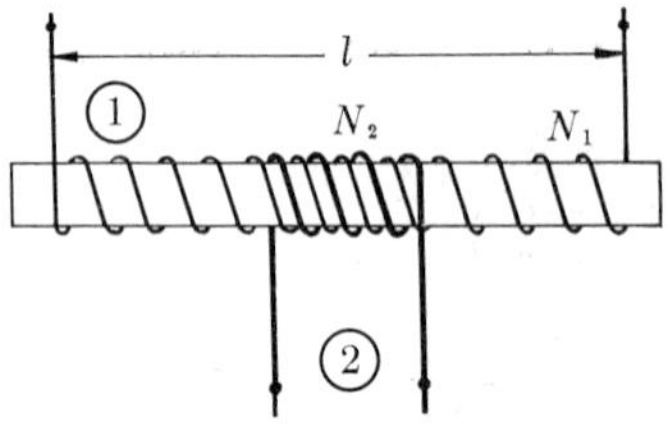

FIG. 13-2. Small coil 2 wound about the center of a much longer coil 1.

(2) What is the induced emf in circuit 2 when the current in circuit 1 changes at the rate of 10 amp/sec?

$$\mathcal{E}_2 = -M \frac{di_1}{dt}$$

$$= -25.1 \times 10^{-6} \times 10 = -251 \text{ microvolts.}$$

13-2 Self-inductance. In the preceding instances of induced emf, the magnetic flux linking the circuit in which the emf was induced has been set up by some external source. But whenever there is a current in a circuit, this current sets up a field which itself links with the circuit and which will vary when the current varies. Hence any circuit in which there is a varying current has induced in it an emf because of the variation of its own field. (Fig. 13-3.) Such an emf is called a *self-induced electromotive force*. For example, as the rheostat slider in Fig. 13-3 is moved back and forth, the flux through the loop varies and induces an emf in the circuit. As in the case of mutual inductance, it is convenient to relate the induced emf to the changing current rather than to the changing flux.

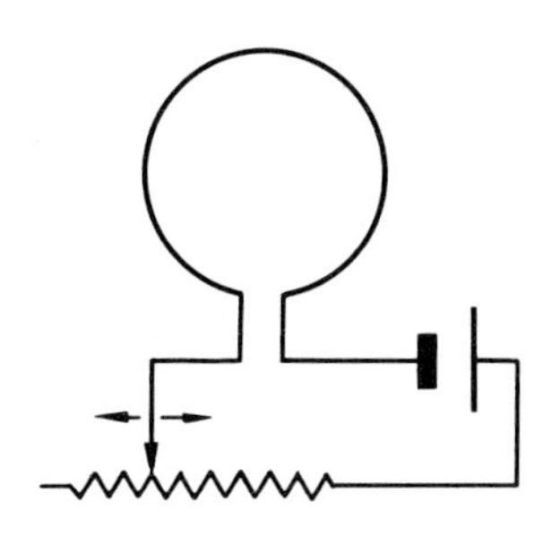

FIG. 13-3. Circuit to show self-inductance.

The number of flux lines linking any given circuit, due to a current in the circuit itself, will depend on the geometry of the circuit, that is, its shape, size, number of turns, etc. Regardless of the geometry, however, the flux density at any point is directly proportional to the current producing it[1] and therefore the flux will also be proportional to the current. Hence one can write

$$\Phi = Ki$$

where K is a geometrical factor, constant for any given circuit. If the circuit has N turns and all of the flux links with each turn, it follows that

$$\mathcal{E} = -N\frac{d\Phi}{dt} = -NK\frac{di}{dt},$$

or, if NK is represented by a single letter L,

$$\boxed{\mathcal{E} = -L\frac{di}{dt}.} \tag{13-5}$$

The constant L is called the *coefficient of self-inductance*, or simply the *self-inductance* of the circuit. Its mks units, like those of mutual inductance, are volts per ampere-per-second or henrys. *The self-inductance*

[1] Except if ferromagnetic materials are present. See Chap. 15.

of a circuit is one henry if an emf of one volt is induced in the circuit when the current in the circuit changes at the rate of one ampere per second.

As with mutual inductance, an alternate point of view consists of equating

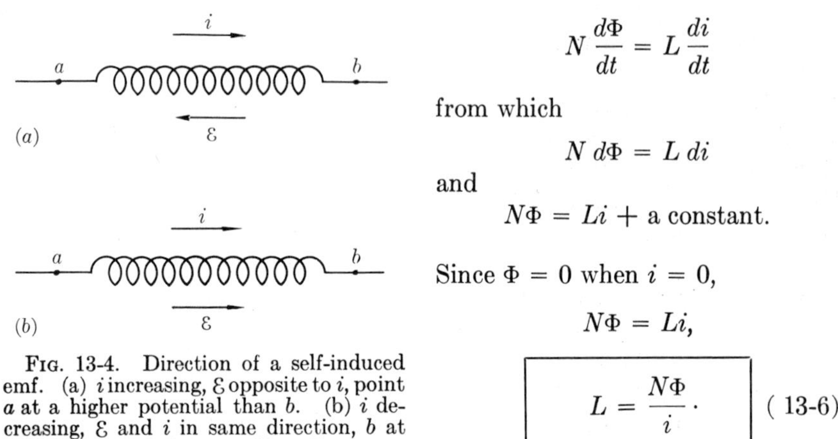

$$N\frac{d\Phi}{dt} = L\frac{di}{dt}$$

from which

$$N\,d\Phi = L\,di$$

and

$$N\Phi = Li + \text{a constant.}$$

Since $\Phi = 0$ when $i = 0$,

$$N\Phi = Li,$$

$$\boxed{L = \frac{N\Phi}{i}.} \tag{13-6}$$

FIG. 13-4. Direction of a self-induced emf. (a) i increasing, $\mathcal{E}$ opposite to i, point a at a higher potential than b. (b) i decreasing, $\mathcal{E}$ and i in same direction, b at higher potential than a. ($R = 0$).

The self-inductance of a circuit may therefore be defined as *the number of flux-linkages per unit current* (in mks units, as the number of weber-turns per ampere). The self-inductance of a circuit is *one henry if a current of one ampere in the circuit results in a flux linkage of one weber-turn.*

A circuit or part of a circuit which has inductance is called an *inductor.* An inductor is represented by the symbol —ᴏᴏᴏᴏᴏᴏᴏᴏᴏ—

The direction of a self-induced emf is found from Lenz's law. The "cause" of the emf is an increasing or decreasing current. If the current is increasing, the direction of the induced emf is opposite to that of the current. If the current is decreasing, the emf and the current are in the same direction. Thus it is the *change* in current, not the current itself, which is "opposed" by the induced emf. See Fig. 13-4.

Examples: (1) Compute the self-inductance of a toroidal winding of mean length l, cross section A, having N turns.

The flux density within the volume enclosed by the winding is

$$B = \mu_0 \frac{Ni}{l}$$

and the flux is

$$\Phi = BA = \mu_0 \frac{ANi}{l}.$$

Since all of the flux links with each turn,

$$L = \frac{N\Phi}{i} = \mu_0 \frac{AN^2}{l} \cdot \left\{\begin{matrix}\text{volts per ampere-per-sec} \\ \text{weber-turns per amp} \\ \text{henrys}\end{matrix}\right\}$$

Let $l = 1$ meter, $A = 10\ \text{cm}^2 = 10^{-3}\ \text{m}^2$, $N = 1000$ turns. Then

$$L = \frac{12.57 \times 10^{-7} \times 10^{-3} \times 10^{6}}{1}$$

$$= 1.26 \times 10^{-3}\ \text{henrys}$$

$$= 1.26\ \text{millihenrys.}$$

(2) If the current in the coil above is increasing at the rate of 10 amperes per second, find the magnitude and direction of the self-induced emf.

$$\mathcal{E} = -L\frac{di}{dt}$$

$$= -1.26 \times 10^{-3} \times 10 = -12.6\ \text{millivolts.}$$

Since the current is increasing, the direction of this emf is opposite to that of the current.

13-3 Growth of current in an inductive circuit. An inductor in which there is an increasing current becomes a seat of emf whose direction is opposite to that of the current. As a consequence of this back emf the current in an inductive circuit will not rise to its final value at the instant when the circuit is closed, but will grow at a rate which depends on the inductance (and resistance) of the circuit. There is a close analogy between the increase of current in an inductor and the increase of charge on a capacitor. (See Sec. 8-5.)

Fig. 13-5 shows a resistanceless inductor and a non-inductive resistor in series between two points maintained at a constant potential difference V_{ab}. Let i represent the current in the circuit at some instant after the connections are made. We have

$$V_{ab} = \Sigma iR - \Sigma\mathcal{E},$$

$$\mathcal{E} = -L\frac{di}{dt}.$$

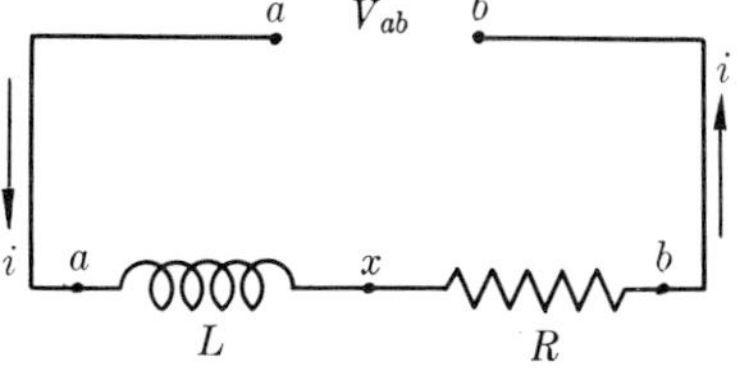

FIG. 13-5. Constant potential difference across an inductor and a resistor in series.

Combining these equations, we get

$$\frac{di}{dt} + \frac{R}{L}i - \frac{V_{ab}}{L} = 0 \qquad (13\text{-}7)$$

and therefore

$$\boxed{i = \frac{V_{ab}}{R}(1 - \epsilon^{-Rt/L})} \qquad (13\text{-}8)$$

Since V_{ab}/R equals the final current in the circuit, say I, one may also write

$$i = I\,(1 - \epsilon^{-Rt/L}). \tag{13-9}$$

A graph of Eq. (13-9) is given in Fig. 13-6. As with a capacitor-resistor combination, an infinite time is required for the current to attain its final value, but for any R/L ratio encountered in practice a very short time suffices for the current to reach essentially its final value.

The *time constant* of the circuit, or the time for the current to increase to within $\dfrac{1}{\epsilon}$th of its final value, is

$$t = \frac{L}{R}.$$

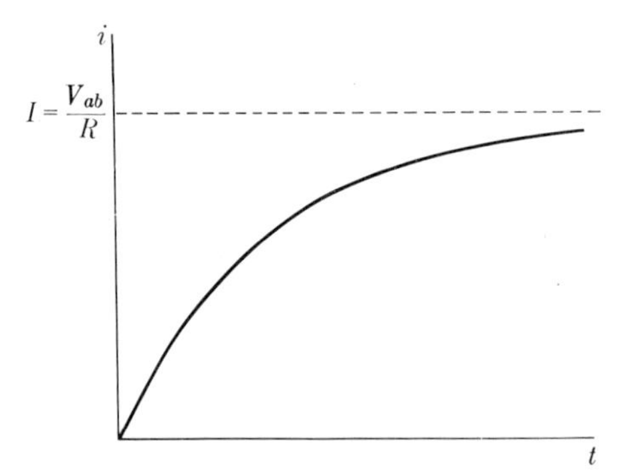

FIG. 13-6. Growth of current in a circuit containing inductance and resistance.

Hence for a circuit with a given resistance, this time is longer the larger the inductance and vice versa. Thus although the graph of i vs t has the same general shape whatever the inductance, the current rises rapidly to its final value if L is small, and slowly if L is large.

The potential difference between the terminals of the inductor in Fig. 13-5 is

$$V_{ax} = \Sigma iR - \Sigma \mathcal{E} = L\frac{di}{dt}.$$

Differentiating Eq. (13-9) to obtain di/dt, we find

$$V_{ax} = V_{ab}\,\epsilon^{-Rt/L} \tag{13-10}$$

The full line voltage V_{ab} thus appears across the inductor at the instant the circuit is closed ($t = 0$), and the potential difference between its terminals decreases exponentially to zero as the current increases.

For simplicity, the inductance and resistance of the circuit were considered as separate units in the preceding discussion. Actually, any inductor will have some resistance associated with it. Points a and b in Fig. 13-5 may then be treated as the terminals of the inductor, and V_{ab} as the potential difference between its terminals. The potential differences v_{ax} and v_{xb} separately cease to have a meaning. The same expression is obtained for the current in the circuit. If there is external resistance in addition to that associated with the inductor, the term R in Eq. (13-9) refers to the equivalent resistance of the entire circuit.

Example. An inductor of inductance 3 henrys and resistance 6 ohms is connected to the terminals of a battery of emf 12 volts and of negligible internal resistance. (a) Find the initial rate of increase of current in the circuit. (b) Find the rate of increase of current at the instant when the current is 1 ampere. (c) What is the instantaneous current 0.2 sec after the circuit is closed?

(a) From Eq. (13-7),

$$\frac{di}{dt} = \frac{V_{ab}}{L} - \frac{R}{L}i.$$

The initial current is zero. Hence the initial rate of increase of current is

$$\frac{di}{dt} = \frac{V_{ab}}{L} = \frac{12}{3} = 4 \text{ amp/sec.}$$

(b) When $i = 1$ amp,

$$\frac{di}{dt} = \frac{12}{3} - \frac{6}{3} \times 1$$
$$= 2 \text{ amp/sec.}$$

(c) From Eq. (13-8),

$$i = \frac{V_{ab}}{R}(1 - \epsilon^{-Rt/L})$$
$$= \frac{12}{6}(1 - \epsilon^{-6\times .2/3})$$
$$= 2(1 - \epsilon^{-.4})$$
$$= 2(1 - .672)$$
$$= 0.65 \text{ amp.}$$

13-4 Energy associated with an inductor. When the switch in Fig. 13-7 is closed, the current in the circuit increases as in Fig. 13-6 from zero to a final value V_{ab}/R. At an instant when the current is i and its rate of increase is di/dt,

$$V_{ab} = V_{ax} + V_{xb} = L\frac{di}{dt} + Ri.$$

Hence the power input to the circuit at this instant is

$$P = iV_{ab}, \quad \text{or}$$

$$P = Li\frac{di}{dt} + i^2R \cdot \quad (13\text{-}11)$$

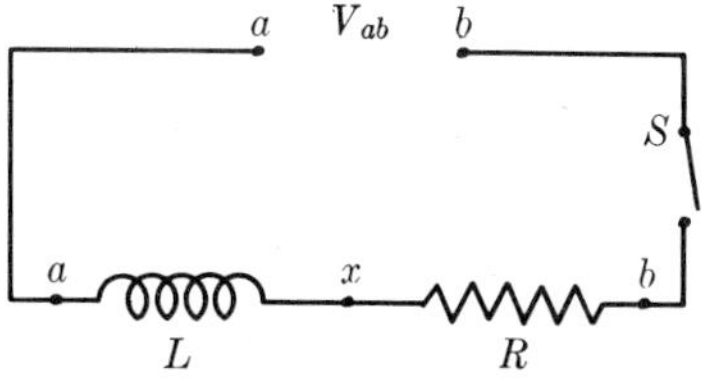

FIG. 13-7. At any instant after the switch S is closed, $V_{ab} = L\frac{di}{dt} + Ri$.

The last term on the right is the power input to (or the rate of development of heat in) the resistor. The first is the power input to the inductor. Hence energy is supplied to the inductor during the time that the current in it is increasing. When the current has reached its final steady value, $di/dt = 0$ and the power input to the inductor ceases. The energy that has been supplied to the inductor is used to establish the magnetic field around the inductor where it is "stored" as potential energy. When the switch is opened, the magnetic field collapses and the energy is returned to the circuit. It is this release of energy which maintains the arc often seen when a switch is opened in an inductive circuit.

The energy associated with the magnetic field of a current-carrying inductor can be evaluated as follows. The instantaneous power input to the inductor, from Eq. (13-11), is

$$P = \frac{dW}{dt} = Li\frac{di}{dt}.$$

Hence

$$dW = Li\,di,$$

and

$$W = \int dW = \int_0^I Li\,di,$$

or

$$\boxed{W = \tfrac{1}{2}LI^2,} \tag{13-12}$$

where W is the energy supplied during the time that the current increases from zero to I. The same amount of energy is released when the current falls to zero (but see Sec. 17-2 for a more complete analysis).

The inductance and resistance of the circuit in Fig. 13-7 are shown as separate units. The same analysis applies to a coil of resistance R and self-inductance L.

13-5 Inductors in series. Inductors, like other electrical devices, are often connected in series or parallel or in more complex networks. The *equivalent* self-inductance of any network is defined as the ratio of the total induced emf (self plus mutual) between the terminals of the network, to the rate of change of current responsible for the emf. We shall consider here only the case of inductors in series.

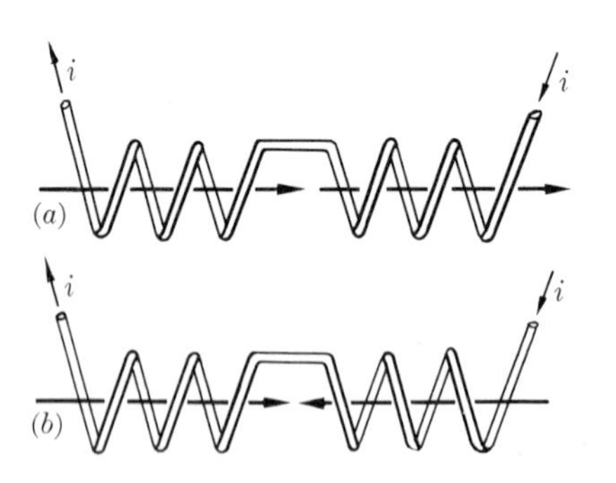

FIG. 13-8. Two inductors in series. (a) Fluxes in same direction. (b) Fluxes in opposite direction.

For the special case of two coils having self-inductances L_1 and L_2 respectively, and between which the mutual inductance is M, the equivalent self-inductance of the two in series can be readily deduced as follows. Let the coils be placed as in Fig. 13-8 so that the flux linking each coil, due to the current in the other, is in the same direction as the flux due to the current in the coil. Then if the current varies, all induced emf's, both self and mutual, will be in the same direction.

$$\begin{aligned}\text{emf in coil 1} &= \text{self-induced emf} + \text{mutual induced emf.}\\ &= L_1 \frac{di}{dt} + M \frac{di}{dt},\\ \text{emf in coil 2} &= L_2 \frac{di}{dt} + M \frac{di}{dt},\\ \text{net emf} &= (L_1 + L_2 + 2M) \frac{di}{dt},\end{aligned}$$

and from its definition the equivalent self-inductance is

$$L = L_1 + L_2 + 2M. \tag{13-13}$$

If one of the coils is reversed so that the flux linking each coil due to the current in the other is opposite in direction to the coil's own flux, the self- and mutual-induced emf's in each coil will be in opposite directions. Then

$$\text{net emf} = \left[L_1 \frac{di}{dt} - M \frac{di}{dt}\right] + \left[L_2 \frac{di}{dt} - M \frac{di}{dt}\right],$$

and

$$L = L_1 + L_2 - 2M. \tag{13-14}$$

Evidently, if the geometry of the circuit is such that none of the flux associated with either coil links with the other, the mutual inductance is zero and the equivalent self-inductance is simply the sum of the individual self-inductances.

Eqs. (13-13) and (13-14) indicate the possibility of constructing a *variable inductor* by connecting in series two coils, one of which may be rotated with respect to the other. Thus while L_1 and L_2 are constant, M varies with the angular position of the movable coil. A common construction is to wind the two coils on spherical forms, one rotatable within the other. The (equivalent) self-inductance of the device is then continuously variable over the range from $L_1 + L_2 + 2M$ to $L_1 + L_2 - 2M$.

The symbol [variable inductor symbol] represents a variable inductor.

One experimental method for measuring the mutual inductance of two coils is to measure the equivalent self-inductance of the two in series, first with their fluxes aiding and again with fluxes opposing. Calling these values L' and L'' respectively, it follows that

$$M = \frac{L' - L''}{4}. \tag{13-15}$$

If two coils are wound on the same iron core, as in a transformer, or if two closely-wound coils are placed side by side, practically all of the flux set up by either coil links with all of the turns of the other. Now by definition

$$L_1 = \frac{N_1\Phi_1}{i_1}, \quad L_2 = \frac{N_2\Phi_2}{i_2},$$

$$M = \frac{N_1\Phi_{12}}{i_2} = \frac{N_2\Phi_{21}}{i_1}.$$

But $$\Phi_{12} = \Phi_2, \quad \text{and} \quad \Phi_{21} = \Phi_1,$$

as will be evident from the significance of the subscripts.

Hence $$M = \frac{N_1\Phi_2}{i_2}, \quad \text{and}$$

$$M = \frac{N_2\Phi_1}{i_1}.$$

After multiplying these equations and rearranging terms we get

$$M^2 = \frac{N_1\Phi_1}{i_1} \times \frac{N_2\Phi_2}{i_2} = L_1L_2.$$

Hence $$M = \sqrt{L_1L_2}. \tag{13-16}$$

The mutual inductance of the two coils is the geometric mean of their self-inductances.

In practice, self- and mutual-inductances are usually measured with the aid of a modified Wheatstone bridge network, using alternating rather than direct current. (See any standard text on electrical measurements.) The mutual inductance of two circuits may also be measured with the aid of a ballistic galvanometer. Let Fig. 13-9 represent the two circuits whose mutual inductance is to be measured. When switch S is closed, the current in circuit 1 increases rapidly from zero to a value I_1 which can be measured

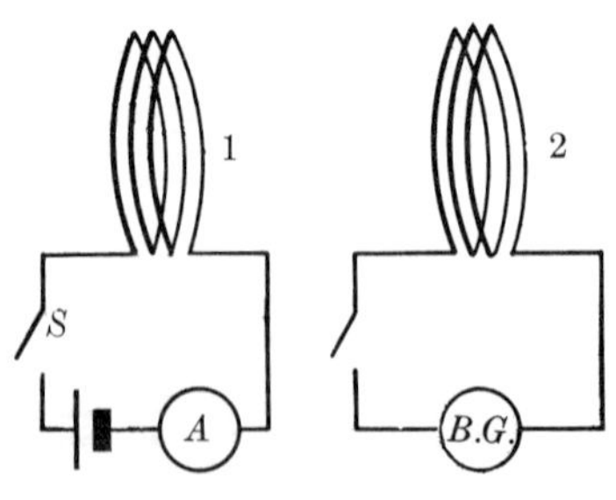

FIG. 13-9. Apparatus for measuring the mutual inductance of two circuits.

by ammeter A. During the growth of this current an emf e_2 was induced in circuit 2. The current in circuit 2 at any instant is therefore

$$i_2 = \frac{e_2}{R_2},$$

where R_2 is the total resistance of circuit 2.
But

$$e_2 = N_2 \frac{d\Phi}{dt} = M \frac{di_1}{dt} + L_2 \frac{di_2}{dt}$$

Hence

$$i_2 = \frac{N_2}{R_2} \frac{d\Phi}{dt} = \frac{M}{R_2} \frac{di_1}{dt} + \frac{L_2}{R_2} \frac{di_2}{dt},$$

$$i_2\,dt = \frac{N_2}{R_2}\,d\Phi = \frac{M}{R_2}\,di_1 + \frac{L_2}{R_2}\,di_2$$

and

$$q = \int_0^\infty i_2\,dt = \frac{N_2}{R_2} \int_0^{\Phi_{21}} d\Phi = \frac{M}{R_2} \int_0^{I_1} di_1 + \frac{L_2}{R_2} \int_0^0 di_2$$

$$= \frac{N_2}{R_2}\,\Phi_{21} = \frac{M}{R_2}\,I_1,$$

where Φ_{21} is the final flux linking circuit 2 when i_1 has reached its final value I_1. The limits of the last integral are both zero, since the initial and final currents in circuit 2 are zero. Therefore, if the maximum angle of swing is observed and the galvanometer has previously been calibrated, both Φ_{21} and M can be computed.

Problems—Chapter 13

(1) A circular coil A of 50 turns of fine wire, 4 cm² in cross-sectional area, is placed at the center of a circular coil B 20 cm in radius and having 100 turns. The axes of the coils coincide. (a) What is the mutual inductance of the coils? (b) What is the induced emf in coil A when the current in coil B is decreasing at the rate of 50 amp/sec? (c) What is the rate of change of flux through coil A at this instant? (d) Draw a diagram and indicate clearly the relative directions of the current in coil B and the induced emf in coil A.

(2) In Fig. 13-10, A and B are two small closely wound coils whose planes are parallel to one another and perpendicular to the line OO. (a) What angle θ should the line joining the centers of the coils make with OO in order that the mutual inductance between the coils shall be a minimum? The distance r is large compared with the radii of the coils. (b) For a given separation r, at what value of θ is the mutual inductance a maximum?

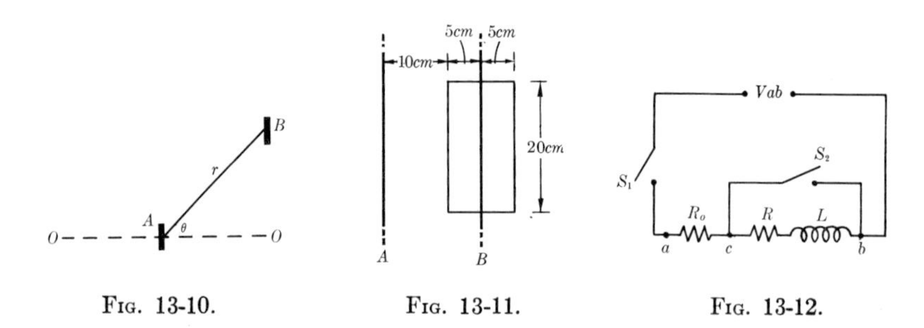

FIG. 13-10. FIG. 13-11. FIG. 13-12.

(3) A closely wound rectangular coil, 10 cm by 20 cm, has 100 turns. The coil is near a long straight wire forming a portion of a closed circuit, with the long sides of the coil parallel to the wire. The remainder of the second circuit is far removed from the coil. Find the mutual inductance of the circuits, (a) when the straight wire is 10 cm from the nearer side of the coil as in A, Fig. 13-11, and (b) when the wire is in position B in Fig. 13-11.

(4) The self-inductance of a closely wound coil of 100 turns is 5 millihenrys. What is the flux through the coil when the current in it is 10 milliamperes?

(5) Prove that the time constant of an inductor is L/R.

(6) The resistance of a 10-henry inductor is 200 ohms. The inductor is suddenly connected across a potential difference of 10 volts. (a) What is the final steady current in the inductor? (b) What is the initial rate of increase of current? (c) At what rate is the current increasing when its value is one-half the final current? (d) At what time after the circuit is closed does the current equal 99% of its final value? (e) Compute the current at the following times after the circuit is closed: 0, 0.025 sec, 0.05 sec, 0.075 sec, 0.10 sec. Show the results in a graph.

(7) An inductor of resistance R and self-inductance L is connected in series with a non-inductive resistor of resistance R_0 to a constant potential difference V_{ab} (Fig. 13-12). (a) Find the expression for the potential difference V_{cb} across the inductor at any time t after switch S_1 is closed. (b) Let $V_{ab} = 20$ volts, $R_0 = 50$ ohms, $R = 150$ ohms, $L = 5$ henrys. Compute a few points, and construct graphs of V_{ac} and V_{cb} over a time interval from zero to twice the time constant of the circuit.

(8) After the current in the circuit of Fig. 13-12 has reached its final steady value the switch S_2 is closed, thus short-circuiting the inductor. What will be the magnitude and direction of the current in S_2, 0.01 sec after S_2 is closed?

(9) Refer to the inductor in the example at the end of Sec. 13-3. (a) What is the power input to the inductor at the instant when the current in it is 0.5 amp? (b) What is the rate of development of heat at this instant? (c) What is the rate at which the energy of the magnetic field is increasing? (d) How much energy is stored in the magnetic field when the current has reached its final steady value?

(10) Refer to Example 1 at the end of Sec. 13-1. Let the resistance of coil 2 be 10 ohms. If the terminals of this coil are connected to a ballistic galvanometer of resistance 20 ohms, how many coulombs are displaced past a point in the galvanometer circuit when the current in the solenoid is suddenly increased from zero to 5 amp?

CHAPTER 14

MAGNETIC PROPERTIES OF MATTER

14-1 Introduction. In the preceding chapters we have discussed the magnetic fields set up by moving charges or by currents in conductors, when the charges or conductors are in air (or, strictly speaking, in a vacuum). Everyone knows, however, that pieces of technical equipment such as transformers, motors, and generators, which make use of the magnetic fields set up by a current, always incorporate iron or an iron alloy in their structure, both for the purpose of increasing the magnetic flux and for confining it to a desired region. Furthermore, by the use of permanent magnets, as in galvanometers or permanent magnet speakers, magnetic fields can be produced without any apparent circulation of charge.

We therefore turn next to a consideration of the magnetic properties which make iron and a few other ferromagnetic materials so useful. We shall find that magnetic properties are not confined to ferromagnetic materials but are exhibited (to a much smaller extent, to be sure) by *all* substances. From the standpoint of the electrical engineer the ferromagnetic materials are of the greatest interest, but a study of the magnetic properties of other materials is of importance also, affording as it does another means of gaining an insight into the nature of matter in general.

The existence of the magnetic properties of a substance can be demonstrated by supporting a small rod-shaped specimen at its center of gravity by a fine thread and placing it in the magnetic field of a powerful electromagnet. Everyone knows that if the specimen is of iron or one of the *ferromagnetic* substances it will align itself in the direction of the magnetic field. Not as familiar is the fact that any substance whatever will be influenced by the field, although to an extent which is extremely small compared with a substance like iron. Some substances will, like iron, set themselves with their long dimension parallel to the field, while others will come to rest with their long dimension perpendicular to the field. The first type are called *paramagnetic,* the second, *diamagnetic.* All substances, including liquids and gases, fall into one or another of these classes. Liquids and gases must, of course, be enclosed within some sort of container, and due allowance must be made for the properties of the container as well as those of the medium (usually air) in which the specimens are immersed.

14-2 Origin of magnetic effects. Our formulation of the electrical properties of a substance in Chap. 7 was based on a specimen in the form of a flat slab, inserted in the field between oppositely-charged parallel plates. The electric field is wholly confined to the region between the plates if their separation is small, and therefore a flat slab between the plates will completely occupy all points of space at which an electric field exists. A specimen of this shape is not as well suited for a study of magnetic effects, however, since lines of magnetic induction are closed lines and there is no way of producing a magnetic field which is confined to the region between two closely-spaced surfaces. The magnetic field within a closely spaced toroidal winding, however, is wholly confined to the space enclosed by the winding. We shall accordingly use such a field on which to base our discussion of magnetic properties, the specimen being in the form of a ring on whose surface the wire is wound. Such a specimen is often called a *Rowland ring* after J. H. Rowland, who made much use of it in his experimental and theoretical work on electricity and magnetism. The winding of wire around the specimen is called the *magnetizing winding,* and the current in the winding, the *magnetizing current.*

The magnetic flux density within the space enclosed by a toroidal winding *in vacuum* is (see Sec. 11-6)

$$B = \mu_0 \frac{Ni}{l}. \qquad (14\text{-}1)$$

Suppose now that the same coil is wound on a Rowland ring, and that a second winding is placed on the ring as in Fig. 14-1, with its terminals connected to a ballistic galvanometer. The flux density within the ring may be measured, as explained in Sec. 12-8, by opening switch S and quickly reducing the magnetizing current to zero. If this is done, the flux density computed from the ballistic galvanometer deflection will not agree with that computed from Eq. (14-1). If the core is made of a ferromagnetic material, the measured flux density will be tremendously larger,

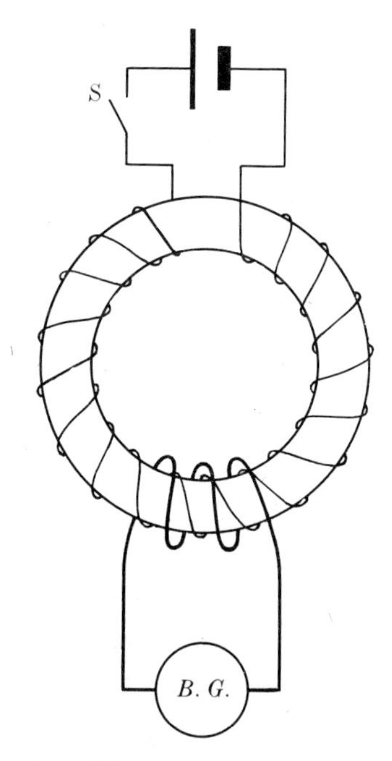

FIG. 14-1. Magnetic specimen in the form of a ring with a toroidal winding.

if of a paramagnetic material very slightly larger, and if of a diamagnetic material it will be very slightly smaller than the calculated value. The differences in the two latter cases are, in fact, so small that this method is not a practical one for investigating these substances, but its principle is simpler than that of other methods so we shall ignore experimental difficulties.

The excess or deficiency of flux in the core, over and above that due to the magnetizing current, arises from the magnetic effects of the electrons which form a part of the structure of all atoms. There is every reason to believe (the effects we are now considering are a part of the experimental evidence) that these electrons are revolving about their parent nuclei, or, if the situation is not quite as simple as that statement implies, they produce magnetic fields which are the same as if they were. Furthermore, each electron is rotating or "spinning" about an axis passing through it. The magnetic properties of the iron atom are due almost entirely to the fact that it has an excess of four uncompensated electron spins, that is, there are four more electrons spinning one way than there are spinning the other way. Because of the circulation or spin of its electrons every atom has associated with it a magnetic field.

Except for the ferromagnetic materials, which can form permanent magnets, no substance exhibits magnetic effects except when situated in an external magnetic field. It follows that in its normal state the electronic currents in a piece of material just cancel one another's magnetic effects. In some substances the cancellation is complete within each atom. In others there may be an outstanding magnetic effect in each atom, but in any sizable group of atoms the currents or spins are so oriented that their magnetic effects cancel. An external magnetic field, however, will alter the normal orientation of the electronic circuits or spins, or alter the rate of revolution of the electrons.

Any atom in which there is a net circulation of charge is called a *magnetic dipole.* If a substance composed of such atoms is in a magnetic field each of its atoms is influenced by the field in the same way as is a current-carrying turn of wire. That is, a torque is exerted on the electronic currents by the field in such a way as to line them up with their planes perpendicular to the field and with the induction within the plane of each turn in the same direction as that of the external field. The observed increase in flux density in a ring of paramagnetic or ferromagnetic material, over its value in vacuum, is due to the flux contributed by the electronic currents within the material.

The tendency toward uniform alignment of the magnetic dipoles with their own fields parallel to the external field is opposed by the thermal agitation of the atoms which tends toward a random arrangement. The higher the temperature the greater the trend toward randomness and the smaller the aligning effect. A paramagnetic or ferromagnetic substance is therefore analogous to a dielectric whose atoms are permanent *electric* dipoles.

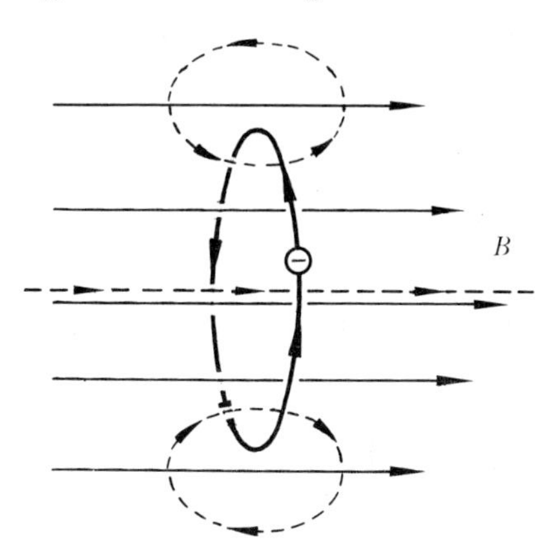

FIG. 14-2. A single electronic circuit in an external magnetic field.

The explanation of diamagnetism is based on the phenomenon of induced emf's. Fig. 14-2 represents a single electronic circuit whose plane is perpendicular to the direction of an external field. (The term "external" means the field exclusive of that set up by the electron under consideration.) When viewed along the direction of this field (left to right in Fig. 14-2) *positive* charge is circulating clockwise in the path, and the circulating charge sets up a flux, shown by dotted lines, which within the plane of the path is from left to right.

Now suppose the external field is increased, say by increasing the current which is producing it. An emf will be induced in the loop because of the change in flux through it, the direction of the emf according to Lenz's law being such as to *oppose* the *increase* in flux. Since the revolving electron is setting up a field in the same direction as the external field, the induced emf causes it to revolve more slowly than before. The increase in flux is therefore smaller than it would otherwise be. Notice that the action on the revolving electron is the same in principle as that on the electrons accelerated in a betatron.

For simplicity, the current loop in Fig. 14-2 was shown with its plane perpendicular to the external field and its own flux in the same direction as the external flux. It can be shown that, regardless of the orientation of an electronic orbit, the same effect will take place; any increase in external flux is opposed by the reaction on the revolving electron.

It follows, then, that the flux density in the core of a toroid surrounded by a magnetizing winding in which there is a current, is always reduced below its value in vacuum by this phenomenon of induced emf's and, if this were the only effect present, all substances would be diamagnetic. However, if the atoms of a substance are magnetic dipoles, that is, if there is a preponderance of electronic currents in one direction over those in the other, then paramagnetic (alignment) effects are superposed on the diamagnetic. If the paramagnetism exceeds the diamagnetism, the substance is paramagnetic. If the paramagnetism is small or absent altogether, the substance is diamagnetic.

14-3 Equivalent surface currents. Fig. 14-3(b) is a sectional view of the core of a Rowland ring. The small circles represent the electronic circuits or spins. Except at the periphery of the section, every portion of each circuit is adjacent to another circuit in which, at the points of contact, the current is in the opposite direction. Hence at interior points the currents mutually cancel one another. The outer portions of the outermost circuits, however, are uncompensated and therefore the whole network of electronic circuits is equivalent to a current around the periphery of the section as in Fig. 14-3(c).

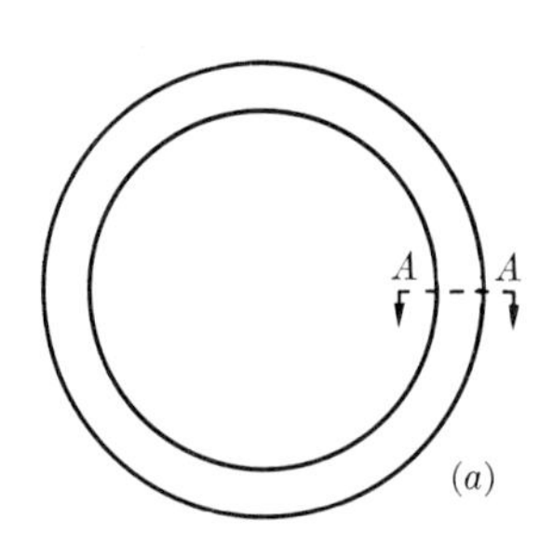

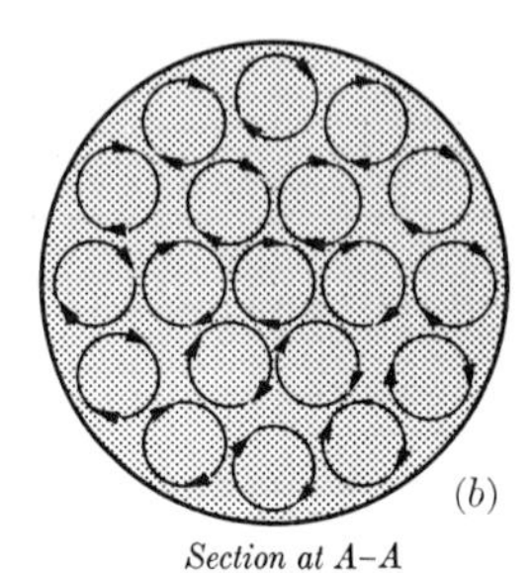

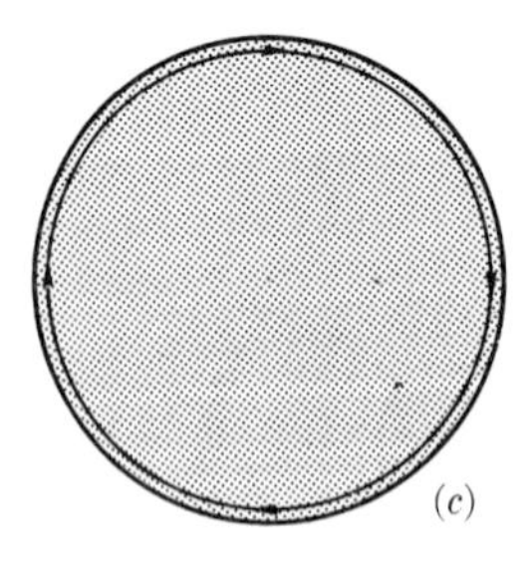

Fig. 14-3. (a) Rowland ring. (b) Electronic currents in section at A-A. (c) Equivalent surface currents.

Fig. 14-3 should be compared with Fig. 7-3, which illustrates how polarization charges cancel one another throughout the volume of a dielectric, but leave uncompensated layers of induced surface charges at opposite faces of the dielectric. The surface current of Fig. 14-3(c) is analogous to these surface charges. The number of ampere-turns of the surface currents can be added to (or subtracted from) those of the magnetizing winding, or the currents can be thought of as setting up a flux of their own which adds to (or subtracts from) the flux set up by the magnetizing current.

14-4 Magnetic susceptibility, permeability, and magnetic intensity. The magnetic flux density B at any point is the resultant of that due to currents in conductors, and to the equivalent surface currents in magnetized matter. We may imagine the surface of a magnetized body to be wound with wire, carrying at every point a current equal to the equivalent surface current. Then if dl_s is the length of an element of this fictitious winding and i_s is the surface current it carries, the flux density at any point is

$$B = \frac{\mu_0}{4\pi}\int \frac{i\,dl \sin\theta}{r^2} + \frac{\mu_0}{4\pi}\int \frac{i_s\,dl_s \sin\theta}{r^2}, \tag{14-2}$$

where the first term includes the current in any actual conductors, and the second, the equivalent currents at the surfaces of magnetized bodies.

In the special case of a closely-wound Rowland ring, where the magnetizing current and the equivalent surface currents have the same geometry, integration of the equation above leads to

$$B = \mu_0 \frac{Ni}{l} + \mu_0\left(\frac{Ni}{l}\right)_s, \tag{14-3}$$

where Ni/l is the number of ampere-turns per unit length in the windings, and $(Ni/l)_s$ the number of ampere-turns per unit length of the equivalent surface currents. The last term in Eqs. (14-2) and (14-3) will be negative if the material of the ring is diamagnetic.

We now define a second magnetic field vector, H, which plays much the same part in magnetic phenomena that the vector D does in dielectric phenomena. For our present purposes, H may be defined by the equation

$$H = \frac{1}{4\pi}\int \frac{i\,dl \sin\theta}{r^2}. \tag{14-4}$$

That is, H is computed in the same way as is the flux density B, with the difference that the proportionality constant μ_0 does not appear in its definition, and in evaluating it only the currents in actual conductors are to be included and *not* any equivalent surface currents. *Note carefully that Eq.* (14-4) *is not the complete definition of H.* See Sec. 15-7 for a further discussion.

The magnitude of H at any point is called the magnetic intensity, and may be expressed in *amperes per meter* as is readily seen from its definition. Like magnetic induction, the magnitude and direction of H can be represented by lines called *lines of magnetic force.* The direction of the magnetic intensity vector at any point is tangent to the line of magnetic force passing through the point, and the number of lines per unit area perpendicular to their direction is made numerically equal to the magnitude of H. Notice that as far as its units are concerned, magnetic intensity, expressed in amperes per meter, is the analogue of electric intensity, expressed in volts per meter.

Evaluation of Eq. (14-4) for the windings of a Rowland ring gives for the magnitude of H at any point within the ring

$$H = \frac{Ni}{l}. \tag{14-5}$$

That is, *in this special case*, the magnetic intensity equals the number of ampere-turns per meter in the magnetizing windings.

Let us return now to Eq. (14-3),

$$B = \mu_0 \frac{Ni}{l} + \mu_0 \left(\frac{Ni}{l}\right)_s.$$

The first term on the right is the flux density set up by the current in the magnetizing windings; the second is the flux density due to the equivalent surface currents. Since the surface currents are brought about by the current in the magnetizing windings, the flux density set up by the surface currents will depend on the number of ampere-turns per unit length in the magnetizing windings and therefore on the magnitude of H. We define a property of the material called its *magnetic susceptibility* and represented by χ, as *the ratio of the flux density due to the surface currents, to the magnetic intensity H.*[1]

$$\boxed{\chi = \frac{\mu_0 (Ni/l)_s}{H}, \qquad \mu_0 (Ni/l)_s = \chi H} \tag{14-6}$$

But

$$H = \frac{Ni}{l}$$

so

$$\chi = \mu_0 \frac{(Ni/l)_s}{Ni/l}$$

Notice that the definition of magnetic susceptibility is analogous to that of dielectric susceptibility. (See Sec. 7-3.) The units of magnetic susceptibility are the same as those of μ_0, webers/amp-meter or henrys/meter, since both $(Ni/l)_s$ and H are in amperes per meter. Some representative values are listed in Table 14-1. The "magnetic susceptibility of a vacuum" is, of course, zero, since equivalent surface currents can only

[1] However, see the discussion following Eq. (14-16).

TABLE 14-1

SUSCEPTIBILITIES OF PARA- AND DIAMAGNETIC MATERIALS

Material	Temp, °C	χ (h/m)
Aluminum	18	1.03×10^{-11}
Alum, ammonium, iron	−258	948
Alum, ammonium, iron	17	48.7
Bismuth	18	−2.12
Carbon (diamond)	20	−0.78
Carbon (graphite)	20	−5.6
Carbon dioxide (gas)	20	−0.68
Cerium	18	24
Copper	18	−0.14
Cupric chloride	18	15
Ferric chloride	20	135
Lead	18	0.19
Mercury	18	−0.30
Nitrogen (gas)	20	−0.54
Oxygen (liquid)	−219	487
Oxygen (gas)	20	167
Silver	18	−0.32
Zinc	18	−0.25

exist in magnetized matter. The magnetic susceptibilities of diamagnetic materials are negative.

At constant temperature, and for relatively small values of H, the magnetic susceptibilities of para- and diamagnetic materials are constant, independent of H. In other words, the equivalent surface currents are proportional to H. The magnetic susceptibilities of ferromagnetic materials are not constant but vary widely as H varies. See Sec. 15-1.

Eq. (14-3) can now be put in the following form, replacing Ni/l by H and $\mu_0(Ni/l)_s$ by χH.

$$B = \mu_0 H + \chi H$$

$$= (\mu_0 + \chi)H. \tag{14-7}$$

Let us now represent the term $\mu_0 + \chi$ by the symbol μ.

$$\boxed{\mu = \mu_0 + \chi.} \tag{14-8}$$

Then Eq. (14-7) takes the simple form

$$\boxed{B = \mu H.} \tag{14-9}$$

The quantity μ is called the *permeability* of the material. Its units are evidently the same as those of μ_0 and χ, henrys meter. In empty space, where $\chi = 0$, $\mu = \mu_0$. The proportionality constant μ_0 can therefore be described as "the permeability of empty space" or "the permeability of a vacuum."

It is often convenient to define still another quantity, represented by K_m and called the *relative permeability* of a material, as the ratio of the permeability of the material to that of a vacuum.

$$K_m = \frac{\mu}{\mu_0}. \tag{14-10}$$

Relative permeability is sometimes represented by the symbol μ_r. Of course it is a pure number and has the same numerical value in any system of units. Relative permeability can be expressed in terms of magnetic susceptibility by combining Eqs. (14-8) and (14-10), giving

$$K_m = 1 + \frac{\chi}{\mu_0}. \tag{14-11}$$

Relative permeability, K_m, is the magnetic analogue of dielectric coefficient, K_e (see Sec. 7-3). The name *magnetic coefficient* would be highly appropriate for K_m, but we shall call it relative permeability to conform with current usage.

The relative permeability of a vacuum is evidently unity, since $\chi = 0$, or $\mu = \mu_0$, in a vacuum. The relative permeabilities of para- and ferromagnetic materials are greater than unity, those of diamagnetic materials are less than unity. However, the magnetic susceptibilities of para- and diamagnetic materials are so small (see Table 14-1) that their relative permeabilities are practically equal to unity. Only for ferromagnetic materials is the difference appreciable.

The magnetic properties of a material are completely specified if any one of the three quantities, magnetic susceptibility χ, relative permeability K_m, or permeability μ, are known. The three are related by the defining equations

$$\mu = \mu_0 + \chi,$$

$$K_m = \frac{\mu}{\mu_0} = 1 + \frac{\chi}{\mu_0}.$$

The only reason for introducing all three is to simplify the form of certain common equations.

The permeability of a material, μ, and the permeability of a vacuum, μ_0, are seen to be analogous to the permittivity of a medium, ϵ, and the permittivity of a vacuum, ϵ_0.

The dielectric coefficient, K_e, which is equal to ϵ/ϵ_0, could also be described as the *relative permittivity.*

Examples. (1) The magnetic susceptibility of iron ammonium alum (see Table 14-1) is 948×10^{-11} henry/meter. Compute its relative permeability and its permeability.

$$\chi = 948 \times 10^{-11} \text{ h/m}$$

$$K_m = 1 + \frac{\chi}{\mu_0}$$

$$= 1 + \frac{948 \times 10^{-11}}{12.57 \times 10^{-7}}$$

$$= 1 + .00754$$

$$= 1.00754 \text{ (no units)}$$

$$\mu = \mu_0 + \chi$$

$$= 12.57 \times 10^{-7} + .0948 \times 10^{-7}$$

$$= 12.66 \times 10^{-7} \text{ h/m}$$

or

$$\mu = K_m \mu_0$$

$$= 1.00754 \times 12.57 \times 10^{-7}$$

$$= 12.66 \times 10^{-7} \text{ h/m}$$

(2) A Rowland ring, made of iron, of mean circumferential length 30 cm and cross section 1 cm², is wound uniformly with 300 turns of wire. Ballistic galvanometer measurements made with a search coil around the ring as in Fig. 14-1, show that when the current in the windings is 0.032 amperes, the flux in the ring is 2×10^{-6} webers. Compute: (a) the flux density in the ring, (b) the magnetic intensity, (c) the number of ampere-turns per meter of the equivalent surface currents and compare with those of the magnetizing winding, (d) the permeability, the relative permeability, and the magnetic susceptibility of the material of the ring.

(a) Flux density $B = \dfrac{\Phi}{A}$

$$= \frac{2 \times 10^{-6}}{10^{-4}}$$

$$= 2 \times 10^{-2} \text{ w/m}^2.$$

(b) Magnetic intensity $H = \dfrac{Ni}{l}$

$$= \frac{300 \times .032}{.30}$$

$$= 32 \text{ amp-turns/m}.$$

(c) From Eqs. (14-6) and (14-7),

$$\left(\frac{Ni}{l}\right)_s = \frac{B}{\mu_0} - H$$

$$= \frac{2 \times 10^{-2}}{12.57 \times 10^{-7}} - 32$$

$$= 1.59 \times 10^4 - 32$$

$$= 15{,}900 \text{ (equivalent) amp-turns/m,}$$

compared with only 32 amp-turns/m in the magnetizing windings.

(d) Permeability $\mu = \dfrac{B}{H}$

$$= \frac{2 \times 10^{-2}}{32}$$

$$= 6250 \times 10^{-7} \text{ h/m}.$$

Relative permeability $K_m = \dfrac{\mu}{\mu_0}$

$$= \frac{6250 \times 10^{-7}}{12.57 \times 10^{-7}}$$

$$= 498. \text{ (no units)}$$

Magnetic susceptibility

$$\begin{aligned}\chi &= \mu - \mu_0 \\ &= 6250 \times 10^{-7} - 12.57 \times 10^{-7} \\ &= 6240 \times 10^{-7}\ \text{h/m},\end{aligned}$$

or,

$$\begin{aligned}\chi &= \mu_0\,(K_m - 1) \\ &= 12.57 \times 10^{-7}\,(498 - 1) \\ &= 6240 \times 10^{-7}\ \text{h/m},\end{aligned}$$

or,

$$\begin{aligned}\chi &= \frac{\mu_0\,(Ni/l)_s}{H} \\ &= \frac{12.57 \times 10^{-7} \times 1.59 \times 10^4}{32} \\ &= 6240 \times 10^{-7}\ \text{h/m}.\end{aligned}$$

We shall next compute the self-inductance of a toroidal winding on a Rowland ring of permeability μ. From Eqs. (14-5) and (14-9), the flux density in the ring is

$$B = \mu H = \mu \frac{Ni}{l}.$$

The total flux across any section of the ring is

$$\Phi = BA,$$

and the self-inductance is

$$L = \frac{N\Phi}{i} = \frac{NBA}{i} = \mu \frac{AN^2}{l} = K_m \mu_0 \frac{AN^2}{l}. \tag{14-12}$$

Since the self-inductance of the same winding in vacuum, say L_0, is

$$L_0 = \mu_0 \frac{AN^2}{l} \tag{14-13}$$

we see that the self-inductance is altered by a factor of K_m when the coil is wound on a material substance. However, as is evident from Table 14-1, unless the ring is ferromagnetic the self-inductance is practically equal to that in vacuum. Note also that since the relative permeability of a ferromagnetic material is not a constant, the self-inductance of a coil on a ferromagnetic core is not constant either.

When Eq. (14-12) is divided by Eq. (14-13) we obtain

$$K_m = \frac{L}{L_0}, \tag{14-14}$$

and the relative permeability of a substance may be defined as the ratio of the self-inductance of a toroid whose core is composed of the substance, to the self-inductance of the same toroid in vacuum. This definition is exactly analogous to that of the dielectric coefficient K_e of a substance in terms of the capacitance of a capacitor (Eq. 8-9)

14-5 Magnetization. The equivalent currents on the surface of a magnetized body are only one aspect of the influence of a magnetic field on the body. Just as the *polarization* P of a dielectric interprets the induced surface charges in terms of the displacement of charge throughout the volume of the dielectric, so the *magnetization* of a material interprets the surface currents in terms of the rearrangement of internal electronic currents.

The *magnetic moment* of a current loop has been defined as the product $\mu_0 ia$, where i is the current in the loop and a is its area. Let us assume for simplicity that all the electronic current loops in a magnetized material have the same magnetic moment $\mu_0 ia$ and that there are n such loops per unit volume, all aligned with their planes in the same direction. The magnetic moment per unit volume within the material is then equal to the product of the magnetic moment of each loop and the number of loops per unit volume. This product is called the *intensity of magnetization* or simply the *magnetization* and is represented by the letter $\mathcal{I}$.

Magnetization is a vector quantity, whose direction is the same as that of the magnetic flux density due to the electronic current loops.

The magnetic moment per unit volume of a uniformly magnetized body can also be expressed in terms of the equivalent surface currents. Consider a short element of a Rowland ring of circumferential length Δs (Fig. 14-4). Let ΔN_s represent the number of turns of the equivalent surface currents in this length. Then by proportion,

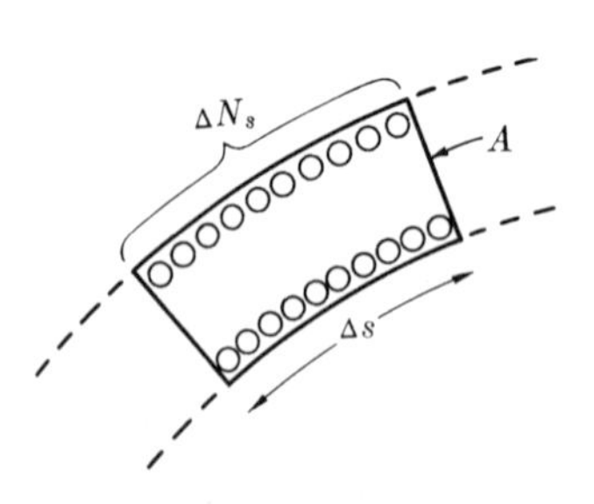

FIG. 14-4. Equivalent surface currents in a short element of a Rowland ring.

$$\frac{\Delta N_s}{\Delta s} = \frac{N_s}{l},$$

$$\Delta N_s = \frac{N_s \Delta s}{l}.$$

The magnetic moment of each turn is $\mu_0 i_s A$, so the total magnetic moment of the element is

$$\Delta N_s \times \mu_0 i_s A = \mu_0 \frac{N_s i_s A \, \Delta s}{l}.$$

The volume of the element is $A \, \Delta s$, and the magnetic moment per unit volume is therefore

$$\mathcal{J} = \frac{\mu_0 \dfrac{N_s i_s A \, \Delta s}{l}}{A \, \Delta s}$$

or

$$\mathcal{J} = \mu_0 \left(\frac{Ni}{l}\right)_s. \tag{14-15}$$

That is, the magnetic moment per unit volume equals the product of μ_0 and the equivalent ampere-turns per unit length. If we wish to consider the influence of the external field as a surface effect, as in the preceding section, we describe it in terms of equivalent surface currents. If we wish to consider it as a volume effect we speak of intensity of magnetization or magnetic moment per unit volume. The two quantities are directly proportional.

Magnetization can be expressed in terms of susceptibility. From the definition of the latter,

$$\mu_0 \left(\frac{Ni}{l}\right)_s = \chi H,$$

and hence

$$\mathcal{J} = \chi H, \qquad \chi = \frac{\mathcal{J}}{H}. \tag{14-16}$$

Eq. (14-16) is actually the correct general definition of magnetic susceptibility, rather than Eq. (14-6). In the special case of a Rowland ring, the magnetization is constant throughout the material of the ring and is directly proportional to the equivalent ampere-turns per unit length. In more general cases the equivalent currents may not be wholly confined to the surface of a body, and the magnetization may vary from point to point. The magnetic susceptibility, at any point, is defined as the ratio of the magnetization at that point to the magnetic intensity at the point.

One of the many equations relating B and H is

$$B = \mu_0 H + \chi H,$$

and since

$$\mathcal{J} = \chi H,$$

$$\boxed{B = \mu_0 H + \mathcal{J}.} \tag{14-17}$$

The first term on the right is the contribution to B of the magnetizing current, the second is the contribution of the internal electronic currents.

Example. Find the magnetization in the iron core of the Rowland ring in the preceeding example.

In that example we found

$$B = 2 \times 10^{-2}\ \text{w/m}^2, \quad H = 32\ \text{amp-turns/m}.$$

Then

$$\begin{aligned} \mu_0 H &= 12.57 \times 10^{-7} \times 32 \\ &= 0.00402 \times 10^{-2}\ \text{w/m}^2. \end{aligned}$$

That is, this is the flux density due to the magnetizing current, or the flux density that would have existed in vacuum.

$$\begin{aligned} \mathcal{J} &= B - \mu_0 H \\ &= 2 \times 10^{-2} - 0.00402 \times 10^{-2} \\ &= 2 \times 10^{-2}\ \text{w/m}^2, \end{aligned}$$

and practically all of the flux is due to the magnetization, or the electronic currents.

SUMMARY

The method followed in this chapter has been to develop general formulas by reasoning based on a special case (the Rowland ring). The steps in the derivation are summarized below, where formulas of general applicability have been indicated by enclosing them in a box. Formulas not so enclosed are correct for the special case of a Rowland ring. Equations are numbered as in the text.

The flux density B in a Rowland ring is the sum of the flux densities set up by the current in the windings, and by the equivalent surface currents.

$$B = \mu_0 \frac{Ni}{l} + \mu_0 \left(\frac{Ni}{l}\right)_s. \tag{14-3}$$

Define magnetic intensity H by the equation

$$H = \frac{1}{4\pi}\int \frac{i\,dl \sin\phi}{r^2}. \tag{14-4}$$

(See Sec. 15-7 for a more general definition.)

For the special case of a Rowland ring, Eq. (14-4) integrates to

$$H = \frac{Ni}{l}. \tag{14-5}$$

Define magnetic susceptibility χ by the equation

$$\mu_0(Ni/l)_s = \chi H. \tag{14-6}$$

(See Sec. 14-5 for a more general definition.)

Combine Eqs. (14-3), (14-5), and (14-6) to get

$$\boxed{B = \mu_0 H + \chi H = (\mu_0 + \chi)H.} \tag{14-7}$$

Define permeability by the equation

$$\boxed{\mu = (\mu_0 + \chi).} \tag{14-8}$$

Then from Eqs. (14-7) and (14-8),

$$\boxed{B = \mu H.} \tag{14-9}$$

Define relative permeability by the equation

$$\boxed{K_m = \frac{\mu}{\mu_0} = 1 + \frac{\chi}{\mu_0}.} \qquad \text{(14-10)} \quad \text{(14-11)}$$

Define magnetization $\mathcal{I}$ as *magnetic moment per unit volume.* For the special case of a Rowland ring,

$$\mathcal{I} = \mu_0(Ni/l)_s. \tag{14-15}$$

Combine Eqs. (14-6) and (14-15) to get

$$\boxed{\mathcal{I} = \chi H} \tag{14-16}$$

Note that Eq. (14-16), rather than Eq. (14-6), is the general definition of χ).

Combine Eqs. (14-7) and (14-16) to get

$$\boxed{B = \mu_0 H + \mathcal{I}.} \tag{14-17}$$

Problems—Chapter 14

(1) Compute the relative permeabilities of (a) bismuth; (b) liquid oxygen; and (c) nitrogen (gas), from the data in Table 14-1.

(2) Suppose the Rowland ring in Example 2 in Sec. 14-4 had been constructed of (a) aluminum; (b) silver. With a current of 0.032 amp in the magnetizing winding, compute for each material the number of ampere-turns/meter in the equivalent surface currents. Compare with those in the iron ring.

CHAPTER 15

FERROMAGNETISM

15-1 Ferromagnetism. The flux density in a Rowland ring of iron may be hundreds or even thousands of times as great as that due to the magnetizing current alone. Furthermore, the flux density B is not a linear function of the magnetic intensity H or, in other words, the permeability μ is not a constant. To complicate matters even further, the permeability depends on the past history (magnetically speaking) of the iron, a phenomenon known as *hysteresis:* In fact, a flux may exist in the iron even in the absence of any external field; when in this state the iron is called a *permanent magnet.*

Any substance which exhibits the properties above is called *ferromagnetic.* Iron, nickel, and cobalt are the only ferromagnetic elements, but a number of alloys whose components are not ferromagnetic also show these effects.

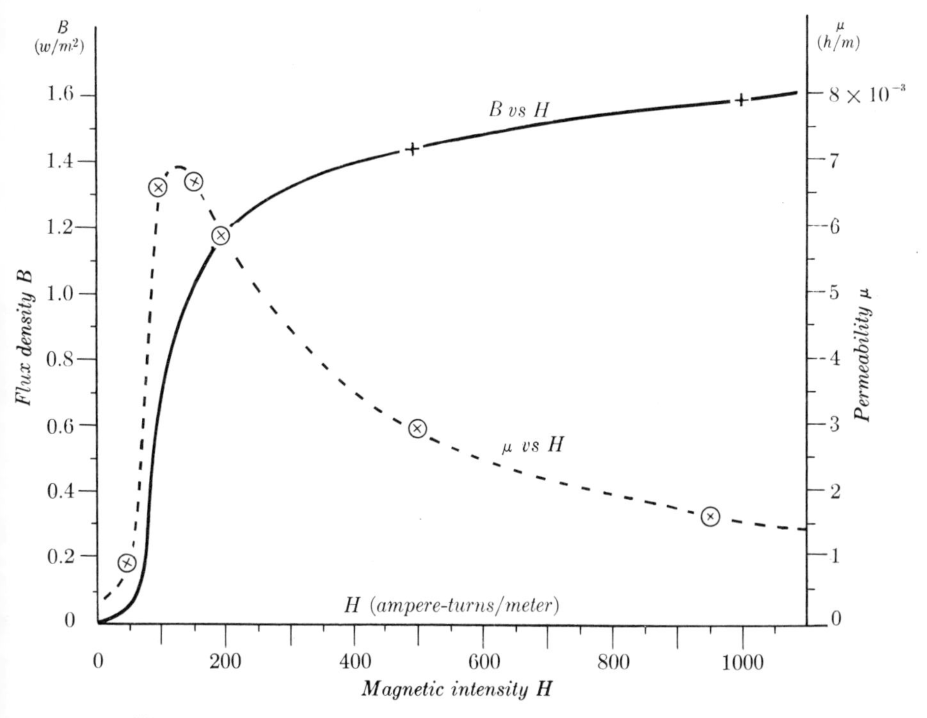

FIG. 15-1. Magnetization curve and permeability curve of annealed iron.

Because of the complicated relation between the flux density B and the magnetic intensity H in a ferromagnetic material, it is not possible to express B as an analytic function of H. Instead, the relation between these quantities is either given in tabular form or is represented by a graph of B vs H. The graph is called the *magnetization curve* of the material.

The magnetization curve of a specimen of annealed iron is shown in Fig. 15-1 in the curve labelled B vs H. The permeability μ, equal to the ratio of B to H, can be found at any point of the curve by dividing the flux density B, at the point, by the corresponding magnetic intensity H. For example, when $H = 150$ amp-turns/m, $B = 1.01$ w/m² and

$$\mu = \frac{B}{H} = \frac{1.01}{150} = .00675 \text{ h/m}.$$

The permeability at any point of the curve is equal to the slope of a line drawn from the origin to the point, due allowance being made for any difference in the scales of B and H. It is evident that the permeability is not a constant. The curve labelled μ vs H in Fig. 15-1 is a graph of μ as a function of H.

TABLE 15-1

MAGNETIC PROPERTIES OF ANNEALED IRON

Magnetic intensity	Flux density	Permeability	Magnetization	Relative permeability	Magnetic susceptibility
H $\frac{\text{amp-turns}}{\text{meter}}$	B $\frac{\text{webers}}{\text{m}^2}$	$\mu = B/H$ $\frac{\text{henrys}}{\text{meter}}$	$\mathcal{J} = B - \mu_0 H$ $\frac{\text{webers}}{\text{m}^2}$	$K_m = \mu/\mu_0$	$\chi = \mu - \mu_0$ $\frac{\text{henrys}}{\text{meter}}$
0	0	$3,100 \times 10^{-7}$	0	250	$3,100 \times 10^{-7}$
10	0.0042	4,200	0.0042	330	4,200
20	.010	5,000	.010	400	5,000
40	.028	7,000	.028	560	7,000
50	.043	8,600	.043	680	8,600
60	.095	16,000	.095	1270	16,000
80	.45	56,000	.45	4500	56,000
100	.67	67,000	.67	5300	67,000
150	1.01	67,500	1.01	5350	67,500
200	1.18	59,000	1.18	4700	59,000
500	1.44	28,800	1.44	2300	28,800
1,000	1.58	15,800	1.58	1250	15,800
10,000	1.72	1,720	1.71	137	1,710
100,000	2.26	226	2.13	18	213
800,000	3.15	39	2.15	3.1	26

Table 15-1 covers a wider range of values for the same specimen, and also includes values of the relative permeability K_m, the magnetization $\mathcal{I}$, and the susceptibility χ. It will be seen from the table that when H is small, practically all of the flux is due to the magnetization (or the equivalent surface currents). Beyond the point where H is of the order of magnitude of 1000 amp-turns/m there is but little further increase in the magnetization and the susceptibility decreases markedly. In this region the iron is said to become *saturated.* Further increases in B are due almost wholly to increases in the magnetizing current.

15-2 The Curie temperature. The permeability of ferromagnetic materials decreases with increasing temperature, and the relative permeability falls practically to unity at a temperature (different for different materials) called the *Curie temperature.* Above the Curie temperature, iron is paramagnetic but not ferromagnetic. In Fig. 15-2 there is shown the temperature variation of the relative permeability of a sample of iron. H is kept constant at 800,000 amp-turns/m. The Curie temperature is about 760°C for this sample.

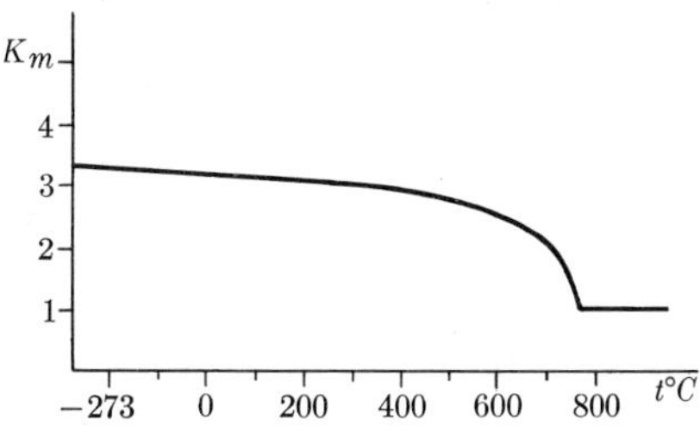

FIG. 15-2. Relative permeability of a sample of iron under a constant magnetic intensity of 800,000 ampere-turns/meter. The Curie temperature is about 760°C.

15-3 Hysteresis. A magnetization curve such as that in Fig. 15-1 expresses the relation between the flux density B in a ferromagnetic material and the corresponding magnetic intensity H, *provided the sample is initially unmagnetized and the magnetic intensity is steadily increased from zero.* Thus in Fig. 15-3, if the magnetizing current in the windings of an unmagnetized ring sample is steadily increased from zero until the magnetic intensity H corresponds to the abscissa Oe, the flux density B is given by the ordinate Of. If starting from the same unmagnetized state, the magnetic intensity is first increased from zero to Og and then

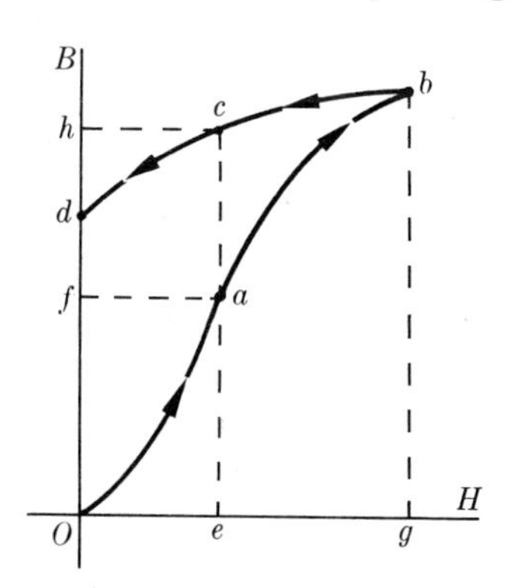

FIG. 15-3. Graph showing hysteresis.

decreased to Oe, the magnetic state of the sample follows the path $Oabc$. The flux density, when the magnetic intensity has been reduced to Oe, is represented by the ordinate Oh rather than Of. If the magnetizing current is now reduced to zero, the curve continues to point d, where the flux density is Od.

The flux density in the sample is seen to depend not on the magnetic intensity alone, but on the magnetic history of the sample as well. It has a magnetic "memory," so to speak, and "remembers" that it has been magnetized to point b even after the magnetizing current has been cut off. At point d it has become a *permanent magnet.* This behavior of the material, as evidenced by the fact that the B-H curve for decreasing H does not coincide with that for increasing H, is called *hysteresis.* The term means literally, "to lag behind."

In many pieces of electrical apparatus, such as transformers and motors, masses of iron are located in magnetic fields whose direction is continually reversing. That is, the magnetic intensity H increases from zero to a certain maximum in one direction, then decreases to zero, increases to the same maximum but in the opposite direction, decreases to zero, and continues to repeat this cycle over and over. The flux density B within the iron reverses also, but in the manner indicated in Fig. 15-4, tracing out a closed curve in the B-H plane known as a *hysteresis loop.* Fig. 15-3 may be considered as the start of such a hysteresis loop.

Positive values of H or B in Fig. 15-4 indicate that the respective directions of these quantities are, say, clockwise in a Rowland ring, while negative values mean that the directions are counterclockwise. The magnitude and direction of H are determined solely by the current in the winding, while those of B depend on the magnetic properties of the sample and its past history.

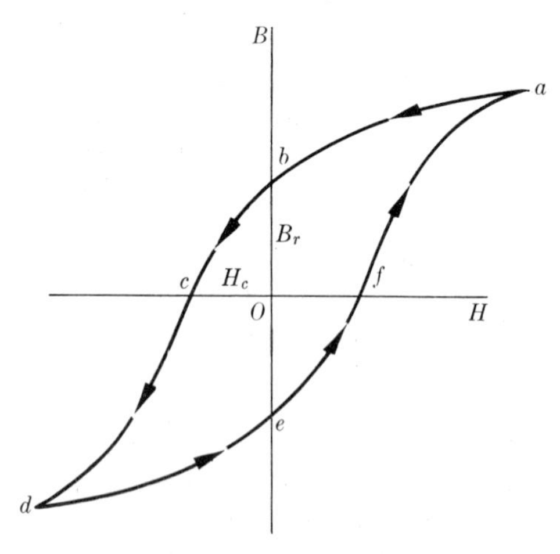

FIG. 15-4. Hysteresis loop.

The ordinate Ob or Oe in Fig. 15-4 represents the flux density remaining in the specimen when the magnetic intensity has been reduced to zero. It is called the *retentivity* or *remanence* of the specimen and is designated by B_r. The abscissa Oc or Of represents the reversed magnetic intensity needed to reduce the flux density to zero after the specimen has been magnetized to saturation in the opposite direction, and it is called the *coercive force* or the *coercivity*, H_c. What is the permeability at points b, c, e, and f?

The magnetization curve Oab in Fig. 15-3, shows the sample initially unmagnetized. One may wonder how this can be accomplished, since cutting off the magnetizing current does not reduce the flux density in the material to zero. A sample may be demagnetized by reversing the magnetizing current a number of times, decreasing its magnitude with each reversal. The sample is thus carried around a hysteresis curve which winds more and more closely around the origin. See Fig. 15-5.

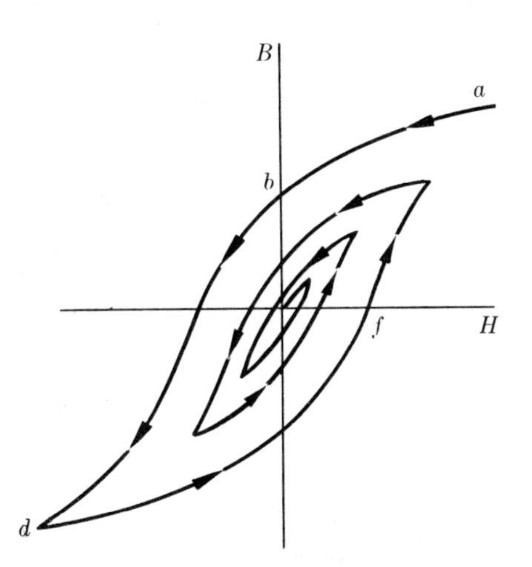

FIG. 15-5. Successive hysteresis loops during the operation of demagnetizing a ferromagnetic sample.

Hysteresis effects introduce a similar difficulty in measuring the flux in a sample. In the absence of hysteresis, the flux falls to zero when the magnetizing current is cut off, and a ballistic galvanometer connected to a search coil around the specimen indicates the flux previously present in the specimen. But since B does not become zero when H is reduced to zero, ballistic galvanometer deflections only indicate the *changes* in flux corresponding to changes in the magnetizing force. Hence in practice the complete hysteresis loop must be traced out in stepwise fashion, measuring the changes in flux accompanying changes in magnetizing current. For a more complete description, reference may be had to a text on electrical measurements.

It is evidently desirable that a material for permanent magnets should have both a large retentivity so that the magnet will be "strong" and a large coercive force so that the magnetization will not be wiped out by stray external fields. Some typical values are given in Table 15-2. Alnico 5 is one of the most recently developed alloys for permanent magnets. Its superiority to carbon steel, which was the material used for many years, is evident.

One consequence of the phenomenon of hysteresis is the production of heat within a ferromagnetic material each time the material is caused to traverse its hysteresis loop. This heat results from a kind of internal

TABLE 15-2

RETENTIVITY AND COERCIVE FORCE OF PERMANENT MAGNET MATERIALS

Material	Composition (percent)	B_r (w/m)	H_c amp-turns/m
Carbon steel	98 Fe, 0.86 C, 0.9 Mn	0.95	3.6×10^3
Cobalt steel	52 Fe, 36 Co, 7 W, 3.5 Cr, 0.5 Mn, 0.7 C	0.95	18×10^3
Alnico 2	55 Fe, 10 Al, 17 Ni, 12 Co, 6 Cu	0.76	42×10^3
Alnico 5	51 Fe, 8Al, 14 Ni, 24 Co, 3 Cu	1.25	44×10^3

friction as the magnetic domains within the material (see Sec. 15-4) are reversed in direction. It can be shown that the heat developed per unit volume, in each cycle, is proportional to the area enclosed by the hysteresis loop. Hence if a ferromagnetic material is to be subjected to a field which is continually reversing its direction (the core of a transformer, for example) it is desirable that the hysteresis loop of the material shall be narrow to minimize heat losses. Fortunately, iron or iron alloys are available which combine high permeability with small hysteresis loss.

To derive the expression for the heat developed per unit volume per cycle, consider a Rowland ring of mean length l and cross section A, having a winding of N turns around its circumference. If the current in the winding is i, then

$$H = \frac{Ni}{l}.$$

Let the current be increased from i to $i + di$ in a time dt. The increase di in the current causes an increase dH in the magnetic intensity, an increase dB in the flux density, and an increase $d\Phi = A\,dB$ in the flux. Because of this increase in flux, an induced emf appears in the winding, of magnitude

$$\mathcal{E} = N\frac{d\Phi}{dt}.$$

The power input during this time, exclusive of i^2R losses in the winding, is

$$P = \mathcal{E}i.$$

The energy supplied is

$$dW = Pdt$$

and the total energy supplied in one cycle is

$$W = \oint dW,$$

where the symbol $\oint$ means that the integration is to be carried around the closed hysteresis loop in the B-H diagram. Combining the preceding equations one finds

$$W = Al \oint H\, dB.$$

But $\oint H\, dB$ is simply the area enclosed by the hysteresis loop, and since $Al = V$, the volume of the ring, we have

Heat per unit volume per cycle = Area of hysteresis loop

If B is expressed in webers per square meter and H in ampere-turns/meter, the heat is in joules per cubic meter, per cycle.

In general it is not possible to express H as any simple function of B, so that the integral cannot be evaluated analytically. Hence the hysteresis loop is plotted and its area measured with a planimeter, due allowance being made for any difference in the scales of B and H.

15-4 The domain theory. Recent studies of ferromagnetism have shown that there exist in a ferromagnetic substance small regions called *domains.* The domains are of microscopic size but large enough to contain from 10^{12} to 10^{15} atoms. Within each domain the magnetic moments of all the spinning electrons are parallel to one another. In other words, each domain is magnetized to saturation. The directions of the magnetic moments of the domains, in iron, are parallel to one or another of the crystal axes, but in unmagnetized iron there is a random variation in direction from one domain to another so that the net magnetic moment of any sizable specimen is zero. (Fig. 15-6(a).)

Several distinct types of change may take place in the domains when the specimen is in an external field. In weak fields the changes consist of, first, rotations of the directions of magnetization of the domains more nearly into parallelism with the external field, and second, movement of the domain boundaries. Domains in which the magnetization is nearly

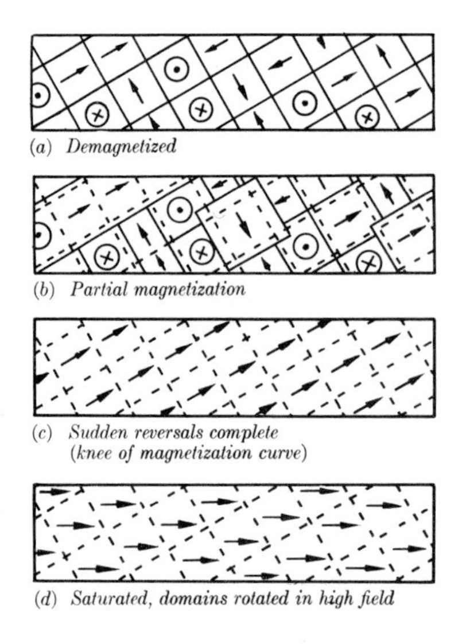

FIG. 15-6. Diagram illustrating changes in domain structure in a single crystal, in an increasing external field directed from left to right. Domains are shown as cubes, for convenience; they are believed to be long and narrow.

parallel to the external field increase in size at the expense of adjacent domains in which the magnetization makes larger angles with the external field. (Fig. 15-6(b).)

In stronger fields, where the magnetization curve rises more steeply, whole domains suddenly rotate by 90° or 180° into parallelism with the crystal axis which most nearly coincides with the direction of the external field. (Fig. 15-6(c).)

Finally, in strong fields, the magnetization in all of the domains rotates continuously into parallelism with the external field (Fig. 15-6(d)) and the entire specimen becomes saturated.

The *Barkhausen effect,* which is most pronounced along the steeply rising portion of the magnetization curve, furnishes evidence that the magnetization proceeds in discontinuous steps as the domains rotate by large angles. If a rod of ferromagnetic material is surrounded by a search coil connected to an audio amplifier, and the rod is placed in an external field which can be steadily increased or decreased, a crackling sound is heard from a speaker connected to the amplifier. As each domain in turn changes direction, it induces a sudden short rush of current through the search coil and these surges are heard as noise from the speaker.

15-5 Magnetic poles. A ring sample of magnetic material simplifies the presentation of most of the concepts associated with the magnetic properties of matter for the reason that the magnetic field of such a ring is confined entirely to its interior. We consider next the magnetization of a body whose field extends to the region surrounding it. The most common examples are a compass needle, or a bar or horseshoe magnet. These, however, are rather complex problems and we shall therefore discuss first the magnetization of a homogeneous isotropic sphere in a uniform field.

When placed in a field which, before the introduction of the sphere, was uniform, the sphere becomes magnetized in a way similar to that in which a dielectric sphere becomes polarized in an electric field. In Fig. 15-7(a) the solid lines are the lines of induction of the original uniform field and the dotted lines are the lines of induction due to the magnetized sphere. (That is, the induction contributed by the electronic circuits within the sphere or by its equivalent surface currents.) Fig. 15-7 should be compared carefully with Fig. 7-4 in Sec. 7-2. Note the difference; the lines of the induced electric field in Fig. 7-4 originate on positive and terminate on negative charges and hence are directed from right to left within the sphere, while the lines of induction in Fig. 15-7(a) are closed lines whose direction within the sphere is from left to right.

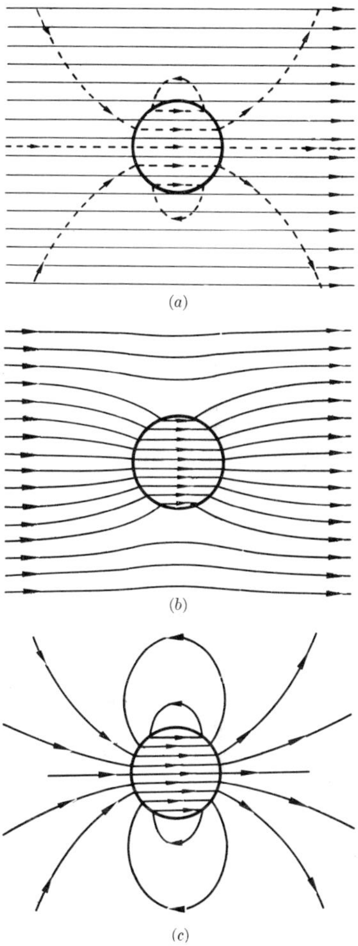

(a)

(b)

(c)

FIG. 15-7. Magnetization of a ferromagnetic sphere in a uniform magnetic field.

Fig. 15-7(b) shows the actual field after the insertion of the sphere, that is, the resultant of the sphere's field and the original field. At *external* points the field is modified in the same way as is the field around a polarized dielectric sphere in an electric field (compare Fig. 7-4(c)); at *internal* points, however, the induction B is increased in the magnetic case while in the electric case the electric intensity E is diminished.

If the sphere in Fig. 15-7 is ferromagnetic, then because of hysteresis effects it will retain some of its magnetization when removed from the field and it becomes a *permanent magnet.* The lines of induction around and within such a magnetized sphere are shown in Fig. 15-7(c).

Since the magnetic field outside the magnetized sphere is geometrically identical with the electric field outside a polarized dielectric sphere, one *might* attribute the external magnetic field of the sphere to fictitious induced "magnetic charges" on its surface. In fact, before the relations between electrical and magnetic effects were understood as well as they are today, it was assumed that the magnetism actually was due to magnetic charges, or, as they were called, magnetic *poles*. "North" magnetic poles correspond to positive charges, "south" magnetic poles to negative charges. In spite of the fact that magnetic charges are no longer believed to exist, it is nevertheless convenient in many instances to make use of the magmetic pole concept and to compute the field of a magnetized body in terms of its poles rather than in terms of its equivalent surface currents.

The term corresponding to "quantity of electric charge" is "magnetic pole strength." The unit of pole strength, in the mks system, turns out to be *one weber*, as will be explained shortly. Pole strength is represented by the letter m, which corresponds to q or Q for quantity of charge. The magnetized sphere in Fig. 15-7 has magnetic poles distributed over its entire surface, although with a surface density which is not uniform. North poles are distributed over the hemisphere from which lines of induction emerge, south poles over the hemisphere where lines of induction enter.

The pole strength of a magnetized body can be expressed in terms of the magnetization within the body. The pole strength, m, within any volume, is equal to the negative of the surface integral of the magnetization $\mathcal{J}$ over the surface bounding the volume, a relation exactly analogous (except for the minus sign) to Gauss's law for the electric displacement D and the charge q.

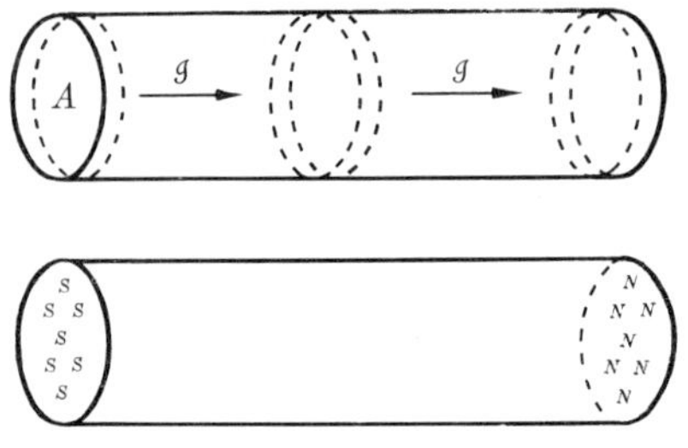

FIG. 15-8. Magnetic poles on a cylindrical magnet.

$$m = -\int \mathcal{J} \cos\theta \, dA. \quad (15\text{-}1)$$

In this equation, a positive value of m indicates a north pole, and $\mathcal{J}$ is considered positive if directed outward.

To make the definition more concrete, consider a cylindrical permanent bar magnet, within which we shall assume the magnetization to be the same at all points and directed from left to right. The surface integral of $\mathcal{J}$ over any short cylinder lying wholly within the magnet, such as the one near its center in Fig. 15-8, is zero, since $\mathcal{J}$ is directed inward over one face and outward over the other. Hence there are no poles in the interior of the magnet.

Over a short cylinder at the right end of the magnet, however, the surface integral is $-\mathcal{J}A$ at the left face but is zero at the right face, since

$\mathcal{I} = 0$ outside the magnet. That is,

$$-\int \mathcal{I} \cos \theta \, dA = -(-\mathcal{I}A) = \mathcal{I}A = m,$$

and since m is positive, a north pole of strength $m = \mathcal{I}A$ is located at the right end of the magnet. By the same reasoning, a south pole of equal strength is found at the left end of the magnet.

The *pole strength per unit area* at either end, m/A, is,

$$\frac{m}{A} = \mathcal{I}.$$

That is, the pole strength per unit area equals the magnetization (in this special case). Pole strength per unit area is the magnetic analogue of electric charge per unit area or surface density of charge, and is related to magnetization in the same way that surface density of charge is related to polarization in the special case of a flat sheet of dielectric. Since $\mathcal{I}$ is expressed in webers/m^2, pole strength per unit area is also in webers/m^2 and hence pole strength itself is expressed in webers and the unit of pole strength is one weber, as was stated earlier.

15-6 The magnetic field of the earth. To a first approximation, the magnetic field of the earth is the same as that outside a uniformly magnetized sphere. Fig. 15-9 represents a section through the earth. The heavy vertical line is its axis of rotation, and the *geographic* north and south poles are lettered N_G and S_G. The direction of the (presumed) in-

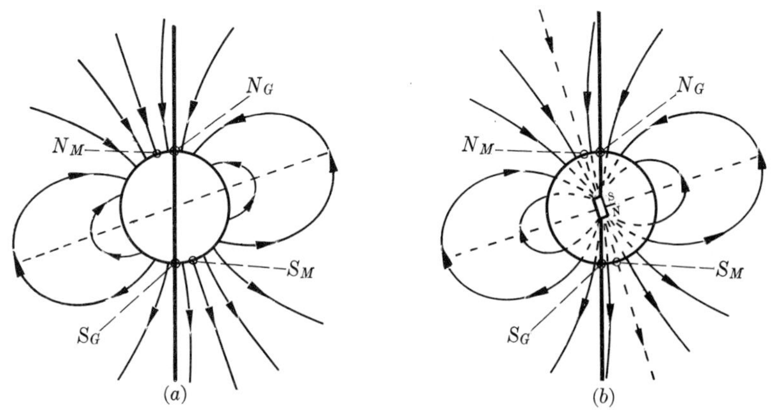

FIG. 15-9. Simplified diagram of the magnetic field of the earth.

ternal magnetization makes an angle of about 15° with the earth's axis. The dotted line indicates the plane of the magnetic equator, and the letters N_M and S_M the so-called magnetic north and south poles. Note carefully that lines of induction emerge from the earth's surface over the entire southern magnetic hemisphere and enter its surface over the entire northern magnetic hemisphere. Hence if we wish to attribute the earth's field to magnetic poles we must assume, on the basis of this hypothesis about the earth's internal magnetic state, that *north* magnetic poles are distributed over the entire *southern* magnetic hemisphere, and *south* magnetic poles over the entire *northern* magnetic hemisphere. This can be very confusing. The north and south magnetic poles, considered as points on the earth's surface, are simply those points where the field is vertical. The former is located at Lat. 70° N, Long. 96° W.

It is interesting to note that the same field at *external* points would result if the earth's magnetism were due to a short bar magnet near its center as in Fig. 15-9(b), with the south pole of the magnet pointing towards the north magnetic pole. The field within the earth is different in the two cases but for obvious reasons experimental verification of either hypothesis is impossible.

Except at the magnetic equator, the earth's field is not horizontal. The angle which the field makes with the horizontal is called the *angle of dip* or the *inclination*. At Cambridge, Mass. (about 45° N lat.), the magnitude of the earth's field is about 5.8×10^{-5} w/m² and the angle of dip about 73°. Hence the horizontal component at Cambridge is about 1.7×10^{-5} w/m² and the vertical component about 5.5×10^{-5} w/m². In northern magnetic latitudes the vertical component is directed downward, in southern magnetic latitudes it is upward. The angle of dip is, of course, 90° at the magnetic poles.

The angle between the horizontal component and the true north-south direction is called the *variation* or *declination*. At Cambridge, Mass., the declination at present is about 15° W, that is, a compass needle points about 15° to the west of true north.

The magnetic field of the earth is not as symmetrical as one might be led to suspect from the idealized drawing of Fig. 15-9. It is in reality very complicated, the inclination and the declination varying irregularly over the earth's surface, and also varying with the time.

15-7 General definition of magnetic intensity. It is a familiar fact that a body can be magnetized by being brought near a permanent magnet, just as it can be magnetized by sending a current through a winding of wire around it. In setting up the equations expressing the magnetic state of a specimen in the form of a Rowland ring, it was convenient to introduce the quantity H which could be thought of as the "cause" of the magnetization. H was defined by the equation

$$H = \frac{1}{4\pi} \int \frac{i\,dl \sin\theta}{r^2}, \tag{15-2}$$

but it was stated that this definition was not complete. It will now be realized why this is so, since the "cause" of the magnetization of any body may be other magnetized bodies as well as currents in conductors. Hence a term must be added to the definition of H to include the effects of such bodies.

The magnetic state of a magnetized body can be expressed in three different ways. We can give (1) the magnetization at all points within the body, (2) the equivalent surface currents, or (3) the strength of the magnetic poles. If any one of these is known, the other two may be computed. For example, if the magnetization $\mathcal{I}$ within a cylinder of length l and cross section A is uniform, then the equivalent surface current per unit length around the curved surface is, from Eq. (14-15),

$$\left(\frac{Ni}{l}\right)_s = \frac{\mathcal{I}}{\mu_0},$$

and the pole strength per unit area at the end faces is

$$\frac{m}{A} = \mathcal{I}.$$

The magnetic effects produced by a magnetized body can be expressed in terms of any one of these three aspects of its magnetic state. It is customary to use the third, that is, the poles of the magnetized body.

If the poles are sufficiently small, or sufficiently distant from the point at which H is to be computed, they may be considered as "point" poles. The additional term in H is then computed from the pole strengths and distances by the equation

$$H = \frac{1}{4\pi\mu_0} \Sigma \frac{m}{r^2} \text{ (vector sum)} \tag{15-3}$$

More generally, if the poles are distributed over surfaces or throughout volumes, H is computed by the equation

$$H = \frac{1}{4\pi\mu_0} \int \frac{dm}{r^2} \text{ (vector sum).}$$

That is, the H-field of magnet poles is an inverse square field like the E-field of an electric charge. The direction of H is away from a north pole and toward a south pole. The complete definition of magnetic intensity is therefore

$$H = \left[\frac{1}{4\pi}\int\frac{i\,dl\sin\theta}{r^2} + \frac{1}{4\pi\mu_0}\int\frac{dm}{r^2}\right] \text{(vector sum)}. \qquad (15\text{-}4)$$

Example. In Fig. 15-10, a long straight conductor perpendicular to the plane of the paper carries a current i. A bar magnet having point poles of strength m at its ends lies in the plane of the paper. What is the magnitude and direction of the magnetic intensity H at point P?

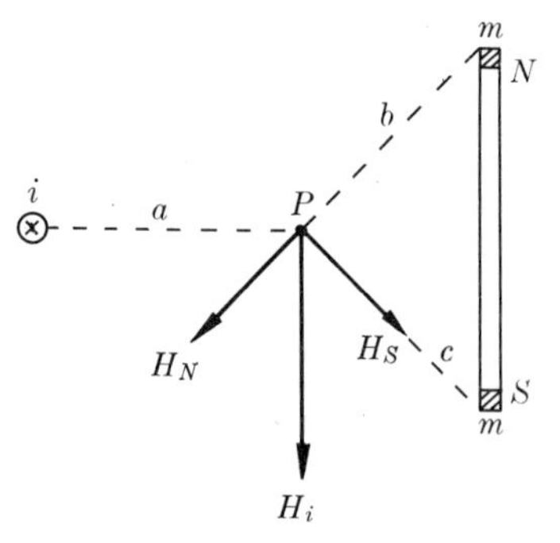

FIG. 15-10. Magnetic intensity at a point in space due to magnet poles and to a current in a wire.

The vectors H_i, H_N, and H_S represent the components of H due respectively to the current, and to the N and S poles of the magnet. Consider first H_i. The flux density B at point P, due to the current i, is

$$B = \frac{\mu_0}{2\pi}\frac{i}{a},$$

and since in empty space $H = \dfrac{B}{\mu_0}$,

$$H_i = \frac{1}{2\pi}\frac{i}{a}.$$

The same result would, of course, be obtained by integration of the first term of Eq. (15-4).

The components H_N and H_S are respectively

$$H_N = \frac{1}{4\pi\mu_0}\frac{m}{b^2},$$

$$H_S = \frac{1}{4\pi\mu_0}\frac{m}{c^2}.$$

The resultant of these three vectors is the magnetic intensity H at the point P.

15-8 Magnetization of a bar. We can now understand why the magnetization of a bar is such a complex phenomenon. Suppose a bar or rod of iron is placed in a magnetic field which, before the introduction of the bar, is uniform. (The field near the center of a long solenoid, for example.) The magnitude of H at every point, before the bar is inserted, is given by the first term in Eq. (15-4), since no poles are present. Furthermore, if the current in the solenoid is kept constant, this component of H remains constant after the bar is inserted. Its magnitude is given by

$$H_i = \frac{Ni}{l}.$$

As soon as the bar is inserted it becomes magnetized and north and south poles appear at its ends. The lines of force of H_m, the component of H produced by the poles, are shown in Fig. 15-11 by dotted lines. It will be seen that the direction of this component, at points within the bar, is opposite to the direction of H_i. Furthermore, the magnitude of H_m is not the same at all points but is large near the poles and much weaker in the center of the bar. The resultant magnetic intensity therefore varies widely from point to point of the bar, the latter does not become uniformly magnetized, and its poles are not confined to its end faces as in the idealized example in Fig. 15-8. An exact analysis of the problem can not be carried out analytically, since it involves the shape of the magnetization curve of the iron. The poles will, however, be more or less localized in regions near the ends of the bar and their contribution to the magnetic intensity at points near the center of the bar will be small. Therefore the magnetic intensity at the center is determined chiefly by the current in the solenoid and can be computed from a knowledge of this current and the number of turns per unit length. If the length of the bar is greater than about 10 times its diameter, the reversed or "demagnetizing" field at its center, due to its poles, can in general be neglected.

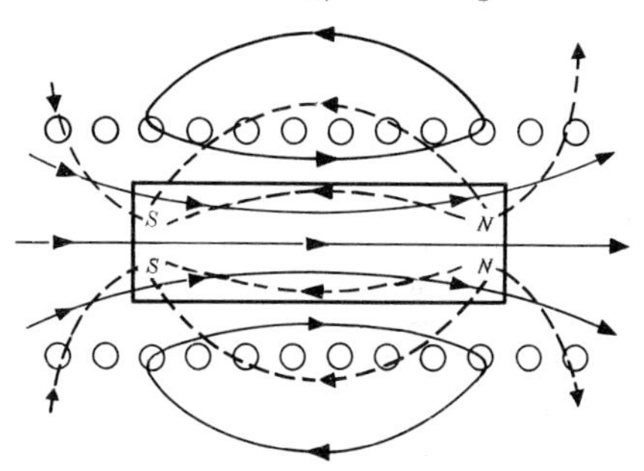

FIG. 15-11. Lines of magnetizing force H, around an iron bar in a solenoid. Solid lines represent H_i, the field of the solenoid, dotted lines represent H_m, the field of the magnetized bar.

It will now be evident why a Rowland ring was chosen to introduce the discussion of magnetization, since there are no poles present and the H-field is due entirely to the current in the magnetizing windings.

15-9 Torque on a bar magnet. A bar magnet, or a compass needle, aligns itself, if free to turn, with its axis parallel to the direction of an external field in which it is placed. The torque aligning it with the field is often attributed to forces exerted on the poles of the magnet. Let us reconcile this point of view with that of equivalent surface currents.

In Fig. 15-12(a) there is shown a solenoid of length l and cross section A, having N turns and carrying a current i. The solenoid is in an external magnetic field of flux density B at right angles to l, and the system is in vacuum or in air. The total torque on the N turns of the solenoid is

$$\tau = BNiA.$$

The direction of the torque is such as to turn the solenoid into a position in which the lines of induction through each turn, due to the current

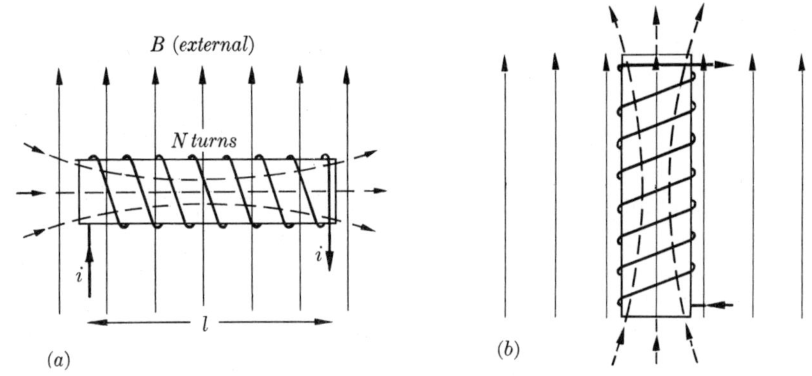

FIG. 15-12. The torque acting on a solenoid placed in a magnetic field is such as to rotate it to the position (b).

in the turn, are in the same direction as the external induction. The equilibrium position is shown in Fig. 15-12(b).

Fig. 15-13 represents a bar magnet, with its equivalent surface currents and its N and S poles, also in an external field at right angles to the magnet. If $(Ni)_s$ represents the number of equivalent ampere turns of the surface currents, the torque on the magnet is

$$\tau = B(Ni)_s A.$$

But we have shown that

$$\mathcal{I} = \mu_0 \left(\frac{Ni}{l}\right)_s = \frac{m}{A}.$$

Hence

$$(Ni)_s = \frac{ml}{\mu_0 A}, \qquad (15\text{-}5)$$

and therefore

$$\tau = B \frac{ml}{\mu_0 A} A = \frac{B}{\mu_0} ml.$$

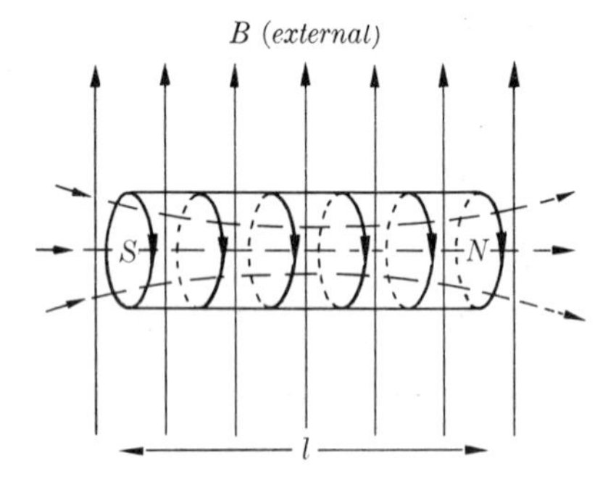

FIG. 15-13. Bar magnet with its equivalent surface currents and its N and S poles.

Since the system is in vacuum or in air,

$$\frac{B}{\mu_0} = H,$$

and

$$\tau = Hml. \tag{15-6}$$

(Note that B and H refer to the *external* field and not the magnet's own field. The magnet cannot exert a torque on itself.)

This equation can be interpreted as follows. Suppose that on each pole of the magnet, of pole strength m, there is exerted a force

$$F = Hm, \tag{15-7}$$

this force being in the same direction as H on a north pole, and opposite to H on a south pole (Fig. 15-14). Since the poles are separated by a distance l, the torque on the magnet is

$$\tau = Fl = Hml,$$

which is the same as Eq. (15-6).

We may therefore consider the torque as due either to (a) equivalent surface currents, and compute it from the equation

$$\tau = B(Ni)_s A,$$

or to (b) forces on the poles of the magnet, and compute it from the equation

$$\tau = Hml.$$

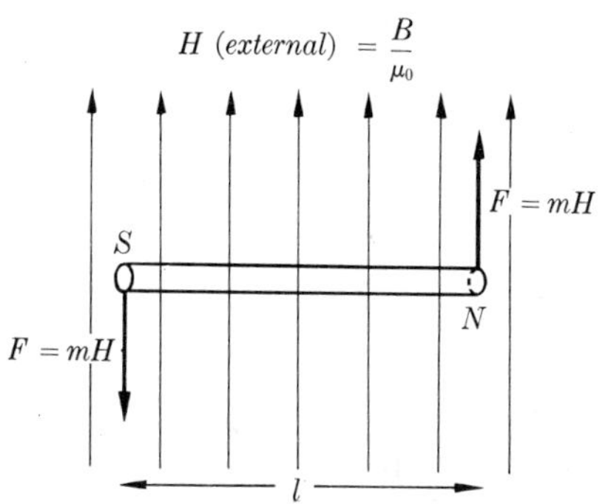

FIG. 15-14. Torque on a bar magnetic interpreted in terms of forces acting on the poles.

One more relation should be developed at this point for completeness. The magnetic intensity at a point P, at a distance r from a point pole of strength m, is, from Eq. (15-3),

$$H = \frac{1}{4\pi\mu_0}\frac{m}{r^2}.$$

The force on a second point pole of strength m' at P is, from Eq. (15-7)

$$F = Hm'.$$

Hence the force exerted on one pole by the other is

$$F = \frac{1}{4\pi\mu_0}\frac{mm'}{r^2}. \tag{15-8}$$

If the poles are alike the force is a repulsion; if unlike, the force is an attraction.

This equation has precisely the same form as Coulomb's law for the force between point charges, and it is often taken as the starting point from which one develops all "magnetic" equations. However, since we now believe that the magnetic effects of magnetized matter are actually due to its revolving or spinning electrons and that the "poles" of a magnetized body are only a convenient mathematical artifice, it seems preferable to base the magnetic equations on the general law of force between moving charges.

15-10 Magnetic moment. The magnetometer. Let us write Eq. (15-5) in the form

$$ml = \mu_0 (Ni)_s A.$$

The term on the right is, by definition, the total magnetic moment of the equivalent surface currents of a bar magnet. Hence the magnetic moment of the magnet can also be expressed as the product of pole strength and separation, ml. In this form it is completely analogous to the dipole moment p of an electric dipole,

$$P = ql,$$

and it is now evident why the term *dipole* was applied to the pair of electric charges in the first place. If we introduce a single symbol, say M, for the magnetic moment, then

$$M = ml,$$

and Eq. (15-6) can be written

$$\tau = HM, \quad M = \frac{\tau}{H}. \tag{15-9}$$

In the preceding discussion we have assumed the poles to be point poles, concentrated at the ends of the magnet. But whether its poles are concentrated or not, any bar magnet will experience a certain torque τ at a point where the magnetic intensity is H. We can therefore use Eq. (15-9),

$$M = \frac{\tau}{H},$$

as an experimental method of determining the *effective* or *equivalent* magnetic moment of any bar magnet. That is, the effective magnetic moment M may be *defined* as the ratio of the torque on the magnet to the external magnetic intensity. Conversely, if the magnetic moment M of a magnet is known, an unknown magnetic intensity can be measured by measuring the torque on the magnet. This method is known as the *magnetometer method* for measuring magnetic intensity, or for comparing the magnetic intensity at two points.

Fig. 15-15 is a top view of a bar magnet in a horizontal field H. The magnet is pivoted at its center, either suspended by a supporting fiber or pivoted from below as is a compass needle. The support is assumed to exert no torque on the magnet.

When the magnet is displaced from its equilibrium position by an angle θ, the restoring torque is

$$\begin{aligned}\tau &= -mH \times l \sin\theta \\ &= -HM \sin\theta.\end{aligned}$$

The minus sign is introduced since we have a *restoring* torque, and the torque by definition equals $-HM \sin\theta$ even if the poles are not concentrated at the ends of the magnet.

If θ is small we may replace $\sin\theta$ by θ.

$$\tau = -HM\theta.$$

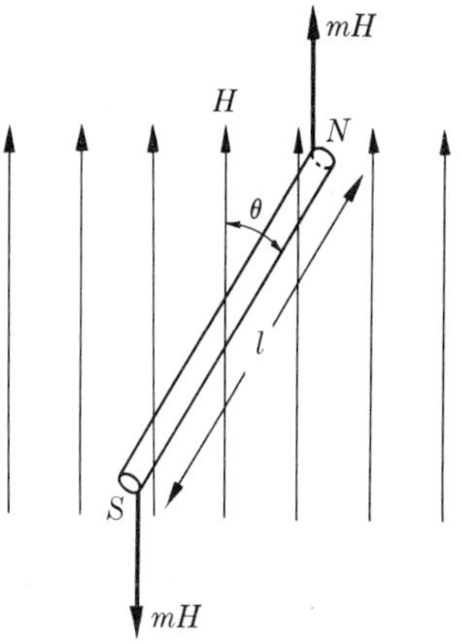

FIG. 15-15. Bar magnet pivoted at its center, vibrating in a uniform magnetic field.

This is the condition necessary for angular harmonic motion, and the magnet therefore oscillates about its equilibrium position with a period

$$T = 2\pi\sqrt{\frac{I_0}{HM}},$$

where I_0 is the moment of inertia of the magnet about its center of mass. The magnetic intensity H can therefore be computed if the other factors are known.

If we measure the periods T_1 and T_2 at two points where the magnetic intensities are H_1 and H_2 respectively, then

$$\frac{T_1}{T_2} = \sqrt{\frac{H_2}{H_1}},$$

and the *ratio* of H_2 to H_1 can be found without knowing I_0 or M.

15-11 The magnetic circuit. As we have seen, every line of induction is a closed line. Although there is nothing in the nature of a flow along these lines, it is useful to draw an analogy between the closed paths of the flux lines and a closed conducting circuit in which there is a current. The region occupied by the magnetic flux is called a *magnetic circuit,* of which a Rowland ring is the simplest example. When the windings on such a ring are closely spaced over its surface, practically all of the flux lines are confined to the ring (Fig. 15-16(a)). Even if the winding is concentrated over only a small portion of the ring as in Fig. 15-16(b) the permeability of the ring is so much greater than that of the surrounding air that most of the flux is still confined to the material of the ring. The small part which returns via an air path is called the *leakage flux* and is indicated by dotted lines.

If the ring contains an air gap as in Fig. 15-16(c) there will be a certain amount of spreading or "fringing" of the flux lines at the air gap, but again most of the flux is confined to a well-defined path. This magnetic circuit may be considered to consist of the iron ring and the air gap "in series."

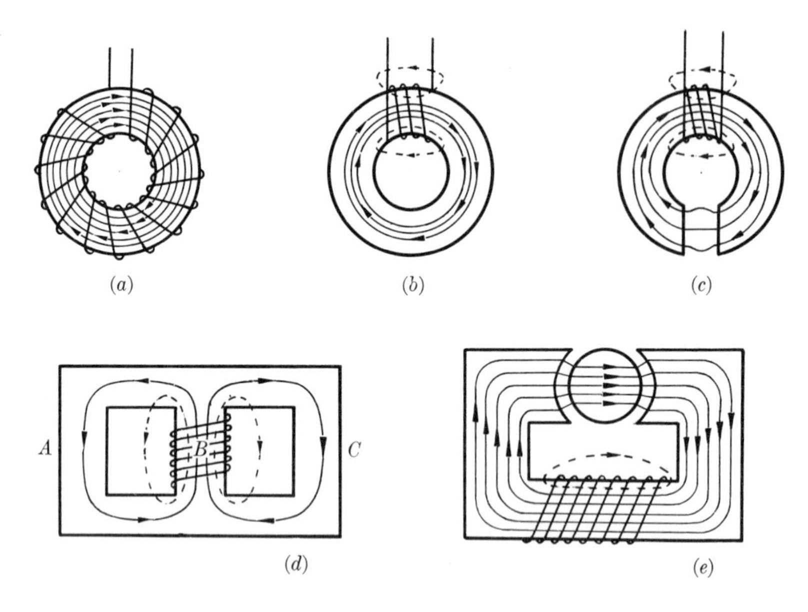

FIG. 15-16. Various magnetic circuits. (a) Rowland ring completely wound. (b) Rowland ring partially wound, showing leakage. (c) Fringing in an air gap. (d) Leakage in a transformer core. (e) Fringing and leakage in the core of a motor or generator.

Fig. 15-16(d) shows a section of a common type of transformer core. Here the magnetic circuit is divided, and sections A and C may be considered to be "in parallel" with one another, and in series with section B.

Fig. 15-16(e) is the magnetic circuit of a motor or generator. The two air gaps are in series with the iron portion of the circuit.

One of the important problems in the design of apparatus in which there is a magnetic circuit is to compute the flux density which will result from a given current in a given winding on a given core or, conversely, to design a core and windings so as to produce a desired flux density. It is this problem which we are now to consider, although naturally many of the factors which arise in practice must be ignored for simplicity.

Consider first a closed ring (no air gap) of uniform cross section. We have shown that within the ring

$$B = \mu H = \mu \frac{Ni}{l},$$

and since $\Phi = BA$,

$$\Phi = \mu \frac{NiA}{l},$$

or,

$$\Phi = \frac{Ni}{l/\mu A}. \tag{15-10}$$

Now the resistance R of a conductor of uniform cross section A, length l, and resistivity ρ is given by

$$R = \rho \frac{l}{A},$$

or in terms of the conductivity σ $(\sigma = 1/\rho)$

$$R = \frac{l}{\sigma A}.$$

If such a conductor is connected to the terminals of a seat of emf of negligible internal resistance, the circuit equation becomes

$$I = \frac{\text{emf}}{l/\sigma A}.$$

The form of this equation is the same as that of Eq. (15-10), with current corresponding to magnetic flux, the quantity Ni corresponding to electromotive force, and $l/\mu A$ corresponding to resistance. In view of the close analogy, the numerator in Eq. (15-10) is called the *magnetomo tive force,* and the denominator is called the *reluctance* of the magnetic circuit.

$$\text{Magnetomotive force (mmf)} = Ni, \tag{15-11}$$

$$\text{Reluctance } (\mathcal{R}) = l/\mu A, \tag{15-12}$$

and Eq. (15-10) may be written

$$\Phi = \frac{\text{mmf}}{\mathcal{R}}. \tag{15-13}$$

Magnetomotive force is evidently expressed in *ampere turns* and reluctance in *ampere-turns per weber.* The reluctance of a magnetic circuit is the required number of ampere-turns, per weber of magnetic flux in the circuit.

The advantage of writing the expression for the flux in a magnetic circuit in the form of Eq. (15-13) is most apparent when one considers a circuit containing an air gap. (Or more generally, when the circuit is composed of sections of different permeabilities, lengths, and cross sections.) It turns out that the *equivalent reluctance* of such a circuit may be found in the same way that one finds the equivalent resistance of a network of conductors. For example, a ring containing an air gap corresponds to two resistors in series and the equivalent reluctance of the circuit is the sum of the reluctances of ring and gap. The arms A and C of Fig. 15-16(d) are in parallel and the reciprocal of their equivalent reluctance is the sum of the reciprocals of the reluctances of the arms individually. For a simple "series" magnetic circuit one has

$$\Phi = \frac{\text{mmf}}{\Sigma\mathcal{R}} = \frac{\text{mmf}}{\Sigma(l/\mu A)} = \frac{\text{mmf}}{l_1/\mu_1 A_1 + l_2/\mu_2 A_2 + \cdots} \tag{15-14}$$

where l_1, l_2, etc., are the lengths of the various portions of the circuit, μ_1, μ_2, etc., the corresponding permeabilities, and A_1, A_2, etc., the cross-sectional areas.

The statements above are correct to a good approximation if the leakage flux is small. Further details will be found in any good text on electrical machinery.

Examples. (1) The mean length of a Rowland ring is 50 cm and its cross section is 4 cm². Use the permeability curve of Fig. 15-1 to compute the magnetomotive force needed to establish a flux of 4×10^{-4} webers in the ring. What current is required if the ring is wound with 200 turns of wire?

The desired flux density B is

$$B = \frac{\Phi}{A} = \frac{4 \times 10^{-4}}{4 \times 10^{-4}} = 1 \text{ w/m}^2.$$

From Fig. 15-1, the permeability at this flux density is about 65×10^{-4} h/m. Hence the reluctance $\mathcal{R}$ is

$$\mathcal{R} = \frac{l}{\mu A} = \frac{.5}{65 \times 10^{-4} \times 4 \times 10^{-4}} = 1.92 \times 10^5 \text{ ampere-turns/weber},$$

and since mmf $= \Phi\mathcal{R}$, the required magnetomotive force is

$$\text{mmf} = 4 \times 10^{-4} \times 1.92 \times 10^5 = 77 \text{ ampere-turn}.$$

If the ring is wound with 200 turns the current required is 0.385 ampere.

(2) If an air gap one millimeter in length is cut in the ring, what current is required to maintain the same flux?

The reluctance of the air gap is

$$\mathcal{R} = \frac{l}{\mu_0 A} = \frac{10^{-3}}{12.57 \times 10^{-7} \times 4 \times 10^{-4}} = 20 \times 10^5 \text{ ampere-turns/weber}.$$

Neglecting the small change in length of the iron, its reluctance is the same as before or 1.92×10^5 amp-turns/w. Thus the reluctance of the gap, although only 1 mm long, is ten times as great as that of the iron portion of the circuit. The reluctance of the entire circuit is now $20 \times 10^5 + 1.92 \times 10^5 = 22 \times 10^5$ amp-turns/w. The number of ampere turns required is 880 and the corresponding current is 4.4 amperes.

If it is desired to find the flux or flux density that will be set up by a given mmf in a magnetic circuit having an air gap, the problem cannot be solved as readily. In order to find the reluctance, the permeability must be known and its value depends on the flux density which is just the thing one is attempting to compute. An answer may be obtained, however, by a "cut-and-try" process. A value of flux density is assumed and the corresponding permeability read from a graph or table. The reluctance is then computed using this value of the permeability, and the mmf computed from mmf $= \Phi\mathcal{R}$. If the assumed flux density was correct, this should equal Ni. If the two are not equal, a new value of flux is assumed and the process repeated. After a few trials an experienced computer is able to secure as close an agreement as may be desired.

15-12 Derivation of magnetic circuit equation. In deriving the equation for the flux in a magnetic circuit it will be helpful to begin by generalizing Eq. (11-9) which states that the line integral of the magnetic induction around a closed curve equals μ_0 times the current through the surface bounded by the curve.

$$\oint B \cos \theta \, dl = \mu_0 i.$$

Fig. 15-17 is a section through a Rowland ring. The small circles outside the ring represent the magnetizing windings and the small circles inside the ring the equivalent surface currents. The dotted circle is a closed flux line of length l, and the surface bounded by it is shaded.

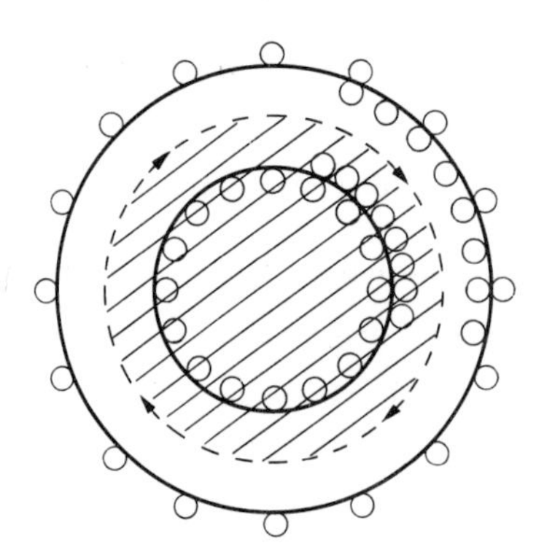

FIG. 15-17. Section through a Rowland ring.

The flux density in the ring is

$$B = \mu_0 \left[\left(\frac{Ni}{l} \right) + \left(\frac{Ni}{l} \right)_s \right],$$

and therefore

$$Bl = \mu_0[(Ni) + (Ni)_s].$$

But Bl, in this special case, equals the line integral of B around the dotted circle, while the sum $[(Ni) + (Ni)_s]$ is the *total* current through the shaded area, magnetizing current plus equivalent surface currents.

In other words, when a closed curve lies within magnetized matter, *the line integral of B around the curve equals μ_0 times the net total current through the enclosed surface, magnetizing current plus equivalent surface currents.*

The magnetic intensity, H, on the other hand, is defined as the field due to the magnetizing current plus the field due to the poles, if any. If the ring is closed as in Fig. 15-17 there are no poles and we have from Eq. (15-4)

$$H = \frac{1}{4\pi} \int \frac{i\,dl \sin\theta}{r^2} = \frac{Ni}{l},$$

Hence

$$\oint H \cos\theta\, dl = \oint \frac{Ni}{l}\, dl = Ni$$

and the line integral of H equals the number of ampere-turns in the windings alone.

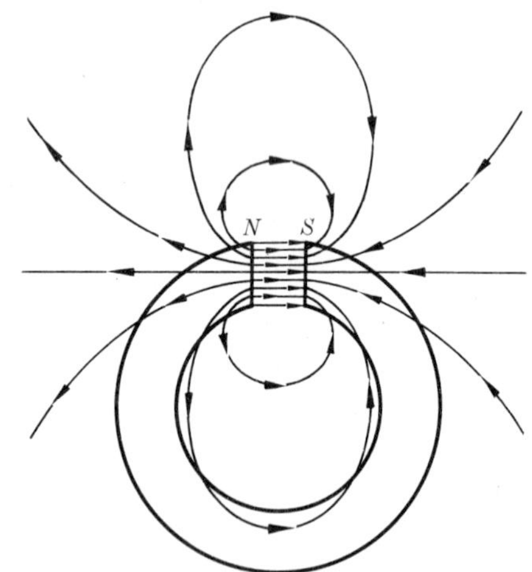

FIG. 15-18. The H-field due to the poles on the faces of a gap in a Rowland ring.

If there is a gap in the ring as in Fig. 15-16(c), then poles appear at the faces of the gap and their field adds vectorially to the field of the magnetizing windings. The H-field around the poles is like the E-field around a pair of + and − charges and is shown in Fig. 15-18. Within the gap this field is clockwise, while within the material of the ring it is counterclockwise. In evaluating the line integral of H around the dotted curve, a positive contribution results from the portion of the curve within the gap and a negative contribution from the portion within the iron. It is not difficult to show that these exactly cancel, i.e., the line integral of that part of H resulting from the magnet poles is zero. (The proof is identical with that which shows that in an electrostatic field, $\oint E \cos\theta\, dl = 0$). Hence the line integral of H around any closed curve equals the net *magnetizing* current through the enclosed surface.

To sum up, then,

$$\oint B \cos\theta\, dl = \mu_0 \times \text{net total current through enclosed surface}$$

$$= \mu_0[Ni + (Ni)_s]. \tag{15-15}$$

$$\oint H \cos\theta\, dl = \text{net magnetizing current through enclosed surface}$$

$$= Ni. \tag{15-16}$$

These equations should be compared with Eqs. (7-10) and (7-13) in Sec. 7-4, which relate to analogous surface integrals in an electric field, when the surface may lie within polarized matter.

We shall now apply Eq. (15-16) to a magnetic circuit. Let us take as our closed curve the boundary of the shaded area in Fig. 15-19. Let l_1 be the length of the boundary lying in the iron, and l_2 the length in the air gap. Because of the presence of poles at the ends of the iron forming the gap, the magnetic intensity in the iron is not equal to that in the gap. Let H_1 be the magnetic intensity in the iron, H_2 that in the air gap, and let μ_1 and μ_2 be the corresponding permeabilities. We shall assume that H_1 and H_2 are uniform in their respective regions. The line integral of H around the boundary of the loop then reduces to

$$H_1 l_1 + H_2 l_2.$$

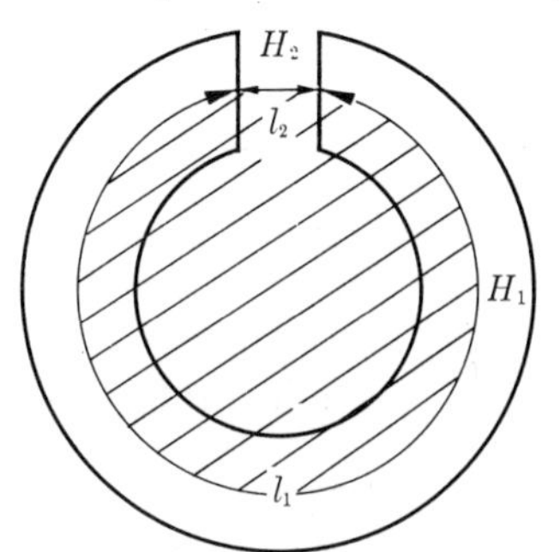

FIG. 15-19. Magnetic circuit with air gap.

Each turn of the magnetizing winding passes once through the shaded area. If there are N turns in the winding the magnetizing current through the shaded area is Ni. Hence from Eq. (15-16) we have

$$H_1 l_1 + H_2 l_2 = Ni \tag{15-17}$$

The quantity B is the flux density due to currents in conductors and to electronic circuits or their equivalent surface currents. The fluxes set up by these types of source are alike in that both are closed lines encircling the currents which give rise to them. The lines of induction in the magnetic circuit are therefore continuous around the circuit and *the flux Φ through any cross section is the same as that through any other.* (This conclusion can be checked experimentally by search coil measurements.) Hence if A_1 and A_2 are the cross sections of iron and air gap, and Φ is the flux in the circuit,

$$B_1 = \frac{\Phi}{A_1}, \quad B_2 = \frac{\Phi}{A_2},$$

and in the iron we have

$$B_1 = \mu_1 H_1, \quad \text{or} \quad H_1 = \frac{B_1}{\mu_1} = \frac{\Phi}{\mu_1 A_1},$$

while in the air gap

$$H_2 = \frac{\Phi}{\mu_2 A_2}.$$

Introducing these values in Eq. (15-17), we obtain

$$\frac{\Phi l_1}{\mu_1 A_1} + \frac{\Phi l_2}{\mu_2 A_2} = Ni$$

or finally

$$\Phi = \frac{Ni}{\dfrac{l_1}{\mu_1 A_1} + \dfrac{l_2}{\mu_2 A_2}}.$$

15-13 Energy per unit volume in a magnetic field. It has been shown that a quantity of energy $\frac{1}{2}Li^2$ is required to establish a current i in an inductor of self-inductance L. It is customary to associate this energy with the magnetic field of the inductor, and to consider it distributed through the field with a certain energy density. The expression for the energy density is readily derived with the aid of a Rowland ring, within which the magnetic field is of constant magnitude, so that by symmetry the energy per unit volume is constant throughout the volume of the ring.

When the current in the windings is i, the energy associated with the field is

$$W = \tfrac{1}{2}Li^2.$$

We have shown that self-inductance may be defined as

$$L = \frac{N\Phi}{i},$$

and that the flux within a toroidal winding is

$$\Phi = \frac{Ni}{l/\mu A} = BA = \mu HA.$$

Combining these equations, and remembering that the volume of the ring is Al, we obtain

$$\begin{aligned}\text{Energy density} = \frac{W}{Al} &= \tfrac{1}{2}BH \\ &= \tfrac{1}{2}\mu H^2 \\ &= \tfrac{1}{2}\frac{B^2}{\mu}.\end{aligned}$$

In the mks system, energy density is expressed in joules per cubic meter. Notice that these expressions are identical in form with those for the energy density in an electric field. (See Sec. 8-8.)

Problems—Chapter 15

(1) From the data in Table 15-1, construct a graph of K_m vs B for annealed iron, over a range of B from zero to 1.5 w/m^2.

(2) Table 15-3 lists corresponding values of H and B for a specimen of commercial hot-rolled silicon steel, a material widely used in transformer cores. (a) Construct graphs (similar to Fig. 15-1) of B and μ as functions of H, in the range from $H = 0$ to $H = 1000$ amp-turns/m. (b) What is the maximum permeability? (c) What is the initial permeability ($H = 0$)? (d) What is the permeability when $H = 800{,}000$ amp-turns/m?

TABLE 15-3

MAGNETIC PROPERTIES OF SILICON STEEL

Magnetic Intensity H $\frac{\text{amp-turns}}{\text{meter}}$	Flux Density B $\frac{\text{webers}}{\text{m}^2}$
0	0
10	0.050
20	.15
40	.43
50	.54
60	.62
80	.74
100	.83
150	.98
200	1.07
500	1.27
1000	1.34
10000	1.65
100000	2.02
800000	2.92

(3) Suppose a cylindrical bar magnet of Alnico 5, 1 cm in diameter and 10 cm long, is permanently and uniformly magnetized to an intensity of magnetization of 1.2 w/m^2. (a) What is the strength of the poles at the ends of the magnet? (b) What is the magnetic moment of the magnet? (c) What is the torque exerted on the magnet when it is suspended in air at right angles to a magnetic field of flux density 1 milliweber/m^2? (d) Compare the magnitude of H in the above field with the coercive force of Alnico 5.

(4) A solenoidal winding of 100 turns of wire is wound on a wooden rod of the same dimensions as the bar magnet in problem 3. What current must be sent through the wire in order that the torque acting on the solenoid in an external field shall equal the torque on the bar magnet?

(5) Sketch the lines of induction of a diamagnetic sphere in a uniform external field.

(6) Refer to Fig. 15-10. Let the length of the magnet be 8 cm and its pole strength $m = 10^{-5}$ weber. Let $a = 6$ cm, $b = 5$ cm, $c = 5$ cm, and let $i = 100$ amp. Find the magnitude of the magnetic intensity H at point P.

(7) A Rowland ring has a cross section of 2 cm^2, a mean length of 30 cm, and is wound with 400 turns. Find the current in the winding that is required to set up a flux density of 0.1 w/m^2 in the ring, (a) if the ring is of annealed iron (Table 15-1), (b) if the ring is of silicon steel (Table 15-3). (c) Repeat the computations above if a flux density of 1.2 w/m^2 is desired.

(8) Refer to example (2) at the end of Sec. 15-11. Find the current required to establish a flux of 4×10^{-4} weber in the ring, if it is of silicon steel (Table 15-3).

(9) Refer to example (2) at the end of Sec. 15-11. Compute the flux in the ring if the current in the windings is 6 amp.

(10) Refer to example (2) at the end of Sec. 15-11. (a) Compute the energy density in the air gap and in the iron. (b) Compute the quantity of energy in the air gap and in the iron.

(11) A copper wire of radius a carries a steady current i. (a) What is the flux density at a point within the material of the wire, at a distance r from its axis? (b) Find the magnetic energy density at points within a thin cylindrical shell of radius r, thickness dr, and length l. (c) Find the total magnetic energy within a portion of the wire of length l. (d) Find the self-inductance of the wire, per unit length, due to the internal flux.

CHAPTER 16

ALTERNATING CURRENTS

16-1 The alternating current series circuit. A coil of wire, rotating with constant angular velocity in a uniform magnetic field, develops a sinusoidal alternating emf as explained in Sec. 12-6. This simple device is the prototype of the commercial alternating current generator, or alternator, the field and armature structure of which are illustrated in Fig. 16-1. A number of pairs of poles are spaced around the inner circumference of the *stator*. As each conductor on the surface of the armature or *rotor* sweeps across the magnetic field, a motional emf is induced in it, in one direction as the conductor passes a north pole and in the opposite direction as it passes a south pole. The induced emf is therefore alternating, the number of complete cycles in each revolution equalling the number of *pairs* of poles. This multipole structure enables a sufficiently high frequency to be attained without an unduly high angular velocity of the rotor. For a more detailed description, a text on A.C. machinery may be consulted.

The induced emf of a commercial alternator may differ slightly from a purely sinusoidal form, but, unless explicitly stated otherwise, we shall assume in this chapter that we have to do with an alternator which maintains between its terminals a sinusoidal potential difference given by

$$v = V_m \sin 2\pi ft, \qquad (16\text{-}1)$$

where v is the instantaneous potential difference, V_m the maximum potential difference, and f is the frequency, equal to the number of revolutions per second of the rotor multiplied by the number of pairs of poles. For many of the A.C. generators in this country, $f = 60$ cycles/sec.

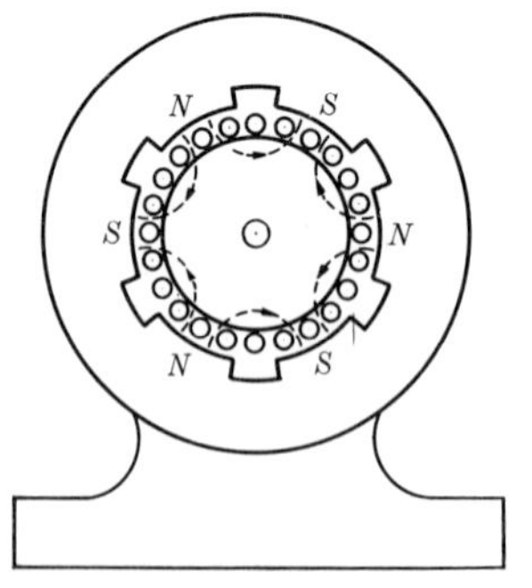

Fig. 16-1. Schematic diagram of a commercial alternating current generator.

We now proceed to investigate the current in a circuit when a sinusoidal alternating potential difference is maintained across its terminals. Let a series circuit composed of a resistor, an inductor, and a capacitor be connected to the terminals of an alternator as in Fig. 16-2. The instantaneous potential difference between a and b equals the sum of the instantaneous p.d.'s. across R, L, and C. That is

$$V_m \sin 2\pi ft = Ri + L\frac{di}{dt} + \frac{q}{C}$$

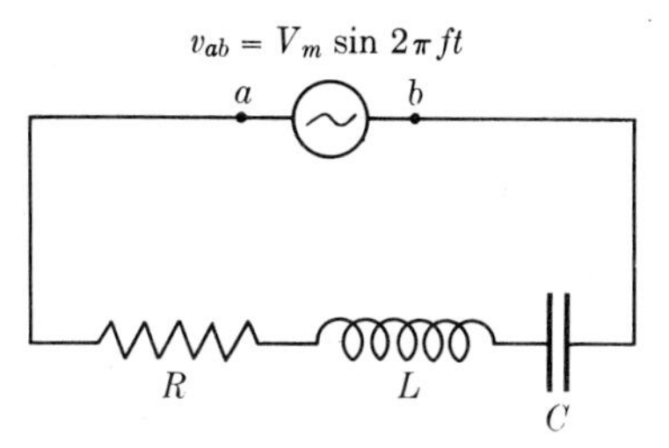

FIG. 16-2. Series circuit of a resistor, an inductor, and a capacitor.

where i, di/dt, and q are respectively the instantaneous current, its rate of change, and the charge on the capacitor. Differentiation of each term with respect to t gives

$$2\pi fV_m \cos 2\pi ft = R\frac{di}{dt} + L\frac{d^2i}{dt^2} + \frac{1}{C}\frac{dq}{dt}.$$

After rearranging terms and replacing dq/dt by i, we obtain

$$L\frac{d^2i}{dt^2} + R\frac{di}{dt} + \frac{1}{C}i = 2\pi fV_m \cos 2\pi ft.$$

For the method of solution of this second order differential equation the reader is referred to any standard textbook such as Phillips' "Differential Equations." The solution is

$$i = \frac{V_m}{\sqrt{R^2 + \left(2\pi fL - \dfrac{1}{2\pi fC}\right)^2}} \sin(2\pi ft - \phi) + A\epsilon^{-bt}$$

where

$$\phi = \tan^{-1}\frac{2\pi fL - \dfrac{1}{2\pi fC}}{R}.$$

This equation appears rather formidable, but its interpretation is not too difficult.

Consider first the term $A\epsilon^{-bt}$. The quantities A and b depend on the circuit constants and on the initial conditions, that is, on the values of V, i, di/dt, and q at the instant when the circuit is closed. However, the term ϵ^{-bt} decreases exponentially with the time and becomes negligible after a sufficient time (usually very short) has elapsed. This term is called a *transient*, and while in practice transient potential differences and currents may become undesirably large, we shall neglect them and consider only the first term which is called the *steady state* solution.

The steady state current, like the terminal voltage, is seen to vary sinusoidally with the time. Its maximum value is

$$I_m = \frac{V_m}{\sqrt{R^2 + \left(2\pi fL - \dfrac{1}{2\pi fC}\right)^2}}.$$

The steady state current may hence be written

$$i = I_m \sin (2\pi ft - \phi). \tag{16-2}$$

The frequency of the current is therefore the same as that of the voltage, but the two differ in phase, or are out of phase, by the angle ϕ.

Let us introduce the following abbreviations:

$$2\pi fL = X_L$$

$$\frac{1}{2\pi fC} = X_C$$

$$2\pi fL - \frac{1}{2\pi fC} = X_L - X_C = X$$

$$\sqrt{R^2 + X^2} = Z.$$

Then

$$I_m = \frac{V_m}{\sqrt{R^2 + (X_L - X_C)^2}} = \frac{V_m}{\sqrt{R^2 + X^2}} = \frac{V_m}{Z}, \tag{16-3}$$

$$\phi = \tan^{-1} \frac{X}{R}. \tag{16-4}$$

The quantity Z is called the *impedance* of the circuit; X, the *reactance;* and X_L and X_C the *inductive reactance* and *capacitive reactance* respectively. Impedances and reactances are expressed in ohms. The general term for a device possessing reactance is a *reactor*.

The maximum current is seen to be related to the maximum potential difference by an equation having the same form as Ohm's law for steady currents, impedance Z corresponding to resistance R.

16-2 Root-mean-square or effective values. The instantaneous value of an alternating current, emf, or potential difference varies continuously from a maximum in one direction through zero to a maximum in the opposite direction, and so on. It has been found convenient to describe these varying quantities by stating their root-mean-square or rms values. (Their average value over any number of complete cycles is obviously zero.) It was shown in Sec. 4-10 that the rms value of a varying current equalled the steady current which would develop heat at the same rate in the same resistance, and for that reason the rms value was called the *effective* value of the varying current. The same term is applied to the rms value of a varying emf or potential difference.

If a quantity varies sinusoidally, its rms or effective value is $1/\sqrt{2} = 0.707$ times its maximum value. For example, when it is stated that the alternating potential difference between the supply mains of a household power line is 110 volts, this means that the rms or effective potential difference is 110 volts, and hence the maximum potential difference is $110 \times \sqrt{2} = 155$ volts.

When the first and last terms of Eq. (16-3) are divided by $\sqrt{2}$ we obtain

$$\frac{I_m}{\sqrt{2}} = \frac{V_m/\sqrt{2}}{Z}, \quad \text{or}$$

$$I_{\text{rms}} = \frac{V_{\text{rms}}}{Z}. \tag{16-5}$$

It will be understood from now on that the letters I, $\mathcal{E}$, or V, without subscripts, refer to the root-mean-square or effective values of the corresponding quantities. Eq. (16-5) will therefore be written

$$I = \frac{V}{Z}.$$

Let us consider next the factors which determine the resistance, reactance, and impedance of a circuit. In the first place, the resistance of a conductor carrying an alternating current may not be the same as the resistance of the same conductor when carrying a steady current. The difference arises from the fact that the current density in a wire carrying alternating current is not uniform over the cross section of the wire, but is greater near the surface, a phenomenon known as "skin effect." The effective cross section of the conductor is therefore reduced and its resistance increased. Skin effect is brought about by self-induced emf's set up by the variations in the internal flux in a conductor and is greater the

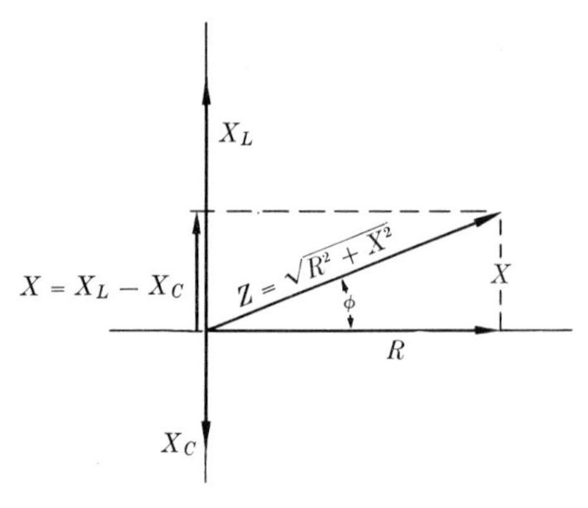

FIG. 16-3. Vector impedance diagram for the series circuit of Fig. 16-2.

higher the frequency. While an important factor at radio frequencies, it can usually be neglected at a frequency of 60 cycles. The actual resistance of a conductor at any frequency is called its *effective resistance* at that frequency. Unless explicitly stated otherwise, we shall ignore skin effects and assume the resistance of a conductor to be independent of frequency.

The reactance of an inductor, $X_L = 2\pi fL$, is proportional both to its inductance and to the frequency. If there is any iron associated with the inductor then, as has been shown, the self-inductance is not a constant, but again for simplicity we shall ignore this variation.

Capacitive reactance, $X_C = 1/2\pi fC$, is inversely proportional both to the capacitance and the frequency.

The relation between the impedance Z of a series circuit, and the values of R, X_L, and X_C, may be represented graphically by treating all of these quantities as vectors. The resistance R is represented by a vector along the positive X-axis, and the reactances X_L and X_C by vectors along the positive and negative Y-axes respectively. The impedance Z is the vector sum or resultant of these three vectors. See Fig. 16-3, which is called the *vector impedance diagram* of the circuit. Fig. 16-3 has been drawn for the case where $X_L > X_C$ and X is positive. If $X_L < X_C$, X is negative and extends downward rather than upward.

Example. A 600-ohm resistor is in series with a 0.5 henry inductor and a 0.2 μf capacitor. Compute the impedance of the circuit and draw the vector impedance diagram (a) at a frequency of 400 cycles, (b) at 600 cycles.

Solution. (a) At 400 cycles

$$X_L = 2\pi \times 400 \times 0.5 = 1256 \text{ ohms},$$

$$X_C = \frac{1}{2\pi \times 400 \times 0.2 \times 10^{-6}} = 1990 \text{ ohms},$$

$$X = X_L - X_C = 1256 - 1990 = -734 \text{ ohms}.$$

$$Z = \sqrt{(600)^2 + (-734)^2} = 949 \text{ ohms}. \text{ [See Fig.16-4(a).]}$$

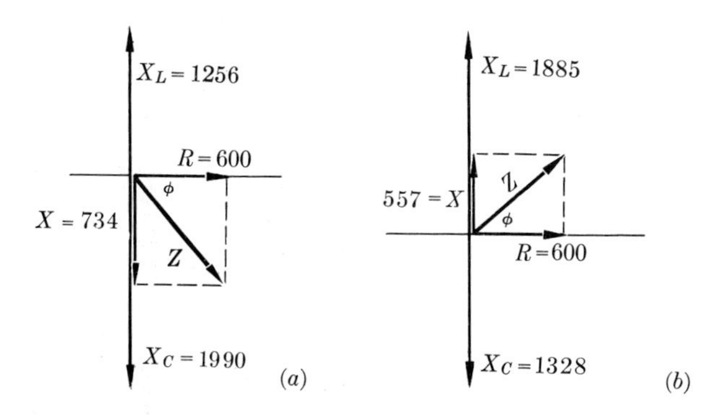

FIG. 16-4. Vector impedance diagrams for Example in Sec. 16-2. (a) $f = 400$ cycles. (b) $f = 600$ cycles.

(b) At 600 cycles

$$X_L = 2\pi \times 600 \times 0.5 = 1885 \text{ ohms},$$

$$X_C = \frac{1}{2\pi \times 600 \times 0.2 \times 10^{-6}} = 1328 \text{ ohms},$$

$$X = X_L - X_C = 1885 - 1328 = 557 \text{ ohms}.$$

$$Z = \sqrt{(600)^2 + (557)^2} = 818 \text{ ohms}. \quad \text{(See Fig. 16-4(b).)}$$

16-3 Phase relations between voltage and current. The effective alternating current in a series circuit equals the ratio of the effective terminal voltage to the impedance of the circuit. We now consider the *phase* of the current in relation to that of the voltage. Eqs. (16-1), (16-2) and (16-4) will be rewritten for convenience.

$$v = V_m \sin 2\pi ft,$$

$$i = I_m \sin (2\pi ft - \phi),$$

$$\phi = \tan^{-1} \frac{X}{R} \cdot$$

The product $2\pi ft$ represents an angle (in radians) and its magnitude at any time t is called the *phase angle* or simply the *phase* of the voltage. Similarly, the quantity $(2\pi ft - \phi)$ is the phase angle or the phase of the current. The current is said to *differ in phase* from the voltage, or to be *out of phase* with the voltage, by an angle ϕ. Since the reactance X may be either positive or negative, the same is true of the angle ϕ. If $X_L > X_C$, then X is positive, ϕ is positive and the current maxima, minima, etc., occur at later times than do those of the voltage. The current is said to *lag* the voltage. On the other hand, if $X_L < X_C$, X is negative, ϕ is negative also, and the current *leads* the voltage.

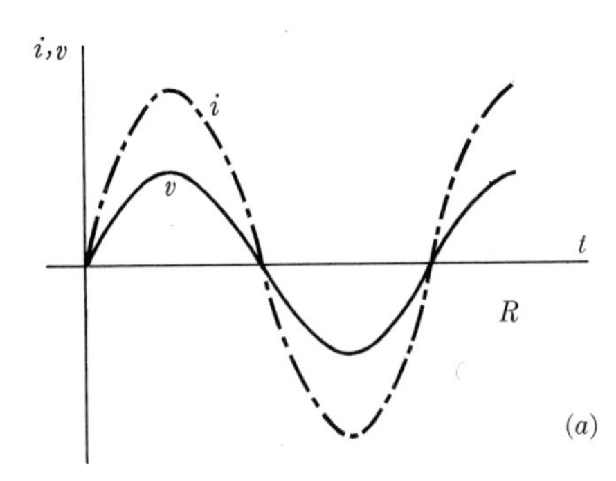

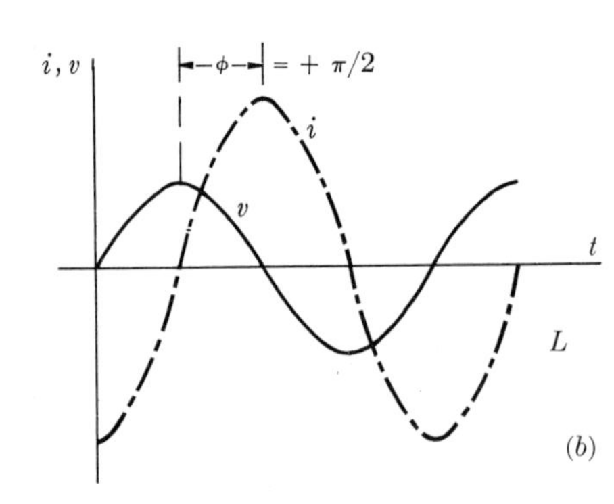

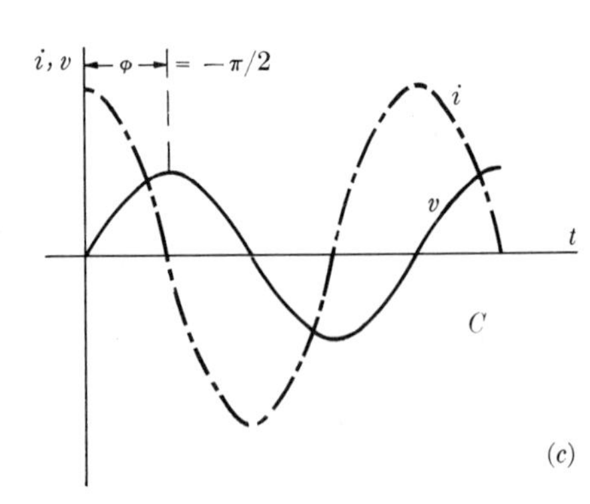

FIG. 16-5. Phase relations between potential difference and current in special circuits. (a) Resistance only. (b) Inductance only. (c) Capacitance only.

The angle ϕ can be found at once from the vector impedance diagram, since $\tan\ \phi = X/R$. See Figs. 16-4(a) and 16-4(b).

As special cases it is evident that if a circuit consists of a pure resistance connected to an alternating potential difference

$$X = 0, \quad Z = R, \quad \phi = 0,$$

and the current and voltage are *in phase*, as in Fig. 16-5(a).

If the circuit consists of a pure inductance,

$$R = 0, \quad Z = X_L, \quad \phi = +\frac{\pi}{2},$$

and the current lags the voltage by $\pi/2$ radians or 90°, as in Fig. 16-5(b).

If the circuit contains capacitance only,

$$R = 0, \quad Z = X_C, \quad \phi = -\frac{\pi}{2},$$

and the current leads the voltage by 90°, as in Fig. 16-5(c).

The *current* in all parts of a series circuit is in the same phase. That is, it is a maximum in the resistor, the inductor, and the capacitor at the same instant, zero in all three at a later instant, a maximum in the opposite direction at a still later instant, and so on.

16-4 Potential difference between points of an A.C. circuit. The rms potential difference between any two points of a series circuit equals the product of the current and impedance of the circuit between the two points, provided there is no seat of emf in the path between the points.

$$V_{ab} = IZ_{ab}.$$

The phase angle ϕ between V_{ab} and I is

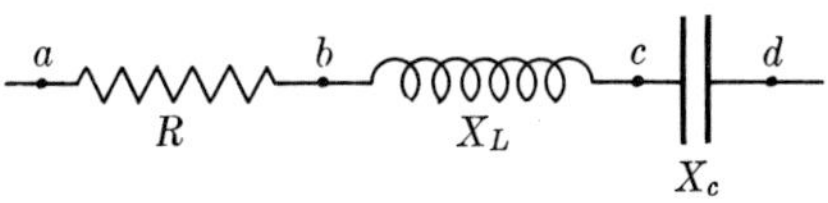

$$\phi = \tan^{-1}\frac{X_{ab}}{R_{ab}}.$$

FIG. 16-6. The potential difference across the circuit V_{ad} is *not* equal to the arithmetic sum of the potential differences V_{ab}, V_{bc}, V_{cd}.

In Fig. 16-6, the impedance Z_{ab} between a and b equals R, since there are no other circuit elements between these points. Hence, $V_{ab} = IR$ and $\phi = \tan^{-1} 0 = 0$. That is, *the potential difference between the terminals of a pure resistance is in phase with the current in the resistance.* Between points b and c, $Z_{bc} = X_L$, $V_{bc} = IX_L$, $\phi = \tan^{-1} \infty = +\pi/2$. *The potential difference between the terminals of a pure inductance leads the current in the inductance by* 90°. Between points c and d, $Z_{cd} = X_c$, $V_{cd} = I\,X_c$, $\phi = \tan^{-1} - \infty = -\pi/2$. *The potential difference between the terminals of a capacitor lags the current in the capacitor by* 90°.

As a numerical example, if the current I in Fig. 16-6 is 5^{a}, $R = 8^{\omega}$, $X_L = 6^{\omega}$, $X_C = 12^{\omega}$, we have

$$V_{ab} = IR = 40^{v}, \; v_{ab} \text{ and } i \text{ in phase},$$

$$V_{bc} = IX_L = 30^{v}, \; v_{bc} \text{ leads } i \text{ by } 90°,$$

$$V_{cd} = IX_C = 60^{v}, \; v_{cd} \text{ lags } i \text{ by } 90°.$$

The impedance of the entire circuit, Z_{ad}, is

$$\sqrt{8^2 + (6 - 12)^2} = 10^{\omega},$$

and the rms potential dfference across the circuit, V_{ad}, is

$$V_{ad} = IZ_{ad} = 5 \times 10 = 50 \text{ volts},$$

although

$$V_{ab} + V_{bc} + V_{cd} = 130^{v}.$$

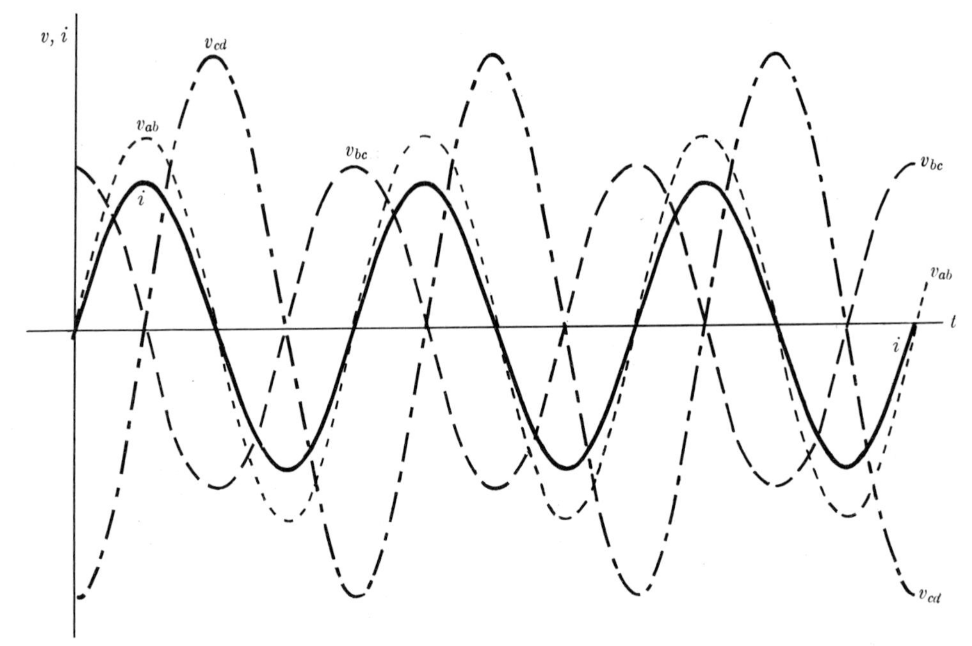

FIG. 16-7. Phase relations between v_{ab}, v_{ac}, v_{cd}, and i for the circuit of Fig. 16-6.

This example illustrates one of the unexpected situations which arise in alternating current circuits, namely, that the sum of the (rms) potential differences across a number of circuit elements in series does *not* equal the (rms) potential difference between the terminals of the group as a whole. This anomaly is readily explained, however, when the phase relations between the individual potential differences are taken into account.

Fig. 16-7 should be studied carefully. The full line represents the instantaneous current, the same at each instant in each part of the circuit, and having a maximum value of $5\sqrt{2}$ amperes. The other three curves represent instantaneous potential differences between a and b, b and c, and c and d, having maximum values of $40\sqrt{2}$, $30\sqrt{2}$, and $60\sqrt{2}$ volts respectively, and bearing the phase relations to i as shown.

The *instantaneous* potential difference v_{ad} is equal to the sum of the instantaneous potential differences v_{ab}, v_{bc}, and v_{cd}. If the three dotted curves are added, the curve obtained will therefore represent the instantaneous potential difference between a and d. This curve is shown in Fig. 16-8. Its maximum value is $50\sqrt{2}$ volts, and its effective value 50 volts, as it should be. It lags the current by 37°, which agrees with the value of ϕ as com-

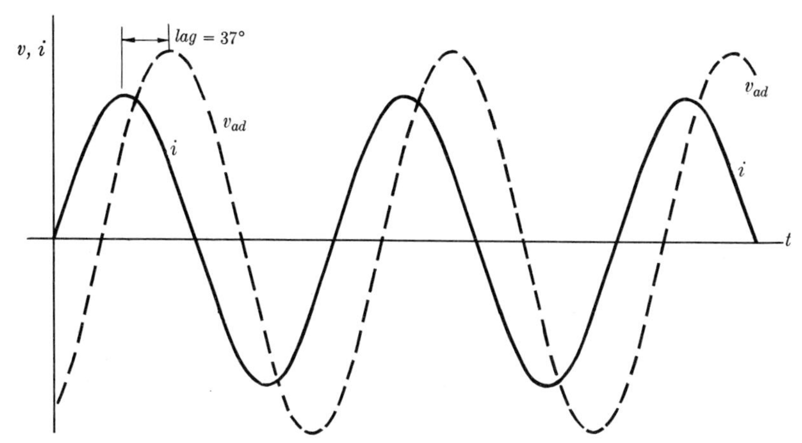

FIG. 16-8. Instantaneous potential difference v_{ad} and current i for the circuit of Fig. 16-6.

puted from

$$\tan \phi = \frac{X_L - X_C}{R} = \frac{6 - 12}{8} = -0.75, \text{ whence } \phi = -37^\circ.$$

16-5 Rotating vector diagrams. The obvious difficulty of drawing and interpreting diagrams such as Fig. 16-7 and 16-8 makes it desirable to introduce some simpler graphical method of representing the phase relations between currents and potential differences in an A.C circuit. This is done as follows. Suppose we wish to represent a sinusoidally varying current of frequency f and maximum value I_m. Construct a vector I_m as in Fig. 16-9 to some convenient scale, and imagine this vector to be rotating counterclockwise about point O at a uniform angular velocity $\omega = 2\pi f$ rad/sec. If the vector is horizontal at time $t = 0$, its projection or component along a vertical axis at any time t will equal the instantaneous current at that time, since from Fig. 16-9 the vertical component is

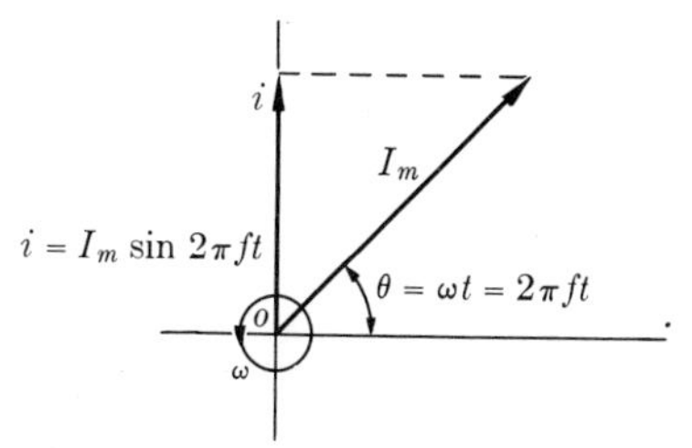

FIG. 16-9. Representation of an alternating current by means of a rotating vector.

$$I_m \sin \theta = I_m \sin 2\pi ft = i.$$

Of course the diagram shows only the value of i at some one instant. The reader must supply the rotation mentally, and follow the fluctuations of i as I_m rotates. It will be rec-

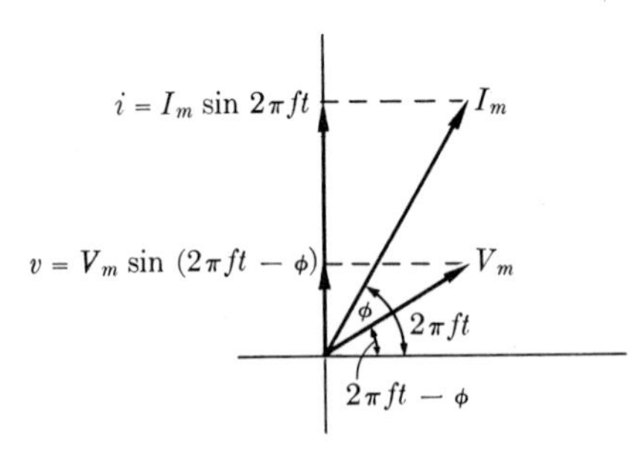

FIG. 16-10. Vector representation of an alternating potential difference and an alternating current.

ognized that this method is equivalent to representing simple harmonic motion by the projection onto a diameter of a point moving in a circle with constant angular velocity.

If we wish to show in the same diagram the instantaneous values of a potential difference $v = V_m \sin (2\pi ft - \phi)$, having the same frequency as the current but lagging the current by an angle ϕ, then a second vector of length V_m is constructed as in Fig. 16-10 to some convenient scale, but displaced clockwise from I_m by an angle ϕ. When both vectors are set into counterclockwise rotation, it is evident that the corresponding instantaneous values of v all occur at a phase angle ϕ later than do those of i.

Taking the circuit of Fig. 16-6 again as an example, let us represent by this rotating vector method the current and the potential differences across parts of the circuit. A single vector I_m, Fig. 16-11, will suffice for the current, since this has the same magnitude and is in the same phase in all parts of the circuit. The potental difference across the resistor is represented by the vector $V_{m_{ab}}$, having the same direction as I_m since v_{ab} and i are in phase, and of length $40\sqrt{2}$ volts. Similarly, vectors $V_{m_{bc}}$ of length $30\sqrt{2}$ volts and $V_{m_{cd}}$ of length $60\sqrt{2}$ volts represent the leading potential difference across the inductor and the lagging potential difference across the capacitor. The components of each of these vectors, and the rotation of the entire diagram, may be supplied mentally by the reader. As the vectors rotate, their vertical components trace out the same variations as do the ordinates of the four curves in Fig. 16-7.

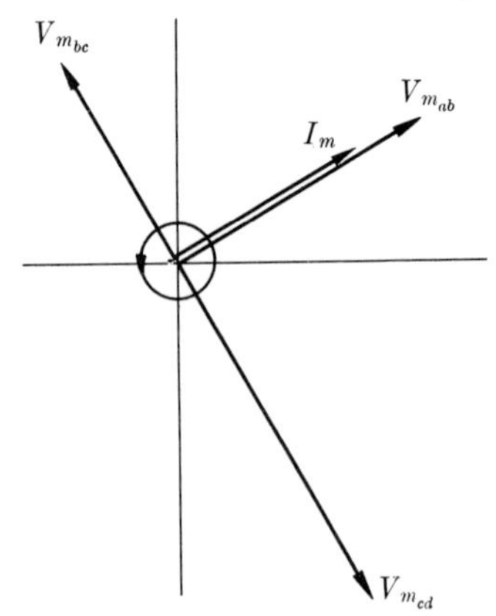

FIG. 16-11. Vector representation of i, V_{ab}, V_{bc}, V_{cd} for the circuit in Fig. 16-6.

What about the potential difference between a and d, Fig. 16-6? The instantaneous value of this potential difference was found in Fig. 16-8 by summing the ordinates of the three dotted curves of Fig. 16-7. Using the rotating vector diagram, it is to be found by summing the vertical components of $V_{m_{ab}}$, $V_{m_{bc}}$, and

$V_{m_{cd}}$. But the sum of the vertical components of these vectors is equal to the vertical component of their vector sum or resultant. Hence if this resultant is found by any convenient method as in Fig. 16-12, its vertical component represents the instantaneous potential difference across the circuit, and the angle ϕ which it makes with I_m represents the phase angle of the circuit as a whole. The resultant itself therefore represents the maximum potental difference between a and d, or $V_{m_{ad}}$.

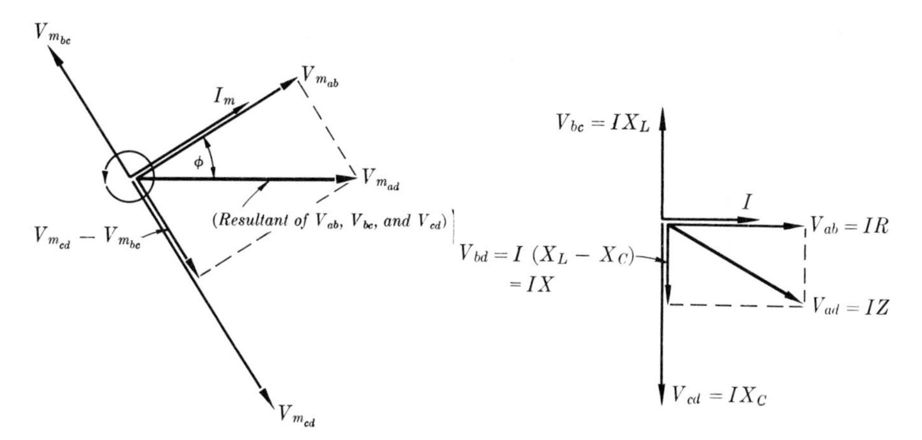

FIG. 16-12. Vector representation of V_{ad} and i for the circuit in Fig. 16-6.

FIG. 16-13. Vector diagram for the circuit in Fig. 16-6, where the vectors represent rms values.

Although the rotating vector diagram is essentially a method of representing instantaneous values, in practice it nearly always suffices to deal with effective values and phase angles only. If then we imagine the scale of Fig. 16-11 or 16-12 to be altered by a factor of $\sqrt{2}$, the same vectors which represent maximum values will, on the new scale, represent effective values, the phase angle ϕ remaining unchanged. Therefore it is standard practice to construct these diagrams with the vectors representing rms, rather than maximum values. If instantaneous values are desired, one has only to imagine all vectors increased in length by a factor $\sqrt{2}$, and the diagram set in rotation.

The particular orientation in which a vector diagram is drawn is quite arbitrary. In a series circuit, one usually begins by drawing the current vector at any convenient angle and constructing the other vectors at the proper relative orientation. The x- and y-axis are usually omitted.

Fig. 16-13 is the same as Fig. 16-12 except that the current vector is horizontal and the scale has been altered so that the vectors represent rms values. Since $V_{ab} = IR$, $V_{bc} = IX_L$, $V_{cd} = IX_C$, and $V_{ad} = IZ$, the voltage vectors are related in exactly the same way as the vectors in a vector impedance diagram such as that of Fig. 16-3. In fact, the vector voltage diagram of a series circuit can be obtained from its vector impedance diagram by multiplication of each resistance, reactance, or impedance vector by the current.

16-6 Circuits in parallel. Consider the parallel grouping of circuit elements illustrated in Fig. 16-14(a). The instantaneous potential difference across both branches is the same, namely, v_{ab}. The current i_1 in the upper (capacitive) branch will lead v_{ab}; that in the lower (inductive) branch will lag. From Kirchhoff's point rule, the *instantaneous* line current i_l equals the sum of the *instantaneous* currents i_1 and i_2. The vector diagram of the circuit is shown in Fig. 16-14(b). Vector V_{ab} represents the common potential difference across each branch, vectors I_1 and I_2 the two currents. I_1 leads V by an angle $\phi_1 = \tan^{-1} X_1/R_1$, I_2 lags V by an angle $\phi_2 = \tan^{-1} X_2/R_2$. The line current is represented by I_l, the vector sum of I_1 and I_2, and ϕ is the phase angle between the line current and V.

The *equivalent impedance* of the circuit is defined as the ratio of V_{ab} to the line current I_l. It is evident that if V_{ab} and the circuit constants are given, the calculation of I_l involves only the calculation of the individual currents in the two branches, their phase angles, and a little trigonometry. Hence the equivalent impedance can be expressed in terms of the resistances and reactances of the branches. The resulting expression is rather formidable and will not be given, although it can readily be derived by reference to Fig. 16-14(b).

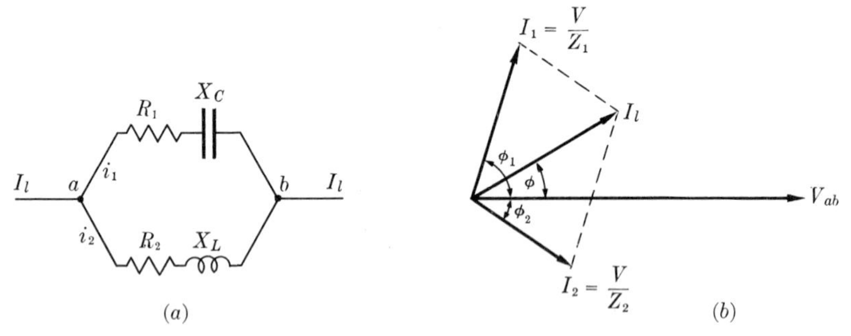

FIG. 16-14. (a) Parallel grouping of circuit elements. (b) Corresponding vector diagram.

16-7 Resonance. An important special case arises in a series circuit containing both inductance and capacitance, whenever L, C, and f have such values that $2\pi fL = 1/2\pi fC$, that is, when $X_L = X_C$. Then $X = 0$,

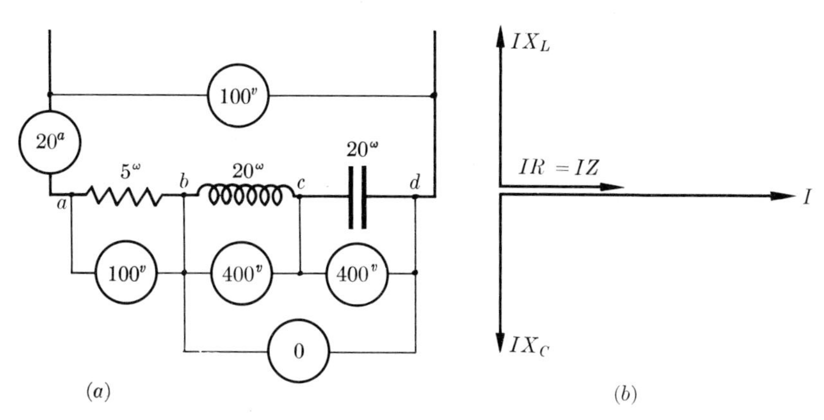

FIG. 16-15. (a) Series resonant circuit. (b) Corresponding vector diagram.

$Z = R$, and $\phi = \tan^{-1} X/R = 0$. The impedance of the circuit is equal simply to its resistance, and the current is in phase with the potential difference across the circuit. The current drawn by a series resonant circuit, if its resistance is small, will be correspondingly large, and the potential differences across the inductor and capacitor may be much larger than that across the entire circuit.

As an example, consider a series circuit in which $R = 5$ ohms, $X_L = 20$ ohms, $X_C = 20$ ohms, connected to an alternating potential difference whose rms value is 100 volts [(Fig. 16-15(a)]. The current in the circuit is

$$I = \frac{V}{Z} = \frac{V}{R} = \frac{100}{5} = 20 \text{ amp.}$$

The potential difference across the resistor is

$$V = IR = 20 \times 5 = 100 \text{ volts.}$$

The potential difference across the inductor is

$$V = IX_L = 20 \times 20 = 400 \text{ volts.}$$

The potential difference across the capacitor is

$$V = IX_C = 20 \times 20 = 400 \text{ volts.}$$

The potential difference across the inductor-capacitor combination (V_{bd}) is

$$V = IX = 0.$$

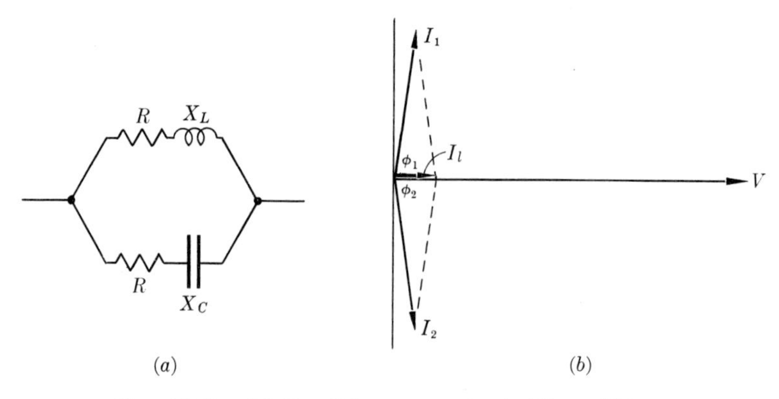

FIG. 16-16. (a) Parallel resonant circuit ($X_L = X_C$). (b) Corresponding vector diagram.

The rotating vector diagram for a series resonant circuit is given in Fig. 16-15(b). It will be seen that the instantaneous potential differences across the inductor and capacitor are 180° out of phase, and while the effective values of each may be large, their resultant at each instant is zero. Hence a voltmeter across b and d, Fig. 16-15(a), reads zero.

Fig. 16-16(a) shows a parallel circuit in which $X_L = X_C$, R being the same in each branch. The rms current is therefore the same in each branch, one of these currents lagging the potential difference and the other leading it by the same angle. If X_L and X_C are very much greater than R, the phase angles are nearly 90° and as is evident from Fig. 16-16(b), the line current I_l is much smaller than the current in either branch. In other words, the equivalent impedance of the circuit as a whole, V/I_l, is very high, much greater than that of either of the parallel branches. Peculiarly enough, the smaller the resistance, the greater the impedance, since as R approaches zero the phase angles approach 90° and the line current approaches zero. The circuit is said to be in *parallel resonance*.

Resonance in power lines is to be avoided, since unduly high currents and potential differences in parts of the circuit will result. In radio circuits, on the other hand, advantage is taken of resonance in the "tuning" process. The antenna circuit of a radio receiving set contains an inductor and a capacitor in series. Broadcasting stations within the range of the set induce in this circuit emf's of frequencies equal to the carrier frequencies of the stations. When the tuning capacitor is adjusted so that the circuit is in resonance with the frequency of any desired station, the current corresponding to that particular frequency is large, and a large potential difference appears across the capacitor. But since at any one capacitor setting the circuit is in resonance for one frequency only, other stations will develop only negligibly small currents and potential differences and hence will not be heard.

The necessary condition for resonance is that

$$2\pi fL = \frac{1}{2\pi fC}$$

from which we find that

$$f = \frac{1}{2\pi}\sqrt{\frac{1}{LC}} = \frac{1}{2\pi}\sqrt{\frac{1/C}{L}}. \qquad (16\text{-}6)$$

It will be recognized that this equation is of the same form as that for the frequency of a mechanical oscillating system, $f = \frac{1}{2\pi}\sqrt{\frac{K}{m}}$. Inductance corresponds to mass or inertia, and capacitance corresponds to the reciprocal of the force constant. In fact, the frequency f given by Eq. (16-6) may be considered to be the "natural" frequency of the electrical circuit (see Sec. 17-1), and the circuit resonates when the driving frequency is equal to this value.

16-8 Power in A.C. circuits. The instantaneous rate at which energy is supplied to an electrical device in which there is an alternating current is equal to the product of the instantaneous potential difference between the terminals of the device and the instantaneous current. The instantaneous power P fluctuates as shown in Fig. 16-17, where the power curve is obtained by multiplying together the curves representing v and i.

$$P = vi \text{ (instantaneous values)}.$$

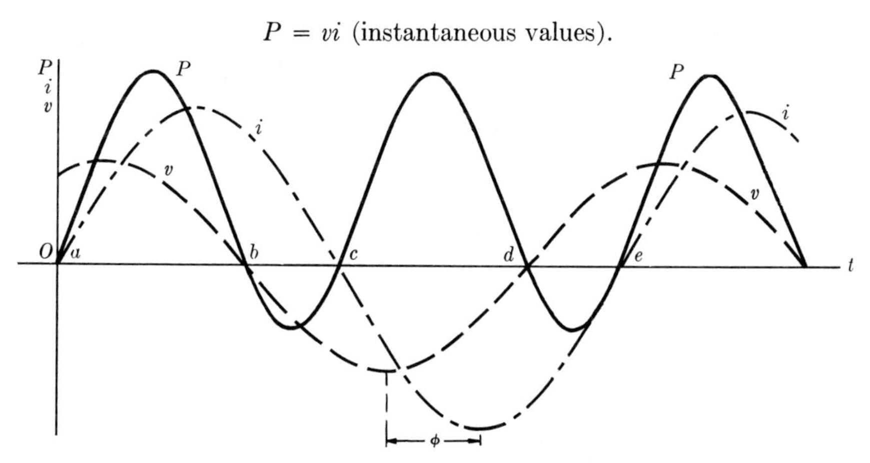

FIG. 16-17. The instantaneous power P is the product of the instantaneous potential difference and the instantaneous current.

It has been shown that energy is supplied to an electrical device when the direction of the current in the device is from the high to the low potential terminal. If the direction of the current is from the low to the high potential terminal the device is supplying energy to the circuit. Between points a and b of Fig. 16-17, where the v and i curves are both positive, the former is the case and energy is supplied to the device. Between points b and c, the potential difference is reversed, the direction of the current remaining unchanged. Hence during this interval the device returns energy to the circuit. Between c and d, both v and i are reversed and energy is again supplied to the device. Where the power curve is positive, then, as between a and b, or c and d, energy is supplied to the device (at a rate equal to the ordinate of the power curve) and where the power curve is negative, as between b and c, energy is returned to the circuit.

The total amount of energy supplied in a time t is represented graphically by the *net* area under the power curve during that time, or analytically by

$$W = \int_0^t P\,dt.$$

The average power equals the total energy supplied divided by the time, or

$$\frac{W}{t} = P_{\text{av}} = \frac{1}{t}\int_0^t P\,dt.$$

When one speaks of "the" power supplied to a device in an A.C. circuit, the average power is meant. The rms power has no significance. We shall therefore drop the subscript from P_{av}.

In general, v and i are out of phase by an angle ϕ. That is,

$$v = V_m \sin \omega t,$$
$$i = I_m \sin (\omega t - \phi).$$

Hence in a time interval of one period,

$$P = \frac{1}{T}\int_0^T V_m I_m \sin \omega t \sin (\omega t - \phi) dt$$
$$= \frac{V_m I_m}{2} \cos \phi, \quad \text{or}$$

$$\boxed{P = VI \cos \phi.} \tag{16-7}$$

That is, the average power supplied to a device in an A.C. circuit equals the product of the effective potential difference times the effective current times the cosine of the angle of lag or lead. The quantity $\cos\phi$ is called the *power factor* of the device (abbreviated p.f.). Depending on the nature of the device, the power factor can have any value between zero (when $\phi = 90°$) and unity (when $\phi = 0°$).

A power factor of zero means that the device consists of a pure reactance, inductive or capacitive. From Eq. (16-7), the average power supplied to such a device is zero, which is evidently correct, since the energy input to a capacitor or inductor goes into building up an electric or magnetic field, and all of this energy is recovered when the field later decreases to zero.[1] During those parts of the cycle when the field is decreasing, the reactor returns energy to the circuit and helps run the generator.

If the circuit contains both resistance and reactance (and no mechanical device such as a motor), then

$$\tan\phi = \frac{X}{R}, \quad \cos\phi = \frac{R}{Z}$$

and Eq. (16-7) becomes

$$P = \frac{VIR}{Z}$$

and since

$$\frac{V}{Z} = I,$$

$$P = I^2R \tag{16-8}$$

as in the D.C. case. Of course if the circuit contains a motor, $\tan\phi$ is not equal to X/R, and Eq. (16-8) does not apply. Eq. (16-7), however, is true for any circuit.

A low power factor (large angle of lag or lead) is undesirable in power circuits because, for a given potential difference, a large current is needed to supply a given amount of power with correspondingly large heat losses in the transmission lines. Since many types of A.C. machinery draw a lagging current, this situation is likely to arise. It can be corrected by connecting a capacitor in parallel with the load. The leading current drawn by the capacitor compensates for the lagging current in the other branch of the circuit. The capacitor itself takes no net power from the line.

[1] Exceptions: (a) If hysteresis is present, all of the energy supplied to the circuit is not recovered. (b) At high frequencies energy is *radiated* from the circuit as electromagnetic waves.

A synchronous motor with its field "over-excited" also draws a leading current and is sometimes used for power factor correction.

The power supplied to (or by) a device in a D.C. circuit may be measured either by an ammeter-voltmeter or by a wattmeter (see page 138). In an A.C. circuit the power cannot be measured by an ammeter and voltmeter since each indicates rms values only and takes no account of phase relations or power factor. A dynamometer type wattmeter, however, automatically includes the correction for power factor and indicates the true power supplied. The same type of wattmeter is used both for D.C. and A.C. measurements.

The instantaneous deflecting torque τ on a dynamometer coil is proportional to the products of the instantaneous currents in the moving and fixed coils. When connected as in a wattmeter, one of these currents is proportional to the line current and the other to the potential difference. That is,

$$\tau = kvi.$$

where k is a proportionality constant. But

$$v = V_m \sin \omega t,$$
$$i = I_m \sin (\omega t - \phi).$$

Hence

$$\tau = kI_mV_m \sin \omega t \sin (\omega t - \phi).$$

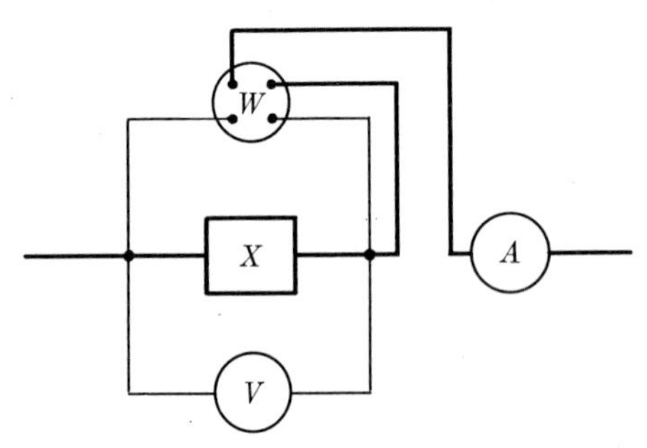

FIG. 16-18. Measurement of the power factor of a circuit by use of A.C. ammeter, voltmeter, and wattmeter simultaneously.

With alternating current, the deflection of the wattmeter is proportional to the *average* torque, or to

$$\bar{\tau} = \frac{k}{T}\int_0^T I_mV_m \sin \omega t \sin (\omega t - \phi)dt$$

and from Eq. (16-7) on page 378

$$\bar{\tau} = kVI \cos \phi = kP$$

and the deflection is proportional to the average power.

If an A.C. ammeter, voltmeter, and a wattmeter are simultaneously connected as in Fig. 16-18, V, I, and P may all be measured independently. This is an experimental method by which the power factor of a device may be measured.

16-9 The transformer. For reasons of efficiency it is desirable to transmit electrical power at high voltages and small currents, with consequent reduction of I^2R heating in the transmission line. On the other hand, considerations of safety and of insulation of moving parts require relatively low voltages in generating equipment and in motors and household appliances. One of the most useful features of A.C. circuits is the ease and efficiency with which voltages (and currents) may be changed from one value to another by means of transformers.

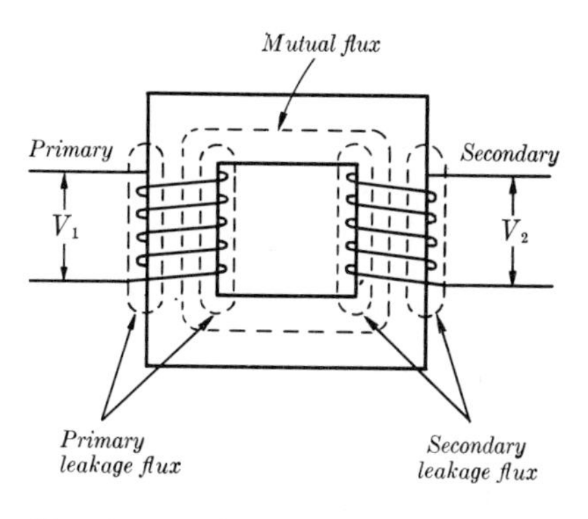

FIG. 16-19. An iron core transformer.

In principle, the transformer consists of two coils electrically insulated from one another and wound on the same iron core. (Fig. 16-19.) An alternating current in one winding sets up an alternating magnetic flux in the core. Most of this flux links with the other winding and induces in it an alternating emf. Power is thus transferred from one winding to the other via the flux in the core. The winding to which power is supplied is called the *primary*, that from which power is delivered is called the *secondary*. Either winding may be used as the primary. The symbol for an iron core transformer is

In any actual transformer the flux lines are not confined entirely to the iron but some of them return through the air as indicated in Fig. 16-19. That part of the flux which links both the primary and secondary windings is called the *mutual* flux. The part linking the primary only is the *primary leakage flux* and the part linking the secondary only is the *secondary leakage flux*.

The power output of a transformer is necessarily less than the power input because of unavoidable losses in the form of heat. These losses consist of I^2R heating in the primary and secondary windings (the copper losses) and hysteresis and eddy current heating in the core (the core losses). Hysteresis is minimized by the use of iron having a narrow hysteresis loop, and eddy currents by laminating the core. In spite of these losses, transformer efficiencies are usually well over 90% and in large installations may reach 99%.

For simplicity we shall first consider an idealized transformer in which there are no losses and no leakage flux. Let the secondary circuit be open. The primary winding then functions merely as an inductor. The primary current, which is small, lags the primary voltage by 90° and is called the *magnetizing* current, I_m. The power input to the transformer is zero. The core flux is in phase with the primary current. Since the same flux links both primary and secondary, the induced emf *per turn* is the same in each. The ratio of primary to secondary induced emf is therefore equal to the ratio of primary to secondary turns, or

$$\frac{\mathcal{E}_2}{\mathcal{E}_1} = \frac{N_2}{N_1}.$$

In the idealized case assumed, the induced emf's $\mathcal{E}_1$ and $\mathcal{E}_2$ are numerically equal to the corresponding terminal voltages V_1 and V_2. Hence by properly choosing the turn ratio N_2/N_1, any desired secondary voltage may be obtained from a given primary voltage. If $V_2 > V_1$, we have a *step-up* transformer; if $V_2 < V_1$, a *step-down* transformer.

The vector diagram of the idealized transformer is given in Fig. 16-20 for a turn ratio of $N_2/N_1 = 2$. The induced emf's in both primary and secondary, since they are proportional to the negative rate of change of flux, will lag 90° behind the flux, but since the induced emf in the primary ($\mathcal{E}_1$) is a back emf, the primary terminal voltage (V_1) is opposite to it in phase. ($V_1 = -\mathcal{E}_1$.)

Consider next the effect of closing the secondary circuit. The secondary current I_2 (Fig. 16-21) and its phase angle ϕ_2 will, of course, depend on the nature of the secondary circuit. It has been assumed in Fig. 16-21 that the load is inductive and hence I_2 lags V_2. As soon as the secondary circuit is closed, some power must be delivered by the secondary (except

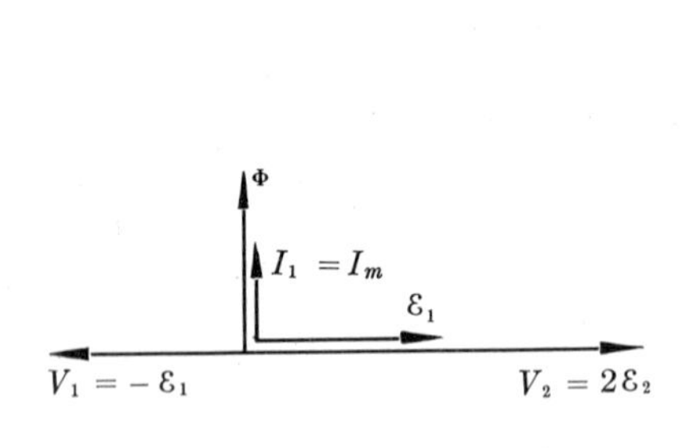

FIG. 16-20. Vector diagram of an idealized transformer under no load. The secondary current is zero, and the primary current is the magnetizing current only.

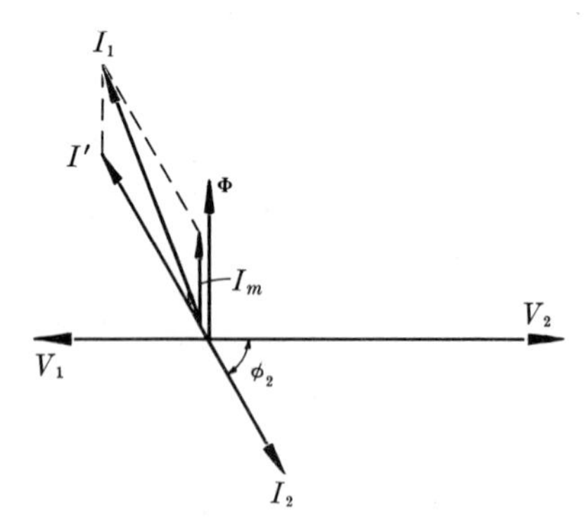

FIG. 16-21. Vector diagram of an idealized transformer under an inductive load.

when $\phi_2 = 90°$) and from energy considerations an equal amount of power must be supplied to the primary. The process by which the transformer is enabled to draw the requisite amount of power is as follows. When the secondary circuit is open, the core flux is produced by the primary current only. But when the secondary circuit is closed, both primary and secondary currents set up a flux in the core. The secondary current, by Lenz's law, tends to weaken the core flux and therefore to decrease the back emf in the primary. But (in the absence of losses) the back emf in the primary must equal the primary terminal voltage which is assumed to be fixed. The primary current therefore increases until the core flux is restored to its original no-load magnitude. The vector I_1' in Fig. 16-21 represents the *change* in primary current that takes place when the secondary delivers the current I_2. It is opposite in phase to the secondary current I_2 and of such magnitude that its magnetomotive force (N_1I_1') is equal and opposite to the magnetomotive force of the secondary current (N_2I_2). That is,

$$N_2I_2 = N_1I_1', \text{ or}$$

$$\frac{I_2}{I_1'} = \frac{N_1}{N_2} \tag{16-9}$$

The resultant primary current, I_1, is the vector sum of I_1' and the magnetizing current I_m. But in practice the magnetizing current is never more than a few percent of the full load current. Hence I_1 and I_1' are practically equal and one may write approximately, from Eq. (16-9)

$$\frac{I_2}{I_1} = \frac{N_1}{N_2}. \tag{16-10}$$

That is, the primary and secondary currents are *inversely* proportional to the primary and secondary turns.

The effect of leakage flux and of resistance of the windings requires that the primary terminal voltage shall be somewhat larger than the primary induced emf and not exactly 180° out of phase with it. Similarly, the secondary terminal voltage is somewhat smaller than, and out of phase with, the secondary induced emf. A text on electrical engineering should be consulted for further details.

16-10 Three-phase alternating current. We shall conclude this chapter with a brief discussion of three-phase alternating current. A coil of wire rotating with constant angular velocity in a uniform magnetic field develops a sinusoidal alternating emf. Suppose that three coils are wound

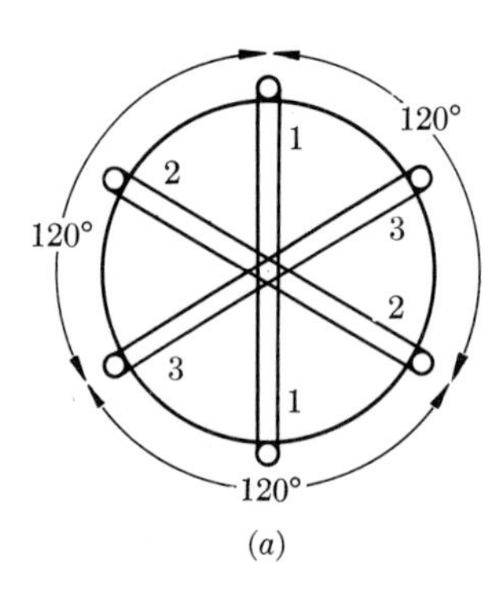

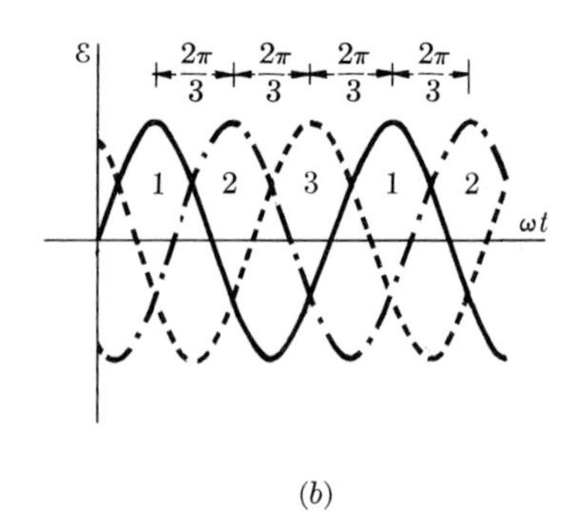

FIG. 16-22. (a) Simple three-phase alternator. (b) Phase relations among the three corresponding emf's.

on an armature as in Fig. 16-22(a), with their planes displaced from one another by 120° or $2\pi/3$ radians. If the armature rotates in a uniform field with constant angular velocity the emf's developed in the three coils are related in phase as in Fig. 16-22(b). The three sinusoidal emf's are numbered to correspond to the three coils in Fig. 16-22(a). The emf in coil 2 lags that in coil 1 by $2\pi/3$ radians, and the emf in coil 3 lags that in coil 2 by the same amount.

It would be perfectly possible to provide each coil with its own set of slip rings and brushes and to utilize each coil as an independent generator supplying single phase A.C. to its own individual circuit. Instead, however, the three coils are interconnected as shown diagrammatically in Fig. 16-23. The three common points a, b, and c are connected to slip rings and the three wires of the external 3-phase circuit leads from brushes making contact with these three rings.[1]

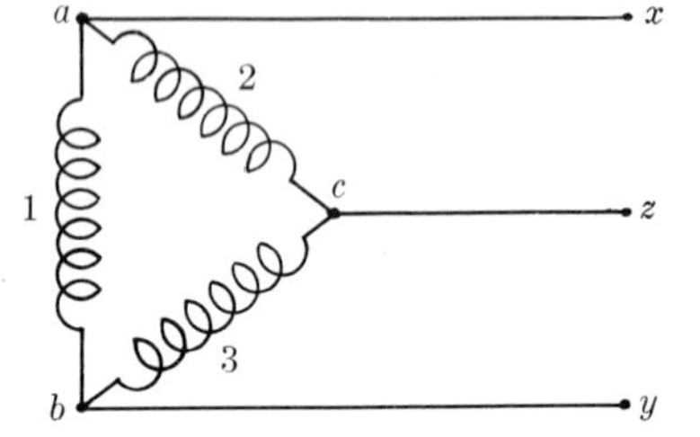

FIG. 16-23. Δ-connection of a three-phase alternator.

Any pair of terminals such as x and y, y and z, or x and z, is a source of single phase A.C. of the same voltage as is developed in any one of the coils. Consider, for example, the terminals x and y. These are connected directly across coil 1, and also across coils 2 and 3 in series. The instantaneous p.d.'s between the terminals of the three coils are equal to the instantaneous emf's. That is,

[1] This is called a Δ-connection. Another hook-up known as a Y-connection is frequently used.

$$V_{ab} = V_{xy} = e_1 = E_m \sin 2\pi ft,$$

$$V_{ca} = V_{zx} = e_2 = E_m \sin (2\pi ft - 2\pi/3),$$

$$V_{bc} = V_{yz} = e_3 = E_m \sin (2\pi ft - 4\pi/3).$$

If we consider points a and b as the terminals of coils 2 and 3 connected in series,

$$\begin{aligned} V_{ab} &= V_{ac} + V_{cb} = -(V_{ca} + V_{bc}) \\ &= -E_m [\sin (2\pi ft - 2\pi/3) + \sin (2\pi ft - 4\pi/3)] \\ &= -E_m [\sin 2\pi ft \cos 2\pi/3 - \cos 2\pi ft \sin 2\pi/3 \\ &\quad + \sin 2\pi ft \cos 4\pi/3 - \cos 2\pi ft \sin 4\pi/3]. \end{aligned}$$

But

$$\cos 2\pi/3 = \cos 4\pi/3, \quad \sin 2\pi/3 = -\sin 4\pi/3.$$

Hence

$$V_{ab} = -E_m[2 \sin 2\pi ft \cos 2\pi/3].$$

Since $\cos 2\pi/3 = -\frac{1}{2}$, this reduces to

$$V_{ab} = E_m \sin 2\pi ft,$$

which is equal to the instantaneous emf in coil 1. Hence coil 1 on the one hand, and the series connection of coils 2 and 3 on the other, are identical seats of emf in parallel and the potential difference between the terminals x and y is the same as that across a single coil. If an A.C. voltmeter across x and y reads 110 volts, it will also read 110 volts across y and z, or across x and z. The voltage across y and z of course lags that across x and y by $2\pi/3$ radians, and that across x and z lags by another $2\pi/3$ radians.

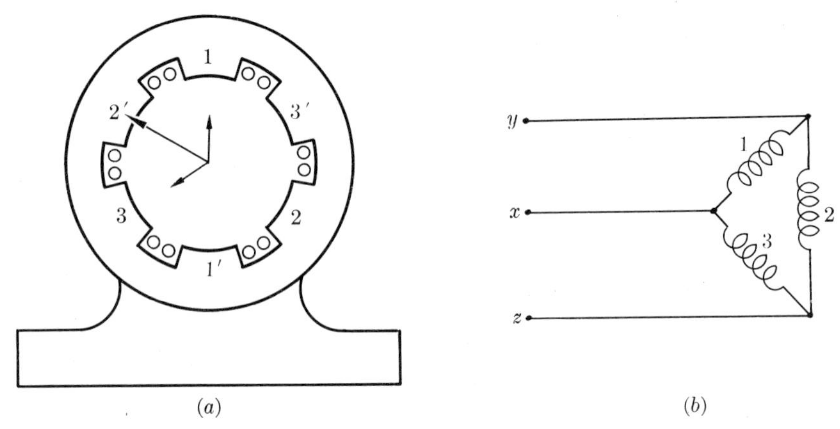

FIG. 16-24. (a) The stator of a three-phase induction motor. (b) Interconnection of the windings of the stator.

We shall describe next the principle of the 3-phase induction motor. Fig. 16-24(a) is a diagram of the stator. The windings around opposite pairs of poles such as those numbered 1 and 1′ are in series, as are the windings of poles 2 and 2′ and 3 and 3′. The windings of the three pairs of poles are interconnected as in Fig. 16-24(b) and connected to the three terminals of a 3-phase A.C. supply. The currents in the three windings are then $2\pi/3$ radians apart in phase.

The poles are so wound that at the instants when the current in any one phase is positive (i.e., above the axis in Fig. 16-22(b)) the unprimed pole in Fig. 16-24(a) is North and the primed pole is South. For example, at an instant when the current in phase 2 is a positive maximum, pole 2 is a North pole of maximum strength and pole 2′ is a South pole of maximum strength. Reference to Fig. 16-22(b) shows that at this instant the currents in phases 1 and 3 are negative. Hence at this instant poles 1 and 3 are South and poles 1′ and 3′ are North.

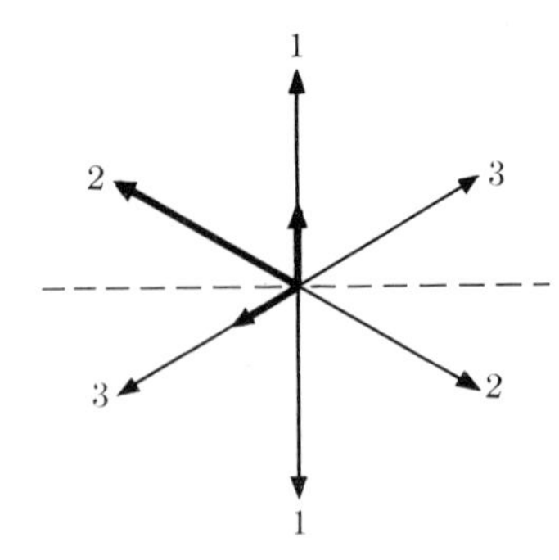

FIG. 16-25. The fields set up by the three pairs of poles of Fig. 16-24(a).

The magnetic field in the space between the poles is the resultant of three fields which differ in *phase* by $2\pi/3$ radians and which also differ in *azimuth* by $2\pi/3$ radians. Thus as shown in Fig. 16-25 the fields set up by the three pairs of poles have the directions indicated and alternate between the indicated maximum values shown by light arrows. The heavy arrows indicate the magnitudes and directions of the three components at the instant when the current in phase 2 is a positive maximum. It can be shown that the resultant field is constant in magnitude but that it rotates in the plane of the diagram with a frequency equal to that of the 3-phase supply.

The magnitude and direction of the resultant field at any instant is computed as follows. Let B_m represent the maximum magnetic flux density due to any one pair of poles, assumed the same for all three pairs. The magnitudes of the instantaneous flux densities are

$$B_1 = B_m \sin 2\pi ft,$$

$$B_2 = B_m \sin\left(2\pi ft - \frac{2\pi}{3}\right),$$

$$B_3 = B_m \sin\left(2\pi ft - \frac{4\pi}{3}\right).$$

The X-component of the resultant flux density is

$$\begin{aligned} B_x &= B_1 + B_2 \cos\frac{2\pi}{3} + B_3 \cos\frac{4\pi}{3} \\ &= B_m \sin 2\pi ft \\ &\quad + B_m \sin\left(2\pi ft - \frac{2\pi}{3}\right)\cos\frac{2\pi}{3} \\ &\quad + B_m \sin\left(2\pi ft - \frac{4\pi}{3}\right)\cos\frac{4\pi}{3} \\ &= \tfrac{3}{2} B_m \sin 2\pi ft. \end{aligned}$$

The Y-component of the resultant flux density is

$$\begin{aligned} B_y &= B_2 \sin\frac{2\pi}{3} + B_3 \sin\frac{4\pi}{3} \\ &= B_m \sin\left(2\pi ft - \frac{2\pi}{3}\right)\sin\frac{2\pi}{3} \\ &\quad + B_m \sin\left(2\pi ft - \frac{4\pi}{3}\right)\sin\frac{4\pi}{3} \\ &= -\tfrac{3}{2} B_m \cos 2\pi ft. \end{aligned}$$

The magnitude of the resultant flux density is

$$B = \sqrt{B_x^2 + B_y^2} = \tfrac{3}{2} B_m,$$

which is constant.

The angle α between the resultant flux density and the Y-axis is

$$\tan \alpha = \frac{B_x}{B_y} = -\tan 2\pi f t.$$

The resultant field therefore rotates with a frequency f equal to that of the 3-phase supply.

When a conductor moves relatively to a magnetic field, a motional emf is induced in it. According to Lenz's law, the side thrust on the induced currents is in such a direction as to oppose the relative motion. Therefore if a conductor is pivoted in the rotating magnetic field within the stator structure in Fig. 16-24(a), the side thrusts on the induced currents in it oppose the relative motion between it and the rotating field. In other words, a torque is exerted on the conductor in such a direction as to force it to rotate with the field and we have a 3-phase induction motor. While a torque is exerted on a conductor of any shape, in practice one uses a rotor having a cylindrical laminated iron core with copper conductors embedded in longitudinal slots in its surface and interconnected by copper rings at the ends of the rotor. Because of the shape of the copper portion it is called a "squirrel-cage" rotor.

One of the chief advantages of this type of motor is that there are no electrical connections to any moving part, and brushes, commutator, or slip rings can be dispensed with. Only the stator, which of course is stationary, is connected to the 3-phase supply. There are no electrical connections whatever to the rotor.

Problems—Chapter 16

(1) (a) What is the reactance of a 1-henry inductor at a frequency of 60 cycles/sec? (b) What is the inductance of an inductor whose reactance is 1 ohm at 60 cycles/sec? (c) What is the reactance of a 1 μf capacitor at a frequency of 60 cycles/sec? (d) What is the capacitance of a capacitor whose reactance is 1 ohm at 60 cycles/sec?

(2) (a) Compute the reactance of a 10-henry inductor at frequencies of 60 cycles/sec and 600 cycles/sec. (b) Compute the reactance of a 10-μf capacitor at the same frequencies. (c) At what frequency is the reactance of a 10-henry inductor equal to that of a 10-μf capacitor?

(3) An inductor of reactance 10 ohms and a capacitor of reactance 25 ohms (both measured at 60 cycles/sec) are connected in series with a 10-ohm resistor across a 100-volt (rms), 60-cycle AC line. (a) Find the voltage across each part of the circuit. (b) What are the expressions for the instantaneous line voltage and line current?

(4) In Fig. 16-26, $I = 5$ amp, $R = 8$ ohms, $X_L = 6$ ohms, $X_C = 12$ ohms. Compute V_{ab}, V_{bc}, V_{cd}, V_{ad}, and the phase angle between line current and line voltage.

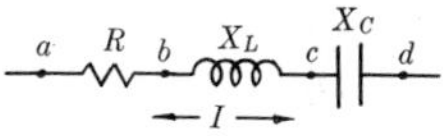

FIG. 16-26.

(5) A 400-ohm resistor is in series with a 0.1-henry inductor and a 0.5-μf capacitor. Compute the impedance of the circuit and draw the vector impedance diagram (a) at a frequency of 500 cycles/sec, and (b) at a frequency of 1000 cycles/sec. Compute in each case the phase angle between line current and line voltage, and state whether the current lags or leads.

(6) (a) A pure resistance and a pure inductance are in series across a 100-volt AC line. An AC voltmeter gives the same reading whether connected across the resistance or the inductance. What does it read? (b) The magnitudes of the resistance and inductance in part (a) are altered so that a voltmeter across the inductance reads 50 volts. What will the voltmeter read when connected across the resistance?

(7) The rms terminal voltage of an AC generator is 100 volts and the so-called angular frequency $\omega = 2\pi f$ is 500 rad/sec. In series across the generator are a 3-ohm resistor, a 50-μf capacitor, and an inductor whose inductance can be varied from 10 to 80 millihenrys. The peak voltage ($V_{\max}$) across the capacitor should not exceed 1200 volts. (a) What is the maximum allowable current in the series circuit? (b) To what value can the inductance be safely increased?

(8) A coil whose resistance is 10 ohms and whose inductance is 15 millihenrys is in series with a 12-ohm resistor and a 200-μf capacitor across a 100-volt, 60-cycle AC line. Find the voltage across the terminals of the coil.

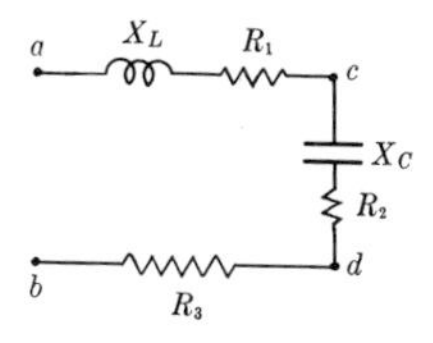

FIG. 16-27.

(9) Points a and b in Fig. 16-27 are the terminals of a 60-cycle AC line. The rms voltage between a and b is 130 volts. If $R_1 = 6$ ohms, $R_2 = R_3 = 3$ ohms, $X_C = 3$ ohms, $X_L = 8$ ohms, compute: (a) the current in the circuit; (b) the voltage between a and c; (c) the voltage between c and d.

(10) The reactances of an inductor and a capacitor are equal at a frequency f_0. Show in the same diagram the reactances of the inductor and the capacitor as a function of frequency, over a frequency range from zero to $2f_0$. Show also the reactance of the two in series.

(11) If the reactances X_L and X_C in problem 4 above are for a frequency of 60 cycles/sec, find the frequency at which the circuit would be in resonance.

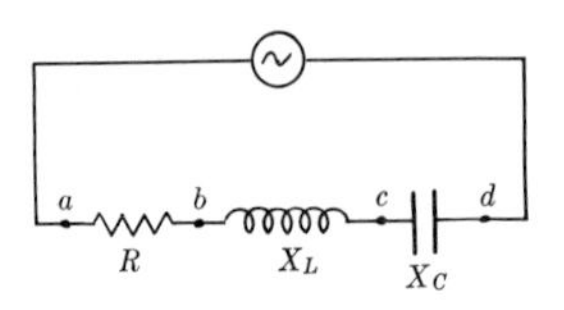

FIG. 16-28.

(12) The terminal voltage of the AC generator in Fig. 16-28 is 150 volts, and $R = 10$ ohms, $X_L = 50$ ohms, $X_C = 50$ ohms. Find the current in the circuit, and V_{ab}, V_{bc}, V_{cd}, V_{bd}, V_{ad}. Construct the rotating vector diagram of the circuit.

(13) A 2-μf capacitor, a 2-henry inductor, and a resistor of resistance R are in series across an AC generator of terminal voltage 100 volts. Compute the current in the circuit for the following values of angular frequency $\omega = 2\pi f$: 0, 200, 400, 500, 600, 800 rad/sec, (a) when $R = 100$ ohms, (b) when $R = 500$ ohms. Construct graphs of the results, plotting I vertically and ω horizontally.

(14) A $6\frac{2}{3}$-μf capacitor is connected in series with a coil to an AC supply of 1.2 volts and of controllable frequency. By adjusting the frequency it is observed that the current reaches its largest effective value of 0.2 amp when the angular frequency $\omega = 2\pi f$ is 50,000 rad/sec. (a) Find the resistance and inductance of the coil. (b) What is the current when $\omega = 150{,}000$ rad/sec? (c) What is the peak voltage across the capacitor at this frequency?

(15) A resistance R, an inductance L, and a capacitance C are in series across an AC generator of constant terminal voltage. (a) At what frequency, in terms of R, L, and C, will the voltage across the inductor be a maximum? (b) At what frequency will the voltage across the capacitor be a maximum?

(16) Compute the power supplied to the circuits in problems 3 and 4 above, using both the equations $P = I^2R$ and $P = VI \cos \theta$.

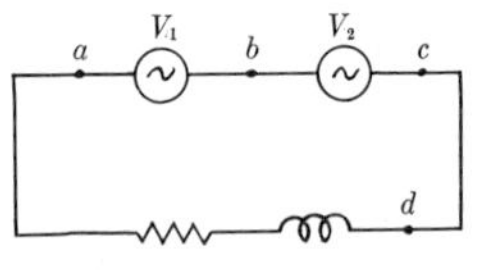

FIG. 16-29.

(17) Two AC generators of the same frequency, V_1 and V_2 in Fig. 16-29, are connected in series and supply 150 watts to a coil. An AC voltmeter reads 40 volts when connected across ab, it reads 30 volts across bc, and 50 volts across ac. An AC ammeter in the line cd reads 5 amp. (a) Show in a diagram how a wattmeter should be connected to measure the power supplied to the coil. (b) What is the phase difference between the voltages of the generators? (c) Find the resistance, reactance, impedance, and power factor of the coil.

(18) An impedance Z is connected across a 100-volt 60-cycle AC line. After a 265-μf capacitor is connected in parallel with the impedance, the current in the line is reduced to one-half its former value and is in phase with the line voltage. Compute the power supplied to the impedance.

(19) In the circuit in Fig. 16-30, $V_{ab} = 100$ volts, $f = 60$ cycles/sec, $C = 26.5\ \mu f$, $L = 0.265$ henry, $R_1 = R_2 = R_3 = 5$ ohms. (a) Find the equivalent impedance between points a and b. (b) Find the current in each resistor. (c) Construct the vector diagram of the circuit.

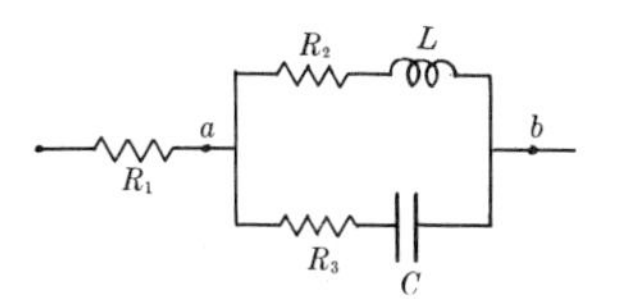

FIG. 16-30.

(20) A circuit draws 330 watts from a 110-volt 60-cycle AC line. The power factor is 0.6 and the current lags the voltage. (a) Find the capacitance of the series capacitor that will result in a power factor of unity. (b) What power will then be drawn from the supply line?

CHAPTER 17

ELECTRICAL OSCILLATIONS AND ELECTROMAGNETIC WAVES

17-1 Electrical oscillations. When the discharge of a capacitor through a resistor was discussed in Sec. 8-5, the effect of inductance in the circuit was ignored. We are now to consider how the discharge is affected by inductance, taking up first a circuit in which the resistance is negligible.

Fig. 17-1(a) represents schematically a charged capacitor, a switch, and an inductor of negligible resistance. At the instant when the circuit is closed, the capacitor starts to discharge through the inductor. At a later instant, represented in Fig. 17-1(b), the capacitor has completely discharged and the potential difference between its terminals (and those of the inductor) has decreased to zero. The current in the inductor has meanwhile established a magnetic field in the space around it. This magnetic field now decreases, inducing an emf in the inductor in the same direction as the current. The current therefore persists, although with diminishing magnitude, until the magnetic field has disappeared and the capacitor has been charged in the opposite sense to its initial polarity as in Fig. 17-1(c). The process now repeats itself in the reversed direction and in the absence of energy losses the charges on the capacitor will surge back and forth indefinitely. This process is called an *electrical oscillation.*

Points a and b in Fig. 17-1(a) can be considered as the terminals of either the capacitor or the inductor. Hence

$$V_{ab} = \frac{q}{C} = L\frac{di}{dt}.$$

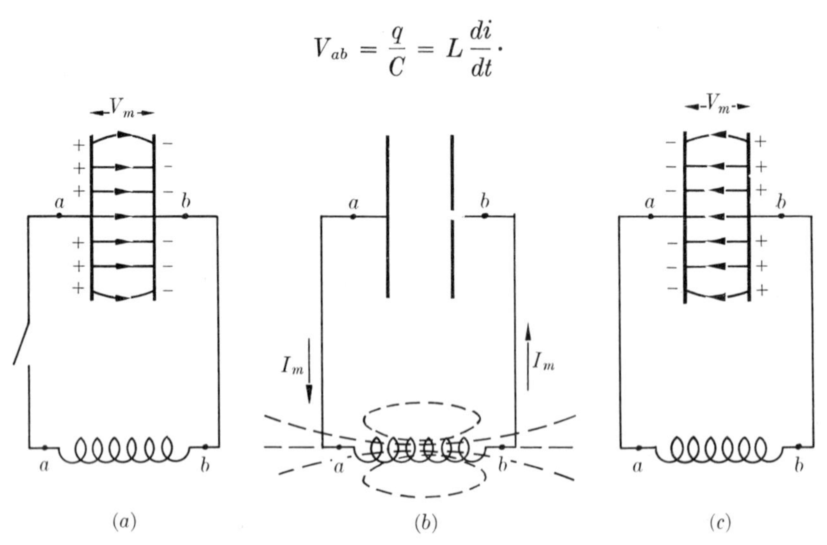

Fig. 17-1. Energy transfer between electric and magnetic fields in an oscillating circuit.

Since $i = -dq/dt$ (the minus sign enters since q is decreasing) one has

$$\frac{di}{dt} = -\frac{d^2q}{dt^2}, \text{ and hence}$$

$$\frac{d^2q}{dt^2} + \frac{1}{LC}q = 0.$$

This will be recognized as the differential equation of simple harmonic motion. Its solution is

$$q = Q \cos 2\pi ft$$

where

$$f = \frac{1}{2\pi}\sqrt{\frac{1}{LC}}$$

and Q is the initial charge on the capacitor. The charge therefore fluctuates or "oscillates" in a manner exactly analogous to the oscillation of a mass suspended from a spring. The frequency f is called the "natural frequency" of the circuit. Note that this frequency is the same as that for which the circuit is in resonance. That is, an electrical circuit, like a mechanical system, esonates when the driving frequency is equal to the natural frequency.

The current in the circuit can be found by differentiation.

$$i = -\frac{dq}{dt} = 2\pi fQ \sin 2\pi ft.$$

The coefficient $2\pi fQ$ is evidently equal to the maximum current, I_m. Hence

$$i = I_m \sin 2\pi ft. \tag{17-1}$$

The current is therefore a sinusoidal alternating one and is 90° out of phase with the charge.

From the energy standpoint, the oscillations of an electrical circuit consist of a transfer of energy back and forth from the electric field of the capacitor to the magnetic field of the inductor, the total energy associated with the circuit remaining constant. Again, this is analogous to the transfer of energy in an oscillating mechanical system from kinetic to potential and vice versa.

The energy of the capacitor at any instant is $\frac{1}{2}Cv^2$, that of the inductor is $\frac{1}{2}Li^2$, hence

$$\tfrac{1}{2}Cv^2 + \tfrac{1}{2}Li^2 = \tfrac{1}{2}CV_m^2 = \tfrac{1}{2}LI_m^2 = \text{constant},$$

where v and i are the instantaneous, V_m and I_m are the maximum potential difference and current respectively.

17-2 Damped oscillations. The analysis of the preceding section will next be extended to include the effect of resistance in the circuit. From Fig. 17-2 we have

$$\frac{q}{C} - L\frac{di}{dt} - iR = 0. \tag{17-2}$$

After replacing i by $-\frac{dq}{dt}$ and $\frac{di}{dt}$ by $-\frac{d^2q}{dt^2}$, Eq. (17-2) becomes

$$\frac{d^2q}{dt^2} + \frac{R}{L}\frac{dq}{dt} + \frac{1}{LC}q = 0. \tag{17-3}$$

This equation is of the same form as the equation of damped harmonic motion of a spring-mass system. (See Principles of Physics I, Sec. 14–8.) If the resistance is not too great ($R_2 < 4L/C$), the charge on the capacitor fluctuates with "damped harmonic motion" and

$$q = \frac{f_0}{f_1}Q\epsilon^{-\frac{Rt}{2L}}\cos(2\pi f_1 t + \phi),$$

where

$$f_0 = \frac{1}{2\pi}\sqrt{\frac{1}{LC}},$$

$$f_1 = \frac{1}{2\pi}\sqrt{\frac{1}{LC} - \frac{R^2}{4L^2}},$$

and

$$\cos\phi = \frac{f_1}{f_0}.$$

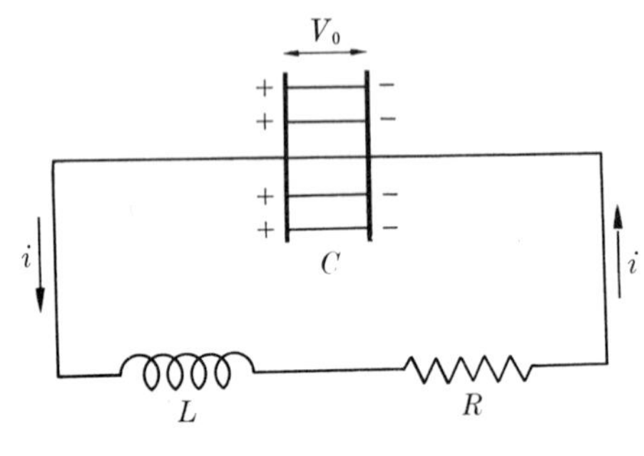

FIG. 17-2.

The frequency of oscillation, f_1, is smaller than the frequency f_0 in the corresponding undamped circuit, i.e., when $R = 0$. The amplitude is not constant but decreases with time because of the presence of the exponential term, $\epsilon^{-\frac{Rt}{2L}}$.

The current may be found from the relation

$$i = -dq/dt.$$

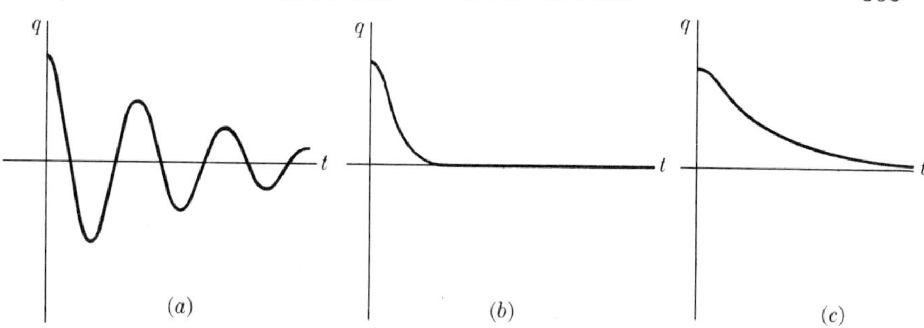

FIG. 17-3. Discharge characteristics in a circuit containing R, L, and C. (a) Oscillation, $R < 2\sqrt{L/C}$. (b) Critical damping, $R = 2\sqrt{L/C}$. (c) Overdamping, $R > 2\sqrt{L/C}$.

This leads to the equation

$$i = I_m \epsilon^{-\frac{Rt}{2L}} \sin 2\pi f_1 t,$$

where

$$I_m = \frac{2\pi f_0^2 Q}{f_1} \tag{17-4}$$

Note that this reduces to Eq. (17-1) when $R = 0$.

If $R = 2\sqrt{L/C}$, the discharge ceases to be oscillatory. The charge on the capacitor decreases asymptotically to zero and the circuit is said to be *critically damped.* If $R > 2\sqrt{L/C}$ the circuit is *overdamped.* The three cases are illustrated in Fig. 17-3.

The multiflash photographs of damped oscillatory motion on page 269 of Principles of Physics I correspond exactly to the discharge of a capacitor through an inductor-resistor combination, for a range of R/L ratios.

17-3 Sustained oscillations. The effect of resistance in an oscillatory circuit is to drain away the energy of the circuit and convert it to heat. Oscillations may be sustained in a circuit if some provision is made for returning energy at the same rate as it is removed. The most common method of doing this at present is to utilize the amplifying properties of a thermionic tube.

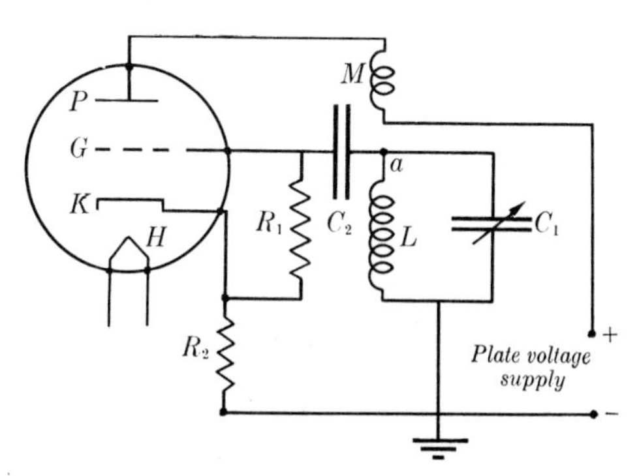

FIG. 17-4. Sustained oscillations in a "tuned-grid" circuit.

A so-called "tuned-grid" circuit is shown in Fig. 17-4. The inductor

L and the capacitor C_1 comprise the tuned circuit. Fluctuations in the potential of point a produce corresponding fluctuations in the potential of grid G, through the capacitor C_2. These variations in grid potential cause variations in the plate current which result in a transfer of energy from the plate voltage supply to the inductor L, through the mutual inductance M between plate coil and tuned circuit. If the circuit constants are properly chosen, energy is fed into the circuit at a rate exactly equal to that at which it is removed.

Cathode and grid are maintained at their proper relative potentials by the iR drops in resistors R_1 and R_2.

17-4 Radiation. In the absence of hysteresis, the energy stored in the electric field of a capacitor, or in the magnetic field of an inductor, is returned to the circuit when the capacitor discharges or the current in the inductor ceases, provided the changes occur slowly. However, if the fields are caused to alternate rapidly, not all of the energy is returned to the circuit but some escapes from it permanently in the form of *electromagnetic radiant energy*. Hence any circuit in which there is an alternating current is a source of electromagnetic waves. More generally, radiant energy is emitted whenever an electric charge is accelerated. An alternating current or an oscillating circuit, is simply a convenient method of accelerating charges.

From the relations which we have already shown to hold between electrical and magnetic effects it can be demonstrated (see the next section) that electromagnetic waves travel with a velocity (in free space) of magnitude 3×10^8 meters per second ($= 186{,}000$ miles per second) which is the same as the velocity of light. In fact, a light beam *is* a train of electromagnetic waves, of much higher frequency (hence shorter wave length) than the waves from an oscillating circuit, but of precisely the same nature.

The velocity, frequency, and wave length of electromagnetic waves in free space are related by the familiar equation

$$c = f\lambda$$

(c is the velocity of light in free space).

The effectiveness of a circuit as a radiator depends to a great extent on its geometry. If the electric field is localized as between a pair of closely-spaced capacitor plates, the rate of radiation of energy is relatively small. As the plates are separated and the field between them has an appreciable magnitude throughout a larger volume, the efficiency of the circuit as a radiator increases. The limiting case is reached when the circuit has been reduced to that of Fig. 17-5—a straight conductor with an alternator of some sort at its center. The two halves of the conductor form the "plates" of the capacitor, the electric field at some instant having the general shape shown by dotted lines. As the polarity of the alternator reverses, this field reverses also. Since charge flows along the conductor with each reversal, the conductor is also surrounded by a magnetic field shown by solid lines which are circles concentric with the conductor. The magnetic field also reverses with reversal of current. The electric field around the conductor is similar to that around an electric dipole (see Fig. 2-8), and the combined electric and magnetic field is like that which would result if the electric moment of the dipole alternated sinusoidally in magnitude according to the relation

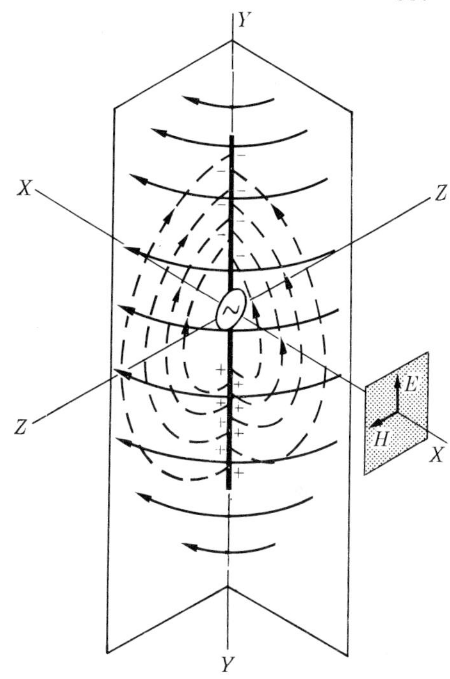

FIG. 17-5. Electric and magnetic fields around an oscillating dipole. Electric field, dotted lines; magnetic field, solid lines.

$$p = p_m \sin 2\pi ft,$$

where p is the instantaneous and p_m the maximum electric moment. Hence the conductor is often referred to as an *oscillating dipole.*

If one looks at the conductor along the X-axis in Fig. 17-5 the electric and magnetic fields within a small area (shown shaded) are at right angles to one another. The lines of electric intensity, E, are vertical or parallel to the Y-axis, the lines of magnetic intensity, H, are parallel to the Z-axis. Close to the conductor these fields are 180° out of phase, since at the instant when the capacitor is fully charged and the electric field a maximum, the current and the magnetic field are both zero. Further from the conductor, and for reasons which are too complex to go into here, the phase relation changes and the fields are in phase with one another. Their spatial relation remains the same, however, the electric field parallel to the conductor and the magnetic field perpendicular to it.

A *wave front* is defined as a surface, at all points of which the vector E (or H) is in the same phase. A line in the direction of propagation, perpendicular to the wave front, is called a *ray*. At a sufficient distance from the conductor the wave fronts may be considered planes. Fig. 17-6 shows a portion of the X-axis at a considerable distance from the conductor. The waves are advancing along this axis toward the right. If the alternating current in the radiator is sinusoidal, the magnitude and direction of the electric field at all points of a plane perpendicular to the X-axis can be represented by the ordinate of the sine curve labelled E in Fig. 17-6. The curve labelled H indicates the magnitude and direction of the magnetic field at all points of planes perpendicular to the X-axis. Both the E and the H curves or, more correctly, the field variations which they represent, are advancing toward the right at a velocity of 3×10^8 m/sec.

The relative directions of the vectors E, H, and the vector V in the direction of propagation, are always related as in Fig. 17-7. A useful method of remembering these relative directions is to imagine the V-axis to be a right-hand threaded rod, on which is a nut carrying the E and H vectors. If the nut is rotated in such a direction as to turn E into the position occupied by H, as indicated by the curved arrow in the figure, the direction of *advance* of the nut is in the direction of propagation.

The wave length of light waves is of the order of 5×10^{-5} cm. Wave lengths in the broadcast band are of the order of 100 meters. Thus the former are very much shorter than the dimensions of most optical apparatus while the latter are much longer than the dimensions of ordinary radio tubes. The development in recent years of apparatus capable of generating ultra-high-frequency electromagnetic waves whose wave length is a few centimeters has opened up a wholly new field of research in which for the first time the dimensions of one's apparatus are of the same order of magnitude as the wave length of the generated or received waves.

It turns out that when wave lengths (or frequencies) in this range are

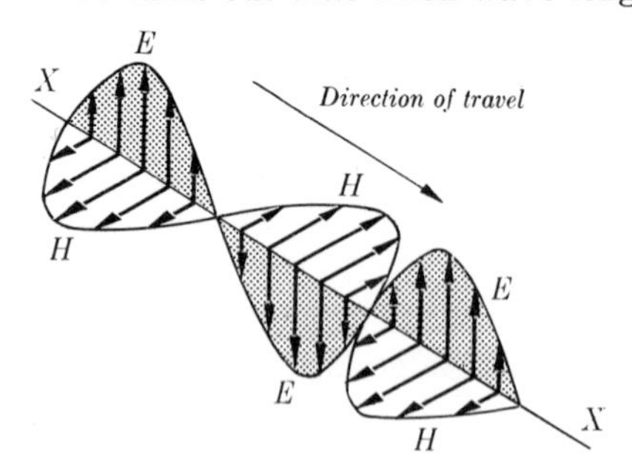

Fig. 17-6. Variations in electric and magnetic intensity in a traveling electromagnetic wave.

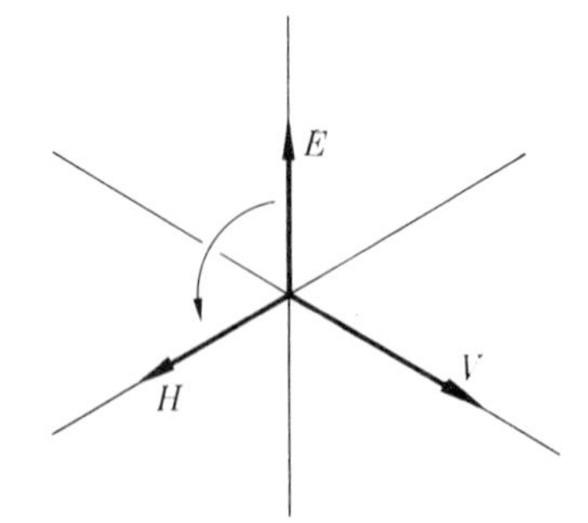

FIG. 17-7. Relative directions of electric intensity E, magnetic intensity H, and velocity of propagation V in a plane polarized electromagnetic wave.

involved, many concepts such as resistance and impedance, which are entirely adequate for handling direct currents, or alternating currents at a frequency of 60 cycles/sec, lose much of their significance. The emphasis shifts from a consideration of the motion of charges in conductors to the variations in the electric and magnetic intensities in the space outside of or enclosed by conductors. While conductors are still used to convey energy from one point to another, the conductor is considered merely as a guide for the electromagnetic waves and is called a *wave-guide.*

17-5 Velocity of electromagnetic waves. The crowning achievement of the electromagnetic theory was its prediction that an oscillating circuit should be a source of electromagnetic waves, traveling with a velocity in exact agreement (within the limits of experimental error) with the velocity observed. The theory was developed by Maxwell in 1864, and the first experimental tests were carried out by Hertz in 1887. A complete exposition of the theory of electromagnetic radiation must naturally be reserved for more advanced texts, but the underlying ideas, and the wave equation, can be developed without too much difficulty. We shall need to make use of the following relations:

(1) The line integral of the electric intensity E around a closed loop equals the negative of the rate of change of magnetic flux through the loop.

$$\oint E \cos \theta \, dl = -\frac{d\Phi}{dt}.$$

If the magnetic field is perpendicular to the plane of the loop and B is the average flux density and A the area of the loop, then

$$\Phi = AB,$$

and since

$$B = \mu H,$$

it follows that

$$\oint E \cos \theta \, dl = -\mu A \frac{dH}{dt}. \tag{17-5}$$

(2) The line integral of the magnetic intensity H around a closed loop equals the current through the loop.

$$\oint H \cos \theta \, dl = i. \tag{17-6}$$

(3) In a region in which the electric field is varying there exists a *displacement current.* The displacement current density is

$$J = \frac{dD}{dt} = \epsilon \frac{dE}{dt}. \tag{17-7}$$

In the derivation of Eq. (17-6), i stood for the conduction current (actual motion of charges) through the plane of the loop. Maxwell took the bold step of asserting that it should hold equally well for displacement currents, and experimentally we find this to be the case. Then since $i = JA$, Eqs. (17-6) and (17-7) may be combined to give

$$\oint H \cos \theta \, dl = \epsilon A \frac{dE}{dt}. \tag{17-8}$$

Eqs. (17-5) and (17-8) are of the same form, one expressing a relation between the line integral of E and the rate of change of H, the other a relation between the line integral of H and the rate of change of E.

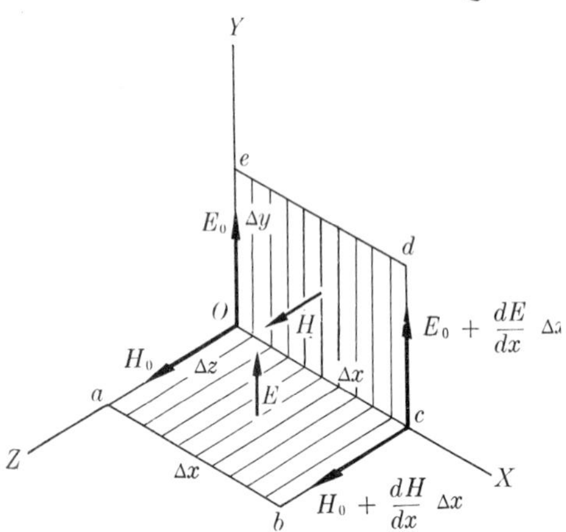

FIG. 17-8. Variations in E and H in an electromagnetic wave.

Now consider the region around an oscillating circuit in which both electric and magnetic fields exist, and where the magnitudes of both are varying from point to point and varying with time as well. At some point O (Fig. 17-8) let a set of coordinate axes be constructed. We have seen that the electric and magnetic fields are at right angles to one another. Let the magnetic field be perpendicular to the X-Y plane, the electric field perpendicular to the X-Z plane.

Construct an infinitesimal rectangle $Oabc$ of sides Δx and Δz, lying in the X-Z plane, and evaluate the line integral of H around this rectangle. Since H is perpendicular to the sides ab and Oc, these contribute nothing to the integral. The line integral along side Oa is $H_0 \, \Delta z$, where H_0 is the magnetic intensity in the Y-Z plane.

Let dH/dx be the rate of change of H with x. Then in the distance Oc or Δx the change in H is the product of the rate of change with distance, times the distance, or

$$\Delta H = \frac{dH}{dx} \Delta x.$$

Hence the magnitude of H at the side bc is

$$H_0 + \Delta H = H_0 + \frac{dH}{dx}\Delta x.$$

The line integral along this side is therefore

$$-\left(H_0 + \frac{dH}{dx}\Delta x\right)\Delta z,$$

the minus sign appearing since $\theta = 180°$ (H and bc are in opposite directions).

Hence the line integral around the whole loop becomes

$$\oint H\cos\theta\, dl = H_0\,\Delta z - \left(H_0 + \frac{dH}{dx}\Delta x\right)\Delta z = -\frac{dH}{dx}\Delta x\,\Delta z.$$

Replacing $\Delta x\,\Delta z$ by A, the area of the loop, and combining with Eq. (17-8) we obtain

$$\frac{dH}{dx} = -\epsilon\frac{dE}{dt}, \tag{17-9}$$

where E is the average value of the electric intensity over the infinitesimal loop, or essentially its value at the "point" at which the loop is constructed.

Consider next the rectangle $\Delta x\,\Delta y$ in Fig. 17-8, lying in the X-Y plane and therefore perpendicular to H. When the line integral of E around this loop is evaluated by method used above, we obtain

$$\frac{dE}{dx} = -\mu\frac{dH}{dt}. \tag{17-10}$$

Let us next eliminate E from Eqs. (17-9) and (17-10) by differentiating the first with respect to x and the second with respect to t. This leads to the relations

$$\frac{d^2H}{dx^2} = -\epsilon\frac{d}{dx}\left(\frac{dE}{dt}\right),$$

$$\frac{d}{dt}\left(\frac{dE}{dx}\right) = -\mu\frac{d^2H}{dt^2}.$$

But

$$\frac{d}{dx}\left(\frac{dE}{dt}\right) = \frac{d}{dt}\left(\frac{dE}{dx}\right).$$

Hence

$$\frac{d^2H}{dt^2} = \frac{1}{\mu\epsilon}\frac{d^2H}{dx^2}. \tag{17-11}$$

By eliminating H in the same way, a similar equation is obtained for E.

$$\frac{d^2E}{dt^2} = \frac{1}{\mu\epsilon}\frac{d^2E}{dx^2}. \tag{17-12}$$

It will be recalled that the differential equation of a transverse wave in a string (see Principles of Physics I, pages 479 and 480) is

$$\frac{d^2y}{dt^2} = \frac{T}{\mu}\frac{d^2y}{dx^2},$$

and the differential equation of a compressional wave in a liquid or gas is

$$\frac{d^2y}{dt^2} = \frac{1}{k\rho_0}\frac{d^2y}{dx^2}.$$

Eqs. (17-11) and (17-12) have the same mathematical form as both of the above differential equations, which means that the field vectors H and E vary in time and in space exactly as do the displacements of a particle in a transverse wave in a string, or in a sound wave. We therefore have a *traveling electromagnetic wave*, which is transverse in the sense that the electric and magnetic intensities are at right angles to the direction of propagation. The wave considered here is of a relatively simple type called *plane polarized*, which means that at all points and at all times the H vectors remain parallel to a definite plane while the E vectors are parallel to a second plane at right angles to the first. No "sources" of the electric or magnetic fields are required other than the wave itself, which is, so to speak, self-sustaining. The changing magnetic field maintains the electric field, as indicated by the equation

$$\oint E \cos\theta \, dl = -\mu A \frac{dH}{dt},$$

and the changing electric field maintains the magnetic field, as indicated by

$$\oint H \cos\theta \, dl = \epsilon A \frac{dE}{dt}.$$

When the differential equation of wave motion is written in the form of Eq. (17-11) or (17-12), the coefficient of the space derivative is the square of the velocity of propagation. For example, the velocity of propagation of a transverse wave in a string is $\sqrt{T/\mu}$, and of a compressional wave in a fluid it is $\sqrt{1/k\rho_0}$. Hence the velocity of an electromagnetic wave is

$$V = \sqrt{\frac{1}{\mu\epsilon}} = \sqrt{\frac{1}{K_m K_e}}\sqrt{\frac{1}{\mu_0\epsilon_0}} \qquad (17\text{-}13)$$

In free space, where $K_m = K_e = 1$, the velocity is represented by c and is given by

$$c = \sqrt{\frac{1}{\mu_0\epsilon_0}}. \qquad (17\text{-}14)$$

The product of the proportionality constants μ_0 and ϵ_0 can be determined by purely electrical, magnetic, and mechanical measurements and definitions. For instance, in the mks system the factor μ_0 is by definition exactly $4\pi \times 10^{-7}$ newton-sec²/coulomb², and the factor ϵ_0, found by an experiment equivalent to measuring the force between two point charges, is $1/4\pi \times 8.9875 \times 10^9$ newton-meters²/coulomb². (The most precise value is from measurements made by Rosa and Dorsey at the National Bureau of Standards.) Hence

$$c = \sqrt{\frac{1}{\mu_0\epsilon_0}} = \sqrt{\frac{4\pi \times 8.9875 \times 10^9}{4\pi \times 10^{-7}}\,\frac{\text{newton-m}^2}{\text{coul}^2} \times \frac{\text{coul}^2}{\text{newton-sec}^2}}$$
$$= 2.9979 \times 10^8 \text{ m/sec.}$$

The best direct experimental measurements of the velocity of light in a vacuum give

$$c = 2.99773 \times 10^8 \text{ m/sec.}$$

The two are in excellent agreement.

Consider next the velocity of light in a material substance such as glass or water. The relative permeability K_m of all transparent substances is so nearly equal to unity that this term can be omitted from Eq. (17-13), which reduces to

$$V = \sqrt{\frac{1}{K_e}}\sqrt{\frac{1}{\mu_0\epsilon_0}},$$

or, since

$$\sqrt{\frac{1}{\mu_0\epsilon_0}} = c,$$

$$V = c\sqrt{\frac{1}{K_e}}.$$

In optics, one defines the *index of refraction*, n, of a material, as the ratio of the velocity of light in free space to its velocity in the material. Hence

$$n = \frac{c}{V} = \sqrt{K_e}, \tag{17-15}$$

and the index of refraction equals the square root of the dielectric coefficient. One must be cautious, however, about inserting in Eq. (17-15) the value of the dielectric coefficient obtained from electrostatic measurements or measurements at low frequencies. The dielectric coefficient of a substance is a measure of the polarization or orientation effects produced by electric fields on the charges within the substance. At very high frequencies such as those of light waves (of the order of 10^{15} cycle/sec) the field alternates in direction so rapidly that the polarization effects cannot build up to their full values before the field reverses. Hence the effective dielectric coefficient at these frequencies is much smaller than at low frequencies. Furthermore, since the ability of the dipoles to follow a disturbance of a given frequency varies with the frequency, it follows that the effective dielectric coefficient, and hence the velocity of a light wave in a material substance, also depends on the frequency, an effect known as *dispersion*.

For example, the dielectric coefficient of water, found from electrostatic measurements, is about 80, and $\sqrt{80} \approx 9$, while the index of refraction of water is 1.3428 for blue light and 1.3308 for red light.

We next derive an important relation between the magnitudes of the electric and magnetic intensities in an electromagnetic wave. Let us assume for simplicity that the wave is sinusoidal. Then from the well-known expression for a sine wave of frequency f, traveling in the positive x-direction with a velocity V, we can write,

$$E = E_m \sin 2\pi f\left(t - \frac{x}{V}\right), \tag{17-16}$$

$$H = H_m \sin 2\pi f\left(t - \frac{x}{V}\right), \tag{17-17}$$

where E_m and H_m are the "amplitudes" or maximum values of E and H.

We have shown that E and H must satisfy Eq. (17-9). By differentiating Eqs. (17-16) and (17-17) with respect to t and x respectively, we obtain

$$\frac{dE}{dt} = 2\pi f E_m \cos 2\pi f\left(t - \frac{x}{V}\right),$$

$$\frac{dH}{dx} = -\frac{2\pi f H_m}{V} \cos 2\pi f\left(t - \frac{x}{V}\right),$$

and when these values are substituted in Eq. (17-9) we get

$$-\frac{2\pi f H_m}{V} \cos 2\pi f\left(t - \frac{x}{V}\right) = -\epsilon \times 2\pi f E_m \cos 2\pi f\left(t - \frac{x}{V}\right).$$

After cancelling, and recalling that $V = \sqrt{1/\mu\epsilon}$, this reduces to

$$\sqrt{\mu}\, H_m = \sqrt{\epsilon}\, E_m.$$

The amplitudes of the electric and magnetic fields thus bear a definite ratio to one another.

If the wave is in empty space,

$$\sqrt{\mu_0}\, H_m = \sqrt{\epsilon_0}\, E_m.$$

17-6 The Poynting vector. Like any other type of wave, a traveling electromagnetic wave transports energy. It can be shown (a more advanced text should be consulted for details) that the instantaneous rate at which energy is transported by an electromagnetic wave, per unit area perpendicular to the direction of propagation, is

$$S = EH, \qquad (17\text{-}18)$$

where E and H are the instantaneous values of the electric and magnetic intensities. In the mks system, S is in joules per second, per square meter, or watts/m², where E is in volts/m and H in amperes/m. Since the direction of propagation of the wave is perpendicular to the plane determined by E and H, the quantity S can be considered as a vector in the direction of propagation. It is known as the *Poynting vector*, after J. H. Poynting (1852–1914).

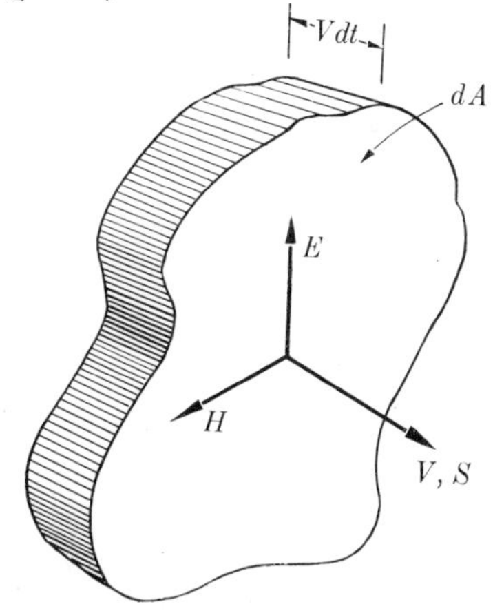

Fig. 17-9.

Eq (17-18) may be justified as follows. Consider a cylindrical volume element (Fig. 17-9) through which there is passing an electromagnetic wave. The cross section of the element, dA, lies in the plane of E and H, and its length is $V\,dt$. We have shown that the energy density in an electric field is $\epsilon\dfrac{E^2}{2}$, and in a magnetic field it is $\mu\dfrac{H^2}{2}$. The electromagnetic energy in the cylinder is therefore

$$dA \times V\,dt \times \left(\epsilon\frac{E^2}{2} + \mu\frac{H^2}{2}\right).$$

The electromagnetic wave travels with velocity V in a direction perpendicular to dA, and if the energy travels with it at the same velocity, all of the electromagnetic energy now in the cylinder will cross the area dA in time dt. Hence the energy crossing the surface, per unit area and per time, is

$$S = \frac{V}{2}\left(\epsilon E^2 + \mu H^2\right). \tag{17-19}$$

We have shown in Sec. 17-5 that

$$\sqrt{\epsilon}\,E = \sqrt{\mu}\,H, \quad \text{or,} \quad \epsilon E^2 = \mu H^2.$$

Hence Eq. (17-19) may be written

$$S = V\epsilon E^2 = V\mu H^2 = EH \tag{17-20}$$

which is the same as Eq. (17-18).

If the wave is a light wave in a transparent dielectric,

$$V = \frac{c}{n}, \quad \epsilon = K_e\epsilon_0 = n^2\epsilon_0, \quad \mu = \mu_0, \quad \text{and}$$

$$S = cn\,\epsilon_0 E^2 = \frac{c}{n}\,\mu_0 H^2 = EH.$$

Finally, in empty space where $n = 1$,

$$S = c\epsilon_0 E^2 = c\mu_0 H^2 = EH.$$

Eq. (17-20) gives the *instantaneous* rate of transport of energy, per unit area. The *time average* rate of transport equals either the product of $V\epsilon$ and the time average value of E^2, or the product of $V\mu$ and the time average value of H^2. If the wave is sinusoidal, the time average value of either E^2 or H^2 is one-half of its maximum value. Hence

$$S_{av} = \tfrac{1}{2}\,V\epsilon E_m^2 = \tfrac{1}{2}\,V\mu H_m^2 = \tfrac{1}{2}\,E_m H_m,$$

and if the wave is in empty space,

$$S_{av} = \tfrac{1}{2}\,c\epsilon_0 E_m^2 = \tfrac{1}{2}\,c\mu_0 H_m^2 = \tfrac{1}{2}\,E_m H_m.$$

The general form of Eq. (17-18) in vector notation is

$$\vec{S} = \vec{E} \times \vec{H},$$

and the relative directions of the three vectors are immediately evident.

17-7 Reflection and refraction. Fresnel's formulae. We shall consider next what happens when an electromagnetic wave, traveling in one medium, strikes the surface of a second medium in which the velocity differs from that in the first. This problem is of the greatest importance since it is the basis for the study of reflection and refraction of light at the surfaces of mirrors, lenses, and prisms, and of the reflection or scattering of the waves in a radar beam. We shall derive the results only for a special case, but the method is the same in all cases.

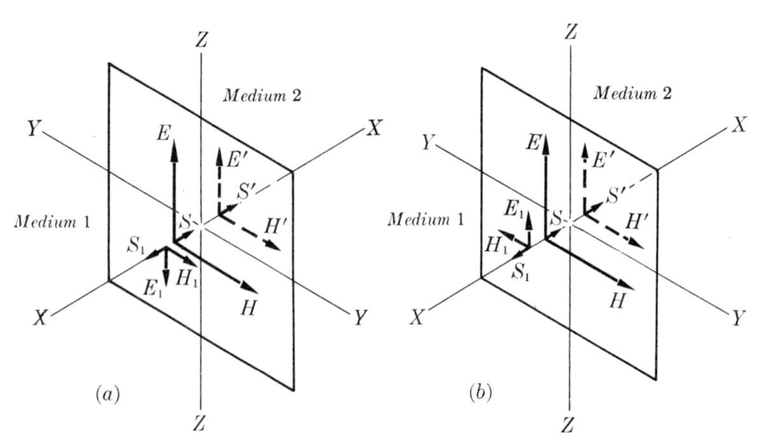

Fig. 17-10. Reflection of an electromagnetic wave at a boundary surface between two media.

In Fig. 17-10, the Y-Z plane represents the boundary between two homogeneous isotropic dielectrics (for example, it might be the surface of a glass prism in air). Medium 1, of dielectric coefficient K_e and index of refraction $n = \sqrt{K_e}$, lies at the left of the plane; medium 2, of dielectric coefficient K_e' and index $n' = \sqrt{K_e'}$, lies at the right. A sinusoidal plane wave traveling in medium 1 is incident normally on the surface from the left. The wave fronts, and the E and H vectors in the incident wave, are parallel to the Y-Z plane. Let us assume the E-vector to be parallel to the Z-axis. The H-vector is then parallel to the Y-axis in the direction shown.

If both media are perfectly transparent, it might appear that the wave would simply pass from medium 1 into medium 2, traveling of course with a different velocity in medium 2. Energy considerations, however, show that something else must happen at the surface. The rate at which energy is incident on the surface, per unit area, is given by the Poynting vector in medium 1,

$$S = \frac{c}{n} \mu_0 H^2.$$

The rate at which energy leaves the surface, per unit area, in the second medium, is

$$S' = \frac{c}{n'} \mu_0 (H')^2.$$

We have shown in Sec. 7-5 that at the boundary surface between two dielectrics in an electric field, the tangential components of E are continuous across the boundary. By the same method used in that section it can be shown that the tangential component of H is also continuous across a boundary. Hence, when an electromagnetic wave strikes the boundary, if the only thing that occurred were that a wave was transmitted across the boundary into medium 2, then of necessity $H = H'$ and the Poynting vectors would *not* be equal since $n \neq n'$.

What actually happens is that, in addition to the *transmitted* wave (also called the *refracted* wave), there originates at the boundary a *reflected* wave, traveling in medium 1 in a direction (in the special case of normal incidence) opposite to that of the incident wave, and of such amplitude that the boundary conditions and conservation of energy are satisfied. Let H_1 and E_1 represent the amplitudes of the magnetic and electric vectors in the reflected wave.

We can see immediately that if the vector H_1 is in the same direction as H, as in Fig. 17-10(a), the vector E_1 must be *opposite* to E since the reflected wave is traveling to the left. (Remember the relations between the directions of E and H and the direction of propagation.) On the other hand, if E_1 is in the same direction as E, as in Fig. 17-10(b), then H_1 and H must be in opposite directions. Therefore to satisfy boundary conditions in case (a),

$$E - E_1 = E', \quad H + H_1 = H', \tag{17-21}$$

and in case (b),

$$E + E_1 = E', \quad H - H_1 = H'. \tag{17-22}$$

Let us assume first that the reflection is as in (a). Since the incident and reflected waves both travel in the first medium,

$$\sqrt{\epsilon}\, E = \sqrt{\mu}\, H, \quad \sqrt{\epsilon}\, E_1 = \sqrt{\mu}\, H_1,$$

while for the refracted wave,

$$\sqrt{\epsilon'}\, E' = \sqrt{\mu'}\, H'.$$

But

$$\mu = \mu' = \mu_0,$$

and

$$\sqrt{\epsilon} = \sqrt{K_e}\, \sqrt{\epsilon_0} = n\, \sqrt{\epsilon_0}, \quad \sqrt{\epsilon'} = \sqrt{K_e'}\, \sqrt{\epsilon_0} = n\, \sqrt{\epsilon_0},$$

so that

$$n\, \sqrt{\epsilon_0}\, E = \sqrt{\mu_0}\, H, \quad n\, \sqrt{\epsilon_0}\, E_1 = \sqrt{\mu_0}\, H_1, \quad n'\, \sqrt{\epsilon_0}\, E' = \sqrt{\mu_0}\, H',$$

or

$$H = n\sqrt{\frac{\epsilon_0}{\mu_0}}\, E, \quad H_1 = n\sqrt{\frac{\epsilon_0}{\mu_0}}\, E_1, \quad H' = n'\sqrt{\frac{\epsilon_0}{\mu_0}}\, E'.$$

When these expressions for the H's are inserted in Eq. (17-21) we get

$$E - E_1 = E',$$

$$nE + nE_1 = n'E'.$$

Simultaneous solution gives

$$E_1 = \frac{n' - n}{n' + n}\, E,$$

$$E' = \frac{2n}{n' + n}\, E.$$

And from the relations between E and H,

$$H_1 = \frac{n' - n}{n' + n}\, H,$$

$$H' = \frac{2n'}{n' + n}\, H.$$

Now if $n' > n$, the term $\dfrac{n' - n}{n' + n}$ is positive and E_1 and H_1, the vectors in the reflected wave, are positive also which means they are in the directions assumed, namely, those in Fig. 17-10(a). If $n' < n$, E_1 and H_1

are negative, that is, opposite to the assumed directions and the reflection is as in case (b). Therefore both types of reflection can occur, the former when $n' > n$, the latter when $n' < n$.

It is left as a problem to show that energy is conserved when the reflected wave is taken into consideration.

The *reflectance* of the surface, represented by r, is defined as the ratio of the rate at which energy leaves the surface in the reflected wave, to the rate at which it is incident on the surface. In other words, it is the ratio of the Poynting vectors of these two waves.

$$r = \frac{\frac{c}{n}\,\mu_0 H_1^2}{\frac{c}{n}\,\mu_0 H^2} = \frac{(n' - n)^2}{(n' + n)^2}. \tag{17-23}$$

For example, suppose that a light wave traveling in air is incident normally on the surface of a glass prism. The index of refraction of air (n) is very nearly 1.00, and that of glass (n') is about 1.50. The reflectance is therefore

$$r = \frac{(1.50 - 1.00)^2}{(1.50 + 1.00)^2} = \frac{.25}{6.25} = 4\%.$$

That is, 4% of the light is reflected and, of course, 96% is transmitted.

When an electromagnetic wave strikes a boundary surface between two dielectrics at other than normal incidence, the problem is somewhat more complicated but the general method of attack is the same. It becomes necessary to consider the boundary conditions that must be satisfied by D and B, as well as those satisfied by E and H. We have shown in Sec. 7-5 that the normal component of D is continuous across a boundary, and the same is easily demonstrated to be true of B. We shall not give a complete discussion of the general case, which can be found in any good text on physical optics, but shall only indicate the factors involved and state the results without proof.

In Fig. 17-11, the Y-Z plane represents the boundary between two dielectrics. Medium 1, of index n, lies at the left of the plane and medium 2, of index n', at the right. A plane wave traveling in medium 1 and indicated by the ray marked "incident ray," is incident on the boundary surface. The wave fronts, and the directions of the E and H vectors in the incident wave, lie in planes perpendicular to the incident ray, which itself lies in the X-Z plane. To avoid confusion these vectors are not shown in the diagram.

Just as in the case of normal incidence, both a refracted and a reflected wave originate at the boundary surface. The directions and relative amplitudes of the incident, refracted, and reflected waves can be found by making use of the boundary conditions that must be satisfied by E, H, D, and B. The polarization of the incident wave also plays an important part, but it is sufficient to consider two special cases, namely, when the incident wave is plane polarized with its E vector (a) perpendicular to the X-Z plane, and (b) parallel to the X-Z plane. When the results for these special cases are known, those for any other may be computed.

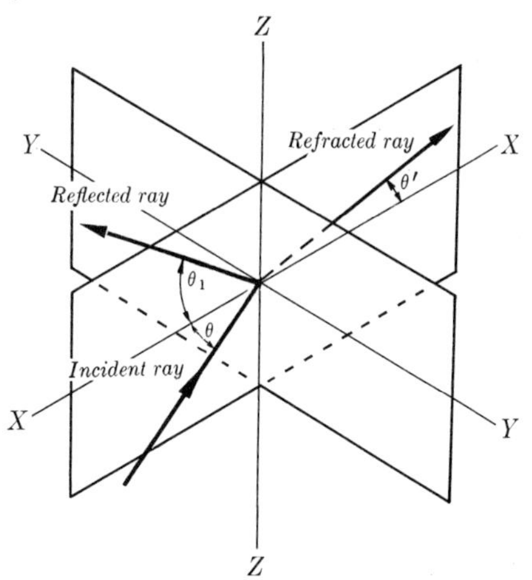

FIG. 17-11. Relation between angles of incidence, reflection, and refraction.

It is found that in all cases the reflected and refracted rays lie in the plane determined by the incident ray and the normal to the boundary surface at the point of incidence. This plane is called the *plane of incidence;* it is the X-Z plane in Fig. 17-11. The angle θ_1 between the reflected ray and the normal is equal to the angle of incidence, θ. The angle θ' between the refracted ray and the normal is given by

$$\frac{\sin \theta'}{\sin \theta} = \frac{\sqrt{K_e}}{\sqrt{K_e'}} = \frac{n}{n'},$$

or

$$n \sin \theta = n' \sin \theta'. \tag{17-24}$$

This relation is known as *Snell's law.*

If the incident wave is plane polarized with its E vector perpendicular to the plane of incidence, the following relations hold between the vectors E_1 and E', in the reflected and refracted waves respectively, and the vector $E_\perp$ in the incident wave.

$$E_1 = -E_\perp \frac{\sin (\theta - \theta')}{\sin (\theta + \theta')}, \tag{17-25}$$

$$E' = E_\perp \frac{2 \sin \theta' \cos \theta}{\sin (\theta + \theta')}. \tag{17-26}$$

If the incident wave is plane polarized with its E vector ($E_{||}$) parallel to the plane of incidence,

$$E_1 = E_{||} \frac{\tan(\theta - \theta')}{\tan(\theta + \theta')}, \tag{17-27}$$

$$E' = E_{||} \frac{2 \sin \theta' \cos \theta}{\sin(\theta + \theta') \cos(\theta - \theta')} \tag{17-28}$$

Eqs. (17-25) to (17-28) are known as *Fresnel's formulae,* after Augustin Fresnel (1788–1827), who was the first to work them out. We shall make use of them later, in optics.

Problems—Chapter 17

(1) The capacitance C in Fig. 17-1 is 300 $\mu\mu$f. (a) What should be the inductance L if the resonant frequency of the circuit is 1 megacycle/sec (10^6 cycles/sec)? (b) How many turns of wire spaced 0.5 mm apart, on a tube 2 cm in diameter, are required for this inductance? Assume that the inductor can be considered a long solenoid.

(2) When switch S in Fig. 17-12 is closed, the steady current in the battery is 10 amp. (a) Explain how oscillations originate in the L-C circuit when the switch is opened. (b) What is the frequency of the oscillations? (c) What is the maximum potential difference across the capacitor?

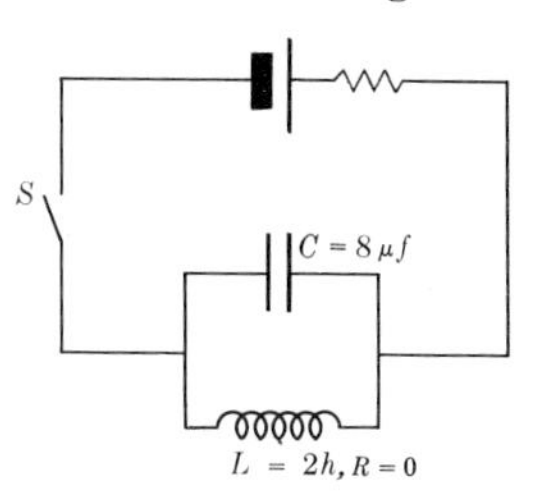

FIG. 17-12.

(3) A 0.01 μf capacitor, a 0.1 henry inductor whose resistance is 1000 ohms, and a switch, are connected in a series circuit. The capacitor is initially charged to a potential difference of 400 volts. The switch is then closed. (a) Find the ratio of the oscillation frequency of the circuit to the frequency if the resistance were zero. (b) How much energy is converted to heat in the first complete cycle? (c) How much energy is converted to heat in the complete train of oscillations? (d) What is the maximum additional resistance that can be inserted in the circuit before the discharge becomes non-oscillatory?

(4) What is the frequency of electromagnetic waves (in free space) of the following wave lengths: 10^{-8} cm (x-rays); 5×10^{-5} cm (yellow light); 10 cm (microwaves); 300 m (broadcast band)?

(5) What is the maximum magnetic intensity in a plane radio wave in which the maximum electric intensity is 100 microvolts/meter?

(6) The average power radiated by a broadcasting station is 10 kilowatts. Assume the power to be radiated uniformly over the surface of any hemisphere with the station at its center. (a) What is the magnitude of the Poynting vector at points on the surface of a hemisphere 10 km in radius? At this distance, the waves can be considered plane. (b) Find the maximum electric and magnetic intensities at points on this hemispherical surface.

(7) A radar transmitter emits a uniform conical beam of radiant energy. The solid angle of the cone is 0.01 steradian. The maximum electric intensity at a distance of 1 km from the transmitter is 10 volts/m. (a) What is the maximum magnetic intensity? (b) What is the total power in the beam?

(8) Assume that a 100-watt incandescent lamp radiates all its energy in a sinusoidal electromagnetic wave, and that the energy is emitted uniformly in all directions. Compute the maximum electric and magnetic intensities at a distance of 2 m from the lamp.

(9) Show that when a plane electromagnetic wave is incident normally on a plane boundary between two different dielectrics, the rate at which energy leaves the surface in the refracted and reflected beams is equal to the rate at which energy is incident on the surface.

(10) What fraction of the energy of a plane electromagnetic wave in air, incident normally on a plane water surface, is reflected, if the frequency of the wave is such that the dielectric coefficient of the water is 80? Compare with the fraction reflected if the wave is of visible light, for which the index of refraction of water is 1.33.

CHAPTER 18

ELECTRONICS

18-1 Elementary particles. At the beginnng of Chapter 1 a brief discussion was given of the elementary particles that form the building blocks of matter. We shall begin this chapter with a summary of present-day information about these particles and others. It is impossible, in the space we can devote to this part of the subject, to do justice to the experimental and theoretical physicists, both in this country and abroad, who in the last thirty or forty years have accumulated a truly impressive body of information about the properties of these particles and have added so much to our knowledge of how the world is put together. A list of reference books for the interested reader is given at the end of the chapter.

Electrons. The discovery of the negative electron dates from the measurement of its charge-to-mass ratio by Thomson in 1897, and the measurement of its charge by Millikan in 1909. These experiments have already been described. The charge of an electron is 1.603×10^{-19} coulombs and its mass is 1/1837 that of a hydrogen atom, or 9.107×10^{-31} kgm.

The positive electron was first observed during the course of an investigation of cosmic rays by Dr. Carl D. Anderson in 1932, in the cloud chamber photograph reproduced in Fig. 18-1. The photograph was made with the cloud chamber in a magnetic field perpendicular to the plane of the paper. A lead plate crosses the chamber and evidently the particle has passed through it. Since the curvature of the track is greater above the plate than below, the velocity is less above than below and the inference is that the particle was moving upward, since it is difficult to see how it could have gained energy going through the lead.

The density of droplets along the path is what would be expected if the particle were an electron. But the direction of the magnetic field and the direction of motion are only consistent with a particle of positive sign. Hence Anderson concluded the track had been made by a positive electron or *positron.* Since the time of this discovery many thousands of such tracks have been photographed and the positron's existence is now definitely established. Its mass is the same as that of a negative electron and its charge is equal but of opposite sign.

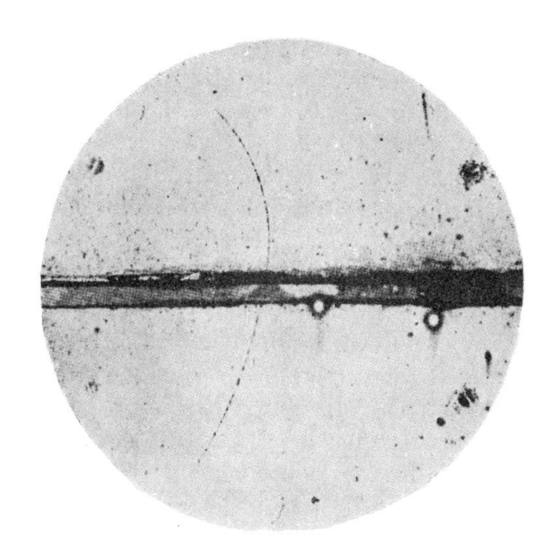

FIG. 18-1. Track of a positive electron traversing a lead plate 6 mm thick. (C. D. Anderson)

Positive electrons have only a transitory existence and do not form a part of ordinary matter. There are two known processes which result in positive electrons. They are ejected from the nuclei of certain artificially radioactive materials, and they spring into existence (along with a negative electron) in a process known as "pair production" in which a γ-ray is simultaneously annihilated. Charge is conserved in the process, since the particles have charges of opposite sign.

The proton. The proton is identical with the nucleus of an atom of hydrogen ${}_1H^1$. It has a positive charge equal in magnitude to the electronic charge. The masses of elementary particles are conveniently expressed in atomic mass units (amu) on a scale in which the mass of an atom of the most abundant isotope of oxygen is arbitrarily set at exactly 16 amu. On this scale the mass of the hydrogen atom is 1.0081 amu and the mass of the proton (the hydrogen nucleus) is this value less the mass of an electron. Hence the mass of a proton is 1.0076 amu or 1.672×10^{-27} kgm.

The neutron. In 1930, W. Bothe and H. Becker in Germany observed that when beryllium, boron, or lithium were bombarded by fast alpha-particles, the bombarded material emitted something, either particles or electromagnetic waves, of much greater penetrating power than the original alpha-particles. Further experiments in 1932, by Irene Curie and F. Joliot in Paris, and by J. Chadwick in England, showed that the emission consisted of uncharged particles now known as neutrons. The neutron mass has been accurately measured and is 1.0090 amu, nearly but not exactly equal to the mass of a proton. The nucleus of every atom, with the single exception of hydrogen $_1H^1$, is a group of protons and neutrons. The great penetrating power of neutrons arises from the fact that they have no charge and hence are not affected by the electric fields around electrons and nuclei.

The mesotron. The mesotron (or meson) was also discovered by Anderson, in 1936, and, like the positron, its existence was inferred from tracks in cloud chamber photographs. The mass of the mesotron has not been accurately measured, but it seems to be of the order of magnitude of 180 times that of an electron. Its charge is equal to the electronic charge, and both + and − mesotrons exist. Until 1945 mesotrons were observed only as a component of cosmic radiation, but they have now been produced in the laboratory with the help of the new 100 million volt betatron. Like the positron, the mesotron is a short-lived creature and does not form a part of ordinary matter.

The neutrino. The neutrino is the most elusive of the elementary particles. No direct evidence of its existence has yet been found. It is postulated to take care of conservation of energy, momentum, and angular momentum in certain atomic processes where otherwise conservation of these quantities would not hold. It is of unknown mass and is uncharged.

Photons. In their interaction with matter, electromagnetic waves behave not like continuous electric and magnetic fields, but like compact bundles of energy called photons. Thus a photon is not a particle in the ordinary sense, but its properties represent one aspect of electromagnetic waves. The energy of a photon is proportional to the frequency of the wave to which it corresponds, a relation first proposed by Planck in 1900 and extended by Einstein and others in the years that followed. The relation between the energy and frequency of a photon is

$$W = hf,$$

where W is the energy, f the frequency, and h a universal constant called Planck's constant. The photon aspect of radiant energy becomes apparent only when the frequency is near to or greater than that of visible light. Thus one speaks of photons of light, x-ray photons, or γ-ray photons.

A photon has mass and momentum but, of course, no electric charge.

18-2 Thermionic emission. The vacuum diode. The phenomenon of conduction in a metal has been described as a relatively slow drift of the free electrons within it, the drift velocity being superposed on the random thermal agitation. The thermal motion can be ignored in the conduction process but it plays an important role in the phenomenon of thermionic emission, that is, the liberation of free electrons from the surface of a heated conductor.

The early theories of thermionic emission assumed that the speeds of free electrons in a metal were distributed in the same way as are the speeds of atoms in a gas, or according to the Maxwell-Boltzmann distribution law. It is now thought that the speed distribution is of a different character, called a Fermi-Dirac distribution. The two functions are shown in Fig. 18-2. The Fermi-Dirac function predicts a considerably larger number of high-speed electrons than the Maxwell distribution.

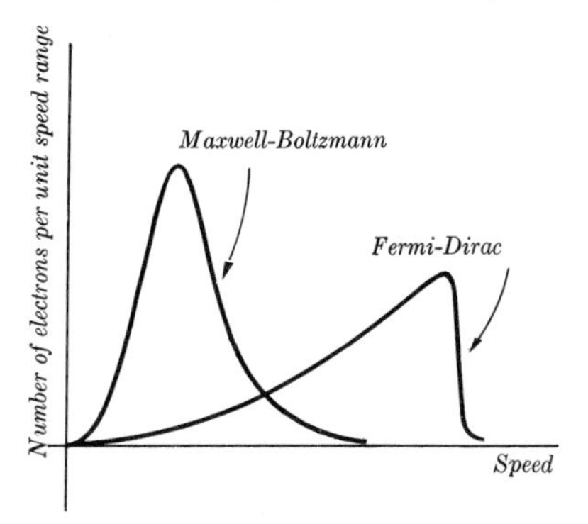

FIG. 18-2. Maxwell-Boltzmann and Fermi-Dirac distribution functions.

It might be thought that an electron coming up to a metal surface from within would simply continue going, break through the surface, and escape. There is, however, a certain minimum energy, different for different metals, which an electron must have in order to be able to get out of the metal. This energy is usually expressed in electron-volts (an electron-volt is the energy of an electron that has been accelerated from rest through a potential difference of 1 volt. 1 ev $= 1.6 \times 10^{-19}$ joules) and its magnitude is referred to as the "height of the potential barrier" at the metal surface. For example, the height of the potential barrier in tungsten is 13.5 electron-volts. At low temperatures very few electrons have this amount of energy, but as the temperature is raised the energies of the electrons increase and the number capable of escaping increases rapidly.

The escape of electrons from the surface of an isolated metallic conductor leaves the conductor positively charged. The electrons which have escaped are therefore attracted by the conductor and form a "cloud" of negative charge outside its surface. This cloud is called a *space charge.* If a second nearby conductor is at a higher potential than the first, the electrons in the cloud are attracted to it, and as long as the potential difference between the conductors is maintained there will be a steady drift

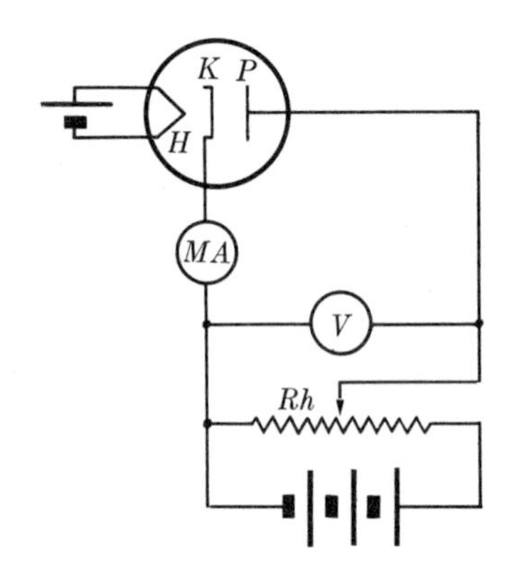

FIG. 18-3. Circuit for measuring plate current and plate voltage in a vacuum diode.

of electrons from the emitter or *cathode* to the other body, which is called the plate or *anode*.

In the common thermionic tube the cathode and anode (and often other electrodes as well) are enclosed within an evacuated glass or metal container, leads to the various electrodes being brought out through the base or walls of the tube. We shall first describe the characteristics of a simple thermionic tube in which the only electrodes are the cathode and anode. Such a tube is called a *diode*.

The diode is shown schematically in Fig. 18-3. The cathode and plate are represented by K and P. The cathode is often in the form of a hollow cylinder, which is heated by a fine resistance wire H within it. Electrons emitted from the outer surface of the cathode are attracted to the plate, which is a larger cylinder surrounding the cathode and coaxial with it. The electron current to the plate is read on the milliammeter MA, which is preferably connected as shown rather than to the plate directly, since the cathode is usually at or near ground potential while the anode may be several hundred volts above ground. The potential difference between plate and cathode can be controlled by the slide wire and read on voltmeter V.

If the potential difference between cathode and anode is small (a few volts) only a few of the emitted electrons reach the plate, the majority penetrating a short distance into the cloud of space charge and then returning to the cathode. As the plate potential is increased, more and more electrons are drawn to it, and with sufficiently high potentials (of the order of a hundred volts) all of the emitted electrons arrive at the plate. Further increase of plate potential does not increase the plate current, which is said to become *saturated*.

A graph of plate current, I_p, vs plate potential V_p is shown in Fig. 18-4(a). Notice that I_p is not zero even when V_p is zero. This is because the electrons leave the cathode with an initial velocity and the more rapidly moving ones may penetrate the cloud of space charge and reach the plate even with no accelerating field. In fact, a *retarding* field is necessary to prevent their reaching the plate, an effect which may be used to measure their velocities of emission.

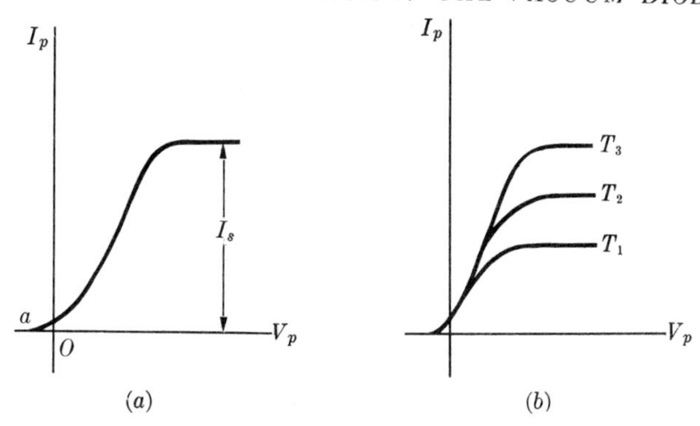

FIG. 18-4. (a) Plate current-plate voltage characteristic of a diode. (b) Plate current curves at three different cathode temperatures. $T_3 > T_2 > T_1$.

The saturation current I_s in Fig. 18-4(a) is equal to the emission current from the cathode, and for a given tube its magnitude depends markedly on the cathode temperature. Fig. 18-4(b) shows three plate current curves at three different temperatures, where $T_3 > T_2 > T_1$. The relation between saturation current and temperature was first derived by O. E. Richardson and later in a slightly different form by S. Dushman. The Dushman equation is

$$J_s = AT^2\epsilon^{-\phi/kT} \tag{18-1}$$

where J_s is the saturation current density at the cathode surface. A is a constant characteristic of the emitting surface, T is the Kelvin temperature of the emitter, k is the Boltzmann constant, and ϕ is the *work function* of the surface, a quantity related to the height of the potential barrier. (It is difficult to describe the phenomenon both accurately and simply.) The work function of a pure tungsten surface is 4.52 electron-volts; the height of its potential barrier is 13.5 electron-volts.

The work function ϕ of a surface may be considerably reduced by the presence of impurities. A small amount of thorium, for example, reduces the work function of pure tungsten by about 50%. Since the smaller the work function the larger the current density at a given temperature (or the lower the temperature at which a given emission can be attained), most vacuum tubes now use cathodes having composite surfaces.

There is a relatively simple relation known as the Langmuir-Child equation which gives the minimum plate voltage required to draw saturation current in a diode. To derive this equation we first consider the effect of the electron cloud, or space charge, on the potential distribution between the cathode and plate of a diode. Assume cathode and plate

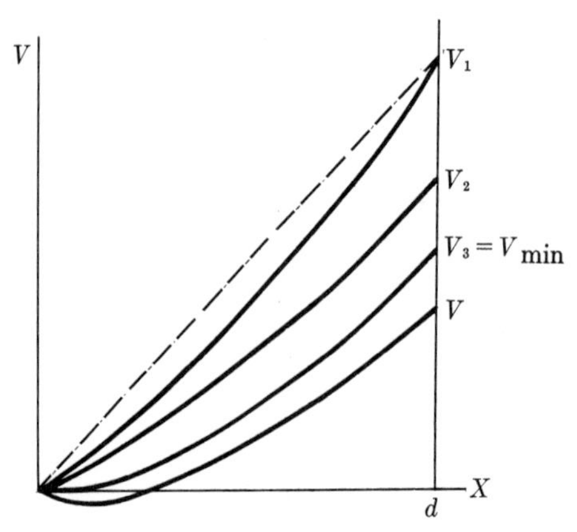

FIG. 18-5. Potential distribution in cathode-anode interspace of a vacuum diode, for various plate voltages.

to be parallel planes separated by a distance d. Let us construct the graphs of Fig. 18-5, in which the X-axis is a line perpendicular to the cathode, and the potential at any point along the line is plotted vertically. The cathode is at ground potential.

First let the plate be at a positive potential V_1, and assume that the cathode is not heated and there is no space charge. We then have essentially a parallel plate capacitor in vacuum and the potential increases linearly from zero to V_1 as shown by the dotted line. Now let us heat the cathode so that electrons are emitted, but assume that the plate potential V_1 is more than enough to draw saturation current. The space charge, since it is negative, will lower the potential at points in the cathode-anode interspace to a form indicated by the solid line from the origin to V_1. If the plate potential is now decreased successively to V_2 and V_3, the potential distribution changes as shown. At the potential V_3, the curve has a horizontal tangent at the origin. If the potential is reduced to V_4, the curve dips below the X-axis.

It will be recalled that the electric intensity equals the negative of the potential gradient,

$$E = -\frac{dV}{dx}. \tag{18-2}$$

Examination of the curves in Fig. 18-5 will show that as long as the plate potential is equal to or greater than V_3 the direction of the electric intensity is from the plate toward the cathode at all points. Hence an electron is subject to an accelerating force toward the anode at all points, and every electron emitted from the cathode reaches the anode. This is merely another way of stating that if the plate potential is equal to or greater than V_3 we are drawing saturation current. But if the potential is less than V_3 the curve of V vs x starts off from the origin with a negative slope and instead of an accelerating field there is a retarding field at the cathode surface. Hence the more slowly moving electrons are turned back by this field and cannot reach the plate, and the plate current begins to decrease when the plate potential drops below V_3. The magnitude of

V_3 is therefore the minimum plate voltage required to draw saturation current, and is characterized by the fact that at this particular voltage, which we shall designate as V_{min}, both V and dV/dx are zero at the origin.

With the plate potential at this particular value V_{min}, let V represent the potential at any distance x from the cathode, and let v be the velocity of an electron at this distance. We shall assume for simplicity that all electrons leave the cathode with zero initial velocity. Further details will be found in any good text on electronics. Then if e is the electronic charge,

$$Ve = \tfrac{1}{2}mv^2, \tag{18-3}$$

$$v = \sqrt{\frac{2Ve}{m}}. \tag{18-4}$$

Let n represent the number of electrons per unit volume in the cloud of space charge, at any distance x from the cathode. The current density J_s (which of course is the same at all planes perpendicular to the X-axis and under the assumed conditions equals the saturation current density given by Eq. (18-1)) is then from Eq. (4-4)

$$J_s = nev. \tag{18-5}$$

The product ne is the charge per unit volume, or the charge density ρ. The equation above can therefore be written

$$J_s = \rho v. \tag{18-6}$$

When this is combined with Eq. (18-4) we get

$$\rho = J_s\sqrt{\frac{m}{2Ve}}. \tag{18-7}$$

This expression for the charge density can now be inserted in Poisson's equation (Eq. 3-35),

$$\frac{d^2V}{dx^2} = \frac{1}{\epsilon_0}\rho, \tag{18-8}$$

to give

$$\frac{d^2V}{dx^2} = \frac{1}{\epsilon_0}J_s\sqrt{\frac{m}{2Ve}}. \tag{18-9}$$

To solve this differential equation we first multiply both sides by $2dV/dx$.

$$2\frac{dV}{dx}\left(\frac{d^2V}{dx^2}\right) = \frac{2}{\epsilon_0}J_s\sqrt{\frac{m}{2Ve}}\frac{dV}{dx}. \tag{18-10}$$

But

$$2\frac{dV}{dx}\left(\frac{d^2V}{dx^2}\right) = \frac{d}{dx}\left(\frac{dV}{dx}\right)^2.$$

Hence Eq. (18-10) becomes

$$\frac{d}{dx}\left(\frac{dV}{dx}\right)^2 = \frac{2}{\epsilon_0}J_s\sqrt{\frac{m}{2Ve}}\frac{dV}{dx}, \tag{18-11}$$

or, after cancelling dx's,

$$d\left(\frac{dV}{dx}\right)^2 = \frac{2}{\epsilon_0}J_s\sqrt{\frac{m}{2e}}V^{-1/2}\,dV. \tag{18-12}$$

Integration of this equation gives

$$\left(\frac{dV}{dx}\right)^2 = \frac{4}{\epsilon_0}J_s\sqrt{\frac{m}{2e}}V^{1/2} + C. \tag{18-13}$$

To determine the integration constant C we make use of the fact that $dV/dx = 0$ when $V = 0$, i.e., at the cathode. Hence $C = 0$ also.

Extraction of the square root of both sides of Eq. (18-13) gives

$$V^{-1/4}\,dV = \sqrt{\frac{4}{\epsilon_0}J_s\sqrt{\frac{m}{2e}}}\,dx, \tag{18-14}$$

which integrates to

$$\frac{4}{3}V^{3/4} = \sqrt{\frac{4}{\epsilon_0}J_s\sqrt{\frac{m}{2e}}}\,x + C, \tag{18-15}$$

and since $V = 0$ when $x = 0$, this integration constant is zero also.

When $x = d$, $V = V_{\text{min}}$, so finally

$$V_{\text{min}}^{3/4} = \frac{3}{4}\sqrt{\frac{4}{\epsilon_0}J_s\sqrt{\frac{m}{2e}}}\,d. \tag{18-16}$$

This is one form of the desired expression for the minimum voltage required to draw saturation current. It is more common to solve it for the saturation current density J_s.

$$J_s = \frac{4\epsilon_0}{9}\sqrt{\frac{2e}{m}}\frac{V_{\text{min}}^{3/2}}{d^2}. \tag{18-17}$$

This is the usual form of the Langmuir-Child equation.

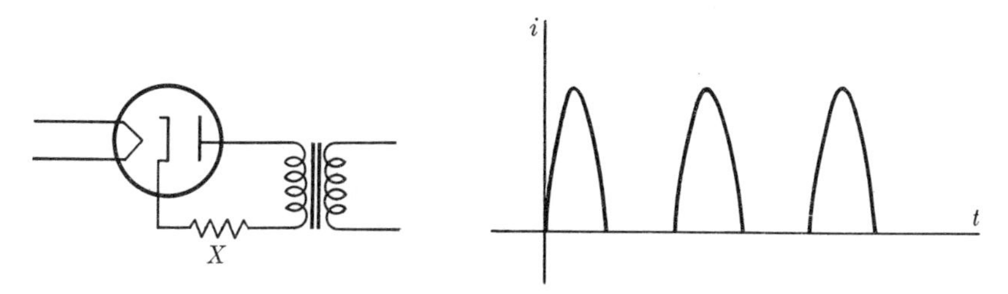

FIG. 18-6. Circuit for obtaining half-wave rectified alternating current.

The abscissa Oa in Fig. 18-4(a) represents the negative plate potential (relative to the cathode) at which the plate current is reduced to zero. If the plate potential is made still more negative, the plate current remains zero. All of the electrons emitted by the cathode are driven back by the reversed field between cathode and plate, and there is no supply of electrons to flow in the other direction since the plate itself is not heated. The diode is thus a "one-way" conductor and behaves like a check-valve in a pipe line, allowing charge to flow through it in one direction but not in the other. It may therefore be used to *rectify* an alternating current. That is, if an alternating voltage is applied between cathode and plate as in Fig. 18-6 the diode will conduct only during the half-cycle in which the plate is positive. One thus obtains in the cathode-plate circuit (say at point X) a *half-wave rectified* current.

The circuit in Fig. 18-7 utilizes a diode with two plates (equivalent to two diodes) and *full-wave rectified* current is obtained at X.

18-3 Multi-electrode vacuum tubes. Lee de Forest, in 1907, discovered that if a third electrode, called a *grid*, is inserted between the cathode and plate of a thermionic tube, the potential of this electrode exerts much more control over the plate current than does the potential of the plate itself. The grid is usually an open mesh or a helix of fine wire which allows

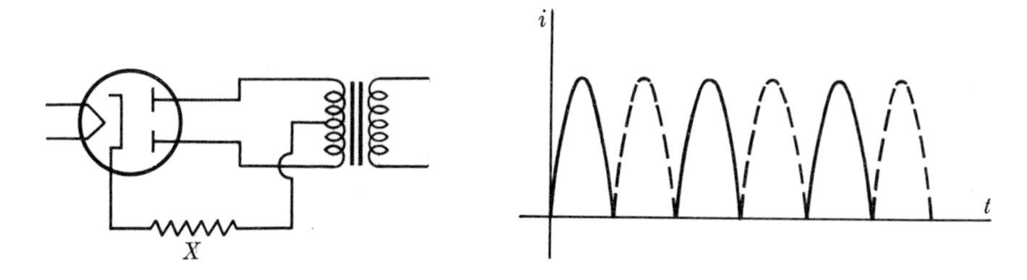

FIG. 18-7. Circuit of a full-wave rectifier.

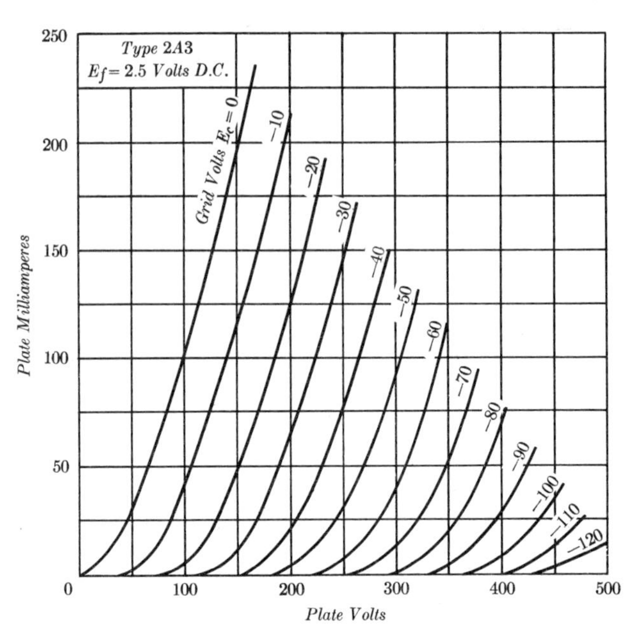

FIG. 18-8. Average plate characteristics of a 2A3 triode.

most of the electrons to pass through its openings. Since small variations in grid potential produce relatively large variations in plate current, the three-electrode tube can be used as an *amplifier*.

Soon after de Forest's introduction of the grid it was found that a fourth electrode, in the space-charge region, would add still further desirable characteristics to the thermionic tube. At present tubes with as many as eight electrodes are in common use.

We shall consider briefly some of the features of a 3-element tube, or triode. Commercial tubes are invariably operated at potentials much lower than would be required to draw saturation current. The currents to various electrodes cannot be represented by any simple equation and are best expressed by graphs. In particular, the current to an electrode is rarely a linear function of the potential of that electrode. The vacuum tube therefore does not obey Ohm's law and is a *nonlinear* circuit element.

The plate current i_b, in a triode, depends both on the plate potential e_b and the control grid potential e_c, both measured relative to the cathode. In Fig. 18-8, plate current is plotted vertically and plate voltage horizontally. The curves give plate current as functions of plate voltage when the control grid voltage is kept constant at the values shown. For example, when the grid voltage is −30 volts and the plate voltage is 250 volts, the plate current is 150 milliamperes.

The interrelations between various electrode potentials and currents are commonly expressed in terms of *tube factors.* The most useful of these are the *amplification factor* μ, the *transconductance* g_m, and the *plate conductance* g_p (or its reciprocal, the *plate resistance* r_p). The unit of conductance, the reciprocal of resistance, is one *mho*, a word whose etymology should be obvious. The *micromho* is often a unit of more convenient size.

The amplification factor μ is defined as the negative ratio of the change in plate voltage to a change in grid voltage when the plate current is kept constant. It measures the effectiveness of the control grid voltage, relative to that of the plate voltage, on the plate current. Strictly speaking, the changes should be infinitesimal but we may approximate them by small but finite changes and write for the amplification factor

$$\mu = -\frac{\Delta e_b}{\Delta e_c} \; (i_b = \text{const}). \tag{18-18}$$

The plate conductance g_p is defined as the ratio of a small change in plate current to a small change in plate voltage, at constant grid voltage. Since the plate resistance r_p is the reciprocal of the plate conductance,

$$r_p = \frac{1}{g_p} = \frac{\Delta e_b}{\Delta i_b} \; (e_c = \text{const}). \tag{18-19}$$

The transconductance g_m is defined as the ratio of a small change in plate current to a small change in grid voltage, at constant plate voltage.

$$g_m = \frac{\Delta i_b}{\Delta e_c} \; (e_b = \text{const}). \tag{18-20}$$

18-4 The cathode ray oscillograph. One of the most useful of recent electronic developments is the cathode ray oscillograph, whose essential features are illustrated in Fig. 18-9. The tube itself is of glass and is

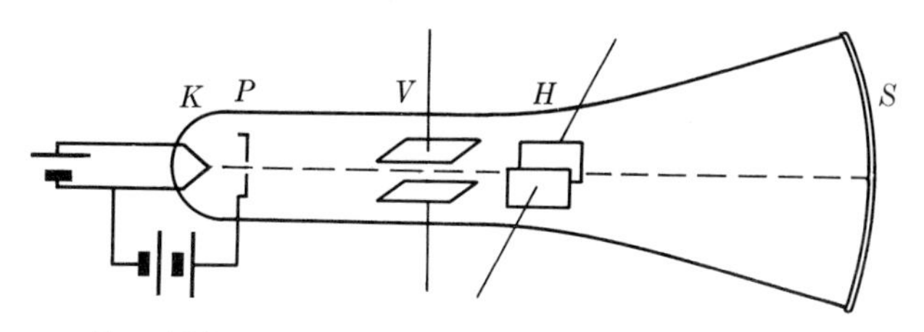

FIG. 18-9. Simplified drawing of a cathode ray tube.

highly evacuated. Electrons (cathode rays) are emitted by a heated filament K and accelerated toward an anode P maintained at a potential of from several hundred to several thousand volts above the filament. Most of the electrons strike the anode but a narrow beam passes through a small hole in the anode and continues on to the screen S which is coated on its inner surface with a substance that emits visible light (or fluoresces) when bombarded by electrons.

The electron beam may be deflected in a horizontal or vertical direction by an electric field between the pairs of deflecting plates marked H and V. (In some tubes the deflection is produced by magnetic fields set up by coils outside the tube.) The extremely small inertia of the electrons enables the electron stream to follow, with practically no time lag, the variations in the electric or magnetic fields which deflect it.

A grid structure (not shown) between cathode and anode controls the electron current and hence the intensity of the spot on the screen. When the tube is used as a television receiver the spot is caused to sweep rapidly over the screen while its intensity is controlled by the grid in accordance with variations in brightness of points in the object being televised. Persistence of vision (and a small persistence of screen fluorescence) creates the illusion of an image.

18-5 The photoelectric effect. In the thermionic emission of electrons from metals, the energy needed by an electron to surmount the potential barrier at the metal surface is furnished by the energy of thermal agitation. Electrons may also acquire enough energy to escape from a metal, even at low temperatures, if the metal is illuminated by light of sufficiently short wave length. This phenomenon is called the *photoelectric effect*. It was first observed by Heinrich Hertz in 1887, who noticed that a spark would jump more readily between two spheres when their surfaces were illuminated by the light from another spark. Hallwachs investigated the effect more fully the following year and it usually goes by his name.

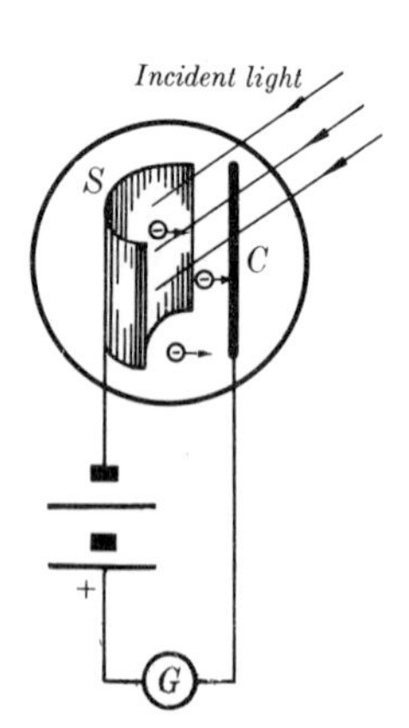

Fig. 18-10. Schematic diagram of photocell circuit.

A modern type phototube is shown schematically in Fig. 18-10. A beam of light, indicated by the arrows, falls on a photosensitive surface S. Electrons emitted by the surface are drawn to the collector C, normally maintained at a positive potential with respect to the emitter. Emitter and collector are enclosed in an evacuated container. The photoelectric current can be read on the galvanometer G.

It is found that with a given material as emitter, the wave length of the light must be shorter than a critical value, different for different surfaces, in order that any photoelectrons at all may be emitted. This critical wave length, or the corresponding frequency, is called the *threshold frequency* of the particular surface. The threshold frequency for most metals is in the ultraviolet (critical wave length 200 to 300×10^{-5} cm) but for potassium and caesium oxide it lies in the visible spectrum (400 to 700×10^{-5} cm).

Just as in the case of thermionic emission, the photoelectrons form a cloud of space charge around the emitter S. That some of the electrons are emitted with an initial velocity is shown by the fact that even with no emf in the external circuit a few electrons penetrate the cloud of space charge and reach the collector, causing a small current in the external circuit. The velocity of the most rapidly moving electrons can be deduced by measuring the reversed voltage (negative potential of collector) which is required to reduce the current to zero.

A remarkable feature of photoelectric emission is the relation between the number and maximum velocity of escaping electrons on one hand, and the intensity and wave length of the incident light on the other. Surprisingly enough, it is found that the maximum velocity of emission is *independent of the intensity* of the light, but does depend on its wave length. It is true that the photoelectric current increases as the light intensity is increased, but only because more electrons are emitted. With light of a given wave length, no matter how feeble it may be, the maximum velocity of the photoelectrons from a given surface is always the same, provided, of course, that the frequency is above the threshold frequency.

The explanation of the photoelectric effect was given by Einstein in 1902, although his theory was so radical that it was not generally accepted until 1906, when it was confirmed by experiments performed by Millikan. Extending a proposal made two years earlier by Planck, Einstein postulated that a beam of light consisted of small bundles of energy which are now called light quanta or photons. The energy W of a photon is proportional to its frequency f,

$$W = hf$$

where h is a universal constant called Planck's constant whose value is 6.62×10^{-27} erg-seconds. When a photon collides with an electron at or just within the surface of a metal, it may transfer its energy to the electron. This transfer is an "all-or-none" process, the electron getting all of the photon's energy or none at all. The photon then simply drops out of existence. The energy acquired by the electron may enable it to penetrate the potential barrier at the surface of the metal, if it is moving in the right direction.

In penetrating the potential barrier the electron loses energy in amount ϕ (the work function of the surface). Some electrons may lose more than this if they start at some distance below the metal surface, but the maximum energy with which an electron can emerge is the energy gained from a photon minus the work function. Hence the maximum kinetic energy of the photoelectrons ejected by light of frequency f is

$$\tfrac{1}{2}mv_{\max}^2 = hf - \phi. \tag{18-2}$$

This is Einstein's photoelectric equation, and it was in exact agreement with Millikan's experimental results.

The currents obtainable with vacuum phototubes are extremely small, of the order of a few microamperes per lumen of light flux. These currents may be increased by a factor of 5 to 10 if a small quantity of gas is left in the tube. The electrons ionize the gas and more current-carriers are available.

A different type of phototube has recently been developed, known as the *barrier-layer* or *photovoltaic cell.* This is the type commonly used in photographic exposure meters. Unlike the true photoelectric tube, which requires an external emf, the photovoltaic cell generates its own emf. In one type, a thin layer of cuprous oxide is deposited on a copper disk. Light falling on the layer of oxide penetrates the latter and photoelectrons are ejected from it into the metallic copper, leaving the oxide positively charged and giving the copper a negative charge.

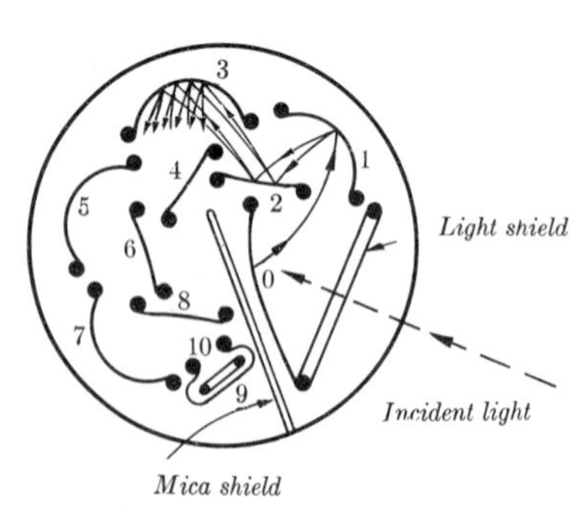

FIG. 18-11. An electron multiplier tube.

The *electron multiplier phototube,* a recent addition to the phototube family, makes use of the phenomenon of *secondary emission* to increase by a factor of several hundred thousand the electron current from a photoelectric surface. The structure of one such tube is shown schematically in Fig. 18-11. Light incident on surface 0 ejects electrons which are accelerated to electrode 1, since the latter is at a higher potential. The bombardment of surface 1 by these electrons causes the ejection of *secondary* electrons whose number may be two or three times as great as the number incident. These electrons are in turn accelerated to electrode 2, where the current is multiplied by another factor of two or three, and so on. In the tube shown in Fig. 18-11 the ratio of the secondary to the primary electron current is about 1.6 times, and since the tube contains nine electrodes, the initial photoelectric current is multiplied by a factor of $(1.6)^9$ or about 230,000 times. Thus a photoelectric current of 10 microamperes per lumen at the first surface is multiplied to 2,300,000 microamperes per lumen, or 2.3 *amperes* per lumen, at the anode.

18-6 The x-ray tube. X-rays are produced when rapidly moving electrons, which have been accelerated through potential differences of some tens or hundreds of thousands of volts, are allowed to strike a metal target. They were first observed by Wilhelm K. Roentgen (1845–1923) in 1895 and are also called "Roentgen rays."

X-rays are of the same nature as light or any other electromagnetic wave and, like light waves, they are governed by quantum relations in their interaction with matter. One may hence speak of x-ray photons or quanta, the energy of such a photon being given by the familiar relation

$$W = hf.$$

Wave lengths of x-rays range from 10^{-10} to 10^{-6} cm.

At present, practically all x-ray tubes are of the Coolidge type, invented by W. D. Coolidge of the General Electric laboratories in 1913. A diagram of a Coolidge tube is given in Fig. 18-12. A thermionic cathode and an anode are enclosed in a glass tube which has been pumped down to an extremely low pressure, so that electrons emitted from the cathode can travel directly to the anode with only a small probability of a collision

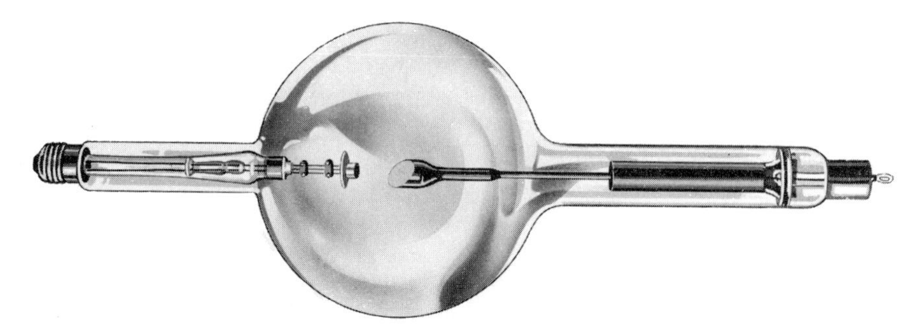

FIG. 18-12. Coolidge type x-ray tube.

on the way, reaching the anode with a speed corresponding to the full potential difference across the tube. X-radiation is emitted from the anode surface as a consequence of its bombardment by the electron stream.

It will be seen that x-ray production is an inverse phenomenon to photoelectric emission. In the latter, electrons acquire kinetic energy from light waves or photons. In the former, x-ray waves or photons are produced at the expense of electronic kinetic energy.

There appear to be two distinct processes going on when x-rays are emitted. Some of the electrons are stopped by the target and their kinetic energy is converted directly to x-radiation. Others transfer their energy in whole or in part to the atoms of the target, which retain it temporarily as "energy of excitation" but very shortly emit it as x-radiation. The latter is characteristic of the material of the target while the former is not.

18-7 Conduction in gases. Unlike a metal or an electrolyte, a gas free from external influences contains no free charges to serve as carriers in an electric field. Gases may be rendered conducting, however, in a variety of ways, in all of which some of the molecules become ionized by the detachment of one or more of their outer electrons. Some of these electrons may then attach themselves to neutral molecules forming negative ions, so that in an ionized gas both positive and negative ions and free electrons are usually present.

In order to ionize a molecule, energy must be supplied to it in some way. This may come about as a result of a collision with a rapidly moving molecule, ion, or electron, or by interaction with a quantum of radiation (a photon). The term *ionizing agent* is applied to the means by which ionization is brought about. Some common ionizing agents are the following:

Cosmic rays. Cosmic rays, at the earth's surface, are a mixture of rapidly moving positive and negative electrons, mesotrons, possibly protons, and photons of high energy. At sea level cosmic rays produce from 1 to 2 ions per cm^3 of air per second. Cosmic rays will penetrate matter for hundreds of feet, so it is impossible to shield a gas from them. A few ions are therefore always present in any sample of gas and a gas is always slightly conducting. In fact, it was a search for the cause of this small residual conductivity of gases which led to the discovery of cosmic radiation.

Radioactivity. Radioactive substances are present in minute amounts in all of the components of the earth's crust. The α, β, and γ rays emitted by them result in a small ionization of the earth's atmosphere.

Photoelectric emission Electrons can be liberated from the atoms of a gas by light, just as they are liberated from metallic surfaces in the photoelectric effect.
X-rays. X-rays are photons, like light, and ionize a gas through which they pass.
Ionization by collision. Ionization by collision is the chief means of producing ions in most instances of gaseous conduction. Positive and negative ions and electrons are accelerated by the electric field in the gas and may acquire sufficient energy to ionize a neutral molecule when a collision takes place.

All of the processes above result in the *production* of ions or free electrons. The number of ions per unit volume in a conducting gas does not increase indefinitely, however, since ions are removed by recombination, by diffusion to the walls, or they are attracted to the electrodes by the electric field. A steady state is eventually reached when the rate of production of ions equals the rate of removal.

A current in a gas is referred to as a *discharge.* Discharges may be classified as self-maintaining or nonself-maintaining. In the former type the discharge itself provides a continuous supply of ions; in the latter the supply is dependent on some external source.

A common example of nonself-maintaining discharge is afforded by the *gas-filled phototube*, which differs from the vacuum phototube only in that it contains a gas (usually argon or neon) at low pressure. Electrons liberated from the cathode by light incident on it are accelerated toward the anode and, if the accelerating field is sufficient, they will in a short distance acquire enough kinetic energy to ionize a gas molecule with which they collide.

The positive ion thus formed moves toward the cathode, while both electrons accelerate toward the anode and after again gaining energy from the field each produces another electron-ion pair. The number of free charges per unit volume thus increases exponentially with distance from the cathode and is much larger than it would be if the photoelectrons were the only source of supply. The current with a given illumination is increased by a factor of 5 to 10 times.

A number of other effects probably play a part also, among them the secondary emission of electrons at the cathode caused by the positive ions which bombard it.

The discharge is nonself-maintaining, since it must be initiated by photoelectrons liberated from the cathode. When the light is cut off, the tube ceases to conduct.

One unexpected characteristic of such a discharge is that the current

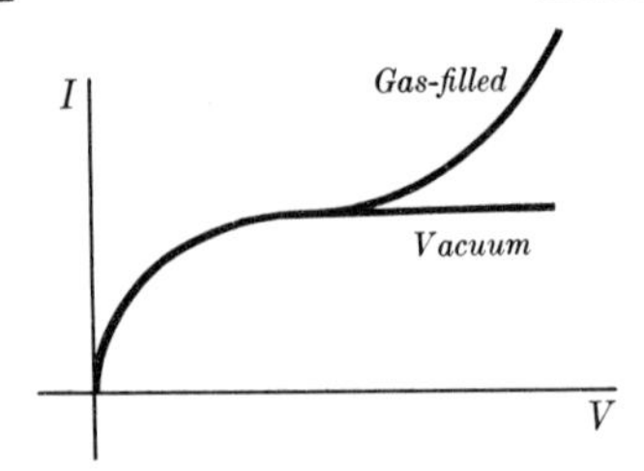

FIG. 18-13. Characteristic curves of gas-filled and vacuum photocells.

increases if the distance between the electrodes is increased, the light intensity and the potential gradient being unchanged. This arises from the fact that as the path length between electrodes is increased the number of collisions made by an electron between cathode and anode increases also.

Fig. 18-13 is a current-voltage graph for a commercial gas filled phototube. The corresponding graph for a vacuum type tube is included for comparison.

We shall consider next the *glow discharge*, which is the type of discharge in a neon sign or a fluorescent lamp. Glow discharges take place at low pressures (a few mm of mercury) but are self-maintaining. Fig. 18-14 represents a tube with electrodes sealed in at its ends and containing a gas at low pressure. Assume the potential difference between the electrodes to be gradually increased from zero. Because of the few ions formed by cosmic rays or other means, the gas will be very slightly conducting even at low voltages. The first part of its current-voltage characteristic, below the *critical voltage* V_{crit}, is like that of the gas-filled photo-tube. Along this part of the curve from O to a, the discharge is nonself-maintaining.

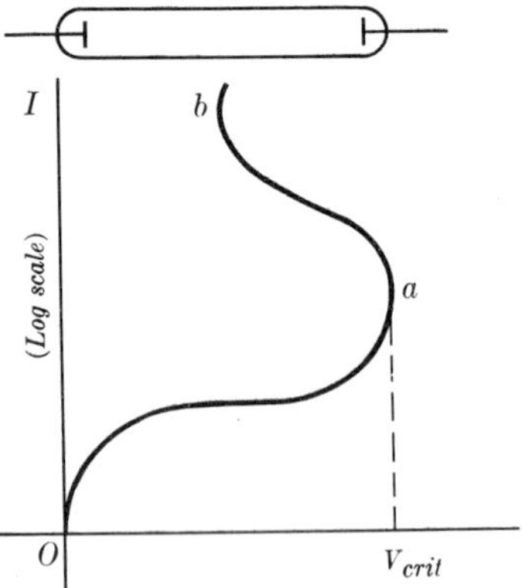

FIG. 18-14. Current-voltage characteristic of gas discharge tube.

When the critical voltage is reached the current increases very rapidly to point b, while at the same time the voltage across the tube decreases. The gas in the tube becomes luminous, whence the term glow discharge.

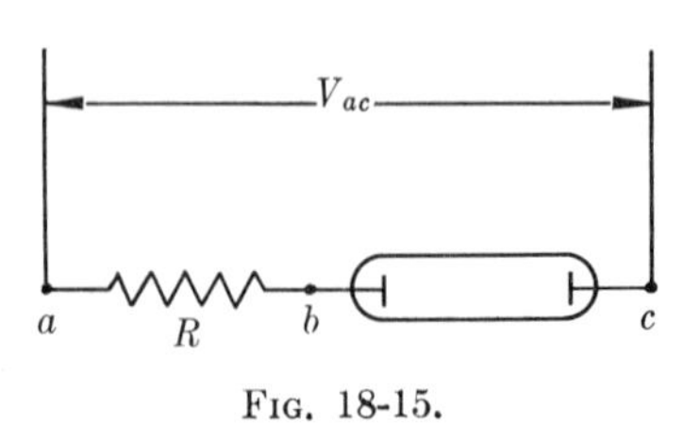

FIG. 18-15.

The fact that the voltage across the tube increases and then decreases requires a word of explanation. For concreteness, suppose the tube is connected in series with a fixed resistor R as in Fig. 18-15, where V_{ac} is controllable at will. The voltage across the tube is $V_{bc} = V_{ac} - V_{ab} = V_{ac} - IR$. As V_{ac} is gradually increased from zero, the critical voltage across the tube will be reached when $V_{bc} = V_{\text{crit}} = V_{ac} - IR$, where I is the current at a in Fig. 18-14. If now V_{ac} is held at this value, then as I increases from point a to point b, the voltage drop IR increases and V_{bc} *decreases.*

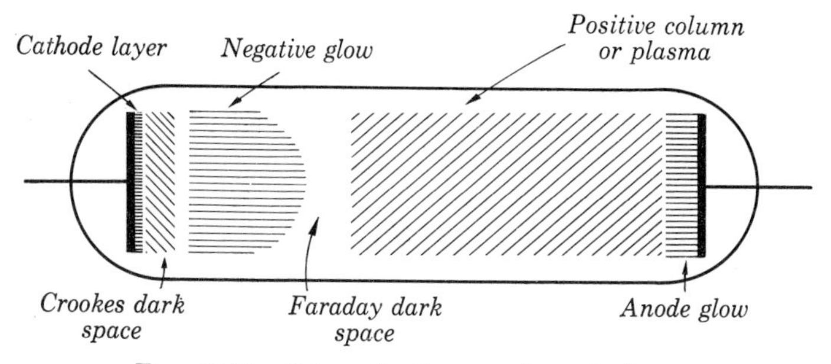

FIG. 18-16. Schematic diagram of gas discharge.

A marked difference in the appearance of the gas near the cathode and anode will be noted after the glow discharge sets in. Fig. 18-16 is a schematic drawing. The surface of the cathode is covered with a thin layer of luminous gas called the *cathode layer.* Immediately beyond it is a relatively nonluminous region called the *Crookes dark space* (after Sir William Crookes, one of the early workers in this field). Beyond the Crookes dark space is a second luminous region, blue if the discharge is in air, and called the *negative glow.* Following this is another relatively dark region, the *Faraday dark space,* and beyond this is the *positive column* or *plasma* which occupies the remainder of the tube. The positive column is often broken up into alternate luminous and nonluminous portions called *striations.*

The discharge is rendered self-maintaining chiefly by processes taking place in the Crookes dark space and at the cathode surface. Positive ions, formed in the Crookes dark space by electrons from the cathode, are accelerated toward the cathode and when they strike its surface secondary electrons may be released. If each electron emitted produces enough positive ions to liberate another secondary electron, a steady state exists and the discharge is self-maintaining.

Most of the light emitted by the gas comes from the positive column. Light is emitted by a gas molecule on returning to its normal state after one of its outer electrons has been removed or raised to a higher "energy level." This light is characteristic of the gas in the tube, and in general much of it lies in the ultraviolet. Fluorescent lamps are gaseous discharge tubes whose inner walls are covered with a material which absorbs ultraviolet light and re-emits light in the visible spectrum. The luminous efficiency of such lamps is considerably greater than that of incandescent lamps.

Problems—Chapter 18

(1) In what ratio is the saturation current density from a tungsten cathode increased when the temperature is raised from 2500° K to 3000° K?

(2) In what ratio would the saturation current density from a cathode operating at 2500° K be increased if the work function of the cathode surface were reduced from 4ev to 2ev?

(3) The electrodes in a certain vacuum diode are parallel plates 1 cm square, spaced 0.3 cm apart. (a) What is the minimum plate voltage required to draw a saturation current of 60 ma? (b) What voltage would be required if the spacing were doubled?

(4) From the plate characteristics of the 2A3 triode in Fig. 18-8, find the amplification factor, the plate resistance, and the transconductance, at the point $e_b = 250$ volts, $e_c = -45$ volts. Compare with the manufacturer's figures of: $\mu = 4.2$, $r_p = 800$ ohms, $g_m = 5250$ micromhos.

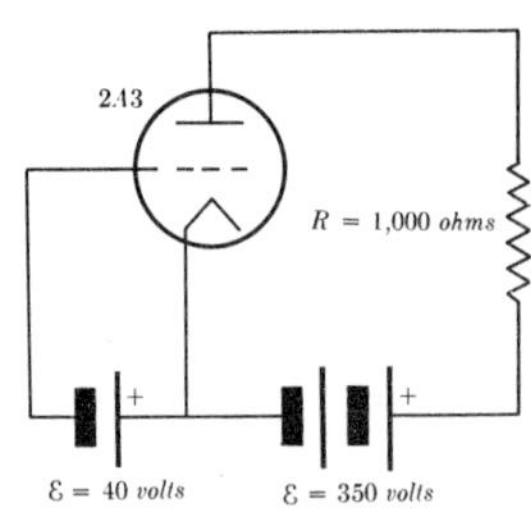

FIG. 18-17.

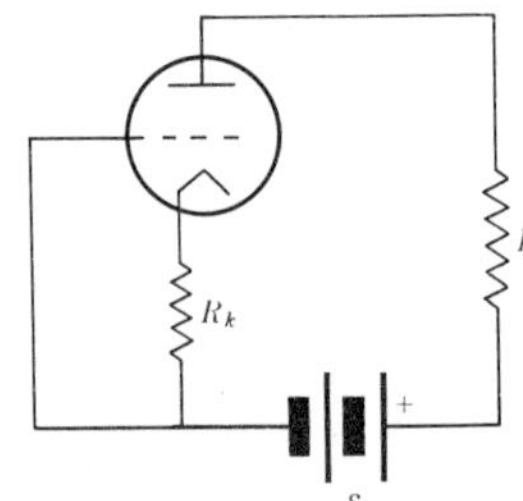

FIG. 18-18.

(5) Find the plate current drawn by the 2A3 triode in Fig. 18-17.

(6) The plate current drawn by the triode in Fig. 18-18 is 60 ma. The potential of the grid is 30 volts negative with respect to the cathode. What is the magnitude of the grid biasing resistor, R_k? The current to the grid is negligible.

SUGGESTED BOOKS FOR COLLATERAL READING

TAYLOR, LLOYD. *Physics, the Pioneer Science.* Houghton Mifflin.

PAGE AND ADAMS. *Principles of Electricity.* D. Van Nostrand.

FRANK, N. H. *Introduction to Electricity and Optics.* McGraw-Hill.

HARNWELL, G. P. *Principles of Electricity and Electromagnetism.* McGraw-Hill.

ABRAHAM-BECKER. *Theory of Electricity.* Blackie.

SMYTHE, W. R. *Static and Dynamic Electricity.* McGraw-Hill.

STRATTON, J. A. *Electromagnetic Theory.* McGraw-Hill.

LAWS, F. A. *Electrical Measurements.* McGraw-Hill.

REICH, H. J. *Theory and Application of Vacuum Tubes.* McGraw-Hill.

M.I.T. Electrical Engineering Department. *Texts in Electrical Engineering* (3 vols.). John Wiley & Sons.

FARADAY, MICHAEL. *Experimental Researches in Electricity and Magnetism.* E. P. Dutton.

LINDSAY and MARGENAU. *Foundations of Physics.* John Wiley & Sons.

WHITTAKER, E. T. *History of the Theories of Aether and Electricity.* Longmans.

ANSWERS TO PROBLEMS

CHAPTER 1

(a) 1.9 dynes.
(b) 9.7 dynes.

3. 81 dynes.

5. (a) 360 newtons.
(b) 5.4×10^{28} m/sec^2.

CHAPTER 2

1. (a) 4.8×10^{-15} newton.
(b) 5.3×10^{15} m/sec^2.

3. (a) 1.4×10^4 newtons/coulomb.
(b) 27° above horizontal.

5. (a) Zero.
(b) 10×10^3 newtons/coulomb, toward the right.
(c) 5×10^3 newtons/coulomb, 77°, first quadrant.
(d) 6.3×10^3 newtons/coulomb, vertically upward.

7. (a) 3.6×10^6 newtons/coulomb.
(b) 6.3×10^{17} m/sec^2.

9. (a) $E_r = 115 \times 10^4$ newtons/coulomb, $E_\theta = 0$.
(b) $E_r = 0$, $E_\theta = 58 \times 10^4$ newtons/coulomb.
(c) $E_r = 100 \times 10^4$ newtons/coulomb, $E_\theta = 29 \times 10^4$ newtons/coulomb.

11. (a) E (newtons/coulomb): 0, 1.3, 1.5, 1.5, 1.2, 0.7.
(b) $a = \sqrt{2}\, b$.

15. 8.85×10^{-13} coulomb.

17. $q(1 - \cos\beta)/\epsilon_0$.

19. 41°.

21. Two electrons.

CHAPTER 3

1. (a) 1600 volts.
(b) 16×10^{-7} joule.
(c) 16×10^{-7} joule.

3. Toward points of higher potential.

5. (a) $v = \sqrt{2qV/m}$.
(b) 5.93×10^5.
(c) 2500 volts.
(d) 0.17% of velocity of light.

7. (a) 8.9×10^6 m/sec.
(b) 10.3×10^6 m/sec.

9. (a) 1800 volts.
(b) 1200 volts.
(c) 960 volts.
(d) 1290 volts.

11. (a) 9×10^{-7} joule.
(b) 19.3×10^{-7} joule.
(c) Zero.

13. $E_{Oa} = 0$, $E_{Ob} = 1000$ volts/m,
$E_{Oc} = 2000$ volts/m,
$E_{Od} = 3000$ volts/m,
$E_{Oe} = 4000$ volts/m.

15. $E_r = \dfrac{2a\cos\theta}{r^3} + \dfrac{b}{r^2}, \quad E_\theta = \dfrac{a\sin\theta}{r^3}$.

17. 37×10^{-6} volt.

21. (a) 11.5 volts.
(b) 100 volts/m, 3.45 volts.

23. (a) V (statvolts): 50, 50, 50, 33, 20, 10.
(b) E (statvolts/cm): 0, 0, 0 (just inside), 25 (just outside), 11, 4, 1.
(c) 15,000 volts.

25. (a) $\rho = 2.95 \times 10^{-9}x^{-2/3}$.
(b) 86 electrons/mm^3.

27. 1.340×10^{-8} coulomb, 0.660×10^{-8} coulomb, 6070 volts.

CHAPTER 4

1. 1.24×10^{21} electrons.

3. 1.3×10^5 electrons/mm^3.

5. (a) 2 ma
(b) 0.09μa

7. 0.067 volts/m.

9. (a) 4 amp, 20 amp.
(b) 50 volts, 0.5 volt.
(c) 5 ohms.

11. 94×10^{-8} ohm-meter.

13. (a) 0.32 ohm.
(b) 8 volts.

15. 0.15 ohm.

17. (a) 1.1 megohms, 0.55 megohm.
(b) 91 μa.
(c) 18×10^{-6} amp/m^2.
(d) 91 μa.

19. (b) 50.5° C.

21. (a) 22 ohms.
(b) 5.5 amp.
(c) 157 cal/sec.
(d) 550 watts.

23. (a) 605 coulombs.
(b) 121 amp.
(c) 130 amp.

25. (a) I_m/π.
(b) $I_m/2$

CHAPTER 5

1. (a) From d to a; from a to d.
(b) 2 coul/sec.
(c) $V_a = 10$ volts, $V_b = 10$ volts, $V_c = 0$, $V_d = 0$.
(d) $V_{ad} = 10$ volts.
(e) $V_{bc} = 10$ volts.
(f) Zero.

3. (a) R (ohms): ∞, 5, 2, 1, 0.5, 0.2, 0.
(b) 24 amp.
(c) V_{ab} (volts): 24, 20, 16, 12, 8, 4, 0.
(d) 88 volts, aiding first battery.
(e) 40 volts, opposing first battery.
(f) $V_{ab} = -4$ volts, $V_{ab} = 28$ volts.
(g) Yes. No.

5. (a) 3.3 amp.
(b) $V_a = 3.3$ volts, $V_b = -14$ volts, $V_c = -7.3$ volts, $V_d = -10$ volts.
(c) $V_{ab} = 17.3$ volts, $V_{dc} = -2.7$ volts, $V_{cd} = 2.7$ volts.

CHAPTER 5 (continued)

7. (a) 200 volts.
(b) Zero.
(c) 1500 ohms.
(d) 185 volts.

9. $\mathcal{E} = 1.52$ volts, $r = 0.1$ ohm.

11. (b) The two are equal.

13. $V_{ea} = -9$ volts, $V_{fc} = -8.7$ volts, $V_{gd} = 6.7$ volts.

17. (a) 2.7 ohms.
(b) 1 amp, 4 volts.

19. 0.044 ohm.

21. 8 ohms, 72 volts.

23. $R_1 = \dfrac{n-1}{n} R_G$, $R_2 = \dfrac{1}{n-1} R_G$.
$R_1 = 81$ ohms, $R_2 = 10$ ohms, 1 ma, 0.1 ma.

25. $\mathcal{E}_1 = 18$ volts, $\mathcal{E}_2 = 7$ volts, $V_{ab} = 13$ volts.

27. *Note:* resistance in upper branch = 4 ohms.
(a) 2 amp, from b to a.
(b) A seat of emf directed from b to a. Any $\mathcal{E}$ and r such that $\mathcal{E} - 4r = 18$ volts.

29. (a) $R_0 = \dfrac{R(R_1 + R_2) + 2R_1R_2}{2R + R_1 + R_2}$.
(b) 2.75 ohms, 3 ohms.

31. (a) 0.5 amp.
(b) 4 amp.
(c) 108 volts.
(d) 60 watts.
(e) 48 watts.
(f) 540 watts.
(g) 71%.

33. (a) 11.5 kw, 0.5 kw, 11 kw.
(b) 11.1 kw, 0.12 kw.

35. (b) 40 ohms, 126 ohms; 71 ohms, 310 ohms, 480 ohms, 260 ohms.
(c) Positive for tungsten, negative for carbon.
(d) 96 watts; 40 watts; 47 watts.
(e) 0.33 amp; 38%, 92%; 40-watt lamp is brighter.

CHAPTER 6

1. (a) −3.6 mv, 0, 2.6 mv, 0, −4.8 mv.
(b) −5.6 mv, −2 mv, 0.6 mv, −2 mv, −6.8 mv.

3. 11.6 mv, 1.21 mv.

CHAPTER 7

1. 31×10^{-12} coul²/newton-m², 22×10^{-12} coul²/newton-m².
(b) 20 μcoul/m².
(c) 13.3 μcoul/m², 15.0 μcoul/m².

3. (a) 7.5×10^5 volts/m, 5.6×10^5 volts/m.

5. 6×10^6 volts/m, 35.4 μcoul/m².

CHAPTER 8

1. 750 μf.

5. (a) q (μcoul): 0, 400, 630, 870, 1000.
(b) i (μa): 100, 60, 37, 14, 0.
(c) 10 sec.
(d) 6.9 sec.

CHAPTER 8 (continued)

7. (a) 0.0122 sec.
(b) 78 volts, 136 volts, 178 volts.

9. (a) 38×10^{-8} coul.
(b) 7600 volts.
(c) 1.43×10^{-3} joule.
(d) 1.35×10^{-3} joule.

11. (a) 2×10^{-2} coul.
(b) 800 volts.
(c) 8 joules.
(d) 2 joules.

13. (a) 1000 μcoul, 2000 μcoul, 1000 volts.
(b) 333 μcoul, 667 μcoul, 333 volts.

15. 490 μcoul, 250 μcoul, 90 μcoul, 1 amp.

17. $C/2$.

19. (a) 14.2 μcoul/m², 8.8 μcoul/m².
(b) 14.2 μcoul/m², 8.8 μcoul/m².
(c) 5.4 μcoul/m².
(d) 3000 volts.

21. (a) $Q^2/8\pi^2\epsilon l^2 r^2$.
(b) $Q^2 dr/4\pi\epsilon lr$.
(c) $(Q^2 \ln b/a)/4\pi\epsilon l$.
(d) $2\pi\epsilon l/(\ln b/a)$.

23. 286,000 times as much energy from the battery.

CHAPTER 9

1. (a) 0.24 weber, 24×10^6 maxwells.
(b) Zero.
(c) 0.24 weber.

3. (a) 2.9×10^7 m/sec.
(b) 4.3×10^{-8} sec.
(c) 8.7×10^6 volts.

5. (a) 1.7×10^8 m/sec.
(c) 0.48 m.

7. (a) 0.16 newton.
(b) Zero, 0.11 newton, 0.16 newton, 0.08 newton.

9. IBa.

11. (a) 0.013 w/m², upward.
(b) 0.023 w/m², toward the left. Neutral equilibrium.

CHAPTER 10

1. 30°.

3. 2985 ohms, 12,000 ohms, 135,000 ohms.
3000 ohms, 15,000 ohms, 150,000 ohms.

5. 4.97 ohms.

7. 4.2×10^5 ohms.

9. (a) 200 volts.
(b) 6 watts.

11. (a) 150 ohms, 300 ohms.
(b) 1.55 volts, 13.2 ohms.

13. 5.92 volts, 4.94 volts, 3.94 volts.

15. (a) Fractional error $= \dfrac{1}{1 + \dfrac{R_V}{R}}$.

(b) Fractional error $= \dfrac{R_A}{R}$.

Use (a) if R is small, (b) if R is large.

17. 0.0277 ohm, 0.25 ohm, 2.5 ohms.

CHAPTER 11

1. (a) 20×10^{-6} w/m².
(b) 7.1×10^{-6} w/m².
(c) 20×10^{-6} w/m².
(d) Zero.
(e) 5.4×10^{-6} w/m².

3. 2.4×10^{-7} w/m².

5. 72×10^{-5} newton.

7. $\dfrac{\mu_0 ib}{2\pi} \ln \dfrac{a+b}{a}$.

9. B(mw/m²): 2.5, 2.0, 0.89, 0.38, 0.22,

11. $\dfrac{\mu_0 ia^2}{2}\left(\dfrac{1}{a^3} + \dfrac{(a^2+b^2)^{3/2}}{1}\right)$.

13. $\mu_0 Ni/4r$.

15. B(mw/m²): 6.2, 6.1, 5.4, 3.1, 0.92, 0.16, 0.06.

17. (a) $\dfrac{6.4 \times 10^{-4}}{r}$ w/m².
(b) 3.25×10^{-6} w.

19. (a) $\dfrac{\mu_0}{4\pi}\dfrac{q_1 v_1}{a^2}$.
(b) $\dfrac{\mu_0}{4\pi}\dfrac{q_1 v_1 q_2 v_2}{a^2}$, toward the left.
(c) Zero.
(d) Zero.
(e) Linear and angular momentum are both conserved if one includes the linear and angular momentum of the electromagnetic field.

CHAPTER 12

1. (a) 1 volt, from B toward A.
(b) 1.25 newtons.
(c) 5 watts.

3. 36.8 μv, point A.

5. (a) R to L.
(b) R to L.
(c) L to R.

7. 0.375 volt.

9. 2.3×10^{-4} coul.

CHAPTER 13

1. (a) 6.28×10^{-6} henry.
(b) 3.14×10^{-4} volt.
(c) 6.28×10^{-6} w/sec.

3. (a) 2.77×10^{-6} henry.
(b) Zero.

7. (a) V_{cb}
$$= \frac{V_{ab}}{R+R_0}\left(R + R_0 \epsilon^{-\frac{(R+R_0)t}{L}}\right).$$

9. (a) 6 watts.
(b) 1.5 watts.
(c) 4.5 watts.
(d) 6 joules.

CHAPTER 14

1. (a) 0.9999832.
(b) 1.00385.
(c) 0.9999957.

CHAPTER 15

3. (a) 9.4×10^{-5} w.
(b) 9.4×10^{-6} w-m.
(c) 7.5×10^{-3} newton-m.
(d) 1.8% of H_c.

7. (a) 0.045 amp.
(b) 0.012 amp.
(c) 0.15 amp, 0.23 amp.

9. 5.3×10^{-4} w.

11. (a) $\mu_0 Ir/2\pi R^2$.
(b) $\mu_0 I^2 r^2/8\pi^2 R^4$.
(c) $\mu_0 I^2 l/16\pi$.
(d) 5×10^{-8} h/m.

CHAPTER 16

1. (a) 377 ohms.
(b) 2.65 mh.
(c) 2650 ohms.
(d) 2650 μf.

3. (a) 56 volts, 139 volts, 56 volts.
(b) $v = 141 \sin 377t$,
$i = 7.84 \sin (377t + \tan^{-1} 1.5)$.

5. (a) 515 ohms, $\phi = 39°$, current leads.
(b) 505 ohms, $\phi = 38°$, current lags.

7. $I_{\text{rms}} = 21$ amp, $L = 73$ mh.

9. (a) 10 amp.
(b) 100 volts.
(c) 42 volts.

11. 84 cycles/sec.

13. (a) I (amp): 0, .048, 0.22, 1.0, 0.26, 0.10.
(b) 0.047, 0.15, 0.20, 0.16, 0.091.

15. (a) $f = \dfrac{1}{2\pi}\sqrt{\dfrac{1}{LC}\left(\dfrac{1}{1 - \dfrac{R^2C}{L}}\right)}$.

(b) $f = \dfrac{1}{2\pi}\sqrt{\dfrac{1}{LC}\left(1 - \dfrac{R^2C}{L}\right)}$.

17. (a) 90°.
(b) 6 ohms, 8 ohms, 10 ohms, 0.6.

19. (a) 1000 ohms.
(b) 0.1 amp, 1 amp, 1 amp.

CHAPTER 17

1. 107 turns.

3. (a) 0.987.
(b) 6.9×10^{-4} joule.
(c) 8×10^{-4} joule.
(d) 5320 ohms.

5. 2.7×10^{-7} amp/m.

7. (a) 2.7×10^{-2} amp/m.
(b) 2.7 kilowatts.

CHAPTER 18

1. 52 times.

3. (a) 176 volts.
(b) 444 volts.

5. 90 ma.

PHYSICAL CONSTANTS

Quantity	Symbol	Best Experimental Value	Approximate Value for Problem Work	Unit
Electron charge	e	1.6030×10^{-19}	1.6×10^{-19}	coulomb
		4.8025×10^{-10}	4.8×10^{-10}	statcoulomb
		1.6030×10^{-20}	1.6×10^{-20}	abcoulomb
Electron mass	m	9.1066×10^{-31}	9.1×10^{-31}	kilogram
Electron charge-to-mass ratio	e/m	1.7592×10^{11}	1.76×10^{11}	coulomb/kgm
Proton mass	M_p	1.67248×10^{-27}	1.67×10^{-27}	kilogram
Boltzmann constant	k	1.38047×10^{-23}	1.38×10^{-23}	joule/deg K
Planck constant	h	6.624×10^{-34}	6.62×10^{-34}	joule-sec
Velocity of light	c	2.99773×10^{8}	3×10^{8}	meter/sec
Avogadro number	A	6.0228×10^{26}	6×10^{26}	(kgm-mole)$^{-1}$
Electron-volt	ev	1.6030×10^{-19}	1.6×10^{-19}	joule
Permittivity of empty space	ϵ_0	$\dfrac{1}{4\pi \times 8.98776 \times 10^{9}}$	$\dfrac{1}{4\pi \times 9 \times 10^{9}} = 8.85 \times 10^{-12}$	$\dfrac{\text{coulomb}^2}{\text{newton-meter}^2}$ or farad/meter
Permeability of empty space	μ_0	$4\pi \times 10^{-7}$ (exactly)	12.6×10^{-7}	$\dfrac{\text{newton-sec}^2}{\text{coulomb}^2}$ or henry/meter

TABLE OF SYMBOLS

Symbol	Quantity	Unit
A, dA	area	meter^2
α	angle	radian
α	temperature coefficient of resistivity	deg^{-1}
B	magnetic induction, or flux density	weber/meter^2
B_r	retentivity	weber/meter^2
β	angle	radian
C	capacitance	farad, or coulomb/volt
c	velocity of light in free space	meter/sec
χ	magnetic susceptibility	henry/meter
D	displacement	coulomb/meter^2
E	electric intensity	newton/coulomb, or volt/meter
$\mathcal{E}$, e	electromotive force	volt
e	electron charge	coulomb
ϵ	permittivity	farad/meter, or coulomb^2/newton-meter^2
ϵ_0	permittivity of free space	farad/meter, or coulomb^2/newton-meter^2
ϵ	base of natural logarithms	———
η	electric susceptibility	coulomb^2/newton-meter^2
F	force	newton
f	frequency	sec^{-1}
H	magnetic intensity	ampere/meter, or newton/weber
H_c	coercivity	ampere/meter, or newton/weber
I, i	current	ampere, or coulomb/sec
I_{av}	average current	ampere
I_{eff}	effective current	ampere
I_m	maximum current	ampere
I_{rms}	root-mean-square current	ampere
$\mathcal{I}$	Intensity of magnetization, or magnetic moment per unit volume	weber-meter/meter^3, or weber/meter^2
J	current density	ampere/meter^2
K_e	dielectric coefficient	———
K_m	relative permeability	———
L	self inductance	henry, or volt/(ampere/sec)
L, l	length	meter
λ	charge per unit length	coulomb/meter
M	magnetic moment	weber-meter

Symbol	Quantity	Unit
M	mutual inductance	henry, or volt/(ampere/sec)
m	magnetic pole strength	weber
mmf	magnetomotive force	ampere-turn
m	mass	kilogram
μ	permeability	henry/meter, or weber/ampere-meter
μ_0	permeability of free space	henry/meter, or weber/ampere-meter
N	electric flux	(no name for unit)
n	index of refraction	———
p	electric moment, or dipole moment	coulomb-meter
P	polarization, or dipole moment per unit volume	coulomb-meter/meter^3, or coulomb/meter^2
P	power	watt, or joule/sec
ϕ	angle	radian
Φ	magnetic flux	weber
π	Peltier emf	volt
Q, q	quantity of charge	coulomb
r	distance	meter
$\mathcal{R}$	reluctance	ampere-turn/meter
R, r	resistance	ohm, or volt/ampere
ρ	density (mass)	kilogram/meter^3
ρ	resistivity	ohm-meter
ρ	volume density of charge	coulomb/meter^3
S	Poynting vector	watt/meter^2
s, ds	length	meter
σ	conductivity	(ohm-meter)$^{-1}$
σ	surface density of charge	coulomb/meter^2.
σ	Thomson coefficient	volt/deg K
T	temperature	deg K
t	temperature	deg C
T, t	time	second
τ	torque	newton-meter
θ	angle	radian
V, v	potential	volt, or joule/coulomb
V, v	velocity	meter/sec
$V_{ab} \equiv V_a - V_b$	potential difference	volt, or joule/coulomb
W	energy	joule, or newton-meter
X	reactance	ohm
X_C	capacitive reactance	ohm
X_L	inductive reactance	ohm
Z	impedance	ohm

N	0	1	2	3	4	5	6	7	8	9
50	6990	6998	7007	7016	7024	7033	7042	7050	7059	7067
51	7076	7084	7093	7101	7110	7118	7126	7135	7143	7152
52	7160	7168	7177	7185	7193	7202	7210	7218	7226	7235
53	7243	7251	7259	7267	7275	7284	7292	7300	7308	7316
54	7324	7332	7340	7348	7356	7364	7372	7380	7388	7396
55	7404	7412	7419	7427	7435	7443	7451	7459	7466	7474
56	7482	7490	7497	7505	7513	7520	7528	7536	7543	7551
57	7559	7566	7574	7582	7589	7597	7604	7612	7619	7627
58	7634	7642	7649	7657	7664	7672	7679	7686	7694	7701
59	7709	7716	7723	7731	7738	7745	7752	7760	7767	7774
60	7782	7789	7796	7803	7810	7818	7825	7832	7839	7846
61	7853	7860	7868	7875	7882	7889	7896	7903	7910	7917
62	7924	7931	7938	7945	7952	7959	7966	7973	7980	7987
63	7993	8000	8007	8014	8021	8028	8035	8041	8048	8055
64	8062	8069	8075	8082	8089	8096	8102	8109	8116	8122
65	8129	8136	8142	8149	8156	8162	8169	8176	8182	8189
66	8195	8202	8209	8215	8222	8228	8235	8241	8248	8254
67	8261	8267	8274	8280	8287	8293	8299	8306	8312	8319
68	8325	8331	8338	8344	8351	8357	8363	8370	8376	8382
69	8388	8395	8401	8407	8414	8420	8426	8432	8439	8445
70	8451	8457	8463	8470	8476	8482	8488	8494	8500	8506
71	8513	8519	8525	8531	8537	8543	8549	8555	8561	8567
72	8573	8579	8585	8591	8597	8603	8609	8615	8621	8627
73	8633	8639	8645	8651	8657	8663	8669	8675	8681	8686
74	8692	8698	8704	8710	8716	8722	8727	8733	8739	8745
75	8751	8756	8762	8768	8774	8779	8785	8791	8797	8802
76	8808	8814	8820	8825	8831	8837	8842	8848	8854	8859
77	8865	8871	8876	8882	8887	8893	8899	8904	8910	8915
78	8921	8927	8932	8938	8943	8949	8954	8960	8965	8971
79	8976	8982	8987	8993	8998	9004	9009	9015	9020	9025
80	9031	9036	9042	9047	9053	9058	9063	9069	9074	9079
81	9085	9090	9096	9101	9106	9112	9117	9122	9128	9133
82	9138	9143	9149	9154	9159	9165	9170	9175	9180	9186
83	9191	9196	9201	9206	9212	9217	9222	9227	9232	9238
84	9243	9248	9253	9258	9263	9269	9274	9279	9284	9289
85	9294	9299	9304	9309	9315	9320	9325	9330	9335	9340
86	9345	9350	9355	9360	9365	9370	9375	9380	9385	9390
87	9395	9400	9405	9410	9415	9420	9425	9430	9435	9440
88	9445	9450	9455	9460	9465	9469	9474	9479	9484	9489
89	9494	9499	9504	9509	9513	9518	9523	9528	9533	9538
90	9542	9547	9552	9557	9562	9566	9571	9576	9581	9586
91	9590	9595	9600	9605	9609	9614	9619	9624	9628	9633
92	9638	9643	9647	9652	9657	9661	9666	9671	9675	9680
93	9685	9689	9694	9699	9703	9708	9713	9717	9722	9727
94	9731	9736	9741	9745	9750	9754	9759	9763	9768	9773
95	9777	9782	9786	9791	9795	9800	9805	9809	9814	9818
96	9823	9827	9832	9836	9841	9845	9850	9854	9859	9863
97	9868	9872	9877	9881	9886	9890	9894	9899	9903	9908
98	9912	9917	9921	9926	9930	9934	9939	9943	9948	9952
99	9956	9961	9965	9969	9974	9978	9983	9987	9991	9996
100	0000	0004	0009	0013	0017	0022	0026	0030	0035	0039
N	0	1	2	3	4	5	6	7	8	9

N	0	1	2	3	4	5	6	7	8	9
0	. . .	0000	3010	4771	6021	6990	7782	8451	9031	9542
1	0000	0414	0792	1139	1461	1761	2041	2304	2553	2788
2	3010	3222	3424	3617	3802	3979	4150	4314	4472	4624
3	4771	4914	5051	5185	5315	5441	5563	5682	5798	5911
4	6021	6128	6232	6335	6435	6532	6628	6721	6812	6902
5	6990	7076	7160	7243	7324	7404	7482	7559	7634	7709
6	7782	7853	7924	7993	8062	8129	8195	8261	8325	8388
7	8451	8513	8573	8633	8692	8751	8808	8865	8921	8976
8	9031	9085	9138	9191	9243	9294	9345	9395	9445	9494
9	9542	9590	9638	9685	9731	9777	9823	9868	9912	9956
10	0000	0043	0086	0128	0170	0212	0253	0294	0334	0374
11	0414	0453	0492	0531	0569	0607	0645	0682	0719	0755
12	0792	0828	0864	0899	0934	0969	1004	1038	1072	1106
13	1139	1173	1206	1239	1271	1303	1335	1367	1399	1430
14	1461	1492	1523	1553	1584	1614	1644	1673	1703	1732
15	1761	1790	1818	1847	1875	1903	1931	1959	1987	2014
16	2041	2068	2095	2122	2148	2175	2201	2227	2253	2279
17	2304	2330	2355	2380	2405	2430	2455	2480	2504	2529
18	2553	2577	2601	2625	2648	2672	2695	2718	2742	2765
19	2788	2810	2833	2856	2878	2900	2923	2945	2967	2989
20	3010	3032	3054	3075	3096	3118	3139	3160	3181	3201
21	3222	3243	3263	3284	3304	3324	3345	3365	3385	3404
22	3424	3444	3464	3483	3502	3522	3541	3560	3579	3598
23	3617	3636	3655	3674	3692	3711	3729	3747	3766	3784
24	3802	3820	3838	3856	3874	3892	3909	3927	3945	3962
25	3979	3997	4014	4031	4048	4065	4082	4099	4116	4133
26	4150	4166	4183	4200	4216	4232	4249	4265	4281	4298
27	4314	4330	4346	4362	4378	4393	4409	4425	4440	4456
28	4472	4487	4502	4518	4533	4548	4564	4579	4594	4609
29	4624	4639	4654	4669	4683	4698	4713	4728	4742	4757
30	4771	4786	4800	4814	4829	4843	4857	4871	4886	4900
31	4914	4928	4942	4955	4969	4983	4997	5011	5024	5038
32	5051	5065	5079	5092	5105	5119	5132	5145	5159	5172
33	5185	5198	5211	5224	5237	5250	5263	5276	5289	5302
34	5315	5328	5340	5353	5366	5378	5391	5403	5416	5428
35	5441	5453	5465	5478	5490	5502	5514	5527	5539	5551
36	5563	5575	5587	5599	5611	5623	5635	5647	5658	5670
37	5682	5694	5705	5717	5729	5740	5752	5763	5775	5786
38	5798	5809	5821	5832	5843	5855	5866	5877	5888	5899
39	5911	5922	5933	5944	5955	5966	5977	5988	5999	6010
40	6021	6031	6042	6053	6064	6075	6085	6096	6107	6117
41	6128	6138	6149	6160	6170	6180	6191	6201	6212	6222
42	6232	6243	6253	6263	6274	6284	6294	6304	6314	6325
43	6335	6345	6355	6365	6375	6385	6395	6405	6415	6425
44	6435	6444	6454	6464	6474	6484	6493	6503	6513	6522
45	6532	6542	6551	6561	6571	6580	6590	6599	6609	6618
46	6628	6637	6646	6656	6665	6675	6684	6693	6702	6712
47	6721	6730	6739	6749	6758	6767	6776	6785	6794	6803
48	6812	6821	6830	6839	6848	6857	6866	6875	6884	6893
49	6902	6911	6920	6928	6937	6946	6955	6964	6972	6981
50	6990	6998	7007	7016	7024	7033	7042	7050	7059	7067
N	0	1	2	3	4	5	6	7	8	9

NATURAL TRIGONOMETRIC FUNCTIONS

Angle	Sine	Cosine	Tangent	Angle	Sine	Cosine	Tangent
0°	0.000	1.000	0.000				
1°	.018	1.000	.018	46°	.719	.695	1.036
2°	.035	0.999	.035	47°	.731	.682	1.072
3°	.052	.999	.052	48°	.743	.669	1.111
4°	.070	.998	.070	49°	.755	.656	1.150
5°	.087	.996	.088	50°	.766	.643	1.192
6°	.105	.995	.105	51°	.777	.629	1.235
7°	.122	.993	.123	52°	.788	.616	1.280
8°	.139	.990	.141	53°	.799	.602	1.327
9°	.156	.988	.158	54°	.809	.588	1.376
10°	.174	.985	.176	55°	.819	.574	1.428
11°	.191	.982	.194	56°	.829	.559	1.483
12°	.208	.978	.213	57°	.839	.545	1.540
13°	.225	.974	.231	58°	.848	.530	1.600
14°	.242	.970	.249	59°	.857	.515	1.664
15°	.259	.966	.268	60°	.866	.500	1.732
16°	.276	.961	.287	61°	.875	.485	1.804
17°	.292	.956	.306	62°	.883	.470	1.881
18°	.309	.951	.325	63°	.891	.454	1.963
19°	.326	.946	.344	64°	.899	.438	2.050
20°	.342	.940	.364	65°	.906	.423	2.145
21°	.358	.934	.384	66°	.914	.407	2.246
22°	.375	.927	.404	67°	.921	.391	2.356
23°	.391	.921	.425	68°	.927	.375	2.475
24°	.407	.914	.445	69°	.934	.358	2.605
25°	.423	.906	.466	70°	.940	.342	2.747
26°	.438	.899	.488	71°	.946	.326	2.904
27°	.454	.891	.510	72°	.951	.309	3.078
28°	.470	.883	.532	73°	.956	.292	3.271
29°	.485	.875	.554	74°	.961	.276	3.487
30°	.500	.866	.577	75°	.966	.259	3.732
31°	.515	.857	.601	76°	.970	.242	4.011
32°	.530	.848	.625	77°	.974	.225	4.331
33°	.545	.839	.649	78°	.978	.208	4.705
34°	.559	.829	.675	79°	.982	.191	5.145
35°	.574	.819	.700	80°	.985	.174	5.671
36°	.588	.809	.727	81°	.988	.156	6.314
37°	.602	.799	.754	82°	.990	.139	7.115
38°	.616	.788	.781	83°	.993	.122	8.144
39°	.629	.777	.810	84°	.995	.105	9.514
40°	.643	.766	.839	85°	.996	.087	11.43
41°	.656	.755	.869	86°	.998	.070	14.30
42°	.669	.743	.900	87°	.999	.052	19.08
43°	.682	.731	.933	88°	.999	.035	28.64
44°	.695	.719	.966	89°	1.000	.018	57.29
45°	.707	.707	.000	90°	1.000	.000	∞

CONSTANTS AND CONVERSION FACTORS

$\pi = 3.1416$

$\epsilon = 2.7183$

$\log_e 10 = 2.3026$

1 Angström unit = A = 10^{-8} cm
1 micron = 0.001 mm
1 centimeter = 0.39370 in
1 inch = 2.5400 cm
1 foot = 30.480 cm
1 radian = 57.2958 degrees

1 gram = 15.432 grains
1 ounce = 28.350 gm
1 newton = 0.224 lb = 10^5 dynes
1 pound (wt.) = 445,000 dynes

1 atmosphere = 14.697 lb per sq in
1 joule = 10,000,000 ergs
1 calorie = 4.186 joules
1 sq inch = 6.4516 sq cm
1 sq foot = 929.03 sq cm
1 cu inch = 16.387 cu cm
1 liter = 1000 cu cm
1 gallon = 3.785 liters
1 gallon = 231 cu in

1 pound = 453.59 gm
1 kilogram = 2.2046 lb
1 slug = 14.6 kgm

1 foot-pound = 1.3549 joules
1 B.t.u. = 252.00 cal
1 B.t.u. = 778 ft-lb
1 horsepower = 746 watts

Greek Alphabet

Α	α	Alpha
Β	β	Beta
Γ	γ	Gamma
Δ	δ	Delta
Ε	ε	Epsilon
Ζ	ζ	Zeta
Η	η	Eta
Θ	θ	Theta
Ι	ι	Iota
Κ	κ	Kappa
Λ	λ	Lambda
Μ	μ	Mu
Ν	ν	Nu
Ξ	ξ	Xi
Ο	ο	Omicron
Π	π	Pi
Ρ	ρ	Rho
Σ	σ	Sigma
Τ	τ	Tau
Υ	υ	Upsilon
Φ	φ	Phi
Χ	χ	Chi
Ψ	ψ	Psi
Ω	ω	Omega

INDEX

N

O

P

R

S

T

U

V

W

X